Advances in
INORGANIC CHEMISTRY
AND
RADIOCHEMISTRY

Volume 18

CONTRIBUTORS TO THIS VOLUME

K. O. Christe
A. Haas
Joan Mason
Beat Meyer
Jack M. Miller
U. Niemann
C. J. Schack
K. Wade
Robert E. Williams
Gary L. Wilson

Advances in
INORGANIC CHEMISTRY
AND
RADIOCHEMISTRY

EDITORS

H. J. EMELÉUS

A. G. SHARPE

University Chemical Laboratory
Cambridge, England

VOLUME 18

1976

ACADEMIC PRESS  New York San Francisco London

A Subsidiary of Harcourt Brace Jovanovich, Publishers

COPYRIGHT © 1976, BY ACADEMIC PRESS, INC.
ALL RIGHTS RESERVED.
NO PART OF THIS PUBLICATION MAY BE REPRODUCED OR
TRANSMITTED IN ANY FORM OR BY ANY MEANS, ELECTRONIC
OR MECHANICAL, INCLUDING PHOTOCOPY, RECORDING, OR ANY
INFORMATION STORAGE AND RETRIEVAL SYSTEM, WITHOUT
PERMISSION IN WRITING FROM THE PUBLISHER.

ACADEMIC PRESS, INC.
111 Fifth Avenue, New York, New York 10003

United Kingdom Edition published by
ACADEMIC PRESS, INC. (LONDON) LTD.
24/28 Oval Road, London NW1

LIBRARY OF CONGRESS CATALOG CARD NUMBER: 59-7692

ISBN 0-12-023618-4

PRINTED IN THE UNITED STATES OF AMERICA

CONTENTS

LIST OF CONTRIBUTORS vii

Structural and Bonding Patterns in Cluster Chemistry
K. WADE

I.	Introduction	1
II.	The Borane–Carborane Structural Pattern	3
III.	Bonding in Boranes and Carboranes	7
IV.	Some Metal–Carbonyl Clusters	16
V.	Some Mixed Clusters	20
VI.	Hydrocarbon Systems Conforming to the Borane Pattern	35
VII.	Interatomic Distances in Clusters	42
VIII.	Some General Reactions	47
IX.	Other Cluster Systems	50
X.	Conclusion	59
	References	59

Coordination Number Pattern Recognition Theory of Carborane Structures
ROBERT E. WILLIAMS

I.	Introduction	67
II.	Structural Rules	85
III.	Carboranes, Their Analogs and Derivatives	97
IV.	Bridge and Endohydrogens and Relative Lowry-Brønsted Acidity	132
V.	Conclusions	136
	References	137

Preparation and Reactions of Perfluorohalogenoorganosulfenyl Halides
A. HAAS AND U. NIEMANN

I.	Introduction	143
II.	Perfluorohalogenoorganosulfenyl Fluorides	144
III.	Perfluorohalogenoorganosulfenyl Chlorides	146
IV.	Perfluorohalogenoorganosulfenyl Bromides	155
V.	Reactions of Perfluorohalogenoorganosulfenyl Halides and Related Reactions	157
VI.	Characteristics of Perfluorohalogenoorganomercapto Groups	189
	References	190

Correlations in Nuclear Magnetic Shielding. Part I
JOAN MASON

I.	Introduction	197
II.	NMR Measurements and the Periodic Table	198
III.	Theory and Physical Models of Nuclear Magnetic Shielding	202

IV. Absolute Shielding	215
V. Periodicity in Nuclear Magnetic Shielding	218
References	225

Some Applications of Mass Spectroscopy in Inorganic and Organometallic Chemistry

Jack M. Miller and Gary L. Wilson

I. Introduction	229
II. Instrumentation and Sample Handling	231
III. Inorganic and Organometallic Mass Spectra: Practice and Pitfalls	239
IV. Low- and Medium-Resolution Studies	247
V. High-Resolution Studies	268
VI. Metastable-Ion Techniques	270
VII. Coupled Gas Chromatography and Mass Spectrometry Applied to Inorganic and Organometallic Compounds	273
VIII. Conclusion	276
References	276

The Structures of Elemental Sulfur

Beat Meyer

I. Introduction	287
II. Well-Established Allotropes of Sulfur	291
III. Incompletely Characterized Allotropes and Mixtures	308
References	314

Chlorine Oxyfluorides

K. O. Christe and C. J. Schack

I. Introduction	319
II. General Aspects	321
III. Specific Compounds	328
IV. Appendix: Tables of Thermodynamic Properties for Some Chlorine Oxyfluorides	386
References	390

Subject Index	399
Contents of Previous Volumes	406

LIST OF CONTRIBUTORS

Numbers in parentheses indicate the pages on which the authors' contributions begin.

K. O. CHRISTE (319), *Rocketdyne, Division of Rockwell International, Canoga Park, California*

A. HAAS (143), *Chair of Inorganic Chemistry II, Ruhr-University, Bochum, Germany*

JOAN MASON (197), *Open University, Milton Keynes, Buckinghamshire, England*

BEAT MEYER (287), *Chemistry Department, University of Washington, Seattle, Washington, and Inorganic Materials Research Division, Lawrence Berkeley Laboratory, University of California, Berkeley, California*

JACK M. MILLER (229), *Department of Chemistry, Brock University, St. Catharines, Ontario, Canada*

U. NIEMANN (143), *Chair of Inorganic Chemistry II, Ruhr-University, Bochum, Germany*

C. J. SCHACK (319), *Rocketdyne, Division of Rockwell International, Canoga Park, California*

K. WADE (1), *Department of Chemistry, University of Durham, Durham, England*

ROBERT E. WILLIAMS (67), *Chemical Systems Inc., Irvine, California*

GARY L. WILSON (229), *Department of Chemistry, John Abbott College, Ste. Anne de Bellevue, Quebec, Canada*

STRUCTURAL AND BONDING PATTERNS IN CLUSTER CHEMISTRY

K. WADE

Department of Chemistry, University of Durham, Durham, England

I.	Introduction	1
II.	The Borane–Carborane Structural Pattern	3
III.	Bonding in Boranes and Carboranes	7
	A. Localized Bond Treatments	7
	B. Significance of the Numbers of Skeletal Bonding Electron Pairs	10
	C. Summary of Structural and Bonding Pattern	15
IV.	Some Metal–Carbonyl Clusters	16
V.	Some Mixed Clusters	20
	A. Skeletal Electron-Counting Procedures	20
	B. Metallo-boranes and -carboranes	23
	C. Other Mixed Clusters	30
VI.	Hydrocarbon Systems Conforming to the Borane Pattern	35
VII.	Interatomic Distances in Clusters	42
VIII.	Some General Reactions	47
IX.	Other Cluster Systems	50
	A. Bismuth Clusters	50
	B. Some Metal Halide Clusters	51
	C. Gold Clusters	52
	D. Further Metal–Carbonyl Clusters	53
	E. Sulfur Compounds	56
	F. Hydrocarbons	57
X.	Conclusion	59
	References	59

I. Introduction

This is one of two articles in this volume concerned with the borane–carborane structural pattern. In the other (see Williams, this volume, p. 67) Williams has shown how the pattern reflects the coordination number preferences of the various atoms involved. The purpose of the present article is to note some bonding implications of the pattern, and to show its relevance to a wide range of other compounds, including metal clusters, metal–hydrocarbon π complexes, and various neutral or charged hydrocarbons.

Aspects of this theme have been explored in many recent publications (*12, 31b, 43, 96a, 96b, 105a, 128, 137, 158, 161, 163a, 166a, 172a, 173, 173a, 183, 199–205a*) in which boranes, once regarded as obscure chemical rule-breakers, have featured increasingly as pattern-makers for a surprisingly wide range of substances. Arguments derived from borane or carborane chemistry enable us to forecast the probable structures of new metal carbonyl clusters or carbo cations, to devise new syntheses of cluster compounds, to put their redox and acid–base chemistry on a common systematic footing, to predict the likely sites of nucleophilic or electrophilic attack, and to envisage the specific skeletal rearrangements that may accompany such reactions. They also provide us with a type of approach that can profitably be used in other areas of chemistry, where the borane structural pattern itself does not apply.

Boranes and carboranes may be regarded as cluster compounds in the sense defined by Cotton (*48*); they contain a finite group or skeleton of atoms held together entirely, mainly, or at least to a significant extent by bonding directly between those atoms, even though some other atoms may be associated intimately with the cluster. Examples of their structural pattern, however, can be found far beyond the confines of what is normally regarded as cluster chemistry, so this survey includes many systems not commonly referred to as clusters, e.g., cyclic and even acyclic systems, as well as conventional clusters. Because such a wide area of chemistry is involved, it is not possible to be comprehensive, although the examples chosen are intended to illustrate the arguments in sufficient detail, and for a wide enough range of substances, to allow their application to systems not actually covered here. It is hoped that drawing attention to the relationships between areas that had previously been regarded as quite distinct will allow developments in each area to find more rapid general application.

In the following sections, the ways in which the polyhedral structures of boranes and carboranes reflect the numbers of electrons holding them together are outlined, and similar relationships are noted for some metal–carbonyl clusters. Skeletal electron-counting procedures are then given showing that many mixed clusters, containing both transition metals and main group elements, conform to the same structural and bonding pattern. The π complexes formed extensively by transition metals are next shown to be particularly important examples of such mixed clusters. Later sections deal with the question of the sizes (interatomic distances) of borane-type clusters and with some general reaction types. Some other types of cluster are discussed briefly at the end.

II. The Borane–Carborane Structural Pattern

From the early structural studies carried out on boranes, it appeared that most of these compounds adopted structures based on arrangements of their boron atoms that defined fragments of icosahedra. That other triangular-faced polyhedra were important became apparent during the 1960s (*141, 145, 164, 166, 212*), as the structures of key materials such as the borane anions $B_nH_n^{2-}$ and carboranes $C_2B_{n-2}H_n$ were determined, although it was as recently as 1971 that the full structural pattern was first elaborated in a perceptive article by Williams (*213*), who has

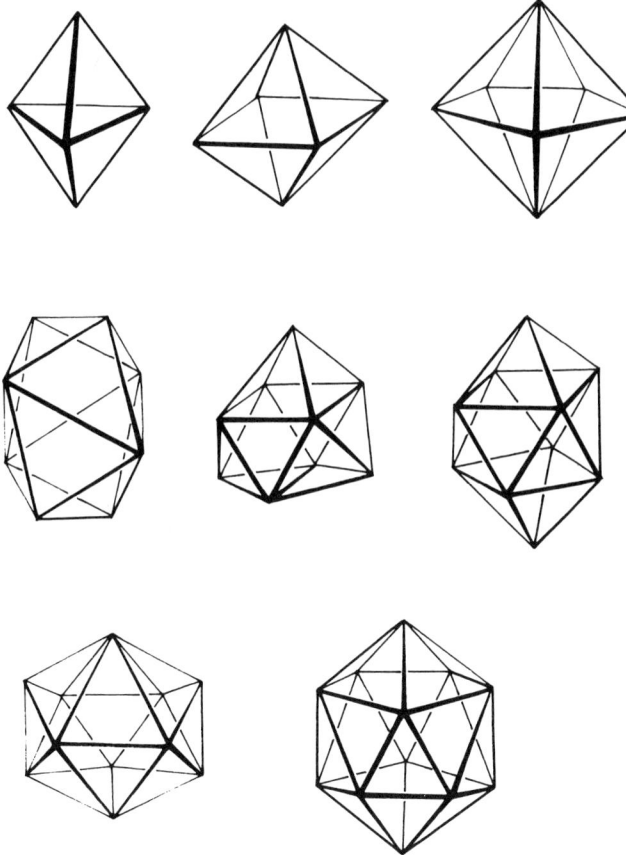

FIG. 1. Polyhedra that form the bases for the structures of *closo*-borane anions $B_nH_n^{2-}$ and carboranes $C_2B_{n-2}H_n$.

profusely illustrated the pattern in the following article of this volume. Here we need concern ourselves only with the main features, outlined in the following.

There are three main structural types, closo, nido, and arachno. Closo structures are adopted by borane anions $B_nH_n^{2-}$ ($n = 6 \rightarrow 12$), carboranes $C_2B_{n-2}H_n$ ($n = 5 \rightarrow 12$), and related isoelectronic species. Their n skeletal boron (or carbon) atoms define the vertices of the triangular-faced polyhedra shown in Fig. 1.

The same polyhedra serve as the basis for the structures of nido and arachno compounds, too, although for these boranes and carboranes the polyhedra are incomplete. Nido structures are adopted by neutral boranes B_nH_{n+4}, carboranes $CB_{n-1}H_{n+3}$, $C_2B_{n-2}H_{n+2}$, $C_3B_{n-3}H_{n+1}$ and $C_4B_{n-4}H_n$, and related ionic species, whose n skeletal boron (or carbon) atoms occupy all but one of the vertices of the appropriate $(n+1)$-vertex polyhedron (see Fig. 2 for examples). Arachno structures are

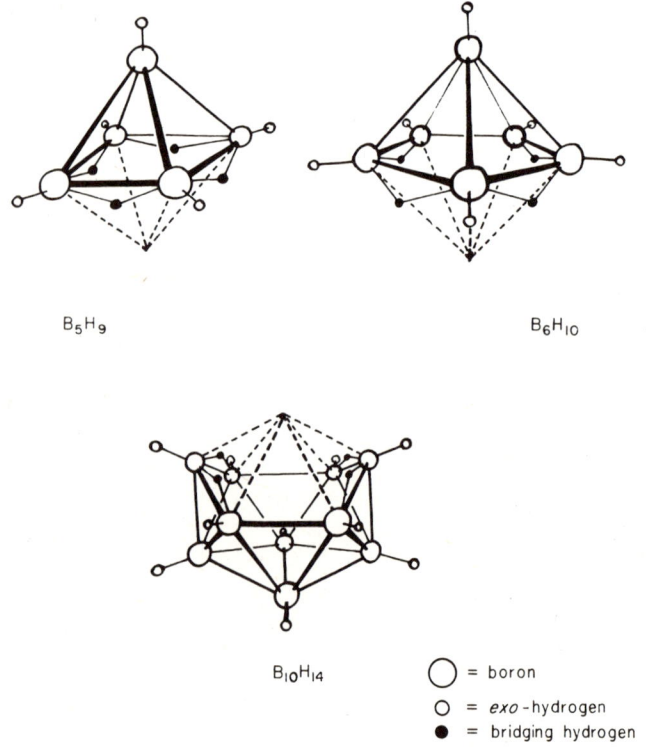

FIG. 2. The *nido*-boranes B_5H_9, B_6H_{10}, and $B_{10}H_{14}$, showing the fundamental polyhedra.

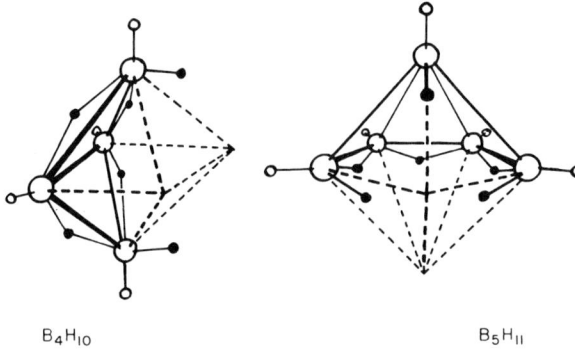

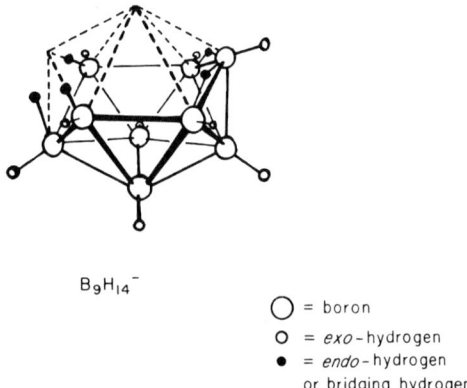

○ = boron
○ = *exo*-hydrogen
● = *endo*-hydrogen
    or bridging hydrogen

FIG. 3. The *arachno*-boranes B_4H_{10}, B_5H_{11}, and $B_9H_{14}^-$, showing the fundamental polyhedra (Williams, this volume, p. 67).

adopted by boranes B_nH_{n+6} and isoelectronic carboranes $C_2B_{n-2}H_{n+4}$, etc. Their n skeletal atoms define all but 2 of the vertices of the appropriate $(n+2)$-vertex polyhedron (Fig. 3). Table I lists formulas of typical *closo*-, *nido*-, and *arachno*-boranes and -carboranes (*12a, 95, 158a, 166a, 181a, 183, 200*, also Williams, this volume).

The positions of the hydrogen atoms in boranes and carboranes are worth noting. In *closo*-borane anions $B_nH_n^{2-}$, the n BH units are aligned so that the B—H bonds point radially outward away from the centers of the B_n polyhedra and are referred to as *exo*-B—H bonds. The CH units of isoelectronic *closo*-carboranes $CB_{n-1}H_n^-$ or $C_2B_{n-2}H_n$ are similarly orientated. When anions $B_nH_n^{2-}$ or $CB_{n-1}H_n^-$ are protonated to form

TABLE I
Typical closo-, nido-, and arachno-Boranes and -Carboranes[a,b]

No. of skeletal bond pairs	No. of polyhedron vertices	Fundamental polyhedron (symmetry)	closo species $B_nH_n^{2-}$	nido species $B_nH_n^{4-}$	arachno species $B_nH_n^{6-}$
6	5	Trigonal bipyramid (D_{3h})	$C_2B_3H_5$	—	$B_3H_8^-$
7	6	Octahedron (O_h)	$B_6H_6^{2-}$ CB_5H_7 $C_2B_4H_6$	B_5H_9 $C_2B_3H_7$	B_4H_{10} —
8	7	Pentagonal bipyramid (D_{5h})	$B_7H_7^{2-}$ $C_2B_5H_7$	B_6H_{10}; $B_6H_{11}^+$ B_5H_{11} $C_xB_{6-x}H_{10-x}$ ($x = 1 \rightarrow 4$)	—
9	8	Dodecahedron (D_{2d})	$B_8H_8^{2-}$ $C_2B_6H_8$ $C_3B_5H_7$	— — —	B_6H_{12}
10	9	Tricapped trigonal prism (D_{3h})	$B_9H_9^{2-}$ $C_2B_7H_9$	B_8H_{12} $C_2B_6H_{10}$	—
11	10	Bicapped Archimedean antiprism (D_{4d})	$B_{10}H_{10}^{2-}$ $CB_9H_{10}^-$ $C_2B_8H_{10}$	$B_9H_{12}^-$ $C_2B_7H_{11}$ —	B_8H_{14} — —
12	11	Octadecahedron (C_{2v})	$B_{11}H_{11}^{2-}$ $CB_{10}H_{11}^-$ $C_2B_9H_{11}$	$B_{10}H_{14}$ CB_9H_{13} $C_2B_8H_{12}$	B_9H_{15} $C_2B_7H_{13}$ —
13	12	Icosahedron (I_h)	$B_{12}H_{12}^{2-}$ $CB_{11}H_{12}^-$ $C_2B_{10}H_{12}$	$CB_{10}H_{13}^-$ $C_2B_9H_{11}^{2-}$ $C_4B_7H_{11}$	$B_{10}H_{15}^-$ $B_{10}H_{14}^{2-}$ —

[a] For the cation $B_6H_{11}^+$, see Ref. (126).
[b] The compound $C_4H_4B_6Me_6$ is not a nido-carborane, but has an adamantane-type structure (30a, 170a).

$B_nH_{n+1}^-$ or $CB_{n-1}H_{n+1}$, the extra hydrogen atom apparently occupies a bridging position either between 2 boron atoms (over an edge of the n-cornered polyhedron) or for 3 atoms (over a polyhedron face). In the (distorted) octahedral carborane CB_5H_7, for example, the extra hydrogen atom is believed, on the basis of a microwave spectroscopic study (150), to be in or over one B_3 octahedral face, which is bounded by longer B—B bonds than the others.

nido-Boranes B_nH_{n+4} contain n exo-BH units orientated like those of the closo species, pointing away from the centers of the fundamental polyhedra, which for these species have $(n + 1)$ vertices. The extra 4 hydrogen atoms generally occupy BHB bridging positions, "stitching-up" the open face where the polyhedron is incomplete (see, e.g., B_5H_9,

B_6H_{10}, and $B_{10}H_{14}$ in Fig. 2), i.e., these BHB bridging hydrogen atoms are located roughly over the centers of incomplete faces of the fundamental polyhedron.

arachno-Boranes, B_nH_{n+6}, too, have n *exo*-BH units pointing outward away from the centers of $(n + 2)$-vertex fundamental polyhedra (Fig. 3). Their extra 6 hydrogen atoms are found in two types of position: either like their counterparts in *nido*-boranes, occupying BHB bridging positions, or in *endo*-B—H positions, terminally attached to particular boron atoms by B—H bonds orientated *tangentially* with respect to the pseudospherical surface of the fundamental polyhedron, which place these *endo*-hydrogen atoms over incomplete edges or faces of the fundamental polyhedra.

The hydrogen atoms of boranes are, thus, of two distinct structural types, for which distinct bonding descriptions are required. One type of hydrogen atom, the exo type, is located over a polyhedron vertex, bound to the skeletal atom occupying that vertex. The other type of hydrogen atom includes both the BHB bridging and endo-terminal BH hydrogens; these are located nearer to the pseudospherical surface of the fundamental polyhedron and are clearly more intimately associated with the skeletal bonding electrons. Alternative ways of treating their bonding are outlined below.

III. Bonding in Boranes and Carboranes

A. Localized Bond Treatments

Although the structural pattern outlined in the foregoing can be rationalized at a simple qualitative level by using a molecular orbital approach to the skeletal bonding of boranes and carboranes (see Section III, B) it is useful to consider first what problems are encountered if one attempts to describe the bonding in terms of localized bonds.

Boranes and carboranes are *electron deficient* (*200*) in the sense that they contain more "bonds"—points of contact between adjacent, covalently bound, atoms—than electron pairs. To describe their bonding in terms of localized bonds, it is, therefore, necessary to resort to three-center or other multicenter bonds. Once the aptness of three-center bonds for describing the bonding in simple boranes had been pointed out by Longuet-Higgins (*146*) (see, for example, the bonding schemes for B_4H_{10}, B_5H_{11}, and B_6H_{10} in Fig. 4), it soon became customary to use two- and three-center bonds and occasionally other multicenter bonds for describing the bonding in electron-deficient compounds in general. For boranes in particular, Lipscomb (*145*) developed an ingenious topo-

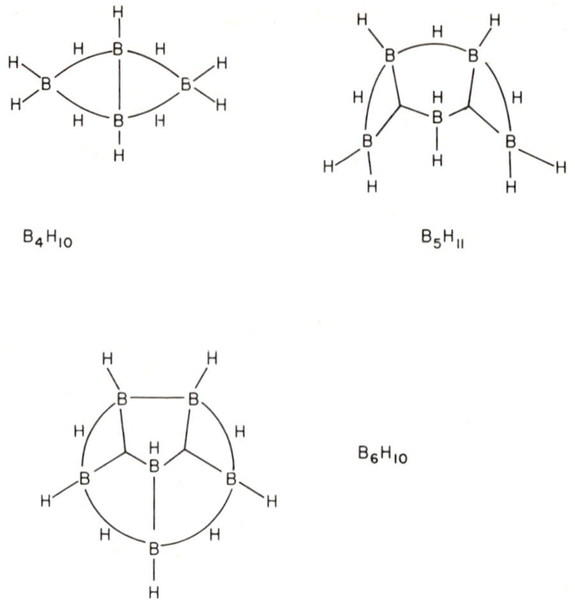

FIG. 4. Localized two- and three-center bond schemes for the boranes B_4H_{10}, B_5H_{11}, and B_6H_{10}.

logical treatment of localized bonds allowing not only known atom networks to be rationalized but also the existence of other networks to be predicted.

Lipscomb's basic assumptions are that each boron atom in a borane is involved in four bonds, and that these bonds can be two-center BH or BB bonds or three-center BHB or BBB bonds. The total number of electron pairs available, and so the total number of bonds, can be calculated simply from the molecular formula, since each boron atom supplies 3 electrons, and each hydrogen atom supplies 1 electron, for bonding. We have already seen that boranes B_nH_{n+m} consist of n exo-BH units and m extra hydrogen atoms. The n BH groups are held together by the $2n$ electrons that they supply for skeletal bonding, together with the m electrons from the extra hydrogens. Thus, there are $(2n + m)$ electrons available for the skeletal bonding that holds the n BH groups together (there are as many skeletal electrons as there are atoms in the molecule), and Lipscomb allocated these in pairs to s three-center BHB bonds, t three-center BBB bonds, y two-center BB bonds, and x two-center BH bonds. The numbers of different types of bond, s, t, y, and x, for a borane B_nH_{n+m} can be shown quite simply to be related by the following equations:

$$x = m - s \qquad t = n - s \qquad 2y = s - x$$

From these equations one can work out what sets of *styx* numbers are possible for known or hypothetical boranes $B_n H_{n+m}$ and so rationalize known atomic networks or predict new ones. For example, the *styx* numbers for the compounds shown in Fig. 4 are B_4H_{10}, 4012; B_5H_{11}, 3203; and B_6H_{10}, 4220.

In the hands of Lipscomb and others (*76, 143, 145a, 153, 189, 190*), such localized bond schemes, particularly when obtained via self-consistent field (SCF) molecular orbital (MO) treatments, have proved particularly valuable for rationalizing the shapes and interatomic distances and estimating the charge distribution in many higher boranes.

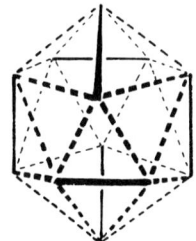

FIG. 5. An apparently satisfactory localized skeletal bond scheme for a hypothetical icosahedral anion $B_{12}H_{12}^{4-}$ (two-center B—B bonds are represented by full lines, three center BBB bonds by triangles with exclusively broken-line edges).

However, used qualitatively, localized bond schemes are less satisfactory for rationalizing the shapes of such relatively symmetrical species as the *closo* anions $B_n H_n^{2-}$, for most of which it is not possible to find a single localized bond arrangement that appears appropriate for the symmetry. For example, the equations of balance for the anion $B_6H_6^{2-}$ require its 6 BH units to be held together by three two-center BB bonds and four 3-centre BBB bonds, a combination which fits the known octahedral symmetry but poorly. Again, the three two-center BB bonds and ten three-center BBB bonds required for anion $B_{12}H_{12}^{2-}$ appear to suit the icosahedral anion far less appropriately than the combination of six two-center BB bonds and eight three-center BBB bonds that could be used for a hypothetical icosahedral anion $B_{12}H_{12}^{4-}$ (Fig. 5) (*200*). In short, for the polyhedral *closo*-borane anions and related carboranes, two- and three-center bond schemes are generally unsatisfactory. They give poor fits to the shapes of known species, and, as in the case of the hypothetical anion $B_{12}H_{12}^{4-}$, lead one to expect the existence of other species than are actually found, e.g., compounds B_nH_{n+4} with closo structures.

B. SIGNIFICANCE OF THE NUMBERS OF SKELETAL BONDING ELECTRON PAIRS

1. Closo Species

The problems outlined in the previous section can be avoided if, instead of allocating the skeletal bonding electron pairs to localized bonds, one simply compares their number with the number of skeletal bonding MO's (199). The closo, nido, and arachno structures of boranes and carboranes can then be seen to reflect the numbers of skeletal bond pairs that are available to hold their skeletal atoms together.

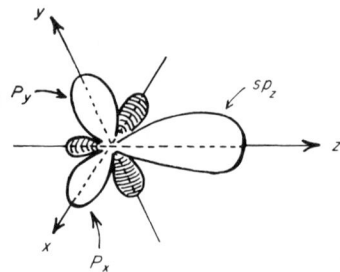

FIG. 6. The radially orientated sp hybrid atomic orbital (AO) and tangentially orientated p AO's that a BH unit can supply for skeletal bonding.

The number of bonding MO's appropriate for a particular structure is best illustrated by considering first the closo-borane anions $B_nH_n^{2-}$. These adopt the polyhedral structures shown in Fig. 1, and each contains n exo-BH units held together by $(n + 1)$ skeletal bond pairs. (Effectively, one skeletal bond pair is supplied by each BH unit; the remaining pair arises from the charge on these closo anions.) If one assigns to the BH bond one sp hybrid orbital on each boron atom, pointing radially outward away from the center of the cluster, there remain three atomic orbitals (AO's) on each boron atom available for cluster bonding (see Fig. 6). One of these is an inward-pointing, radially orientated, sp hybrid orbital; the other two are p orbitals, orientated tangentially with respect to the pseudospherical surface of the cluster. The symmetries of the polyhedra shown in Fig. 1 are such that, in each case, $(n + 1)$ bonding MO's are generated by interactions of the $3n$ AO's available (113–116, 139, 145, 147, 148, 163, 164). For example, an octahedral arrangement of six BH groups generates seven bonding MO's (*147*): a unique orbital of a_{1g} symmetry resulting from in-phase interactions of the six radial sp AO's (see Fig. 7) and two triply degenerate sets of MO's, of symmetries t_{2g} and t_{1u}, to which the tangentially orientated p orbitals make the major contribution.

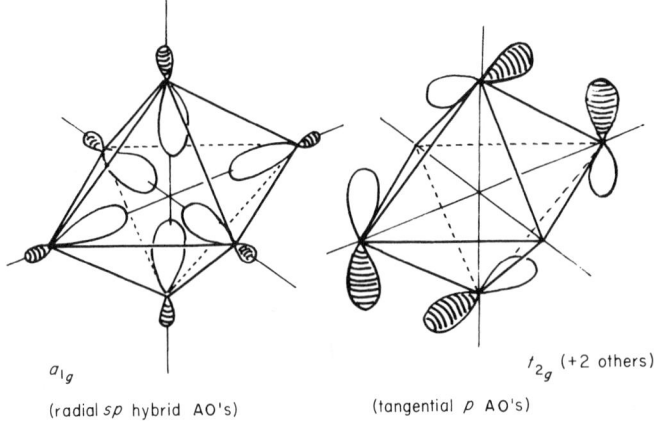

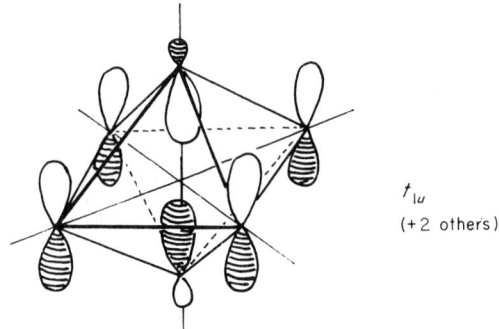

FIG. 7. The skeletal bonding molecular orbitals of $B_6H_6^{2-}$ (*147*). (AO, atomic orbital.)

For the higher borane anions $B_nH_n^{2-}$, although the sets of bonding MO's become progressively more numerous as n increases (*139, 145, 148, 163, 164*), features of the $B_6H_6^{2-}$ MO pattern persist. Of the $(n + 1)$ bonding MO's, there is invariably one, of a_1 or a_g symmetry, representing an in-phase combination of the radially orientated sp hybrid orbitals, to which the tangentially orientated p orbitals make no contribution. The remaining n bonding MO's are primarily polyhedron surface orbitals in that they result from interactions of the $2n$ tangentially orientated p AO's, although some have stabilizing contributions from combinations of radially orientated sp hybrid orbitals, as in the case of the t_{1u} bonding MO's for $B_6H_6{}^{2-}$. These features can be seen in the MO diagram for the icosahedral *closo*-borane anion $B_{12}H_{12}{}^{2-}$ (Fig. 8), a species of particular interest as its existence was predicted, on the basis of an LCAO-MO treatment, by Longuet-Higgins and Roberts (*148*) some years before salts of the anion were actually prepared.

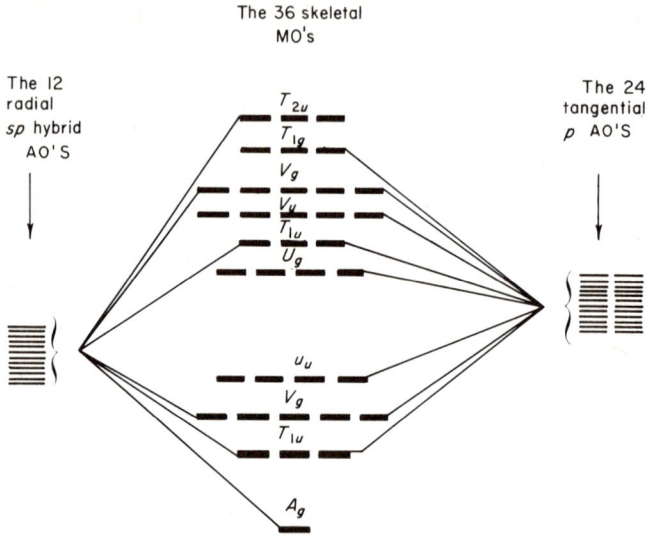

FIG. 8. Relative energy levels and degeneracies of the 13 bonding and 23 antibonding skeletal molecular orbitals (MO's) of the icosahedral anion $B_{12}H_{12}{}^{2-}$ (*148*). (AO, atomic orbital.)

The division of the $(n + 1)$ bonding AO's into two types, a unique radial orbital and n "surface" orbitals, incidentally is also a feature of a free-electron MO treatment (*198*) which has been applied to borane polyhedra, as distinct from the linear combination of atomic orbitals (LCAO)-MO treatments mentioned above.

The polyhedra in Fig. 1 thus represent suitable shapes for cluster species with n skeletal atoms (each of which can furnish three AO's for use in skeletal bonding) and with $(n + 1)$ skeletal bond pairs. Since it is the cluster symmetry that determines the number of bonding MO's, the same polyhedra can serve as the basis for the structures of a whole range of isoelectronic species, including neutral carboranes of formula $C_2B_{n-2}H_n$, bismuth clusters, such as the trigonal-bipyramidal $Bi_5{}^{3+}$, which as Corbett pointed out (43) has six skeletal bond pairs to hold its 5 atoms together if one allocates a "lone pair" to each bismuth atom, and many other systems as outlined below.

For a recent photoelectron spectroscopic study of some *closo*-carboranes $C_2B_{n-2}H_n$ ($n = 5$, 6, 7, or 12) which supports this skeletal bonding treatment, see Ref. (82a).

2. Nido and Arachno species

The structures of *nido-* and *arachno*-boranes and carboranes can be rationalized in a similar manner to those of the closo species if one considers the hypothetical anions, $B_nH_n^{4-}$ and $B_nH_n^{6-}$, from which they are formally derived. Since each BH unit can contribute 2 electrons for skeletal bonding, these two anions contain $(n + 2)$ and $(n + 3)$ skeletal bond pairs, respectively. Nido species related to the hypothetical anions $B_nH_n^{4-}$ thus contain the number of skeletal bond pairs, $(n + 2)$, appropriate for an $(n + 1)$-vertex polyhedral arrangement of their n skeletal atoms, one vertex being left vacant, whereas arachno species formally derived from hypothetical anions $B_nH_n^{6-}$ contain the number of skeletal bond pairs, $(n + 3)$, appropriate for a polyhedron with $(n + 2)$ vertices, two of which are left vacant. Even though not all the vertices of the polyhedra are occupied by skeletal atoms, the symmetries of the nido and arachno fragments ensure in each case that the number of skeletal bonding MO's is one more than the number of polyhedron vertices. (This point is illustrated in Fig. 9 for species formally related to $B_6H_6^{2-}$.)

Significantly, in the neutral *nido*-boranes B_nH_{n+4} (Fig. 2), treated as anions $B_nH_n^{4-}$ to which four H$^+$ ions have been added, the extra 4 hydrogen atoms occupy BHB bridging positions that preserve so far as possible the symmetry of the system, thus also preserving the number of bonding MO's. Similarly, the extra 6 hydrogen atoms of *arachno*-boranes B_nH_{n+6} (Fig. 3) occupy endo-terminal BH or bridging BHB positions that preserve the symmetry of the system. In B_4H_{10}, for example, the extra 6 hydrogen atoms occupy four BHB bridging sites and two endo-terminal BH sites. Similarly, in B_5H_{11} (Fig. 3), the 6 extra hydrogen atoms occupy three BHB bridging and three endo-terminal BH sites. It should be stressed that the electron pairs that, in a localized bond treatment, would be allocated to *endo*-BH or bridging BHB bonds are here included in the *skeletal* electron count.

That the number of skeletal bonding MO's associated with a particular polyhedral arrangement of BH units need not change if 1 or 2 BH units are removed from the closo species is illustrated in Fig. 9 for species based on an octahedron, namely, the closo anion $B_6H_6^{2-}$ and the hypo-

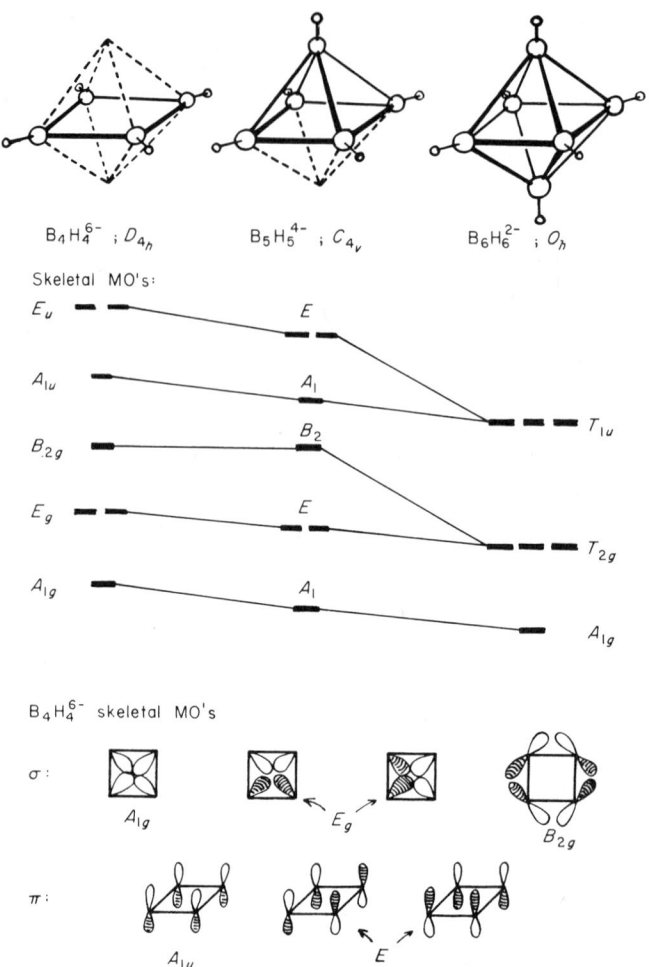

Fig. 9. Orbital correlation diagram for D_{4h} $B_4H_4^{6-}$, C_{4v} $B_5H_5^{4-}$, and O_h $B_6H_6^{2-}$. (MO, molecular orbitals.)

thetical nido and arachno anions $B_5H_5^{4-}$ (C_{4v} symmetry) and $B_4H_4^{6-}$ (D_{4h}). A hypothetical square-planar anion $B_4H_4^{6-}$ (cf the isoelectronic cyclobutadiene dianion $C_4H_4^{2-}$) would have four σ-bonding MO's (of symmetries A_{1g}, B_{1g}, and E_g) and three π-bonding MO's (of symmetries A_{1u} and E_u), which would be stabilized and modified in symmetry, but not added to, if 2 BH^{2+} units were brought up the fourfold axis to complete the octahedron of $B_6H_6^{2-}$. Significantly, the neutral *arachno*-borane B_4H_{10} actually has a C_{2v} arrangement of its 4 skeletal boron atoms (Fig. 3) (formally derived from the octahedral $B_6H_6^{2-}$ by removing 2 cis- rather than trans boron atoms) rather than a D_{4h} square-planar arrangement, probably because the C_{2v} symmetry, but not the D_{4h}, can be preserved for the molecule B_4H_{10}. A square-planar structure, however, appears possible for an anion $B_4H_8^{2-}$ which is already known in the form of the coordination complex $B_4H_8Fe(CO)_3$ (94) (Fig. 10).

The progressive opening of the molecular skeleton that accompanies formal addition of electron pairs to a closo-borane anion $B_nH_n^{2-}$ is intelligible in that a pair of electrons is the contribution made to skeletal bonding by a neutral BH unit. Formally adding BH to $B_nH_n^{2-}$ to convert

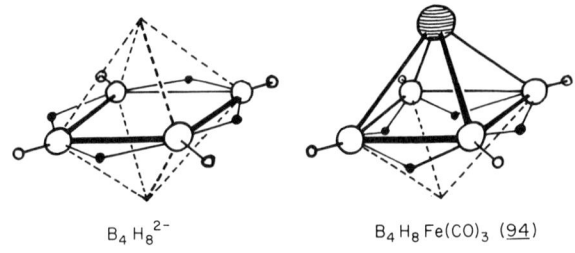

$B_4H_8^{2-}$ $B_4H_8Fe(CO)_3$ (94)

○ = boron; o = exo-hydrogen; • = bridging hydrogen;

⊖ = iron; terminal carbonyls not shown

FIG. 10. Possible structures for the arachno anion $B_4H_8^{2-}$ and its iron complex $B_4H_8Fe(CO)_3$.

it into $B_{n+1}H_{n+1}^{2-}$ has the same effect on the positions of the original n boron atoms as has the addition of a pair of electrons to $B_nH_n^{2-}$ to convert it into $B_nH_n^{4-}$. In each case, the original n boron atoms move to n of the $(n + 1)$ vertices of the next higher triangular-faced polyhedron (199).

C. SUMMARY OF STRUCTURAL AND BONDING PATTERN

Before exploring other areas to which the borane–carborane structural and bonding pattern is relevant, it is useful to summarize the main features, as follows.

1. The triangular-faced polyhedra shown in Fig. 1 form the bases for the structures of boranes and carboranes.
2. The skeletal boron or carbon atoms occupy all, all but one, or all

but two of the vertices of the appropriate polyhedron in closo, nido, or arachno compounds, respectively.

3. Each skeletal boron or carbon atom has a hydrogen atom (or some other singly bonded ligand) terminally attached to it, by a bond radiating outward (*exo*) away from the center of the polyhedron, to which a pair of electrons is allocated.

4. The remaining valence shell electrons (b pairs) are regarded as skeletal bonding electrons.

5. Each skeletal boron or carbon is considered to provide three AO's for skeletal bonding.

6. The symmetries of the polyhedra are such as to generate $(n + 1)$ skeletal bonding MO's from these AO's, where $n =$ the number of polyhedron vertices $[n = (b - 1)]$.

7. Compounds with a skeletal atoms and b skeletal bond pairs adopt closo structures if $b = (a + 1)$, nido structures if $b = (a + 2)$, and arachno structures if $b = (a + 3)$.

IV. Some Metal–Carbonyl Clusters

That transition metal–carbonyl clusters, which contain an apparent abundance of electrons, might have much in common with boranes and carboranes, notorious for their deficiency of electrons, appears at first sight improbable. However, the structural and bonding relationship between them becomes apparent if one considers certain metal–carbonyl clusters for which localized bond treatments are unsatisfactory.

Among the large numbers of metal–carbonyl clusters now known (*1, 33, 127, 138, 170*), species based on triangular, tetrahedral, or octahedral arrays of metal atoms are common. For the first two types, exemplified by the iron-group triangular clusters $M_3(CO)_{12}$ [M = Fe (*208, 210*), Ru (*156*), or Os (*45*)] and the cobalt-group tetrahedral clusters $M_4(CO)_{12}$ [M = Co (*47, 210*), Rh (*211*), or Ir (*211*)] (Fig. 11), the skeletal bonding can be described in terms of localized two-center metal–metal bonds, it being assumed that each metal atom attains an inert gas configuration (*134*). However, many octahedral clusters that contain the wrong number of electrons for similar localized bond descriptions are now known. For example, the octahedral complex $Rh_6(CO)_{16}$ (Fig. 12) contains a pair of electrons too many to allow its skeletal bonding to be described in terms of twelve two-center rhodium–rhodium bonds along the octahedron edges (*46*), although an equivalent orbital approach suggested it had 2 electrons too few (*135*). The same valence problem is posed by other octahedral clusters such as the

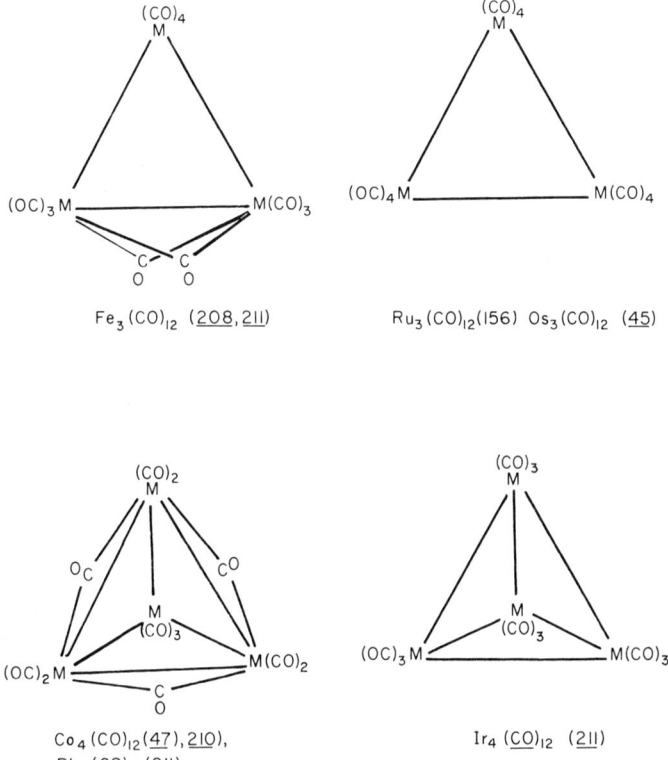

FIG. 11. Triangular and tetrahedral metal clusters.

carbonyl hydride $H_2Ru_6(CO)_{18}$ *(39)* and the carbonyl carbides $Ru_6(CO)_{17}C$ *(125, 182)* and $Ru_6(CO)_{14}C(arene)$ *(125, 157)* (see Fig. 12).

All these molecules contain 86 electrons associated with the valence shells of the 6 metal atoms (the core carbon atoms of the carbides are considered to contribute all their valence shell electrons to these clusters). In the hydride $H_2Ru_6(CO)_{18}$, for example, the metal atoms supply 48 electrons, the carbonyl ligands 36, and the hydrogen atoms 2. Formally, this hydride may be regarded as derived from the anion $[Ru(CO)_3]_6^{2-}$, which, in turn, can be shown to be formally related to the *closo*-borane anion $B_6H_6^{2-}$ as follows *(199, 200)*.

If 18 electron pairs are assigned to the eighteen metal–carbon bonds (3 on each metal), there remain six AO's on each metal atom, and 25 electron pairs, for use in metal–metal bonding. Since well over half the

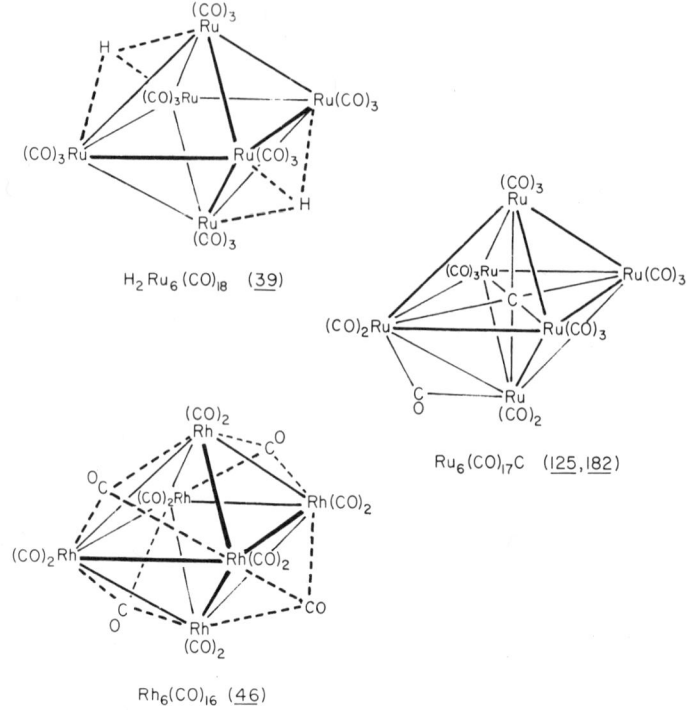

FIG. 12. Some octahedral metal–carbonyl clusters.

available MO's (25 of the 36) will be filled, clearly not all of the electron pairs in them can perform a metal–metal bonding role. Either some must occupy orbitals that are essentially nonbonding, or else the metal–metal antibonding effect of some must nullify the metal–metal bonding effect of others.

For simplicity, one can allocate 3 electron pairs to "nonbonding" orbitals on each metal atom, which leaves three AO's on each metal atom for cluster skeletal bonding, and 7 electron pairs to be accommodated in the bonding MO's these AO's generate. This description is reminiscent of that outlined above (Fig. 7) for $B_6H_6^{2-}$. Each $Ru(CO)_3$ unit of $H_2Ru_6(CO)_{18}$, like each BH unit of $B_6H_6^{2-}$, can contribute three AO's, and 2 electrons, for skeletal bonding (*199, 200*).

Similar bonding descriptions are possible for the other octahedral metal carbonyl clusters. In $Rh_6(CO)_{16}$, for example, 4 electron pairs can be allocated to the four metal–carbon bonds each metal atom is involved in (each metal has two terminal and two triply bridging carbonyl groups attached); 2 more pairs can be allocated to nonbonding orbitals on each metal atom; and 7 electron pairs remain for skeletal bonding.

The relationship between boranes and metal–carbonyl clusters can be extended by considering the compound $Fe_5(CO)_{15}C$, which has the square-based pyramidal structure shown in Fig. 13, with the carbide carbon atom just below the center of the Fe_4 square, clearly contributing all its valence shell electrons to the cluster (24). The metal–carbonyl residue $Fe_5(CO)_{14}^{4-}$ formally left by removal of this carbon as C^{4+} has the nido structure appropriate for a cluster with 5 skeletal atoms and seven skeletal bond pairs.

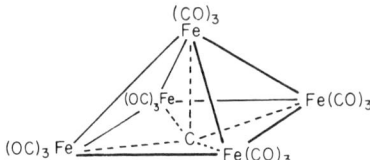

FIG. 13. The *nido*-iron–carbonyl carbide cluster $Fe_5(CO)_{15}C$ (24).

It should be stressed that, in this treatment of metal–carbonyl clusters, the number of nonbonding electron pairs allocated to each metal atom [3 pairs for each ruthenium atom of $H_2Ru_6(CO)_{18}$; 2 pairs for each rhodium atom of $Rh_6(CO)_{16}$] is not arbitrary but is chosen with two objectives in mind: (a) to reduce the number of electrons formally remaining for skeletal bonding to fewer than the number of orbitals remaining, because only then is it realistic to assume that all these electrons can be accommodated in bonding MO's; and (b) to provide a suitable number of electron pairs on each metal atom for metal → carbon dative π bonding to the carbonyl ligands. The so-called nonbonding electrons thus contribute to metal–ligand bonding and, indeed, must also be expected to make a small net metal–metal bonding contribution to the skeletal bonding, as indicated by a recent MO treatment (162) of the carbonyl anion (3) $Co_6(CO)_{14}^{4-}$.

In addition to the closo- and nido-metal–carbonyl clusters already mentioned, with structures based on the octahedron, another interesting category of structure that is found among metal–carbonyl clusters is one in which n skeletal atoms are held together formally by n skeletal bond pairs. These adopt structures based on polyhedra with $(n - 1)$ vertices, as might be expected. The extra metal atom caps one of the triangular faces of the closo residue, where the three vacant orbitals that it can formally furnish for cluster bonding enable it to bond to the 3 metal atoms defining the face without modifying the number of bonding MO's for the rest of the cluster. Examples include the anion $[Rh_7(CO)_{16}]^{3-}$ (4), which has the capped octahedral geometry (Fig. 14) appropriate for its 7 skeletal atoms and seven skeletal bond pairs, and the neutral osmium complex $Os_6(CO)_{18}$ (73, 158), which has a capped trigonal-bipyramidal (or bicapped tetrahedral) structure (Fig. 14) instead of the regular octahedral shape that seemed likely, on a localized bond approach, for an arrangement of six $Os(CO)_3$ units, each capable of forming four two-center bonds.

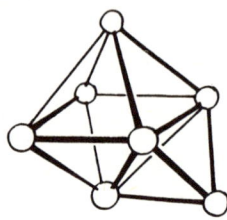 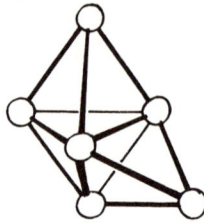

FIG. 14. The capped octahedral skeleton of the anion $[Rh_7(CO)_{16}]^{3-}$ (4) and capped trigonal bipyramidal (bicapped tetrahedral) skeleton of $Os_6(CO)_{18}$ (73, 158).

Many further examples could be added to those already cited to support the close relationship between carbonyl clusters and borane clusters, particularly for species based on an octahedron. These other *closo* octahedral clusters include anions such as $Co_6(CO)_{15}{}^{2-}$ (34), $[Fe_6(CO)_{16}C]^{2-}$ (38, 40), $[Rh_6(CO)_{15}I]^{-}$ (6), $[Ni_6(CO)_{12}]^{2-}$ (31), $[Co_4Ni_2(CO)_{14}]^{2-}$ (8), and $[Rh_{12}(CO)_{30}]^{2-}$, in which two Rh_6 octahedra are linked by a metal–metal bond and two bridging carbonyl groups (2).

V. Some Mixed Clusters

A. Skeletal Electron-Counting Procedures

The foregoing examples show the relevance to metal–carbonyl cluster chemistry of the borane–carborane structural and bonding pattern. Its relevance to other areas of chemistry may be explored readily using a systematic skeletal electron-counting procedure (161, 201).

Tables II and III list the numbers of electrons provided by various potential cluster units, assuming that the skeletal atoms make available three AO's apiece for skeletal bonding, and use their remaining valence shell orbital(s) to bond ligands to the cluster. For example, a main group element E (Table II) such as boron can make three AO's available for cluster bonding if it uses its one remaining valence shell AO (an inert

TABLE II

THE NUMBER OF SKELETAL BONDING ELECTRONS $(v + x - 2)$ THAT MAIN GROUP CLUSTER UNITS CAN CONTRIBUTE

		Typical cluster unit[b]		
v^a	Main group element E	E $(x=0)$	EH, EX $(x=1)$	EH$_2$, EL $(x=2)$
1	Li, Na	[−1]	0	1
2	Be, Mg, Zn, Cd, Hg	0	1	2
3	B, Al, Ga, In, Tl	1	2	3
4	C, Si, Ge, Sn, Pb	2	3	4
5	N, P, As, Sb, Bi	3	4	5
6	O, S, Se, Te	4	5	[6]
7	F, Cl, Br, I	5	[6]	[7]

[a] v = No. of valence shell electrons (periodic group number) of main group element.
[b] x = No. of electrons from ligand(s); X = a 1-electron ligand; L = a 2-electron ligand.

TABLE III

THE NUMBER OF SKELETAL BONDING ELECTRONS $(v + x - 12)$ THAT TRANSITION METAL CLUSTER UNITS CAN CONTRIBUTE[a]

		Typical cluster unit			
v	Transition metal M	M(CO)$_2$ $(x=4)$	M(π-C$_5$H$_5$) $(x=5)$	M(CO)$_3$ $(x=6)$	M(CO)$_4$ $(x=8)$
6	Cr, Mo, W	[−2]	−1	0	2
7	Mn, Tc, Re	−1	0	1	3
8	Fe, Ru, Os	0	1	2	4
9	Co, Rh, Ir	1	2	3	5
10	Ni, Pd, Pt	2	3	4	6

[a] v = No. of valence shell electrons on M; x = No. of electrons from ligands.

gas configuration is assumed) either to accommodate a lone pair or to bond a suitable ligand (as in a group EL). The number of electrons that atom E or group EL supplies for cluster bonding is then $v + x - 2$, where v = the number of electrons in the valence shell of the cluster atom E (its periodic group number) and x = the number of electrons supplied by the ligand L. A transition metal atom M, with five more orbitals in its valence shell, can accommodate more ligands and will furnish $(v + x - 12)$ electrons for skeletal bonding (where x represents the total number of electrons supplied by the ligands on M). For mixed clusters containing both transition metal and main group skeletal atoms, the skeletal electron count may be carried out by summing the valence shell and ligand electrons, and subtracting 2 for each main group atom and 12 for each transition metal atom in the cluster (201). For example, the skeletal electron count for a simple borane B_nH_{n+m} would be $2n + m$, obtained either by adding the contribution from n BH units (2 electrons from each; see Table II) to the contribution from the m extra hydrogen atoms, or

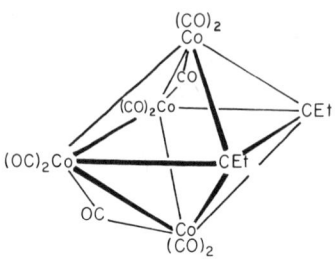

FIG. 15. The closo distorted octahedral structure of the cobalt–carbonyl acetylene complex $Co_4(CO)_{10}C_2Et_2$ (50).

by summing the valence shell $(3n)$ and ligand $(n + m)$ electrons and subtracting 2 for each boron atom $(2n)$.

The application of the method to a cluster containing cobalt and carbon skeletal atoms is illustrated by the cobalt–carbonyl acetylene complex $Co_4(CO)_{10}C_2Et_2$ (50), which can be prepared from the tetrahedral cobalt carbonyl cluster $Co_4(CO)_{12}$ and diethylacetylene (142). Summing the valence shell electrons (4×9 from Co_4; 2×4 from C_2) and ligand electrons (10×2 from the ten carbonyl groups plus 2×1 from the two ethyl groups) and subtracting 12 for each cobalt atom and 2 for each carbon atom, leaves 14 electrons, i.e., 7 pairs, appropriate for a closo-octahedral arrangement of its 6 (Co_2C_2) skeletal atoms. [The same number of skeletal electrons is arrived at by treating the cluster as containing 4 $Co(CO)_2$ units, 2 EtC units, and 2 extra carbonyl ligands.] In the actual structure (Fig. 15) the octahedral skeleton is distorted because of the differing sizes of the cobalt and carbon atoms.

For mixed main group–transition element clusters containing but one transition element, modifying the above electron-counting procedure to include the 6 nonbonding electron pairs on the transition metal has been advocated (128) as a way of bringing the electron count into line with the familiar "18-electron rule" for transition metal complexes. This produces a "$2n + 14$ electron rule" for closo clusters with n skeletal atoms, one a transition metal, and a "$2n + 16$ electron rule" for related nido clusters.

B. METALLO-BORANES AND -CARBORANES

Large numbers of metallo-boranes and -carboranes (*31b, 68a, 94a, 95, 96a, 96b, 172a, 183, 195, 205b*, also Williams, this volume) are now

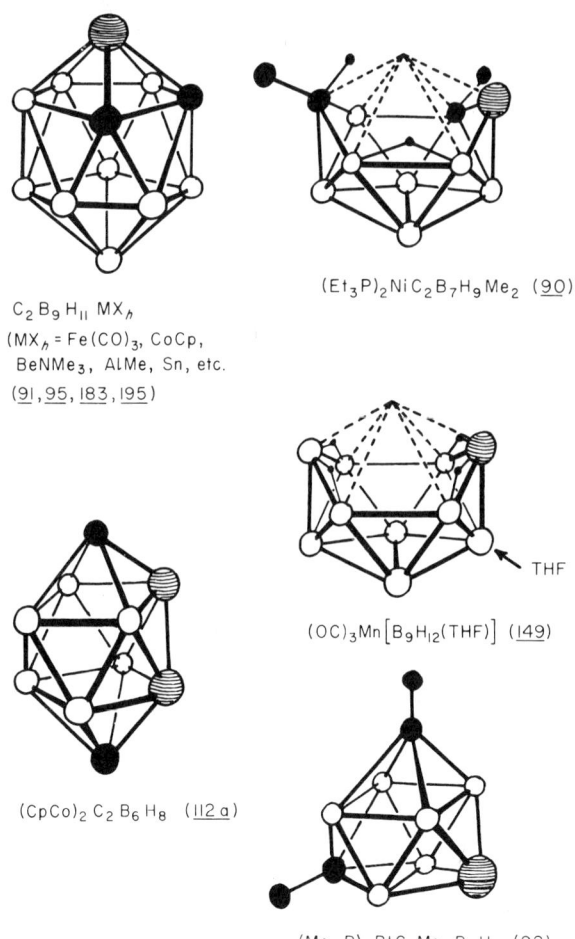

$C_2B_9H_{11}MX_h$
($MX_h = Fe(CO)_3$, CoCp, BeNMe$_3$, AlMe, Sn, etc.
(*91, 95, 183, 195*)

$(Et_3P)_2NiC_2B_7H_9Me_2$ (*90*)

$(OC)_3Mn[B_9H_{12}(THF)]$ (*149*)

$(CpCo)_2C_2B_6H_8$ (*112a*)

$(Me_3P)_2PtC_2Me_2B_6H_6$ (*92*)

FIG. 16. Some metallocarboranes and a metalloborane (metal ligands and exo-hydrogen atoms omitted). (THF, tetrahydrofuran.)

known in which metal atoms occupy vertices of the fundamental polyhedra. Some typical structures are illustrated in Fig. 16, and formulas of some metallo-boranes and -carboranes are listed in Tables IV and V. Reference to Tables II and III shows that such groups as $M(CO)_3$ (M = Fe, Ru, or Os), $M'(h^5\text{-}C_5H_5)$ (M' = Co, Rh, or Ir) and $M''(CO)_2$ (M'' = Ni, Pd, or Pt), like a BH group, can in principle supply three

TABLE IV

Some *closo*-Metallo-boranes and -carboranes Containing at Least 1 Metal Atom in the n-Vertex Polyhedral Cluster

n	Examples (95, 183, 195)
6	$(CpCo)_2B_4H_6$ (159); $CpCoC_2B_3H_5$ (159a); $(OC)_3FeC_2B_3H_5$ (159a)
7	$(OC)_3FeC_2B_4H_6$ (96); $MeMC_2B_4H_6$ (M = Ga, In) (98); $(Ph_3P)_2NiC_2B_4H_6$ (159a); $(OC)_3MnC_3B_3H_5Me$; $(CpCo)_2C_2B_3H_5$ (14, 159a); $[(OC)_3Fe]_2C_2B_3H_5$ (159a)
9	$(R_3P)_2PtC_2B_6H_6R_2'$ (92); $CpCoC_2B_6H_8$ (72, 77); $[(OC)_3MnC_2B_6H_8]^-$; $[CpCoCB_7H_8]^-$; $(CpCo)_2C_2B_5H_7$ (82, 159a); $CpFeC_2B_6H_8$ (77); $[Co(C_2B_6H_8)_2]^-$ (77)
10	$CpCoC_2B_7H_9$ (72, 77); $[Co(C_2B_7H_9)_2]^-$; $(CpCo)_2C_2B_6H_8$ (77, 112a); $CpNi(CpCo)CB_7H_8$ (175); $(CpCo)_3C_2B_5H_7$ (159a)
11	$CpCoC_2B_8H_{10}$ (72, 77); $CpFeC_2B_8H_{10}$ (77); $(C_2B_9H_{11})CoC_2B_8H_{10}$; $(CpCo)_2C_2B_7H_9$ (80, 82); $[Co(C_2B_8H_{10})_2]^-$ (77)
12	$XC_2B_9H_{11}$ [X = CpCo, $(R_3P)_2Pt$ (91); L_2Ni, Ge, Sn, Pb, Tl$^-$, EtAl, [LBe]; $M(C_2B_9H_{11})_2^{x-}$ (M = Fe, $x = 2$; M = Co, $x = 1$; M = Ni, $x = 0$); $M(CB_{10}H_{11})_2^{x-}$ (M = Co, $x = 3$; M = Ni, $x = 2$); $AsCB_{10}H_{11}$; $M(B_{10}H_{10}S)_2^{x-}$ (M = Fe, $x = 2$; M = Co, $x = 1$) (195a); $XB_{10}H_{10}S$ (X = CpCo, L_2Pt) (195a); $(CpCo)_2C_2B_8H_{10}$ (77, 81, 82); $(CpCo)_3C_2B_7H_9$ (80, 82); $CpCoCB_9H_{10}X$ (X = P, As)
13	$CpCoC_2B_{10}H_{12}$ (35, 70, 72); $(CpCo)(CpFe)C_2B_9H_{11}$ (71); $(CpCo)_2C_2B_9H_{11}$ (82)
14	$(CpCo)_2C_2B_{10}H_{12}$

AO's and 2 electrons for cluster bonding, so it is not surprising to find such units in *closo*-metallo-boranes and -carboranes.

Although most known metallocarboranes have only 1 metal atom per polyhedron, the existence of an increasing number of metallocarboranes with 2 or more metal atoms per cluster emphasizes the close relationship between metal clusters and borane clusters. Since they can be synthesized from *closo*-carboranes by replacing BH units by metal–carbonyl or metal–cyclopentadienyl residues, carborane clusters can effectively be used as disposable templates on which to fabricate particular metal clusters (*106*), although alternative, more direct routes to a range of new metal clusters are rapidly being developed.

One particularly interesting category of metallocarborane is that in which a single metal atom is shared between two polyhedra that have a vertex in common. In effect, the metal is sandwiched between two *nido*-carborane residues. Examples are shown in Fig. 17. For such "commo" compounds, the metal can be assumed to contribute three AO's to the skeletal bonding of each polyhedron, when the "$(n + 1)$ rule" for closo clusters is found to be obeyed. For example, the isoelectronic

TABLE V

SOME *nido*-METALLO-BORANES AND -CARBORANES

n^a	Examples
6	$(OC)_3FeB_4H_8$ (*94*); $CpCoB_4H_8$ (*159*)
7	$(OC)_3FeC_2B_3H_7$ (*26, 84, 96*); $(OC)_4FeB_6H_{10}$ (*52*)^b; $R_3MC_2B_4H_7$ (M = Si, Ge)^b; $[(OC)_4Fe(H_3B)B_6H_9]^-$ (*118*) ^{b}
10	$(R_3P)_2PtC_2B_6H_6R_2{}'$ (*92*); $(OC)_3MnB_9H_{12}OC_4H_8$ (*149*)
11	$(Et_3P)_2MC_2B_7H_9Me_2$ [M = Ni (*90*), Pt (*91*)]; $(Et_3P)_2PtB_9H_{11}PEt_3$ (*195a*); $CpCoC_2B_7H_{11}$ (*31c*)
12	$(Et_3P)_2PtHB_9H_{10}S$ (*195a*); $[Zn(B_{10}H_{12})_2]^{2-}$ (*195a*)

^{a} n = No. of vertices on fundamental polyhedron, one of which is left vacant.
^{b} In these compounds the metal atoms occupy edge-bridging positions (not polyhedron vertices).

complexes $[Fe(C_2B_9H_{11})_2]^{2-}$, $[Co(C_2B_9H_{11})_2]^-$, and $Ni(C_2B_9H_{11})_2$ may be regarded as complexes in which two *nido*-$C_2B_9H_{11}{}^{2-}$ anions sandwich the isoelectronic d^6 metal ions Fe^{2+}, Co^{3+}, and Ni^{4+}, respectively (*214*), and these adopt symmetrical structures (Fig. 17) (*36, 37, 53, 174, 220*). Significantly, related complexes with 2 or more extra electrons, such as $[Ni(C_2B_9H_{11})_2]^{2-}$, $[Cu(C_2B_9H_{11})]^-$, and $[Cu(C_2B_9H_{11})_2]^{2-}$ (*215, 216*) have "slipped" structures (Fig. 17) in which the metal is displaced from the fivefold axis of the $C_2B_9H_{11}$ residue toward the 3 boron atoms of the pentagonal face, as if based on a polyhedron with 13 vertices.

The skeletal changes that accompany changes in skeletal electron numbers are reflected in one of the most widely used routes to metallocarboranes, the so-called polyhedral expansion reaction. This reaction, which has been used to considerable effect by Hawthorne and his coworkers (*68–71, 77–79, 82*), involves conversion of a neutral *closo*-

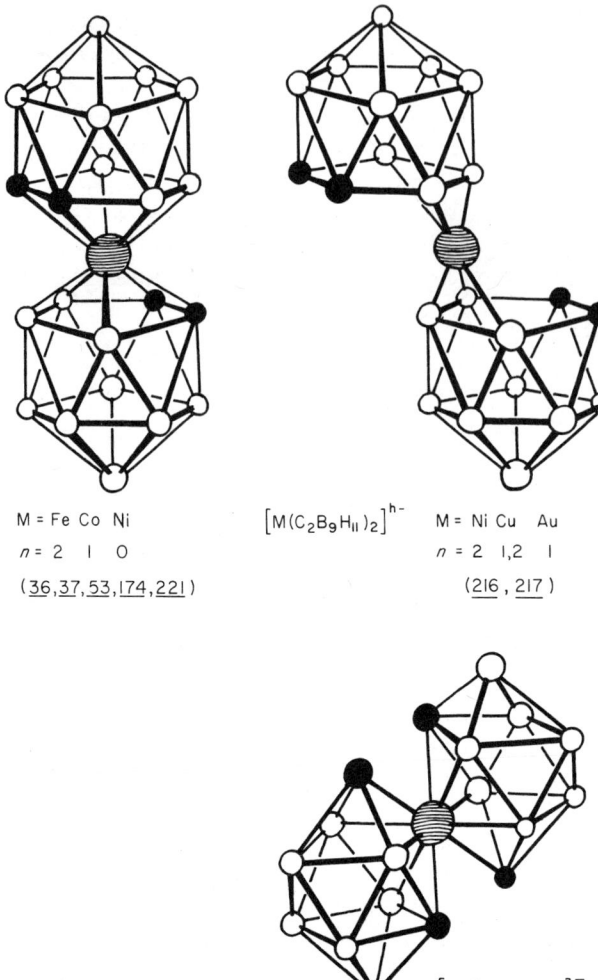

Fig. 17. Metallocarboranes in which the metal is shared between two polyhedra.

carborane into a nido anionic species by adding a pair of electrons to it and then treating it with a suitable metal cation*:

$$C_2B_{n-2}H_n \xrightarrow[\text{THF}]{\text{Na/naphthalene}} C_2B_{n-2}H_n^{2-} \xrightarrow{L_xM^{2+}} L_xMC_2B_{n-2}H_n$$
(closo-) (nido-) (closo-)

* THF, tetrahydrofuran.

Applied to icosahedral carboranes as the starting materials, it leads to metallocarborane based on 13- or 14-vertex polyhedra, e.g. (35, 70),

$$C_2B_{10}H_{12} \xrightarrow[THF]{Na/C_{10}H_8} C_2B_{10}H_{12}{}^{2-} \xrightarrow[CoCl_2]{NaC_5H_5} (C_5H_5)CoC_2B_{10}H_{12}$$
(closo-) (nido-) (closo-; 13 vertices)

$$\downarrow Na/C_{10}H_8$$

$$(C_5H_5)_2Co_2C_2B_{10}H_{12} \xleftarrow[CoCl_2]{NaC_5H_5} [(C_5H_5)CoC_2B_{10}H_{12}]^{2-}$$
(closo-; 14 vertices) (nido-)

The products have structures (35) related to the icosahedron but with 6 atoms instead of 5 in the "tropical" (ortho or meta) rings of atoms (Fig. 18).

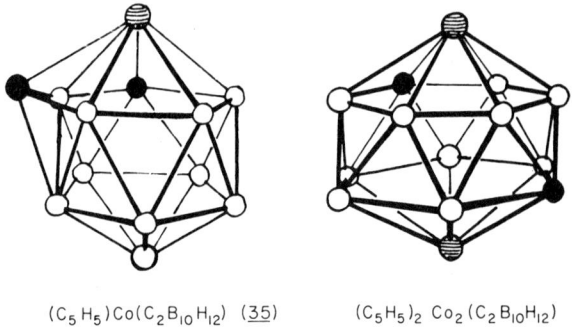

(C₅H₅)Co(C₂B₁₀H₁₂) (35)   (C₅H₅)₂Co₂(C₂B₁₀H₁₂)

FIG. 18. Skeletons of metallocarboranes with 13 and 14 skeletal atoms.

Similar polyhedral expansion reactions have been used to synthesize metallocarboranes containing 3 metal atoms, e.g., the icosahedral $(C_5H_5Co)_3C_2B_7H_9$ (80), or containing two different metals (71, 175), e.g., the bicapped square-antiprismatic $C_5H_5CoCB_7H_8NiC_5H_5$ (175).

Polyhedral expansion is achieved without the need for a preliminary reduction stage in other syntheses of metallocarboranes, e.g., in the preparation of the closo icosahedral platinum compound $(Me_2PhP)_2$-$PtC_2Me_2B_9H_9$ (91) from the closo-carborane $C_2Me_2B_9H_9$ and $(Me_2PhP)_3Pt$ (184), or in the preparation of mono and dimetallocarboranes from the closo-trigonal-bipyramidal $C_2B_3H_5$:

$$C_2B_3H_5 \begin{array}{c} \xrightarrow{Fe(CO)_5} C_2B_3H_5Fe(CO)_3 \xrightarrow{Fe(CO)_5} C_2B_3H_5[Fe(CO)_3]_2 \\ \xrightarrow{C_5H_5Co(CO)_2} C_2B_3H_5CoC_5H_5 \xrightarrow{C_5H_5Co(CO)_2} C_2B_3H_5[CoC_5H_5]_2 \end{array}$$

closo-   closo-   closo-
(trigonal   (octahedra)   (pentagonal bipyramids)
bipyramid)

The recent work of Grimes and his co-workers (*14, 26, 84, 96–96b, 98, 119, 191, 192*) has shown the smaller carboranes to be particularly fertile sources of metal–boron–carbon clusters, of which a few further examples are shown in Fig. 19.

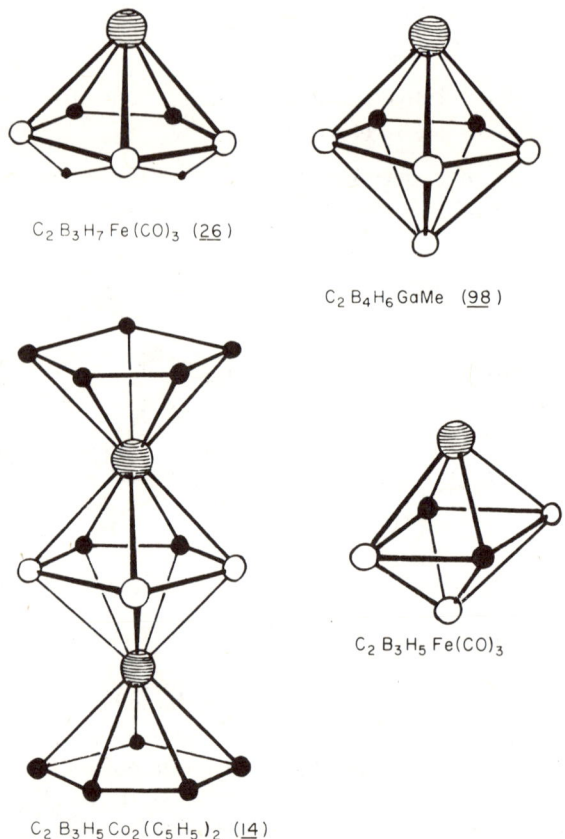

$C_2B_3H_7Fe(CO)_3$ (*26*)

$C_2B_4H_6GaMe$ (*98*)

$C_2B_3H_5Fe(CO)_3$

$C_2B_3H_5Co_2(C_5H_5)_2$ (*14*)

FIG. 19. Metallocarboranes derived from the smaller carboranes. (Terminal hydrogen atoms and carbonyl groups not shown.)

By no means do all metallocarboranes have the metal atoms occupying vertices of the basic polyhedra. Apart from many derivatives in which σ-bonded metal residues occupy exo sites attached to particular skeletal atoms, several metalloboranes and -carboranes are known in which the metal occupies an edge-bridging site, effectively replacing a bridging hydrogen atom of the parent borane. Many are nido species related to B_6H_{10}, for example, the μ-silyl and μ-germyl carboranes,

μ-R$_3$MC$_2$B$_4$H$_7$ (R = H, Me; M = Si, Ge) (*192*) (Fig. 20) and the anion [μ-Fe(CO)$_4$B$_7$H$_{12}$]$^-$, which is formally derived from B$_6$H$_{10}$ by replacing 2 bridging hydrogen atoms by BH$_3$ and Fe(CO)$_4$ units (Fig. 20) (*118*). From its formula, this last anion appears capable of being treated as an

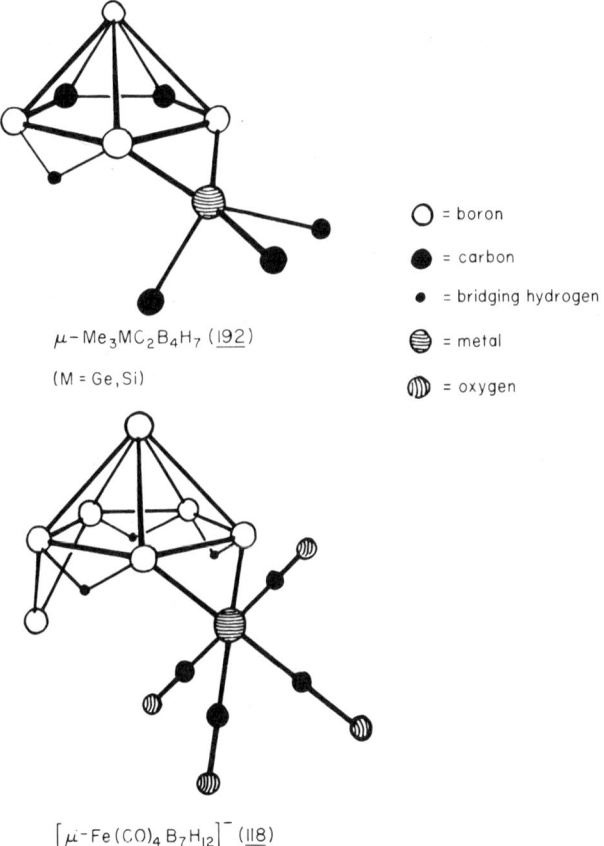

FIG. 20. Metalloboranes with metal atoms in bridging positions. (Terminal hydrogen atoms not shown.)

8-atom cluster (FeB$_7$) with 12 electron pairs available to hold the cluster together, i.e., in Williams's terminology (see following article, this volume) this might be a rare example of a hypho species based on the 11-vertex polyhedron but with 3 vertices unoccupied by skeletal atoms. It seems significant that instead it adopts a structure based on a pentagonal bipyramid with but 1 vertex unoccupied, using its "surplus" electron pairs to bond carbonyl groups or hydrogen atoms to the bridging iron and boron atoms, thereby coordinatively saturating them.

More surprising departures from the normal borane–carborane polyhedral pattern have been found among some metallocarboranes containing platinum (*91, 92*) which adopt nido rather than the expected closo structures. These are the compounds $(Et_3P)_2PtC_2B_7H_7Me_2$ (*91*) and $(R_3P)_2PtC_2B_6H_6R_2'$ (*92*), which have structures based on a very distorted bicapped square antiprism and very distorted tricapped trigonal prism, respectively; the distortions are ones that might have been expected had these compounds contained an extra pair of hydrogen atoms apiece. A closo isomer of the latter compound is known (*92*).

C. OTHER MIXED CLUSTERS

The metal–carbonyl clusters and metalloboranes and -carboranes already mentioned illustrate the range of substances that conform to the

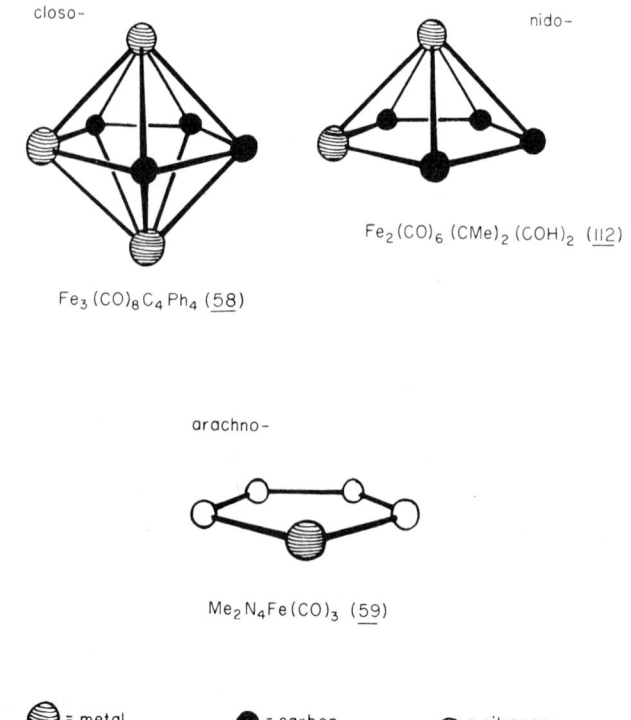

FIG. 21. Skeletons of some mixed clusters with 8 skeletal bond pairs.

borane structural pattern. Some further examples not containing boron but, nevertheless, incorporating both metal and nonmetal skeletal atoms are shown in Figs. 21–23 and listed in Table VI.

Examples with a skeletal electron count of 8 pairs, and with structures formally based on a pentagonal bipyramid, include the *closo*-ferracyclopentadiene complex $Fe_3(CO)_8C_4Ph_4$ (58) and the related nido

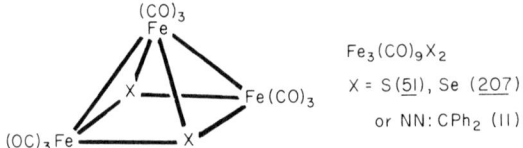

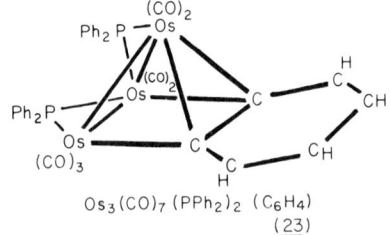

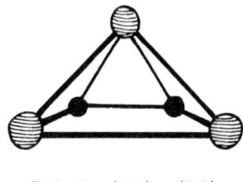

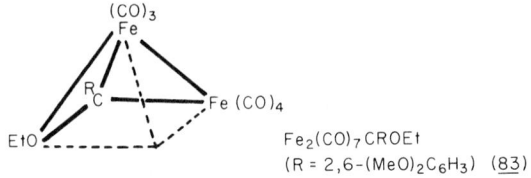

FIG. 22. Some mixed clusters with 7 skeletal bond pairs.

pentagonal-pyramidal complex $[Fe(CO)_3]_2(CMe)_2(COH)_2$ (112). Different types of arachno structure based on the same polyhedron are exemplified by complexes $Fe(CO)_3N_4Me_2$ (55, 59) and $Ir(CO)(PPh_3)_2N_4(C_6H_4F)_2$ (74) (Fig. 21).

The mixed cluster $Co_4(CO)_{10}(C_2Et_2)$ (50) has already been quoted (Fig. 15) as an example of a closo cluster with 6 skeletal atoms and 7 skeletal bond pairs. Nido clusters also formally based on an octahed-

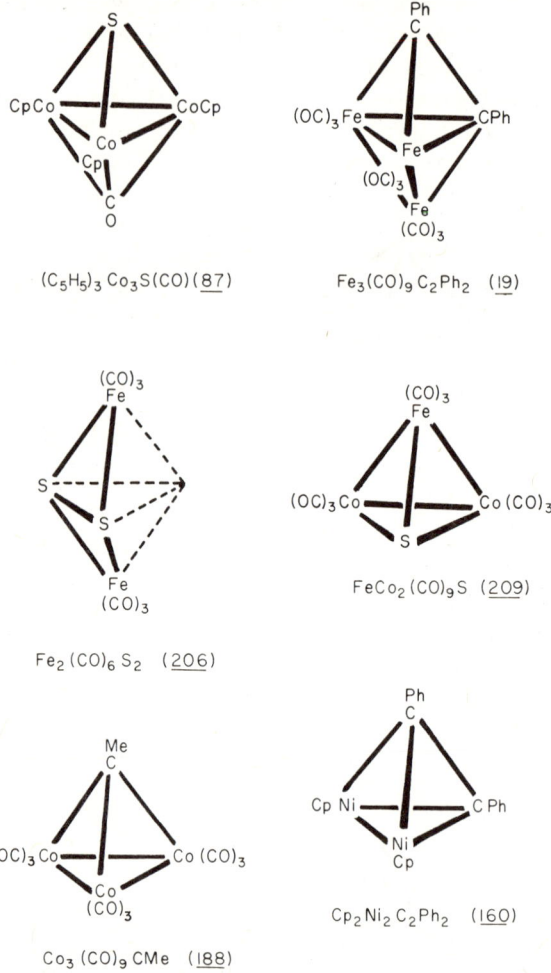

Fig. 23. Some mixed clusters with 6 skeletal bond pairs.

ron include the series of compounds $Fe_3(CO)_9X_2$ [where X = S (51), Se (207), or NN:CPh_2 (11)] all of which have distorted square-pyramidal structures (Fig. 22). Another 5 skeletal atom–7 skeletal bond pair mixed cluster, although less readily recognized as such, is the benzyne complex $Os_3(CO)_7(PPh_2)_2(C_6H_4)$ (23) which has a distorted square-pyramidal arrangement of its Os_3C_2 skeleton (Fig. 22). The electron count for this cluster includes 2 electrons from the $Os_3(CO)_7$ residue, 6 electrons (3

TABLE VI
MIXED CLUSTERS WITH CLOSO, NIDO, OR ARACHNO STRUCTURES[a]

Fundamental polyhedron	b	Closo species ($a = b - 1$)	Nido species ($a = b - 2$)	Arachno species ($a = b - 3$)
Trigonal bipyramid	6	$C_2Ph_2Fe_3(CO)_9$ (19) $Co_3(C_5H_5)_3(CO)S$ (87)	$SFeCo_2(CO)_9$ (209) $MeCCo_3(CO)_9$ (188) $Fe_2(CO)_6S_2$ (206)	$O(CH_2)_2$ $RN(CH_2)_2$ $M(I)M_2(II)(CO)_{12}$[b]
Octahedron	7	$[Co_4Ni_2(CO)_{14}]^{2-}$ (8)	$Fe(_3CO)_9X_2$[c] $Os_3(CO)_7(PPh_2)_2C_6H_4$ (23) $C_2R_2Os_3(CO)_{10}$ (54)	$Fe_2(CO)_7C(OEt)C_6H_3(OMe)_2$ (83)
Pentagonal bipyramid	8	$C_4Ph_4Fe_3(CO)_8$ (58)	$C_4Me_2(OH)_2Fe_2(CO)_6$ (112)	$N_4Me_2Fe(CO)_3$ (55, 59)

[a] $a =$ No. of skeletal atoms; $b =$ No. of skeletal bond pairs. For illustrative structures, see Figs. 15, 21, 22, and 23.
[b] $M(I), M(II) =$ Fe, Ru, Os.
[c] $X = S$ (51), Se (207), or $NN:CPh_2$ (11).

apiece) from the bridging diphenylphosphino ligands, and 6 electrons from the benzyne residue (cf an acetylene C_2R_2). Figure 22 also shows an example of an arachno species formally based on an octahedral fragment, namely, the carbene complex $Fe_2(CO)_7C(OEt)C_6H_3(OMe)_2$ (83).

Mixed clusters formally containing 6 skeletal bond pairs are also common. The acetylene complex $Fe_3(CO)_9C_2Ph_2$ has a distorted trigonal-bipyramidal (closo) arrangement of its Fe_3C_2 skeleton, with apical carbon and iron atoms (Fig. 23) (19). Most clusters with 6 skeletal bond pairs are, however, systems with 4 skeletal atoms arranged at the corners of a tetrahedron. As such, they appear to be n-vertex closo species that contravene the $(n + 1)$ bond pair rule. However, the tetrahedron differs from the triangular-faced polyhedra in Fig. 1 in that its symmetry is such as to generate $(n + 2)$ rather than $(n + 1)$, bonding MO's from the three AO's contributed by each skeletal atom at its corners. Formally, the tetrahedron may be regarded as derivable from the trigonal bipyramid by removal of one apex, so tetrahedral clusters are effectively 4-atom, 6 bond pair nido systems.

The tetrahedron is unique in another respect, in that each vertex is directly linked to only 3 others, so that the three AO's contributed by each skeletal atom can be employed in forming six two-center bonds along the six tetrahedron edges, as illustrated by the mixed clusters $FeCo_2(CO)_9S$ (209), $MeCCo_3(CO)_9$ (188), and $Co_3(C_5H_5)_3(CO)S$ (87). This last compound has the carbonyl group triply bridging the metal atoms, effectively completing the trigonal bipyramid (Fig. 23).

Not all such species conform to the electron count expected for tetrahedral or trigonal-bipyramidal species. For example, the tetrahedral clusters $SCo_3(CO)_9$ (209) and $BuNNi_3(C_5H_5)_3$ (131) formally have 1 electron too many, and the trigonal bipyramidal clusters $S_2Co_3(C_5H_5)_3$ (87) and $S_2Ni_3(C_5H_5)_3$ (87) have 2 and 5 electrons, respectively, in excess of the number appropriate for 5-atom closo species. Such trigonal-bipyramidal clusters have been the subject of a careful study by Dahl and his co-workers, who have shown, by studying the metal–metal distances, that the extra electrons are accommodated in metal–metal *antibonding* MO's (87). These and other examples show that adding extra electrons to a closo cluster, at least one containing metal atoms, may simply increase the *size* of the cluster rather than change its shape to one based on a higher polyhedron.

The remaining structure illustrated in Fig. 23 is that of the 4-atom nido cluster $Fe_2(CO)_6S_2$, with 2 apical iron atoms and 2 equatorial sulfur atoms arranged about an incomplete trigonal bipyramid (206). Similar structures are adopted by various diaza complexes of general formula $Fe_2(CO)_6(RNNR)$ (109).

VI. Hydrocarbon Systems Conforming to the Borane Pattern

Although boranes and hydrocarbons are more notable for their differences than for any similarities, there are several important hydrocarbon systems that adopt structures that conform to the borane pattern (202). They include metal–hydrocarbon π complexes, various aromatic systems, and certain other neutral or charged hydrocarbons. Representative examples are listed in Table VII.

The relationship between borane clusters and metal–hydrocarbon π complexes is illustrated by metal cluster –acetylene complexes such as those already cited, namely, $Co_4(CO)_{10}C_2Et_2$ (50) (Fig. 15) and $Fe_3(CO)_9$-C_2Ph_2 (19) (Fig. 23), whose octahedral and trigonal-bipyramidal skeletons, respectively, are directly analogous to those of the closo-carboranes $B_4H_4C_2H_2$ and $B_3H_3C_2H_2$ (151). Both types of cluster can be prepared by reactions between acetylenes and appropriate homonuclear clusters (metal–carbonyl clusters or boranes). Both types of cluster, when heated, can rearrange to separate the carbon atoms originally triply bonded to each other (18, 95). The dimerization of acetylenes that can take place at metal centers, as in the formation of the ferracyclopentadienyl compound $Fe_3(CO)_8C_4Ph_4$ (58) (Fig. 21), can also take place at borane centers, as in the formation of C-methylated tricarba-nido-hexaboranes $C_3H_2MeB_3H_4$ from B_4H_{10} and HC:CH (97). However, the close bonding relationship between metal cluster–acetylene complexes and carboranes is obscured by the different ways in which their bonding is normally treated. Whereas the bonding in carboranes is not usually discussed in terms of localized two-center bonds, this approach is used for bonding in the carbon skeleton or metal–acetylene complexes (indeed of metal–hydrocarbon π complexes in general) (41); attention is normally focused on the interactions between metal orbitals and hydrocarbon bonding or antibonding π orbitals, the electrons in the C—C σ bonds being ignored. If the C—C σ-bonding electrons are included in the skeletal electron count, then the parallel between boranes and metal–hydrocarbon π complexes becomes apparent.

For example, the pentagonal-pyramidal shape of the C_5M skeleton of pentahaptocyclopentadienyl–metal complexes, as exemplified by $(C_5H_5)Mn(CO)_3$ (Fig. 24), is the nido structure appropriate for a set of 6 atoms held together by eight skeletal bond pairs [3 electrons from each CH unit; 1 electron from the $Mn(CO)_3$ residue]. Two such pentagonal pyramids sharing a common apex are present in ferrocene, $(C_5H_5)_2Fe$ (22) (Fig. 24). Similarly, cyclobutadiene complexes (197) such as $C_4H_4Fe(CO)_3$ (Fig. 24) have the square-pyramidal geometry appropriate for clusters with 5 skeletal atoms and 7 skeletal bond pairs, whereas

TABLE VII

Examples of Closo, Nido, and Arachno Relationships among Hydrocarbons and Their Metal Complexes[a]

Fundamental polyhedron	b	Closo species ($a = b - 1$)	Nido species ($a = b - 2$)	Arachno species ($a = b - 3$)
Trigonal bipyramid	6	$C_2Ph_2Fe_3(CO)_9$ (19)	$C_2Ph_2Co_2(CO)_6$	$C_2H_4Pt(PPh_3)_3$
Octahedron	7	$C_2Et_2Co_4(CO)_{10}$ (50)	$C_2Ph_2Ni_2(C_5H_5)_2$ (160)	$(CH_2)_3$
			C_4Ph_4CoCp (197)	(π-allyl)$Co(CO)_3$
			$C_5H_5^{+b}$	$C_4H_4^{2-}$
Pentagonal bipyramid	8	$C_4Ph_4Fe_3(CO)_8$ (58)	$(C_5H_5)Mn(CO)_3$	(butadiene)$Fe(CO)_3$ (93)
			$C_6Me_6^{2+}$ (117)	$C_5H_5^-$
Hexagonal bipyramid	9	—	$(C_6H_6)_2Cr$	benzene
Dodecahedron	9	—	—	benzvalene, $Me_2AlC_5H_5$ (61)
Heptagonal bipyramid	10	—	$C_7H_7V(CO)_3$ (9)	$C_7H_7^+$

[a] a = No. of skeletal atoms; b = No. of skeletal bond pairs.
[b] References 56, 108, 140, 154, 155, 187, 213.

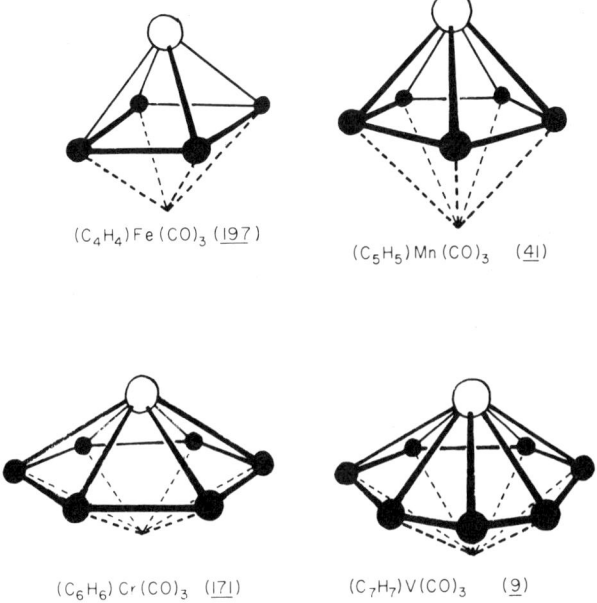

FIG. 24. The nido structures of some metal–hydrocarbon π complexes containing C_nH_n ring systems.

related butadiene complexes such as $(CH_2{:}CHCH{:}CH_2)Fe(CO)_3$ (Fig. 25), with 2 more skeletal electrons but the same number of skeletal atoms, adopt arachno structures based on pentagonal bipyramids. That the metal atom of a butadiene complex occupies an axial, not an equatorial, vertex of the pentagonal bipyramid is apparently in order to allow the 2 extra hydrogen atoms (those bound by skeletal electrons), 1 on

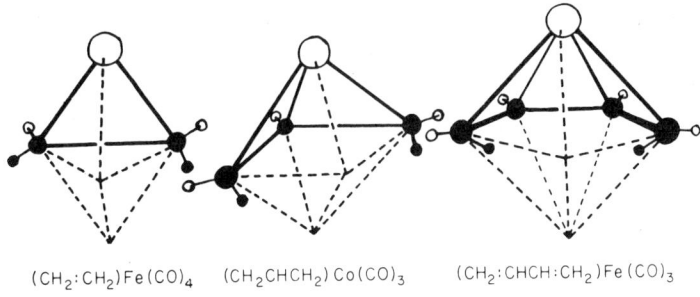

FIG. 25. The arachno structures of some metal–hydrocarbon π complexes. (Carbonyl ligands not shown.)

each methylene group of the butadiene, to be accommodated in endo positions without displacing the remaining methylene hydrogen atoms from their preferred exo positions, directed radially away from the center of the cluster. Significantly, the ^{1}H NMR chemical shift of the endo-hydrogen atoms (9.78τ) *(93)* reveals the influence of their proximity to the metal. A planar C_4Fe skeleton for a complex $(CH_2:CHCH:CH_2)$-$Fe(CO)_3$ which also in principle appears possible for an arachno 8-electron pair system, would only allow an exo orientation of four CH groups if the remaining 2 hydrogens were in C$\cdots$H$\cdots$C or C$\cdots$H$\cdots$Fe bridging positions. Isoelectronic species without these 2 extra hydrogen atoms, e.g. complexes of chelating ligands such as RN:C(R')C(R'):NR or RCOCOR, can and do form planar five-membered C_2N_2M or, C_2O_2M rings in their metal complexes. Similar arguments may be applied to rationalize the nonplanar C_3M skeletons or π-allyl complexes, which, as arachno species based on an octahedron (Fig. 25), might also in principle be planar.

Further examples of organometallic compounds with hydrocarbon residues ranging in complexity from monoolefins or -acetylenes to the cycloheptatrienyl ring are listed in Table VII. Species with 6, 7, or 8 skeletal bonding electron pairs have skeletal structures that can be related to the same fundamental polyhedra (trigonal bipyramid, octahedron, or pentagonal bipyramid, respectively) as can cage boranes or transition metal carbonyl clusters. However, species with 9 or 10 skeletal bond pairs, such as $(C_6H_6)Cr(CO)_3$ *(171)* or π-$C_7H_7V(CO)_3$ *(9)* adopt structures clearly related to the hexagonal or heptagonal bipyramid (Fig. 24) rather than to the D_{2d} dodecahedron or D_{3h} tricapped trigonal prism of $B_8H_8^{2-}$ and $B_9H_9^{2-}$, respectively *(99, 100)* (Fig. 1). Although this shows there is not one unique fundamental polyhedron for each of the 9- and 10-electron pair cases from which skeletal structures are formally derived, it is, nevertheless, apparent that both $(C_6H_6)Cr(CO)_3$ and $(C_7H_7)V(CO)_3$ are *nido* species, i.e., in which the number of skeletal bond pairs exceeds the number of skeletal atoms by 2 and the skeletal structure is clearly recognizable as based on a triangular-faced polyhedron, 1 vertex of which is left vacant. Expressed another way, the carbocyclic systems $C_4H_4^{2-}$, $C_5H_5^-$, C_6H_6, $C_7H_7^+$ (and $C_8H_8^{2+}$) and aromatic ring systems isoelectronic with these form a set of arachno species capable of conversion into *nido* derivatives by coordination to a metal–carbonyl fragment that provides three AO's, but no electrons, for cluster bonding *(202)*.

A corollary of this treatment is that *closo* species appear capable of preparation from these arachno-ring systems by coordination to *two* metal–carbonyl residues in apical positions, on each side of the C_nH_n

ring. Thus, it might be possible to prepare species such as $(CO)_3Mn(C_4H_4)Mn(CO)_3$, $(CO)_3M(C_5H_5)Mn(CO)_3$ (M = Cr, Mo, or W), or $(CO)_3M(C_5H_5)Fe(C_5H_5)M(CO)_3$ as well as coordination polymers such as $[(C_4H_4)M]_n$ (M = Fe, Ru, or Os) or $[(C_5H_5)M]_n$ (M = Mn, Tc, or Re) consisting of octahedra or pentagonal bipyramids, respectively, linked by shared apical metal atoms. The ferracyclopentadienide $Fe_3(CO)_8C_4Ph_4$ (Fig. 21) (58) has a metal–carbonyl residue on each side of an FeC_4 ring and so is of the structural type envisaged, as is the metallocarborane $(C_5H_5)_2Co_2C_2B_3H_5$ (14) (Fig. 19).

The only complex containing a *complete* cyclopentadienyl ring sandwiched between 2 metal atoms that has been the subject of an X-ray crystallographic study is the triple-decker sandwich compound

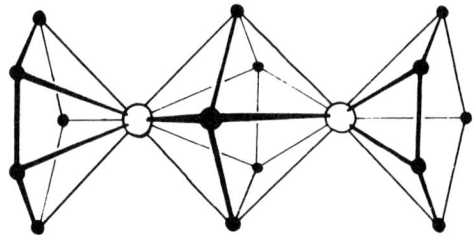

FIG. 26. The triple-decker sandwich structure of cation $[(C_5H_5)Ni(C_5H_5)Ni(C_5H_5)]^+$ (66, 176).

$[(C_5H_5)Ni(C_5H_5)Ni(C_5H_5)]^+BF_4^-$ (Fig. 26) (66, 176) which can be prepared from nickelocene:

$$2(C_5H_5)_2Ni + R^+X^- \rightarrow [(C_5H_5)Ni(C_5H_5)Ni(C_5H_5)]^+X^- + C_5H_5R$$
$$(R = Ph_3C, C_7H_7; \ X = BF_4, PF_6)$$

Like nickelocene itself, this complex has 2 electrons more than the number expected using the electron-counting approach outlined in this article. Interestingly, the iron analog $[(C_5H_5)_3Fe_2]^+$, for which such a triple-decker structure appears appropriate on an electron count, has been detected in mass-spectroscopic studies on ferrocene (179) and on the cyclopentadienyl iron carbonyl tetramer $[C_5H_5FeCO]_4$ (136).

Many main group metal derivatives of unsaturated hydrocarbons adopt structures recognizably nido or arachno in type. Nido structures are, for example, adopted by a range of pentahaptocyclopentadienyl derivatives including $MeBe(C_5H_5)$ (60, 62), $ClBe(C_5H_5)$ (60, 63), $BrBe(C_5H_5)$ (104), $HC\!:\!CBe(C_5H_5)$ (104), $(C_5H_5)_2Be$ (64, 102, 218), $(C_5H_5)_2Mg$ (103), $(C_5H_5)In$ (85, 181), $(C_5H_5)Tl$ (196), $(C_5H_5)_2Sn$ (10), and $(C_5H_5)_2Pb$ (167) (Fig. 27). The pentahapto geometry of the bis(cyclopentadienyls) $(C_5H_5)_2M$ (M = Mg, Sn, or Pb) can formally be reconciled with the borane pattern only if these metals use more than four valence shell orbitals. Bis(cyclopentadienyl)beryllium $(C_5H_5)_2Be$ in the crystal (218) significantly has one pentahapto- and one monohapto-cyclopentadienyl group, as appropriate for the four AO's available in

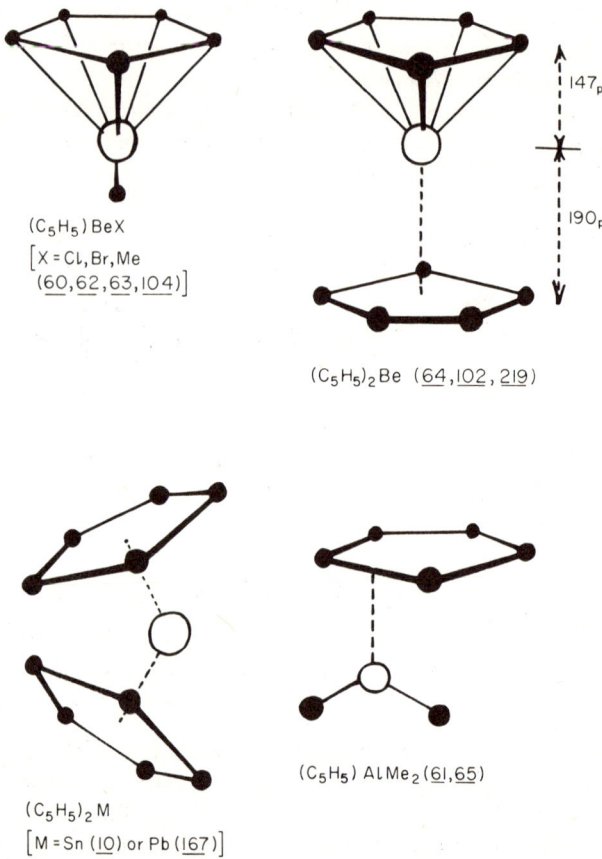

FIG. 27. Cyclopentadienyl derivatives of some main group metals.

the beryllium valence shell. The unsymmetrical sandwich structure (Fig. 27) of this compound in the vapor phase may be interpreted similarly, although it has been pointed out (64, 102) that repulsions between the ring π electrons may not allow a symmetrical ferrocene-type structure to be adopted.

Arachno structures among main group metal–hydrocarbon complexes are represented by the di- or trihaptocyclopentadienylaluminum compound $(C_5H_5)AlMe_2$ *(61, 65)* (Fig. 27) and possibly also by certain lithium derivatives of aromatic hydrocarbons. Compounds such as $Ph_3CLi(Me_2NC_2H_4NMe_2)$ *(29)*, the bis(quinuclidine) adduct of fluorenyllithium *(27)* and the tetramethylethylenediamine adduct of naphthalene dilithium *(28)* have structures in which the lithium atoms are evidently bonded directly to 3 or 4 carbon atoms of the hydrocarbons, although

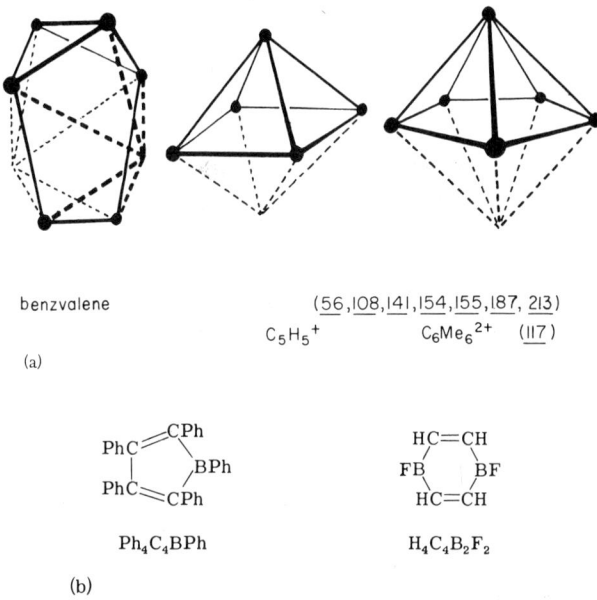

benzvalene   (56,108,141,154,155,187, 213)

(a) $C_5H_5^+$ $C_6Me_6^{2+}$ (117)

Ph_4C_4BPh $H_4C_4B_2F_2$

(b)

FIG. 28. (a) The arachno structure of benzvalene and nido structures of $C_5H_5^+$ and $C_6Me_6^{2+}$. (b) The boron heterocycles Ph_4C_4BPh and $H_4C_4B_2F_2$.

analysis of these complex structures in terms of fragments of 6- or 7-cornered polyhedra is necessarily somewhat arbitrary.

Other hydrocarbon systems that clearly conform to the borane structural pattern include the benzene isomer benzvalene, C_6H_6 (Fig. 28), which has an arachno structure based on a D_{2d} dodecahedron (*cf.* $B_8H_8^{2-}$ in Fig. 1), and the *nido*-carbo cations $C_5H_5^+$ *(56, 108, 140, 154, 155, 187, 213)* and $C_6Me_6^{2+}$ *(117)*, which are believed to adopt the nido-pyramidal structures shown in Fig. 28 instead of planar cyclic structures like the parent *arachno*-aromatic systems, $C_5H_5^-$ and C_6Me_6. Interestingly, the related organoborane, Ph_4C_4BPh, apparently adopts a five-membered borole ring structure (Fig. 28) (*25*, *75*) in which form it is, not surprisingly, highly reactive. It should be possible to prepare a square-pyramidal isomer of this compound.

Another boron heterocycle with a formula apparently appropriate for a nido-pyramidal structure is the compound $H_4C_4B_2F_2$ (*193*) (Fig. 28). The fluorine substituents in this molecule, however, can π-bond to the boron atoms, thereby generating a quinonoid electronic structure. Elsewhere in boron cluster chemistry, the presence of halogen substituents appears to modify the skeletal electron requirements of the cluster.

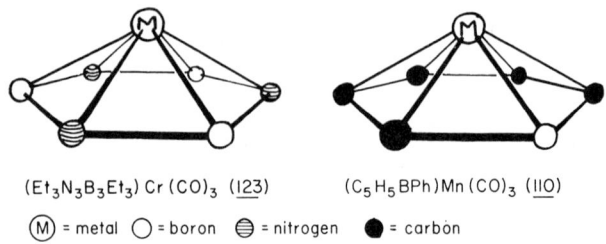

FIG. 29. The nido skeletons of the hexaethylborazine complex $(Et_3N_3B_3Et_3)$-$Cr(CO)_3$ and borabenzene complex $(C_5H_5BPh)Mn(CO)_3$.

For example, the neutral chloride B_8Cl_8 (*169*) and hydride anion $B_8H_8^{2-}$ (*100*) both adopt D_{2d} dodecahedral structures (Fig. 1) despite their differing numbers of skeletal bond pairs, but, significantly, the interatomic distances in the former are longer. Again, both B_9Cl_9 (*122*) and $B_9H_9^{2-}$ (*99*) have tricapped trigonal-prismatic structures, although with different interatomic distances.

Other unsaturated boron heterocycles, such as borazines and borabenzenes, form transition metal complexes with the expected nido geometry, as exemplified by compounds $(Et_3N_3B_3Et_3)Cr(CO)_3$ (*123*) and $(C_5H_5BPh)Mn(CO)_3$ (*110*) (Fig. 29).

VII. Interatomic Distances in Clusters

Whereas the main object of this survey is to explore how the *shapes* of various substances reflect the number of electron pairs holding them together, it is worth considering also how their *sizes* can be rationalized using the electron-counting approach already outlined (*203*).

In a closo-cluster species such as a borane anion $B_nH_n^{2-}$, in which the n skeletal atoms are held together by $(n + 1)$ bond pairs, the interatomic distances—the edge lengths of the B_n polyhedron—can be rationalized if one regards the skeletal atoms as held together by a network of fractional order two-center bonds along the polyhedron edges. The number of electron pairs associated with each edge will be the effective bond order of the two-center link represented by that edge.

The average edge bond order for a particular closo species can be calculated by dividing the total number of skeletal bond pairs $(n + 1)$ by the number of edges, which, for exclusively triangular-faced polyhedra with n vertices, is $(3n - 6)$. This leads directly to the formal boron–boron bond order for those borane anions [$B_6H_6^{2-}$ (*177*) and $B_{12}H_{12}^{2-}$ (*219*)] in which all the polyhedron edges are the same length. The octahedral anion $B_6H_6^{2-}$ has a formal boron–boron bond order of 0.58 and a B—B distance of 169 pm, whereas the icosahedral anion $B_{12}H_{12}^{2-}$ has a formal bond order of 0.43 and a B—B distance of 177 pm.

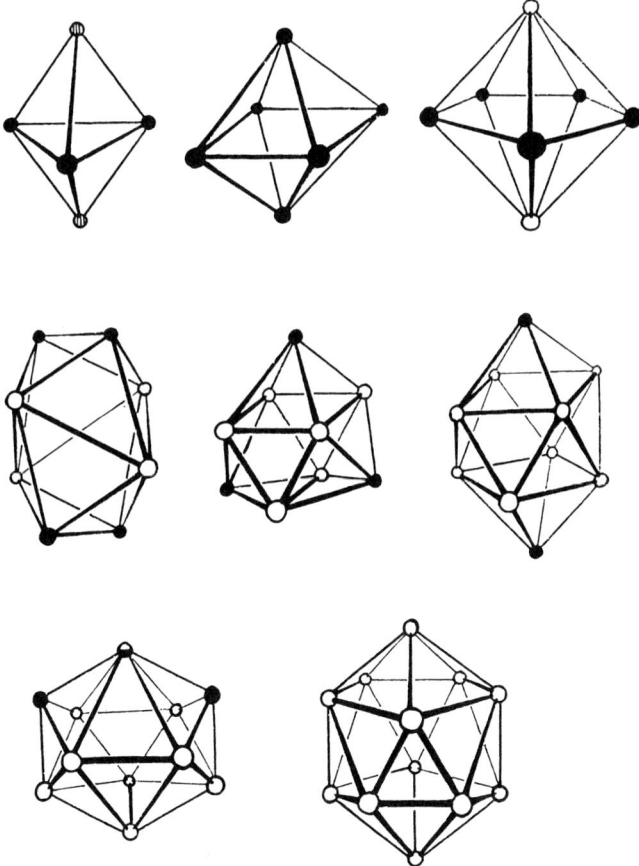

FIG. 30. Skeletal coordination numbers: ◐ = 3; ● = 4; ○ = 5; ◒ = 6.

The less symmetrical polyhedra contain two or more distinct types of edge, linking vertices of similar or different coordination numbers (Fig. 30), and understandably differing in length (see Table VIII) *(57, 99, 100)*. Some insight into these lengths can be obtained by assuming that, for the homonuclear closo clusters $B_8H_8^{2-}$ *(100)*, $B_9H_9^{2-}$ *(99)*, and $B_{10}H_{10}^{2-}$ *(57)* the $(n + 1)$ skeletal electron pairs are distributed evenly among the n skeletal atoms. This gives each skeletal atom $(n + 1)/n$

TABLE VIII

SKELETAL (POLYHEDRON-EDGE) BOND ORDERS FOR ANIONS $B_nH_n^{2-}$

Anion $B_nH_n^{2-}$	Skeletal coordination numbers of linked atoms		Edge bond order[a] $(n + 1)(x_1 + x_2)/nx_1x_2$	Interatomic distance (pm)
	x_1	x_2		
$B_6H_6^{2-}$	4	4	0.58	169 *(177)*
$B_7H_7^{2-}$	4	4	0.57	167[b]
	4	5	0.51	175[b]
$B_8H_8^{2-}$	4	4	0.56	156 *(100)*
	4	5	0.51	174
	5	5	0.45	193
$B_9H_9^{2-}$	4	5	0.50	171 *(99)*
	5	5	0.44	189
$B_{10}H_{10}^{2-}$	4	5	0.50	173 *(57)*
	5	5	0.44	183
$B_{11}H_{11}^{2-}$	4	5	0.49	170[b]
	4	6	0.45	178[b]
	5	5	0.44	179[b]
	5	6	0.40	195[b]
$B_{12}H_{12}^{2-}$	5	5	0.43	177 *(219)*

[a] Estimated, assuming that the skeletal bonding electrons are distributed equally among the skeletal atoms.
[b] Predicted (see text).

electron pairs with which to form bonds to its x skeletal neighbors, where x is the "skeletal coordination number" or order of that vertex (Fig. 30). A skeletal atom with x skeletal neighbors can supply $(n + 1)/nx$ electron pairs to each of the two-center links it forms. The two-center link between 2 skeletal atoms of coordination numbers x_1 and x_2, respectively, will thus contain a total of $[(n + 1)/nx_1 + (n + 1)/nx_2]$ electron pairs, i.e., $(n + 1)(x_1 + x_2)/nx_1x_2$, which is the bond order expected using this approach.

Bond orders calculated on this basis for the various types of two-center link in anions $B_8H_8^{2-}$ *(100)*, $B_9H_9^{2-}$ *(99)*, and $B_{10}H_{10}^{2-}$ *(57)* are given in Table VIII, together with the interatomic distances. The approach leads consistently to the correct qualitative sequence of bond orders for each of these anions—the strongest bonds, and therefore shortest edges, in a particular polyhedron link the atoms of lowest coordination number. However, it underestimates the *extent* to which the various types of bond differ. Judged by their lengths alone, the shortest and longest bonds in $B_8H_8^{2-}$ appear to have orders of about 1.0 and 0.3, respectively. The differences in length are in part, although not entirely, due to the familiar dependence of bond length on coordination number. That the bonds linking atoms of low coordination number are significantly shorter than would be expected on the basis of equal sharing of the skeletal electrons among the skeletal atoms implies that these skeletal atoms have a greater than average share of the skeletal bonding electrons, i.e., the skeletal atoms of lowest coordination number are negatively charged relative to the remainder, whereas atoms of high coordination number are relatively positively charged. These conclusions are consistent with those arrived at by other approaches, e.g., using MO calculations (*100, 145, 164*).

Yet other factors influence the interatomic distances in *nido-* and *arachno-*boranes, for example whether the polyhedron edges border the open face of the polyhedral fragment, and if so whether they are "sewn up" by B---H---B links, when some lengthening of the B----B distance is observed. Nevertheless, for polyhedron edges not subject to such factors, a reasonable consistency of B—B distances can be discerned for species formally based on the same polyhedron. For example, compare the B—B distance in $B_6H_6^{2-}$ (169 pm) with the apical-basal B—B distance of 169 pm in B_5H_9 (*42, 67, 120*) and with the unique short unbridged B—B link in B_4H_{10} (172 pm) (*169*) (Fig. 31). This last link

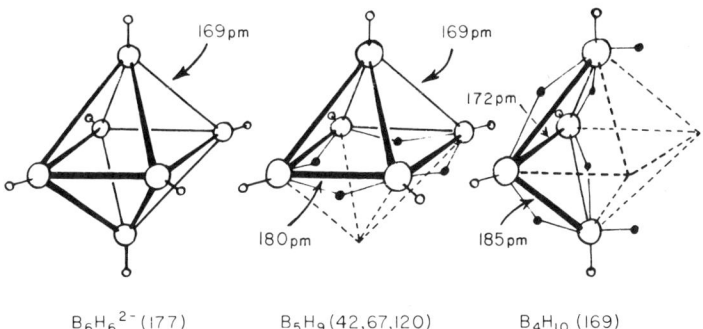

$B_6H_6^{2-}$ (*177*) B_5H_9 (*42,67,120*) B_4H_{10} (*169*)

Fig. 31. Boron–boron interatomic distances in boranes with skeletons based on an octahedron: *closo*-$B_6H_6^{2-}$, *nido*-B_5H_9, and *arachno*-B_4H_{10}.

incidentally is treated as a bond of order 1 in the usual localized bond treatment (Fig. 4). Assuming similar relationships hold for the closo, nido, and arachno series of boranes based on a pentagonal bipyramid, $B_7H_7{}^{2-}$, B_6H_{10} (*172*), and B_5H_{11}, one would predict the following B—B interatomic distances in $B_7H_7{}^{2-}$: approx 167 pm (equatorial-equatorial) and 175 pm (axial-equatorial). Again, comparing $B_{11}H_{11}{}^{2-}$ with $B_{10}H_{14}$ (*145*) leads to the edge lengths predicted in Table VIII for $B_{11}H_{11}{}^{2-}$.

TABLE IX

Carbon–Carbon Bond Distances in Carboranes, Metal–Acetylene π Complexes, and Metal–Cyclobutadienyl and –Cyclopentadienyl π Complexes

No. of vertices on fundamental polyhedron	Structure type	Coordination No. of carbon atoms	C—C distance (pm)	Compound	Ref.
5	Nido	4	135	$Ph_2C_2Ni_2(C_5H_5)_2$	*160*
5	Closo	4, 5	141	$Ph_2C_2Fe_3(CO)_9$	*19*
6	Closo	5	144	$Et_2C_2Co_4(CO)_{10}$	*50*
6	Nido	4	145	$Ph_4C_4Co(C_5H_4CN)$	*197*
6	Closo	5	154	$C_2B_4H_6$	*13*
7	Nido	4	141	$C_2B_3H_7Fe(CO)_3$	*26*
7	Nido	4	142	$(HO)_2Me_2C_4Fe_2(CO)_6$	*112*
7	Nido	4	143	$C_2B_4H_8$	*20*
7	Nido	4	143^a	$C_5H_5MX_n$	*41*
7	Nido	4	143	$C_4B_2H_6$	*168*
7	Closo	5	145	$Ph_4C_4Fe_3(CO)_8$	*58*
7	Closo	5	147	$C_2B_4H_6GaMe$	*98*
10	Nido	5	155	$C_2B_7H_9Me_2$	*121*
12	Closo	6	165^a	$C_2B_{10}H_{12-n}R_n$	*21*

^{a} Average values for compounds of these general formulas.

The C—C and C—B interatomic distances in carboranes can also be related to the coordination numbers of the skeletal atoms. Two factors tend to make these distances shorter than the B—B distances in comparable boranes: the preference of the carbon atoms for sites of low coordination number and the greater electronegativity of carbon than boron, which increases the electron density in the region of the carbon atoms and so strengthens the bonds that they form. Table IX lists some C—C distances for *closo-* and *nido-*carboranes (*13, 20, 21, 26, 98, 121, 168*) and metal–acetylene (*50, 58, 112*) complexes, relating them to the coordination numbers of the atoms concerned. Although it is usual to treat the C—C links of metal–acetylene complexes as triple bonds reduced somewhat in strength by coordination, and the C—C links of carboranes as probably of order less than 1, the data in Table IX show that the formal bond order for both categories of compound lies between 1 and 2 [cf. the bond order in complexes of C_nH_n rings (*197*)].

Despite the resemblance between the shapes of borane clusters and metal clusters, this does not imply that the bond orders in metal clusters are as low as those given in Table VIII for borane anions. Whereas for the borane anions the number of electrons available for skeletal bonding is relatively clear-cut, in metal clusters the assignment of certain electrons to a nonbonding role, as outlined in Section IV, was a convenient device by which to stress their relationship to boranes. As was pointed out in connection with the various 86-electron octahedral clusters such as $H_2Ru_6(CO)_{18}$, these contain just 2 electrons too many to allow their bonding to be described in terms of 12 two-center electron pair bonds along the octahedron edges, and the metal–metal bond order in these clusters is likely to be rather higher than the boron–boron bond order in $B_6H_6^{2-}$. Microcalorimetric studies on the rhodium carbonyl clusters $Rh_4(CO)_{12}$ and $Rh_6(CO)_{16}$ (*29a, 42a*) have nevertheless shown their metal–metal bonds to be weak (less than 30 kcal·mol^{-1}, i.e., about two-thirds of the strength of their metal–carbon bonds).

VIII. Some General Reactions

So far, in this survey, attention has been focused on the shapes and sizes of cluster species, their preparation and reactions having been mentioned only in passing. However, the close link between the shapes of borane-type clusters and the numbers of bonding electrons they contain makes it possible to predict the likely outcome of reactions that may change these numbers.

For example, opening of the cluster skeleton from closo to nido to arachno will accompany the addition of electron pairs, whether by reduction reactions or by addition of Lewis base molecules that do not dislodge other Lewis bases. Cluster closing is expected to accompany the removal of electron pairs, whether by oxidation or by loss of a substituent together with the electron pair by which it was formally bound to the cluster:

$$closo \underset{-2\,\text{electrons}}{\overset{+2\,\text{electrons}}{\rightleftarrows}} nido \underset{-2\,\text{electrons}}{\overset{+2\,\text{electrons}}{\rightleftarrows}} arachno$$

For example, reductive cage opening, closo → nido, is the first step normally used in Hawthorne's polyhedral expansion reactions of carboranes (63–71, 77–80):

$$\underset{closo\text{-}}{C_2B_{n-2}H_n} \xrightarrow{Na/naphthalene} \underset{nido\text{-}}{C_2B_{n-2}H_n{}^{2-}} \xrightarrow{CpCo^{2+}} \underset{closo\text{-}}{C_2B_{n-2}H_nCoCp}$$

Reduction of *closo*-dicarbaboranes, $C_2B_{n-2}H_n$, to the dianionic nido species, $C_2B_{n-2}H_n{}^{2-}$, which are more susceptible to rearrangement reactions, also provides a route to isomers of the starting materials in which the carbon atoms have moved to different polyhedron vertices, e.g. (185, 186),

$$\underset{closo\text{-}}{1,2\text{-}C_2B_{10}H_{12}} \xrightarrow[(+2e^-)]{2Na} \underset{nido\text{-}}{C_2B_{10}H_{12}{}^{2-}} \xrightarrow[(-2e^-)]{H_2O} \underset{closo\text{-}}{1,2\text{-},1,7\text{-},\text{ and }1,12\text{-}C_2B_{10}H_{12}}$$

Similar reactions applied to transition metal–acetylene complexes appear capable of separating the 2 carbon atoms originally linked by the acetylenic triple bond (18). Thermal isomerization of metal–acetylene complexes may achieve the same result, showing how metal clusters can catalyze scrambling reactions of acetylenes, e.g.,

$$R^1C\vdots CR^1 + R^2C\vdots CR^2 \xrightarrow{metal\ cluster} 2R^1C\vdots CR^2$$

The alkaline degradation of *closo*-carboranes, used to prepare smaller carboranes from icosahedral starting materials, also occurs by closo- to nido-, possibly even arachno-, cage opening, e.g. (107),

$$\underset{closo\text{-}}{C_2B_{10}H_{12}} \xrightarrow{OEt^-} \underset{nido\text{-}}{[C_2B_{10}H_{12}OEt]^-} \xrightarrow{OEt^-} \underset{arachno\text{-}}{[C_2B_{10}H_{12}(OEt)_2]^{2-}} \xrightarrow{-HB(OEt)_2} \underset{nido\text{-}}{[C_2B_9H_{11}]^{2-}}$$

Examples of reductive cluster-opening and oxidative cluster-closing reactions are common in the chemistry of metal–hydrocarbon π complexes. For example, bases convert *nido*- (hexa-hapto)arene–manganese tricarbonyl complexes into *arachno*(pentahapto)-π-cyclohexadienyl complexes (129, 130, 217):

This exemplifies a quite general reaction in organometallic chemistry, the cluster-opening conversion of an n-hapto ligand into an $(n-1)$-hapto ligand by addition of a nucleophile X^-. The converse, cluster-closing conversion accompanies removal of X^-.

Reference to Tables II and III shows that cluster expansion reactions are in principle possible if neutral units such as BH, $Fe(CO)_3$, $Co(C_5H_5)$, or $Ni(PR_3)_2$ can be incorporated in an existing polyhedral cluster:

$$\text{An } n\text{-atom cluster} \begin{Bmatrix} \text{closo} \\ \text{nido} \\ \text{arachno} \end{Bmatrix} \xrightarrow{ML_x} \begin{Bmatrix} \text{closo} \\ \text{nido} \\ \text{arachno} \end{Bmatrix} \text{An } (n+1)\text{-atom cluster}$$

where $ML_x = BH$, $Fe(CO)_3$, $Co(C_5H_5)$, $Ni(PR_3)_2$, etc., units that can furnish 1 skeletal atom and 1 skeletal bond pair. It has already been noted that Hawthorne's polyhedral expansion procedure incorporates $Co(C_5H_5)$ units in carborane polyhedra in two stages [reduction to the nido dianion $C_2B_{n-2}H_n{}^{2-}$ followed by addition of $Co(C_5H_5)^{2+}$] (68–71, 77–80, 82). Alternatively, separate stages may not be necessary, as in the formation of the 10-atom clusters $(C_5H_5)_2Co_2C_2B_8H_{10}$ (six isomers) by thermal decomposition of $(C_5H_5)CoC_2B_8H_{10}$ (81), and in the synthesis of the icosahedral metallocarborane $Me_2C_2B_9H_9Pt(PEt_3)_2$ from closo-$Me_2C_2B_9H_9$ and the complex $Pt(PEt_3)_3$, from which one phosphine ligand is readily dislodged (90, 91, 154):

$$closo\text{-}Me_2C_2B_9H_9 + Pt(PEt_3)_3 \xrightarrow{-Et_3P} closo\text{-} Me_2C_2B_9H_9Pt(PEt_3)_2$$

In related reactions, units, such as $Cr(CO)_3$, $Mn(CO)_3{}^+$, $Fe(CO)_3{}^{2+}$, $Co(C_5H_5)^{2+}$, Me_3N, Be^{2+}, or Tl^+, that can supply 1 skeletal atom and three *vacant* AO's (no skeletal electrons) for cluster bonding can be used to fill in the vacant vertices of nido or arachno species, thereby converting them into closo or nido species, e.g. (184, 194),

$$C_6H_6 + M(CO)_6 \xrightarrow{-3CO} (C_6H_6)M(CO)_3 (M = Cr, Mo, \text{ or } W)$$
(*arachno*-) (*nido*-)

$$C_2B_9H_{12}{}^- + Tl^+ \longrightarrow [TlC_2B_9H_{11}]^- + H^+$$
(*nido*-) (*closo*-)

$$2C_6H_6 + Cr \longrightarrow (C_6H_6)_2Cr$$
(*arachno*-) (*nido*-)

The exchange of 1 skeletal atom by another is exemplified by the common "ligand exchange" reactions of metal–hydrocarbon π complexes, e.g. (88)

$$(Ph_4C_4PdBr_2)_2 \xrightarrow{Fe(CO)_5} Ph_4C_4Fe(CO)_3$$
$$(nido\text{-}) \qquad\qquad\qquad (nido\text{-})$$

by the use of the *closo*-thallium species $Tl_2C_2R_2B_9H_9$ in the synthesis of other icosahedral metallocarboranes (184) and by the preparation of the nido-ferraborane $B_4H_8Fe(CO)_3$ from B_5H_9 and $Fe(CO)_5$ (94).

For further examples of the ways in which the reactions of boranes (and related clusters) can be rationalized in terms of changes in their skeletal electron numbers and distribution, see Ref. (173a).

IX. Other Cluster Systems

A. Bismuth Clusters

Various cationic bismuth clusters, Bi_n^{x+}, have been prepared by Corbett and his co-workers (43, 44, 86, 111) who have pointed out the close relationship between these species and borane clusters. For example, reduction of bismuth(III) chloride by bismuth affords a crystalline product of empirical formula Bi_6Cl_7. An X-ray crystallographic study (111) has shown this to contain cations Bi_9^{5+} and anions $BiCl_5^{2-}$ and $Bi_2Cl_8^{2-}$. The cations have tricapped trigonal prism shapes, although very distorted, in a manner suggesting they might be regarded as nido structures based on a bicapped Archimedean antiprism with 1 vacant vertex (assuming there is 1 "lone pair" on each metal atom, a cation Bi_9^{5+} contains 11 skeletal bond pairs). However, the distortion from the ideal D_{3h} symmetry of a tricapped trigonal prism may well be caused by interactions in crystalline "Bi_6Cl_7" between the cations and the adjacent anions, because the same cations, although hardly distorted from D_{3h} symmetry, have been found by X-ray crystallography in bismuth hafnium chloride $Bi^+ Bi_9^{5+} (HfCl_6^{2-})_3$ (86) (prepared from Bi, $BiCl_3$, and $HfCl_4$).

Different bismuth clusters are apparently formed in reactions between bismuth and bismuth trichloride–aluminum trichloride mixtures, from which the salts $(Bi_5)^{3+} (AlCl_4^-)_3$, and $(Bi_8)^{2+} (AlCl_4^-)_2$ have been isolated (43). A trigonal-bipyramidal structure was predicted for Bi_5^{3+} (a closo 6 skeletal bond pair system; cf. $C_2B_3H_5$), and a square antiprismatic structure for Bi_8^{2+}, as appropriate for an arachno 8-atom, 11 skeletal bond pair cluster. Similar polyhedral shapes appear likely for other clusters not only of bismuth but also of other heavy main group metals (the anion Pb_9^{4-}, for example, is isoelectronic with Bi_9^{5+}).

B. SOME METAL HALIDE CLUSTERS

The octahedral metal clusters that have long been familiar features of the lower halide chemistry of niobium, tantalum, molybdenum, and tungsten represent a category of cluster different from those so far considered in that their metal–metal bonding is best treated as involving *four* AO's on each metal (*49, 133, 144, 165, 178*).

There are two important structural types, containing M_6X_8 and M_6X_{12} units, respectively (Fig. 32). The former, typified by $Mo_6Cl_8^{4+}$,

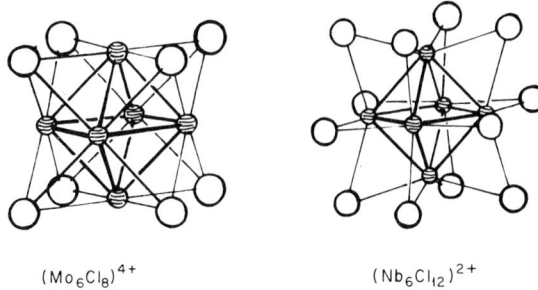

$(Mo_6Cl_8)^{4+}$ $(Nb_6Cl_{12})^{2+}$

FIG. 32. Structures of the octahedral metal–halide clusters $(Mo_6Cl_8)^{4+}$ and $(Nb_6Cl_{12})^{2+}$ (*49, 133, 144, 165, 178*).

have a triply bridging halogen atom X, located over the center of each of the eight octahedral faces. The latter, typified by $Nb_6Cl_{12}^{2+}$, have 12 doubly bridging halogen atoms located over the centers of the twelve octahedral edges. Both types have an effectively square-planar arrangement of halogen atoms about each metal, which contributes four appropriate (e.g., dsp^2 hybrid) orbitals. One further metal orbital (e.g., a pd hybrid) pointing radially outward from the center of the cluster is available for bonding to the exo ligands usually associated with these clusters. This leaves four remaining AO's that point toward the other metals or toward the cluster center. The number of bonding MO's they generate (12 for $Mo_6Cl_8^{4+}$, 8 for $Nb_6Cl_{12}^{2+}$) reflects their orientation (*49*) which, in turn, reflects the orientation of the square-planar MX_4 units with respect to the fourfold axes of the octahedra (Fig. 32). An electron count for the units Mo_6^{12+} and Nb_6^{14+}, which formally constitute the cores of these clusters, shows them to contain 12 and 8 skeletal bond pairs, respectively, as appropriate for closed-shell configurations.

Since an octahedron has twelve edges and eight faces, it is possible to describe the metal–metal bonding in these species in terms of localized two- or three-center bonds, e.g., using twelve two-center octahedron edge bonds for $Mo_6Cl_8^{4+}$, and eight three-center octahedron face bonds for $Nb_6Cl_{12}^{2+}$. Although this may appear to imply much stronger skeletal bonding in the molybdenum cluster, if one takes account of the contribution to skeletal bonding made by the edge- or face-bridging chlorides of these species, then the two types, $Mo_6Cl_8^{4+}$ and $Nb_6Cl_{12}^{2+}$, are seen to be effectively isoelectronic: each contains 20 pairs of electrons associated with the metal–metal or metal–(bridging) halogen bonding, occupying MO's of symmetries A_{1g} (2), A_{2u}, E_g, T_{1u} (2), T_{2g} (2), and T_{2u} *(133)*.

The platinum(II) halides, $PtCl_2$ and $PtBr_2$, adopt hexameric structures Pt_6X_{12} formally analogous to the M_6X_{12} cluster shown in Fig. 32, but with long metal–metal distances (approx. 336 pm). This is not surprising, since the Pt_6^{12+} core on which these halides are formally based contains 24 electron pairs, enough to fill all the cluster antibonding, as well as the bonding orbitals *(200)*.

C. GOLD CLUSTERS

Yet further examples of octahedral metal clusters are to be found in the chemistry of the coinage metals. Borohydride reduction of an ethanolic suspension of tris(*p*-tolyl)phosphinegold nitrate affords, among other products, the cationic cluster $[p\text{-tolyl}_3PAu]_6^{2+}$, which has been isolated as the tetraphenylborate *(16)* and shown by X-ray crystallography to have a slightly distorted octahedral Au_6 skeleton, the phosphine ligands occupying exo positions pointing radially outward. An electron count shows that the gold atoms can use all nine of their valence shell AO's if each atom uses one AO to bond the phosphine ligand, four AO's for nonbonding electron pairs, and the remaining four AO's for cluster bonding, for which 8 electron pairs will then be available. This treatment suggests a closer bonding relationship to clusters such as $Nb_6Cl_{12}^{2+}$ than to $B_6H_6^{2-}$, although a relationship to the latter has been argued *(16)*.

A distorted octahedral arrangement of 6 copper atoms has been found in crystals of the 2-dimethylaminophenylcopper bromide, $[Cu(2\text{-}Me_2\text{-}NC_6H_4)]_4(CuBr)_2 \cdot 1.5C_6H_6$ *(101)*, in which the bromine atoms bridge trans-equatorial edges, whereas the 2-dimethylaminophenyl ligands bridge four octahedron faces. Since both types of ligand perform a bridging role, and so furnish 3 electrons apiece, there are 84 electrons associated with the Cu_6 cluster, which formally has 12 electron pairs available for skeletal bonding if each metal atom uses four AO's for the purpose (cf. $Mo_6Cl_8^{4+}$).

More complicated structures are adopted by some other gold clusters of general formulas $[Au(AuL)_8]^{3+}$ *(15, 16)* and $[Au(AuL)_{10}]^{3+}$ *(5, 16, 17,*

32, 152) (Fig. 33), where L is a 2-electron ligand [the Au_{11} clusters are actually most commonly obtained as neutral species $Au_{11}L_7X_3$, where L is a tertiary phosphine and X is a halogen or pseudohalogen (*32*)]. Both types have a central gold atom surrounded by AuL units at distance (260–290 pm) compatible with metal–metal bonding not only between the central gold atom and the surrounding gold atoms but also between

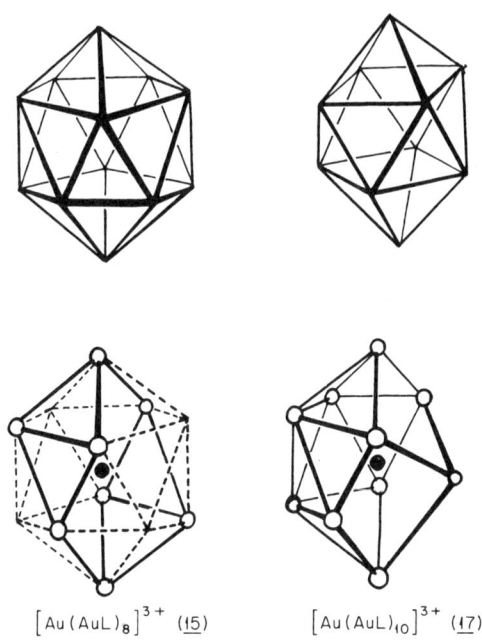

FIG. 33. Skeletons of the gold clusters $(Au(AuL)_8)^{3+}$ and $[Au(AuL)_{10}]^{3+}$ compared with the icosahedron and the bicapped Archimedean antiprism. Central atoms are shaded; peripheral atoms are unshaded.

the surrounding atoms themselves. It, therefore, appears unrealistic to regard their skeletal bonding as involving only interactions between the central atoms and their neighbors, and an ingenious attempt has been made (*16*) to relate these clusters to the borane pattern. However, a simple convincing bonding rationale for these structures has yet to be found.

D. FURTHER METAL–CARBONYL CLUSTERS

In addition to the metal–carbonyl clusters discussed in Sections IV and V, which conform to the borane pattern, there is an increasing number of metal–carbonyl clusters with shapes that have no counterparts in borane chemistry. For example, whereas the nickel–carbonyl dianion $[Ni_3(CO)_3(\mu_2\text{-}CO)_3]_2{}^{2-}$ has an essentially octahedral skeleton (31), as appropriate for a hexanuclear closo cluster formally related to $B_6H_6{}^{2-}$, its platinum analog has a trigonal prismatic skeleton (31a) (Fig. 34). Moreover, it is but one member of a homologous series of

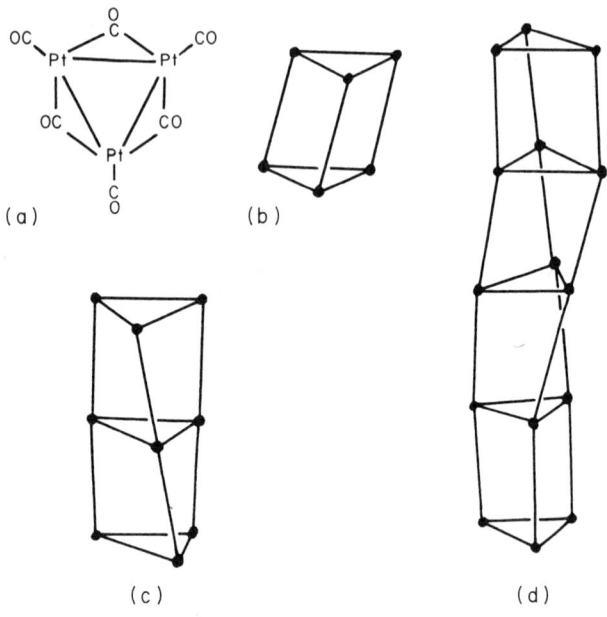

FIG. 34. Structures of the platinum–carbonyl anions $[Pt_3(CO)_3(\mu_2\text{-}CO)_3]_n{}^{2-}$ ($n = 2$, 3, and 5) (31a). (a) The triangular $Pt_3(CO)_3(\mu_2\text{-}CO)_3$ unit from which these anions are built up; (b) $[Pt_6(CO)_{12}]^{2-}$; (c) $[Pt_9(CO)_{18}]^{2-}$; (d) $[Pt_{15}(CO)_{30}]^{2-}$.

platinum–carbonyl anions of general formula $[Pt_3(CO)_3(\mu_2\text{-}CO)_3]_n{}^{2-}$ ($n = 2$, 3, 4, or 5) with column structures (Fig. 34) in which the triangular $Pt_3(CO)_3(\mu_2\text{-}CO)_3$ units are stacked in an eclipsed or nearly eclipsed configuration, forming slightly twisted trigonal prisms (31a). A trigonal prism is, in principle, one of several possible structures for a cluster of 6 units, each capable of forming three bonds, held together by 9 electron pairs, as in prismane (contrast the 10 skeletal bond pairs required for the *tricapped* trigonal-prismatic array of 9 skeletal atoms, as in $B_9H_9{}^{2-}$). That a trigonal prism should form the basis for the structure of $[Pt_3(CO)_3(\mu_2\text{-}CO)_3]_2{}^{2-}$ and its homologs is surprising since a cluster $Pt_6\text{-}(CO)_{12}{}^{2-}$, counting electrons as outlined in Section V, A (Table III), formally contains only 7 skeletal bond pairs.

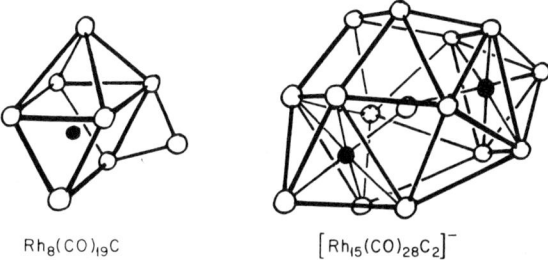

FIG. 35. Skeletal structures of the rhodium–carbonyl carbide complexes $Rh_8(CO)_{19}C$ and $[Rh_{15}(CO)_{28}C_2]^-$ (7).

Among other unusual structures are those of the rhodium–carbonyl carbide clusters $Rh_8(CO)_{19}C$ and $[Rh_{15}(CO)_{28}C_2]^-$ (7) (Fig. 35). The former, for which a closo-dodecahedral structure might have been expected, has a monocapped trigonal prism structure with a central carbon atom and with the remaining metal atom in an edge-bridging position. The latter has a centered tetracapped pentagonal prism structure in which the central metal atom not only has the highly metallic environment of 12 near neighbor metal atoms but also serves as the common vertex of two Rh_6 octahedra, each of which contains a central core carbon atom.

A metal–*nitrosyl* cluster that apparently has no borane counterpart is the anion $[Fe_4S_3(NO)_7]^-$ present in "Roussin's black salt," $Cs^+[Fe_4S_3(NO)_7]^- \cdot H_2O$ (*124*). Its skeleton consists of a trigonal pyramid of iron atoms, the 3 sulfur atoms capping the three pyramid faces (Fig. 36). As a 7-atom cluster formally containing 9 skeletal bond pairs, it might have been expected to adopt a *nido* structure with its Fe_4S_3 skeleton defining all but one of the vertices of a dodecahedron, a structural type that, incidentally and possibly significantly, is missing from the series of *nido*-boranes (see Williams, this volume). This may be because the number of skeletal bond pairs appropriate for such a fragment

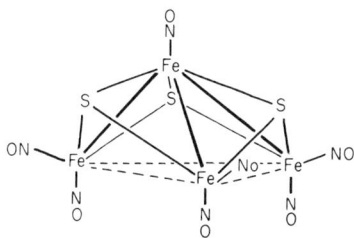

FIG. 36. Structure of anion $[Fe_4S_3(NO)_7]^-$ (*124*).

would not necessarily be the same as the number (9) appropriate for closo-dodecahedral species such as $B_8H_8^{2-}$ or $C_2B_6H_8$. The removal of 1 atom from a dodecahedral cluster would certainly change the symmetry more drastically than would the removal of 2 symmetry-related atoms. In short, closo, nido, arachno relationships are expected to break down when the nido or arachno fragment has a symmetry no longer appropriate to generate the same number of bonding MO's as the parent closo species (cf. the different skeletal bonding requirements of a tricapped trigonal prism—10 bond pairs—and what might be regarded as its hypho derivative a trigonal prism—9 bond pairs). Indeed, the difficulty of preserving the same number of bonding MO's when 3 polyhedron vertices are left vacant is probably one reason why few hypho species are known. Another factor is that, as the excess of electron pairs over skeletal atoms rises, it becomes less appropriate to regard each skeletal atom as contributing as many as three AO's to the cluster: if fewer than three AO's are involved in skeletal bonding, ring or chain structures result.

This last point may be illustrated by considering the bonding in the trinuclear triangular clusters $M_3(CO)_{12}$ (M = Fe, Ru, or Os) (Fig. 11). If one uses the electron-counting procedure summarized in Table III, and assumes that each metal atom uses three AO's for cluster bonding, these clusters formally contain 6 skeletal bond pairs. They can, thus, be regarded as arachno species, and their triangular shapes are, indeed, appropriate for structures based on a trigonal bipyramid with 2 vacant vertices. However, a skeletal *bonding* role cannot be found for all 6 electron pairs formally available, since these would fill more than half of the nine skeletal MO's generated from three AO's per metal. A more realistic treatment of the skeletal bonding is to regard each skeletal atom as using only *two* AO's for cluster bonding, when the number of skeletal bond pairs available (3) is the appropriate number to fill the three skeletal bonding MO's (effectively producing a two-center bond along each edge of the triangle).

E. Sulfur Compounds

Echoes of the borane pattern can be detected in the chemistry of sulfur and the heavier Group VI elements, although many of their relatively electron-rich ring and cage systems can be described quite satisfactorily in terms of localized two-center bonds. For example, tetrasulfur tetranitride, S_4N_4, has a structure (Fig. 37) based on a tetrahedral arrangement of its sulfur atoms, with the nitrogen atoms bridging four of the tetrahedron edges (*89, 180*). If a lone pair is allocated to each sulfur and nitrogen atom, and a bond pair to each two center S—N link,

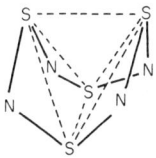

FIG. 37. Structure of tetrasulfur tetranitride, S_4N_4 (*89, 180*).

there remain 6 electron pairs for skeletal bonding in the S_4 tetrahedron, a number that is appropriate if each sulfur atom makes use of two d orbitals (*12, 204*). Elsewhere in sulfur chemistry, the relationship to borane-type clusters is less clear-cut (*12*), although the neighboring Group VI element, selenium, forms square-planar Se_4^{2+} ions (*30*) which may be regarded as tetranuclear arachno species (cf. $C_4H_4^{2-}$ or $B_4H_8^{2-}$).

F. HYDROCARBONS

Benzvalene, C_6H_6, and the carbo cations $C_5H_5^+$ and $C_6Me_6^{2+}$ have already been cited (Section VI) as hydrocarbons of the same structural type as boranes and carboranes, and we have also seen that such aromatic

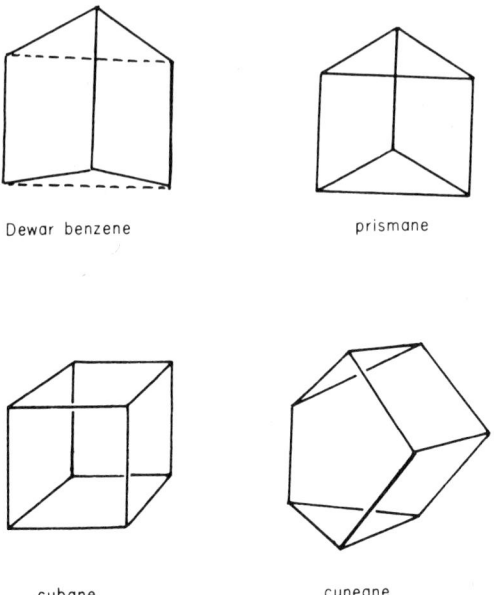

FIG. 38. Skeletons of Dewar benzene, prismane, cubane, and cuneane.

ring systems as benzene and the cycloheptatrienyl cation, $C_7H_7^+$, regarded as arachno systems, effectively extend the series of polyhedra on which borane-type clusters may be based. Quite apart from these, there are many other organic ring or cage systems that serve as models for further structural types, particularly for systems with a higher skeletal electron-to-atom ratio than is common in borane chemistry. They range in type from other benzene isomers such as Dewar benzene and prismane to cages such as cubane, C_8H_8, and its isomer cuneane (Fig. 38) and even more complicated structures. These and other hypothetical cluster shapes have been discussed by Lipscomb (145) and King (137), and the ways in which the skeletons of *electron-precise* species—species for which two-center bond descriptions are adequate—open up as more electron pairs are added have been discussed by Mingos (161).

Two systems, which it is tempting to relate to the 11-vertex octadecahedron of Fig. 1, are worth brief mention. One is the carbo cation $[C_8Me_8H]^+$ believed to have a structure (Fig. 39) formed by bringing a unit CH^+ up to the open face of the double bonds of octamethylnorbornadiene (105, 132). When compared with the octadecahedron, as in Fig. 39, this carbo cation has the CMe_2 unit in an edge-bridging position,

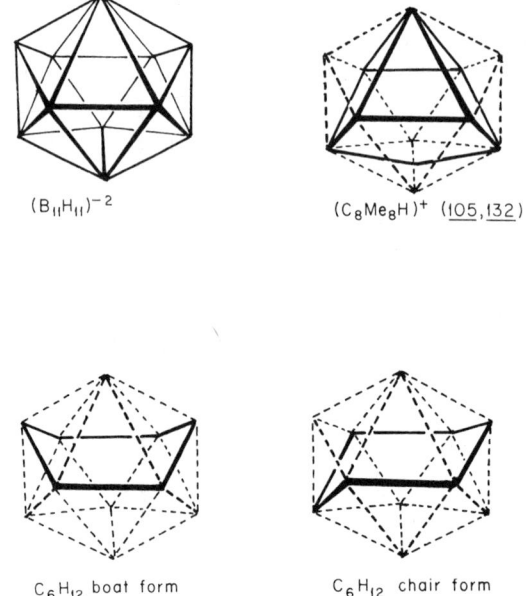

FIG. 39. Skeletal structures of the carbo cation $(C_8Me_8H)^+$ and the boat and chair forms of cyclohexane compared with the octadecahedron of $(B_{11}H_{11})^{2-}$.

and 4 polyhedron vertices are left vacant. (As an octanuclear species with 12 skeletal bond pairs, $C_8Me_8H^+$ might, in principle, have been expected to adopt a hypho structure based on an octadecahedron, but with three vertices vacant.) Figure 39 also shows how the skeletal structures of both boat and chair forms of cyclohexane, C_6H_{12}, interestingly can be related to the octadecahedron, with, however, 5 vacant vertices. However, treating a ring of 6 methylene CH_2 units, each capable of forming *two* bonds and furnishing *2* electrons, as a cluster of the borane type, unnecessarily complicates rather than clarifies the bonding description, although it preserves the structural distinction between the axial (*endo-*) hydrogen and equatorial (*exo-*)hydrogen atoms of C_6H_{12} in its chair configuration.

X. Conclusion

The main object of this survey has been to demonstrate the wide range of substances which, like boranes and carboranes, may be regarded as clusters of n skeletal atoms, each providing three AO's for skeletal bonding, held together by from n to $(n + 3)$ bond pairs. Examples include metalloboranes and carboranes, metal–hydrocarbon π complexes, some hydrocarbons, and certain metal clusters. Their closo, nido, or arachno structures reflect their skeletal electron numbers in a manner that allows the structures of new compounds to be inferred, new syntheses to be devised, reactions to be predicted, and interatomic distances to be rationalized. Exceptions to the simple pattern are to be found in the case of some halogenated species (B_8Cl_8, $C_4H_4B_2F_2$, etc.) and certain metallo systems, in which extra electrons may simply expand rather than deform the skeleton. The existence of other cluster types, e.g., systems using four AO's per skeletal atom ($Mo_6Cl_8^{4+}$, $Nb_6Cl_{12}^{2+}$), or not obeying the inert gas rule (S_4N_4) has also been noted briefly. Although the shapes of some recently prepared metal clusters show that further structural patterns have yet to be recognized, these too may be expected to reflect the number of electrons present.

Acknowledgments

I am pleased to acknowledge many fruitful discussions of cluster chemistry with several cluster chemists, particularly Professor P. Chini and Drs. A. J. Banister and R. E. Williams.

References

1. Abel, E. W., and Stone, F. G. A., *Quart. Rev., Chem. Soc.* **23**, 325 (1969).
2. Albano, V. G., and Bellon, P. L., *J. Organometal. Chem.* **19**, 403 (1969).

3. Albano, V. G., Bellon, P. L., Chini, P., and Scatturin, V., *J. Organometal. Chem.* **16**, 461 (1969).
4. Albano, V. G., Bellon, P. L., and Ciani, G. F., *Chem. Commun.* p. 1024 (1969).
5. Albano, V. G., Bellon, P. L., Manassero, M., and Sansoni, M., *Chem. Commun.* p. 1210 (1970).
6. Albano, V. G., Bellon, P. L., and Sansoni, M., *J. Chem. Soc.*, *A* p. 678 (1971).
7. Albano, V. G., Chini, P., Martinengo, S., Sansoni, M., and Strumolo, D., *J. Chem. Soc. Chem. Commun.* p. 299 (1974).
8. Albano, V. G., Ciani, G., and Chini, P., *J. Chem. Soc.*, *Dalton Trans.* p. 432 (1974).
9. Allegra, G., and Perego, G., *Ric. Sci. Parte 2: Sez. A* [2] **1**, 362 (1961); *Chem. Abstr.* **62**, 4724d (1965).
10. Almenningen, A., Haaland, A., and Motzfeldt, T., *J. Organometal. Chem.* **7**, 97 (1967).
11. Baikie, P. E., and Mills, O. S., *Chem. Commun.* p. 1228 (1967).
12. Banister, A. J., *Nature (London) Phys. Sci.* **239**, 69 (1972).
12a. Beall, H., *in* "Boron Hydride Chemistry" (E. L. Muetterties, ed.), p. 302. Academic Press, New York, 1975.
13. Beaudet, R. A., and Poynter, R. L., *J. Chem. Phys.* **53**, 1899 (1970).
14. Beer, D. C., Miller, V. R., Sneddon, L. G., Grimes, R. N., Mathew, M., and Palenik, G. J., *J. Amer. Chem. Soc.* **95**, 3046 (1973).
15. Bellon, P. L., Cariati, K., Manassero, M., Naldini, L., and Sansoni, M., *Chem. Commun.* p. 1423 (1971)
16. Bellon, P. L., Manassero, M., Naldini, L., and Sansoni, M., *J. Chem. Soc., Chem. Commun.* p. 1035 (1972).
17. Bellon, P., Manassero, M., and Sansoni, M., *J. Chem. Soc., Dalton Trans.* p. 1481 (1972).
18. Benn, H., Wilke, G., and Henneberg, D., *Angew. Chem., Int. Ed. Engl.* **12**, 1001 (1973).
19. Blount, J. F., Dahl, L. F., Hoogzand, C., and Hübel, W., *J. Amer. Chem. Soc.* **88**, 292 (1966).
20. Boer, F. P., Streib, W. E., and Lipscomb, W. N., *Inorg. Chem.* **3**, 1666 (1964).
21. Bohn, R. K., and Bohn, M. D., *Inorg. Chem.* **10**, 350 (1971).
22. Bohn, R. K., and Haaland, A., *J. Organometal. Chem.* **5**, 470 (1966).
23. Bradford, C. W., Nyholm, R. S., Gainsford, G. J., Guss, J. M., Ireland, P. R., and Mason, R., *J. Chem. Soc., Chem. Commun.* p. 87 (1972).
24. Braye, E. H., Dahl, L. F., Hübel, W., and Wampler, D. L., *J. Amer. Chem. Soc.* **84**, 4633 (1962).
25. Braye, E. H., Hübel, W., and Caplier, I., *J. Amer. Chem. Soc.* **83**, 4406 (1961).
26. Brennan, J. P., Grimes, R. N., Schaeffer, R., and Sneddon, L. G., *Inorg. Chem.* **12**, 2266 (1973).
27. Brooks, J. J., Rhine, W., and Stucky, G. D., *J. Amer. Chem. Soc.* **94**, 7339 (1972).
28. Brooks, J. J., Rhine, W., and Stucky, G. D., *J. Amer. Chem. Soc.* **94**, 7346 (1972).
29. Brooks, J. J., and Stucky, G. D., *J. Amer. Chem. Soc.* **94**, 7333 (1972).
29a. Brown, D. L. S., Connor, J. A., and Skinner, H. A., *J. Chem. Soc., Faraday Trans. I* **71**, 699 (1975).
30. Brown, I. D., Crump, D. B., Gillespie, R. J., and Santry, D. P., *Chem. Commun.* p. 853 (1968).

30a. Brown, M. P., Holliday, A. K., and Way, G. M., *J. Chem. Soc., Chem. Commun.* p. 850 (1972); p. 532 (1973).
31. Calabrese, J. C., Dahl, L. F., Cavalieri, A., Chini, P., Longoni, G., and Martinengo, S., *J. Amer. Chem. Soc.* **96**, 2616 (1974).
31a. Calabrese, J. C., Dahl, L. F., Chini, P., Longoni, G., and Martinengo, S., *J. Amer. Chem. Soc.* **96**, 2614 (1974).
31b. Callahan, K. P., Evans, W. J., and Hawthorne, M. F., *Ann. N.Y. Acad. Sci.* **239**, 88 (1974).
31c. Callahan, K. P., Lo, F. Y., Strouse, C. E., Sims, A. L., and Hawthorne, M. F., *Inorg. Chem.* **13**, 2842 (1974).
32. Cariati, F., and Naldini, L., *Inorg. Chim. Acta* **5**, 172 (1971).
33. Chini, P., *Inorg. Chim. Acta Rev.* **2**, 31 (1968).
34. Chini, P., and Albano, V. *J. Organometal. Chem.* **15**, 433 (1968).
35. Churchill, M. R., and DeBoer, B. G., *J. Chem. Soc., Chem. Commun.* p. 1326 (1972).
36. Churchill, M. R., and Gold, K., *J. Amer. Chem. Soc.* **92**, 1180 (1970).
37. Churchill, M. R., Gold, K., Francis, J. N., and Hawthorne, M. F., *J. Amer. Chem. Soc.* **91**, 1222 (1969).
38. Churchill, M. R., and Wormald, J. J., *J. Chem. Soc., Dalton Trans.* p. 2410 (1974).
39. Churchill, M. R., Wormald, J., Knight, J., and Mays, M. J., *Chem. Commun.* p. 458 (1970).
40. Churchill, M. R., Wormald, J., Knight, J., and Mays, M. J., *J. Amer. Chem. Soc.* **93**, 3073 (1971).
41. Coates, G. E., Green, M. L. H., and Wade, K., "Organometallic Compounds." Methuen, London, 1967.
42. Cohen, E. A., and Beaudet, R. A., *J. Chem. Phys.* **48**, 1220 (1968).
42a. Connor, J. A., Skinner, H. A., and Virmani, Y., *Faraday Symp. Chem. Soc.* **8**, 18 (1974).
43. Corbett, J. D., *Inorg. Chem.* **7**, 198 (1968).
44. Corbett, J. D., and Rundle, R. E., *Inorg. Chem.* **3**, 1408 (1964).
45. Corey, E. R., and Dahl, L. F., *Inorg. Chem.* **1**, 521 (1962).
46. Corey, E. R., Dahl, L. F., and Beck, W., *J. Amer. Chem. Soc.* **85**, 1202 (1963).
47. Corradini, P., and Sirigu, A., *Ric. Sci.* **36**, 188 (1966).
48. Cotton, F. A., *Quart. Rev., Chem. Soc.* **20**, 389 (1966).
49. Cotton, F. A., and Haas, T. E., *Inorg. Chem.* **3**, 10 (1964).
50. Dahl, L. F., and Smith, D. L., *J. Amer. Chem. Soc.* **84**, 2450 (1962).
51. Dahl, L. F., and Sutton, P. W., *Inorg. Chem.* **2**, 1067 (1963).
52. Davison, A., Traficante, D. D., and Wreford, S. S., *Chem. Commun.* p. 1155 (1972).
53. DeBoer, B. G., Zalkin, A., and Templeton, D. H., *Inorg. Chem.* **7**, 2288 (1968).
54. Deeming, A. J., Hasso, S., and Underhill, M., *J. Organometal. Chem.* **80**, C53 (1974).
55. Dekker, M., and Knox, G. R., *Chem. Commun.* p. 1243 (1967).
56. Dewar, M. J. S., and Haddon, R. C., *J. Amer. Chem. Soc.* **95**, 5836 (1973).
57. Dobrott, R. D., and Lipscomb, W. N., *J. Chem. Phys.* **37**, 1779 (1962).
58. Dodge, R. P., and Schomaker, V., *J. Organometal. Chem.* **3**, 274 (1965).
59. Doedens, R. J., *Chem. Commun.* p. 1271 (1968).
60. Drew, D. A., and Haaland, A., *Chem. Commun.* p. 1551 (1971).
61. Drew, D. A., and Haaland, A., *J. Chem. Soc., Chem. Commun.* p. 1300 (1972).

62. Drew, D. A., and Haaland, A., *Acta Chem. Scand.* **26**, 3079 (1972).
63. Drew, D. A., and Haaland, A. *Acta Chem. Scand.* **26**, 3351 (1972).
64. Drew, D. A., and Haaland, A. *Acta Crystallogr.*, *Sect. B* **28**, 3671 (1972).
65. Drew, D. A., and Haaland, A., *Acta Chem. Scand.* **27**, 3735 (1973).
66. Dubler, E., Textor, M., Osward, H.-R., and Salzer, A., *Angew. Chem., Int. Ed. Engl.* **13**, 135 (1974).
67. Dulmage, W. J., and Lipscomb, W. N., *Acta Crystallogr.* **5**, 260 (1952).
68. Dunks, G. B., and Hawthorne, M. F., *J. Amer. Chem. Soc.* **92**, 7213 (1970).
68a. Dunks, G. B., and Hawthorne, M. F., *in* "Boron Hydride Chemistry" (E. L. Muetterties, ed.), p. 383. Academic Press, New York, 1975.
69. Dunks, G. B., McKown, M. M., and Hawthorne, M. F. *J. Amer. Chem. Soc.* **93**, 241 (1971).
70. Dustin, D. F., Dunks, G. B., and Hawthorne, M. F., *J. Amer. Chem. Soc.* **95**, 1109 (1973).
71. Dustin, D. F., Evans, W. J., and Hawthorne, M. F., *J. Chem. Soc., Chem. Commun.* p. 805 (1973).
72. Dustin, D. F., Evans, W. J., Jones, C. J., Wiersema, R. J., Gong, H., Chan, S., and Hawthorne, M. F., *J. Amer. Chem. Soc.* **96**, 3085 (1974).
73. Eady, C. R., Johnson, B. F. G., and Lewis, J., *J. Organometal. Chem.* **37**, C39 (1972).
74. Einstein, F. W. B., Gilchrist, A. B., Rayner-Canham, G. W., and Sutton, D., *J. Amer. Chem. Soc.* **93**, 1826 (1971).
75. Eisch, J. J., Hota, N. K., and Kozima, S., *J. Amer. Chem. Soc.* **91**, 4575 (1969).
76. Epstein, I. R., Tossell, J. A., Switkes, E., Stevens, R. M., and Lipscomb, W. N., *Inorg. Chem.* **10**, 171 (1971).
77. Evans, W. J., Dunks, G. B., and Hawthorne, M. F., *J. Amer. Chem. Soc.* **95**, 4565 (1973).
78. Evans, W. J., and Hawthorne, M. F., *J. Amer. Chem. Soc.* **93**, 3063 (1971).
79. Evans, W. J., and Hawthorne, M. F., *J. Chem. Soc., Chem. Commun.* p. 611 (1972).
80. Evans, W. J., and Hawthorne, M. F., *J. Chem. Soc., Chem. Commun.* p. 706 (1973).
81. Evans, W. J., and Hawthorne, M. F., *J. Amer. Chem. Soc.* **96**, 301 (1974).
82. Evans, W. J., and Hawthorne, M. F., *Inorg. Chem.* **13**, 869 (1974).
82a. Fehlner, T. P., *Inorg. Chem.* **14**, 934 (1975).
83. Fischer, E. O., Winkler, E., Huttner, G., and Regler, D., *Angew. Chem., Int. Ed. Engl.* **11**, 238 (1972).
84. Franz, D. A., Miller, V. R., and Grimes, R. N., *J. Amer. Chem. Soc.* **94**, 412 (1972).
85. Frasson, E., Menegus, F., and Panattoni, C., *Nature (London)* **199**, 1087 (1963).
86. Friedman, R. M., and Corbett, J. D., *Chem. Commun.* p. 422 (1971); *Inorg. Chem.*, **12**, 1134 (1973).
87. Frisch, P. D., and Dahl, L. F., *J. Amer. Chem. Soc.* **94**, 5082 (1972).
88. Games, M. L., and Maitlis, P. M., *J. Amer. Chem. Soc.* **85**, 1887 (1963).
89. Gleiter, R., *J. Chem. Soc. A* p. 3174 (1970).
90. Green, M., Howard, J., Spencer, J. L., and Stone, F. G. A., *J. Chem. Soc., Chem. Commun.* p. 153 (1974).
91. Green, M., Spencer, J. L., Stone, F. G. A., and Welch, A. J., *J. Chem. Soc., Chem. Commun.* p. 571 (1974).

92. Green, M., Spencer, J. L., Stone, F. G. A., and Welch, A., *J. Chem. Soc., Chem. Commun.* p. 794 (1974).
93. Green, M. L. H., Pratt, L., and Wilkinson, G., *J. Chem. Soc., London* p. 3753 (1959).
94. Greenwood, N. N., Savory, C. G., Grimes, R. N., Sneddon, L. G., Davison, A., and Wreford, S. S., *J. Chem. Soc., Chem. Commun.* p. 718 (1974).
94a. Greenwood, N. N., and Ward, I. M., *Chem. Soc. Rev. (London)* **3**, 231 (1974).
95. Grimes, R. N., "Carboranes," Academic Press, New York, 1970.
96. Grimes, R. N., *J. Amer. Chem. Soc.* **93**, 261 (1971).
96a. Grimes, R. N., *Ann. N.Y. Acad. Sci.* **239**, 180 (1974).
96b. Grimes, R. N., Beer, D. C., Sneddon, L. G., Miller, V. R., and Weiss, R., *Inorg. Chem.* **13**, 1138 (1974).
97. Grimes, R. N., Bramlett, C. L., and Vance, R. L., *Inorg. Chem.* **7**, 1066 (1968); **8**, 55 (1969).
98. Grimes, R. N., Rademaker, W. J., Denniston, M. L., Bryan, R. F., and Greene, P. T., *J. Amer. Chem. Soc.* **94**, 1865 (1972)
99. Guggenberger, L. J., *Inorg. Chem.* **7**, 2260 (1968).
100. Guggenberger, L. J., *Inorg. Chem.* **8**, 2771 (1969).
101. Guss, J. M., Mason, R., Thomas, K. M., van Koten, G., and Noltes, J. G., *J. Organometal. Chem.* **40**, C79 (1972).
102. Haaland, A., *Acta Chem. Scand.* **22**, 3030 (1968).
103. Haaland, A., Lusztyk, J., Novak, D. P., Brunvoll, J., and Starowieyski, K. B., *J. Chem. Soc., Chem. Commun.* p. 54 (1974).
104. Haaland, A., and Novak, D. P., *Acta Chem. Scand., A* **28**, 153 (1974).
105. Hart, H., and Kuzuya, M., *J. Amer. Chem. Soc.* **94**, 8958 (1972).
105a. Hawthorne, M. F., Callahan, K. P., and Wiersema, R. J., *Tetrahedron* **30**, 1795 (1974).
106. Hawthorne, M. F., and Dunks, G. B., *Science* **178**, 462 (1972).
107. Hawthorne, M. F., Young. D. C., Garrett, P. M., Owen, D. A., Schwenn, S. G., Tebbe, F. N., and Wegner, P. A., *J. Amer. Chem. Soc.* **90**, 862 (1968).
108. Hehre, W. J., and Schleyer, P. von R., *J. Amer. Chem. Soc.* **95**, 5837 (1973).
109. Herberhold, M., and Golla, W., *J. Organometal. Chem.* **26**, C27 (1971).
110. Herberich, G. E., and Becker, H. J., *Angew. Chem., Int. Ed. Engl.* **12**, 764 (1973).
111. Hershaft, A., and Corbett, J. D., *Inorg. Chem.* **2**, 979 (1963).
112. Hock, A. A., and Mills, O. S., *Acta Crystallogr,* **14**, 139 (1961).
112a. Hoel, E. L., Strouse, C. E., and Hawthorne, M. F., *Inorg. Chem.* **13**, 1388 (1974).
113. Hoffmann, R., and Gouterman, M., *J. Chem. Phys.* **36**, 2189 (1962).
114. Hoffmann, R., and Lipscomb, W. N., *J. Chem. Phys.* **36**, 2179 (1962).
115. Hoffmann, R., and Lipscomb, W. N. *J. Chem. Phys.* **36**, 3489 (1962).
116. Hoffmann, R., and Lipscomb, W. N. *J. Chem. Phys.* **37**, 2872 (1962).
117. Hogeveen, H., and Kwant, P. W., *J. Amer. Chem. Soc.* **96**, 2208 (1974); *Tetrahedron Lett.* p. 4351 (1974).
118. Hollander, O., Clayton, W. R., and Shore, S. G., *J. Chem. Soc., Chem. Commun.* p. 604 (1974).
119. Howard, J. W., and Grimes, R. N., *Inorg. Chem.* **11**, 263 (1972).
120. Hrostowski, H. J., and Myers, R. J., *J. Chem. Phys.* **22**, 262 (1954).
121. Huffman, J. C., and Streib, W. E., *J. Chem. Soc., Chem. Commun.* p. 665 (1972).
122. Hursthouse, M. B., Kane, J., and Massey, A. G., *Nature (London)* **228**, 659 (1970).

123. Huttner, G., and Krieg, B., *Angew. Chem.*, *Int. Ed. Engl.* **10**, 512 (1971).
124. Johansson, G., and Lipscomb, W. N., *Acta Crystallogr.* **11**, 594 (1958).
125. Johnson, B. F. G., Johnston, R. D., and Lewis, J., *J. Chem. Soc.*, *A* p. 2865 (1968).
126. Johnson, H. D., Brice, V. T., Brubaker, G. L., and Shore, S. G., *J. Amer. Chem. Soc.* **94**, 6711 (1972).
127. Johnston, R. D., *Advan. Inorg. Chem. Radiochem.* **13**, 471 (1970).
128. Jones, C. J., Evans, W. J., and Hawthorne, M. F., *J. Chem. Soc., Chem. Commun.* p. 543 (1973).
129. Jones, D., Pratt, L., and Wilkinson, G. *J. Chem. Soc. London* p. 4458 (1961).
130. Jones, D., and Wilkinson, G., *J. Chem. Soc., London* p. 2479 (1964).
131. Kamijyo, N., and Watanabe, T., *Bull. Chem. Soc. Jap.* **47**, 373 (1974).
132. Kemp-Jones, A. V., Nakamura, N., and Masamune, S. *J. Chem. Soc., Chem. Commun.* p. 109 (1974).
133. Kettle, S. F. A., *Theor. Chim. Acta* **3**, 211 (1965).
134. Kettle, S. F. A., *J. Chem. Soc.*, *A* p. 1013 (1966).
135. Kettle, S. F. A., *J. Chem. Soc.*, *A* p. 314 (1967).
136. King, R. B., *Chem. Commun.* p. 436 (1969).
137. King, R. B., *J. Amer. Chem. Soc.* **94**, 95 (1972).
138. King, R. B., *Progr. Inorg. Chem.* **15**, 287 (1972).
139. Klanberg, F., Eaton, D. R., Guggenberger, L. J., and Muetterties, E. L., *Inorg. Chem.* **6**, 1271 (1967).
140. Kollmar, H., Smith, H. O., and Schleyer, P. v. R. *J. Amer. Chem. Soc.* **95**, 5834 (1973).
141. Köster, R., and Grassberger, M. A., *Angew. Chem.*, *Int. Ed. Engl.* **6**, 218 (1967).
142. Krüerke, U., and Hübel, W. *Chem. Ber.* **94**, 2829 (1961).
143. Laws, E. A., Stevens, R. M., and Lipscomb, W. N., *J. Amer. Chem. Soc.* **94**, 4467 (1972).
144. Libby, W. F., *J. Chem. Phys.* **46**, 399 (1967).
145. Lipscomb, W. N., "Boron Hydrides," Benjamin, New York, 1963.
145a. Lipscomb, W. N., in "Boron Hydride Chemistry" (E. L. Muetterties, ed.), p. 39. Academic Press, New York, 1975.
146. Longuet-Higgins, H. C., *J. Chim. Phys.* **46**, 275 (1949).
147. Longuet-Higgins, H. C. *Quart. Rev., Chem. Soc.* **11**, 121 (1957).
148. Longuet-Higgins, H. C., and Roberts, M. de V., *Proc. Roy. Soc., Ser. A* **230**, 110 (1955).
149. Lott, J. W., Gaines, D. F., Shenhav, H., and Schaeffer, R., *J. Amer. Chem. Soc.* **95**, 3042 (1973).
150. McKown, G. L., Don, B. P., Beaudet, R. A., Vergamini, P. J., and Jones, L. H., *J. Chem. Soc., Chem. Commun.* p. 765 (1974).
151. McNeill, E. A., Gallaher, K. L., Scholer, F. R., and Bauer, S. H., *Inorg. Chem.* **12**, 2108 (1973).
152. McPartlin, M., Mason, R., and Malatesta, L., *Chem. Commun.* p. 334 (1969).
153. Marynick, D. S., and Lipscomb, W. N., *J. Amer. Chem. Soc.* **94**, 1748 (1972).
154. Masamune, S., Sakai, M., and Ona, H., *J. Amer. Chem. Soc.* **94**, 8955 (1972).
155. Masamune, S., Sakai, M., Ona, H., and Jones, A. J., *J. Amer. Chem. Soc.* **94**, 8956 (1972).
156. Mason, R., and Rae, A. I. M., *J. Chem. Soc.*, *A* p. 778 (1968).
157. Mason, R., and Robinson, W., *Chem. Commun.* p. 468 (1968).
158. Mason, R., Thomas, K. M., and Mingos, D. M. P., *J. Amer. Chem. Soc.* **95**, 3802 (1973).

158a. Middaugh, R. L., in "Boron Hydride Chemistry" (E. L. Muetterties, ed.), p. 273. Academic Press, New York, 1975.
159. Miller, V. R., and Grimes, R. N., *J. Amer. Chem. Soc.* **95**, 5078 (1973).
159a. Miller, V. R., Sneddon, L. G., Beer, D. C., and Grimes, R. N., *J. Amer. Chem. Soc.* **96**, 3090 (1974).
160. Mills, O. S., and Shaw, B. W., *J. Organometal. Chem.* **11**, 595 (1968).
161. Mingos, D. M. P., *Nature (London) Phys. Sci.* **236**, 99 (1972).
162. Mingos, D. M. P., *J. Chem. Soc., Dalton Trans.* p. 133 (1974).
163. Moore, E. B., Lohr, L. L., and Lipscomb, W. N., *J. Chem. Phys.* **35**, 1329 (1961).
163a. Muetterties, E. L., in "Boron Hydride Chemistry" (E. L. Muetterties, ed.), p. 1. Academic Press, New York, 1975.
164. Muetterties, E. L., and Knoth, W. H., "Polyhedral Boranes," Dekker, New York, 1968.
165. Müller, H., *J. Chem. Phys.* **49**, 475 (1968).
166. Onak, T., *Advan. Organometal. Chem.* **3**, 263 (1965).
166a. Onak, T., in "Boron Hydride Chemistry" (E. L. Muetterties, ed.), p. 349. Academic Press, New York, 1975.
167. Panattoni, C., Bambieri, G., and Croatto, U., *Acta Crystallogr.* **21**, 823 (1966).
168. Pasinski, J. P., and Beaudet, R. A., *J. Chem. Soc., Chem. Commun.* p. 928 (1973).
169. Pawley, G. S., *Acta Crystallogr.* **20**, 631 (1966).
170. Penfold, B. R., *Perspect. Struct. Chem.* **2**, 71 (1968).
170a. Rayment, I., and Shearer, H. M. M. (personal communication).
171. Rees, B., and Coppens, P., *Acta Crystallogr., Sect. B* **29**, 2516 (1973).
172. Rossman, M. G., Jacobson, R. A., Hirshfeld, F. L., and Lipscomb, W. N., *Acta Crystallogr* **12**, 530 (1959).
172a. Rudolph, R. W., and Chowdhry, V., *Inorg. Chem.* **13**, 248 (1974).
173. Rudolph, R. W., and Pretzer, W. R., *Inorg. Chem.* **11**, 1974 (1972).
173a. Rudolph, R. W., and Thompson, D. A., *Inorg. Chem.* **13**, 2779 (1974).
174. St. Clair, D., Zalkin, A., and Templeton, D. H., *J. Amer. Chem. Soc.* **92**, 1173 (1970).
175. Salentine, C. G., and Hawthorne, M. F., *J. Chem. Soc., Chem. Commun.* p. 560 (1973).
176. Salzer, A., and Werner, H., *Angew. Chem., Int. Ed. Engl.* **11**, 930 (1972).
177. Schaeffer, R., Johnson, Q., and Smith, G. S., *Inorg. Chem.* **4**, 917 (1965).
178. Schneider, R. F., and Mackay, R. A., *J. Chem. Phys.* **48**, 843 (1968).
179. Schumacher, E., and Taubenest, R., *Helv. Chim. Acta.* **47**, 1525 (1964).
180. Sharma, B. D., and Donohue, J., *Acta Crystallogr.* **16**, 891 (1963).
181. Shibata, S., Bartell, L. S., and Gavin, R. M., *J. Chem. Phys.* **41**, 717 (1964).
181a. Shore, S. G., in "Boron Hydride Chemistry" (E. L. Muetterties, ed.), p. 79. Academic Press, New York, 1975.
182. Sirigu, A., Bianchi, M., and Benedetti, E., *Chem. Commun.* p. 596 (1969).
183. Snaith, R., and Wade, K., *MTP Int. Rev. Sci., Inorg. Chem. Ser.* **1 1**, 139 (1972); *Ser. 2* **1**, 95 (1975).
184. Spencer, J. L., Green, M., and Stone, F. G. A., *J. Chem. Soc., Chem. Commun.* p. 1178 (1972).
185. Stanko, V. I., Brattsev, V. A., and Gol'tyapin, Yu. V. *J. Gen. Chem. USSR* **39**, 1142 and 2623 (1969).
186. Stanko, V. I., Brattsev, V. A., Gol'tyapin, Yu. A., Khrapov, V. V., Babushkina, T. A., and Klimova, T. P., *J. Gen. Chem. USSR* **44**, 319 (1974).

187. Stohrer, W. D., and Hoffmann, R., *J. Amer. Chem. Soc.* **94**, 1661 (1972).
188. Sutton, P. W., and Dahl, L. F., *J. Amer. Chem. Soc.* **89**, 261 (1967).
189. Switkes, E., Lipscomb, W. N., and Newton, M. D. *J. Amer. Chem. Soc.* **92**, 3847 (1970).
190. Switkes, E., Stevens, R. M., Lipscomb, W. N., and Newton, M. D., *J. Chem. Phys.* **51**, 2085 (1969).
191. Thompson, M. L., and Grimes, R. N., *J. Amer. Chem. Soc.* **93**, 6677 (1971).
192. Thompson, M. L., and Grimes, R. N., *Inorg. Chem.* **11**, 1925 (1972).
193. Timms, P. L., *J. Amer. Chem. Soc.* **90**, 4584 (1968).
194. Timms, P. L., *Advan. Inorg. Chem. Radiochem.* **14**, 121 (1972).
195. Todd, L. J., *Advan. Organometal. Chem.* **8**, 87 (1970).
195a. Travers, N. F., *MTP Int. Rev. Sci. Inorg. Chem. Ser.* *1* **1**, 79 (1972).
196. Tyler, J. K., Cox, A. P., and Sheridan, J., *Nature (London)* **183**, 1182 (1959).
197. Villa, A. C., Coghi, L., Manfredotti, A. G., and Guastini, C., *Acta Crystallogr., Sect. B* **30**, 2101 (1974).
198. Waddington, T. C., *Trans. Faraday Soc.* **63**, 1313 (1967).
199. Wade, K. *Chem. Commun.* p. 792 (1971).
200. Wade, K., "Electron Deficient Compounds." Nelson, London, 1971; Appleton, New York, 1973.
201. Wade, K., *Inorg. Nucl. Chem. Lett.* **8**, 559 (1972).
202. Wade, K., *Inorg. Nucl. Chem. Lett.* **8**, 563 (1972).
203. Wade, K., *Inorg. Nucl. Chem. Lett.* **8**, 823 (1972).
204. Wade, K., *Nature (London), Phys. Sci.* **240**, 71 (1972).
205. Wade, K., *New Sci.* **62**, 615 (1974).
205a. Wade, K., *Chem. Br.* **11**, 177 (1975).
205b. Wegner, P. A., in "Boron Hydride Chemistry" (E. L. Muetterties, ed.), p. 431. Academic Press, New York, 1975.
206. Wei, C. H., and Dahl, L. F., *Inorg. Chem.* **4**, 1 (1965).
207. Wei, C. H., and Dahl, L. F., *Inorg. Chem.* **4**, 493 (1965).
208. Wei, C. H., and Dahl, L. F., *J. Amer. Chem. Soc.* **88**, 1821 (1966).
209. Wei, C. H., and Dahl, L. F., *Inorg. Chem.* **6**, 1229 (1967).
210. Wei, C. H., and Dahl, L. F., *J. Amer. Chem. Soc.* **91**, 1351 (1969).
211. Wei, C. H., Wilkes, G. R., and Dahl, L. F. *J. Amer. Chem. Soc.* **89**, 4792 (1967).
212. Williams, R. E., *Progr. Boron Chem.* **2**, 51 (1970).
213. Williams, R. E., *Inorg. Chem.* **10**, 210 (1971).
214. Wilson, R. J., Warren, L. F., and Hawthorne, M. F., *J. Amer. Chem. Soc.* **91**, 758 (1969).
215. Wing, R. M., *J. Amer. Chem. Soc.* **89**, 5599 (1967).
216. Wing, R. M., *J. Amer. Chem. Soc.* **90**, 4828 (1968).
217. Winkhaus, G., Pratt, L., and Wilkinson, G., *J. Chem. Soc. London.* p. 3807 (1961).
218. Wong, C.-H., Lee, T.-Y., Chao, K.-J., and Lee, S., *Acta Crystallogr., Sect B* **28**, 1662 (1972).
219. Wunderlich, J. A., and Lipscomb, W. N., *J. Amer. Chem. Soc.* **82**, 4427 (1960).
220. Zalkin, A., Hopkins, T. E., and Templeton, D. H., *Inorg. Chem.* **5**, 1189 (1966).

COORDINATION NUMBER PATTERN RECOGNITION THEORY OF CARBORANE STRUCTURES*

ROBERT E. WILLIAMS
Chemical Systems Inc., Irvine, California

I. Introduction 67
   A. Viewpoints on Carborane Structures 68
   B. History of Coordination Number Pattern Recognition Theory . 69
II. Structural Rules 85
   A. Reappraisal of Deltahedron–Deltahedral Fragment Hypothesis (Rule 1) 85
   B. Structural Preferences of Various Moieties: Prelude to the Exposition of Rules 2, 3, and 4 86
   C. Bridge and Endohydrogen Considerations (Rule 2) . . . 90
   D. Carbon and Other Heteroelement Atom Considerations (Rule 3) 93
   E. Boron Considerations (Rule 4) 94
   F. Primary, Secondary, and Tertiary Expressions of Rules 1, 2, 3, and 4 95
III. Carboranes, Their Analogs and Derivatives 97
   A. *closo*-Carboranes 97
   B. *nido*-Carboranes 100
   C. *arachno*-Carboranes 116
   D. Heteroatom Carborane Analogs 125
   E. One Heteroatom Donating Two Electrons 129
IV. Bridge and Endohydrogens and Relative Lowry-Brønsted Acidity . 132
V. Conclusions 136
   References 137

I. Introduction

In the early 1950s the structures of the more common boranes were still a matter of debate. Although the hypothesis appeared to be losing favor, there remained strong tendencies to anticipate and interpret borane structures as having *mildly nonconforming hydrocarbon* structures. As a number of boranes were verified as having polyhedral fragment configurations the structural thread to hydrocarbon chemistry became weakened. Subsequently, many other boranes and carboranes were

* Presented at the Second International Meeting on Boron Compounds held at Leeds University, Leeds, United Kingdom, 25–29 March 1974.

discovered, and a seemingly endless array of multifarious structural types had to be considered.

A. VIEWPOINTS ON CARBORANE STRUCTURES

Several schools of thought then arose. Members of one school treated each structure as a separate case, more or less succumbing to the thesis of an almost infinite variety of structural parameters. A second school took the view that although carborane structures were complex in nature, such structures could eventually be categorized or collated by improved theoretical treatments which could be expected to become ever more accurate because of constantly improving computer systems. A third school considered the complexity of carborane structures to be not as severe as had appeared at first view, arguing that when a sufficient number of structures would eventually be determined, the fundamental structural precepts would become decipherable and, consequently, amenable to a simplistic empirical organizational format.

Supporters of each view have made substantial contributions. Advocates of the first school have prepared an almost infinite number of derivatives, whereas subscribers to the second school, in the process of theoretically predicting all possible configurations, generated the correct configuration in a surprisingly large percentage of the cases.

This paper subscribes to the third viewpoint, and is based on an empirical approach that involves coordination number pattern recognition (CNPR). It is a simplistic approach, yet it apparently accommodates most if not all carborane and borane structures. For compounds that are still controversial and for compounds that have not yet been discovered or characterized, the CNPR thesis frequently predicts different structures, or at least fewer candidates, than do any of the theoretical treatments.

The structures, relative stabilities, and relative Lowry-Brønsted acidities of carboranes and boranes as well as related anions, Lewis base adducts, and heteroelement analogs are rationalized primarily on the basis of rudimentary coordination numbers. The principal factors, in decreasing order of importance, are (*a*) the various deltahedra and deltahedral fragments, (*b*) the placement of bridge and endohydrogens, (*c*) the placement of carbon and other heteroelements, and (*d*) the resulting coordination number of boron.

Elucidation of the CNPR theory is based on the presumption *that bridge and endohydrogens (BE hydrogens) when present are of primary structural importance within the carboranes and in related species.* At the very least the theory appears to establish as fact that the influence of BE hydrogens on structure has been vastly underrated.

B. History of Coordination Number Pattern Recognition Theory

In 1971, a note (*164*) was published favoring the hypothesis that the carboranes, boranes, their isoelectronic anions, Lewis base adducts, and heteroatom-substituted analogs should be viewed as constructed about the vertices of either the most spherical series of triangular-faceted polyhedra (deltahedra) found to be characteristic of the dicarba-*closo*-carboranes (Fig. 1) or, with one lone exception, fragments of the series of deltahedra produced by the successive removal of the highest coordinated vertices that sequentially define the nido and arachno classes. This position was in conflict with the then prevalent shibboleth that all nido and arachno compounds [except B_5H_9 (I-N5)] had or would prove to have icosahedral fragment structures.

The 1971 hypothesis (*164*) (throughout this paper referred to as rule 1) may now be expanded by the addition of supplementary rules that deal with the placement of BE hydrogens (rule 2), the placement of carbon and other heteroelements (rule 3), and the consequent coordination situation of boron (rule 4).

Various structures within and among the twenty-five different figures are compared throughout this text. To facilitate reference and cross-reference, all the figures are grouped together in a contrived fashion so that, in Figs. 5 through 19, the last digit or last two digits in the molecular identification numbers not only indicate the figure number but also reflect the number of atoms in the various molecular skeletons. The identifying term I-N5 for B_5H_9, for example, reveals that B_5H_9 is the first structure in Fig. 5; the N denotes that it is a nido species.

Before attempting to outline the supplementary material that expands the original hypothesis, it would appear desirable to establish some degree of credibility by reviewing the present author's research and the many interrelated studies by others.

Included in the 1971 reappraisal (*164*) of carborane structures, and contrary to many alternatively suggested structures pervading the literature at that time, were the following predictions:

1. Nonicosahedral structures should be anticipated for seven- and nine-vertex nido species.
2. The parent B_7H_{11} (I-N7) and B_9H_{13} (I-N9) boranes, which uniquely have unfavorably puckered five-membered open faces (see

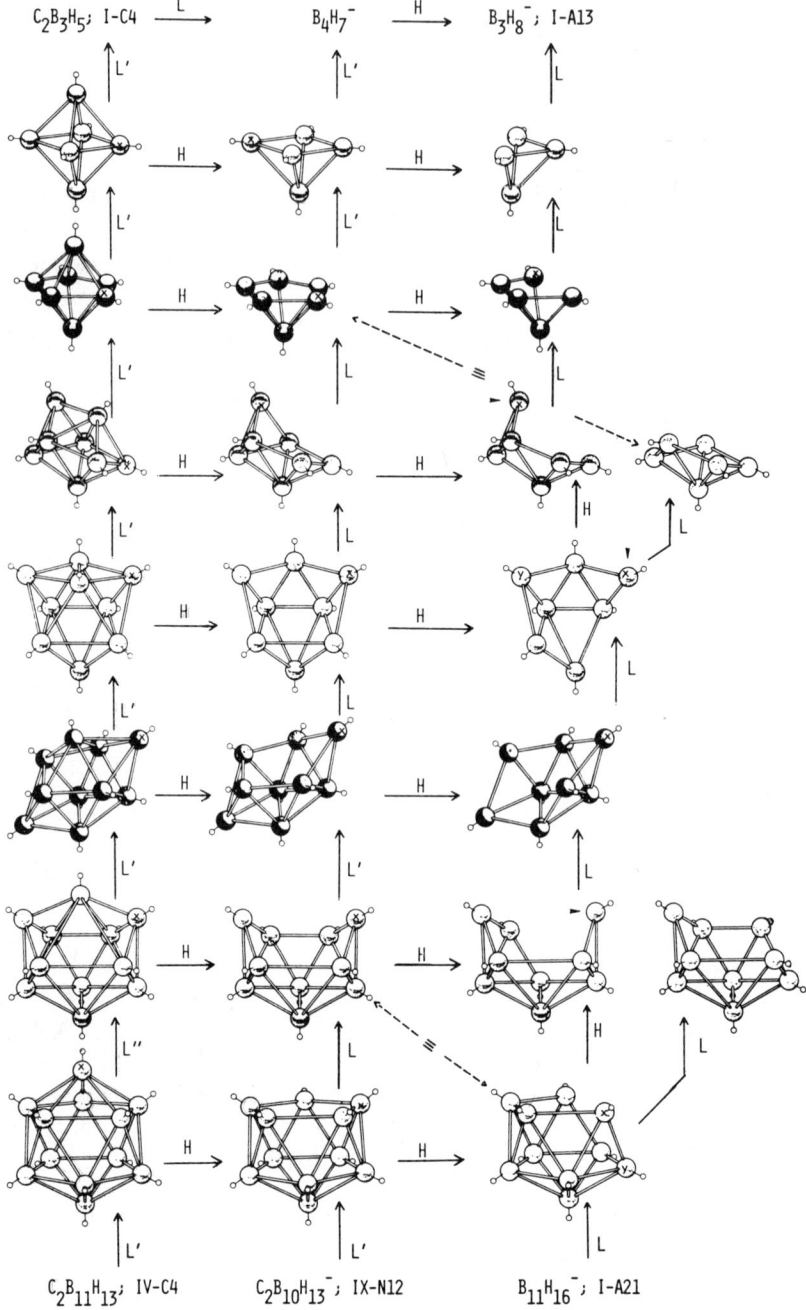

FIG. 1. Deltahedra and deltahedral fragments.

CNPR THEORY 71

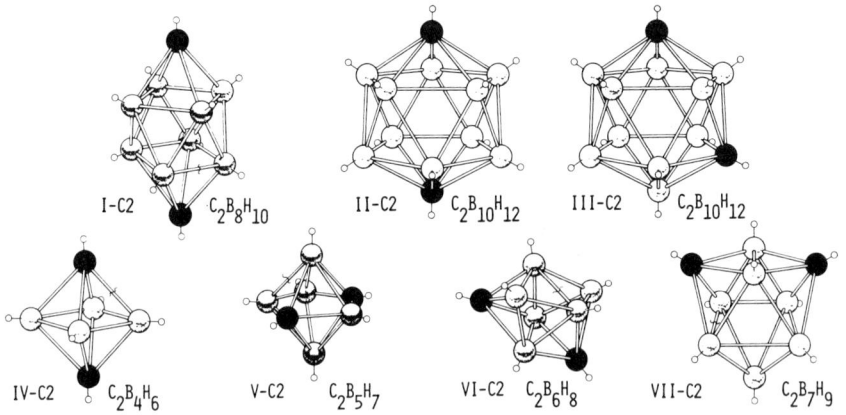

FIG. 2. Stable dicarba-*closo*-carboranes.

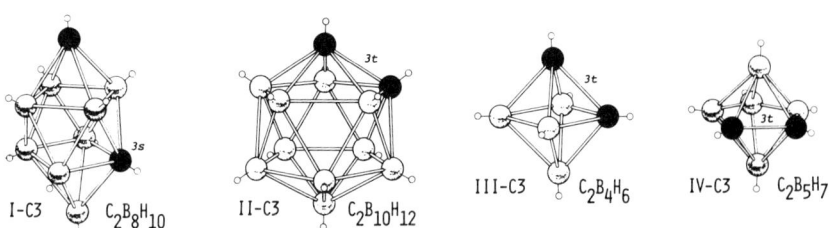

FIG. 3. Rearrangement-prone dicarba-*closo*-carboranes.

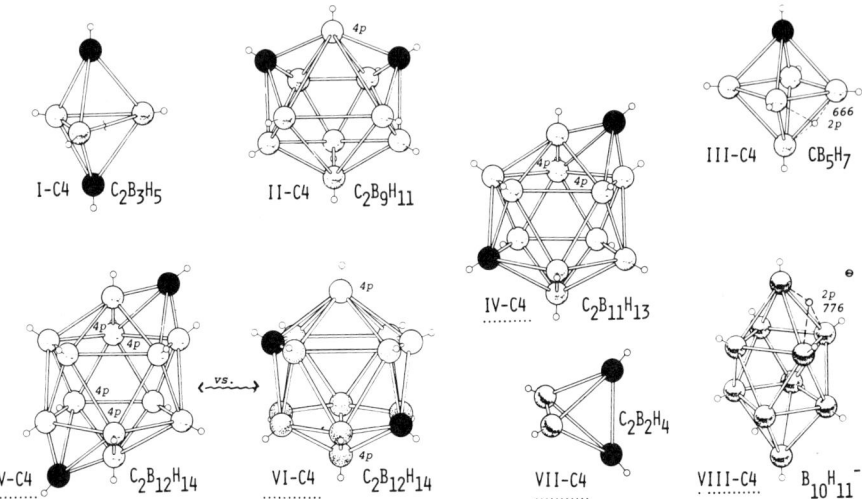

FIG. 4. Unstable and/or unknown *closo*-carboranes.

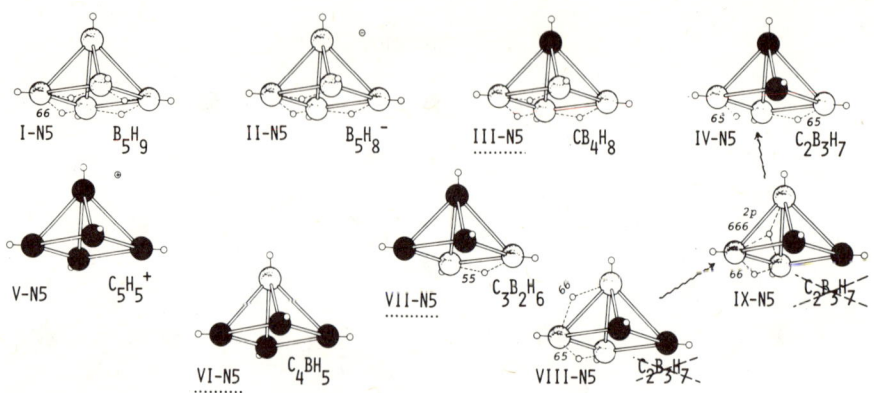

FIG. 5. The B_5H_9 family of nido-carboranes.

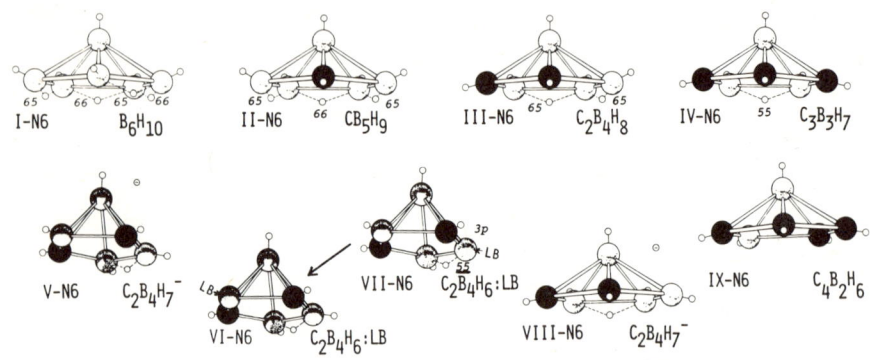

FIG. 6. The B_6H_{10} family of nido-carboranes. (LB, Lewis base.)

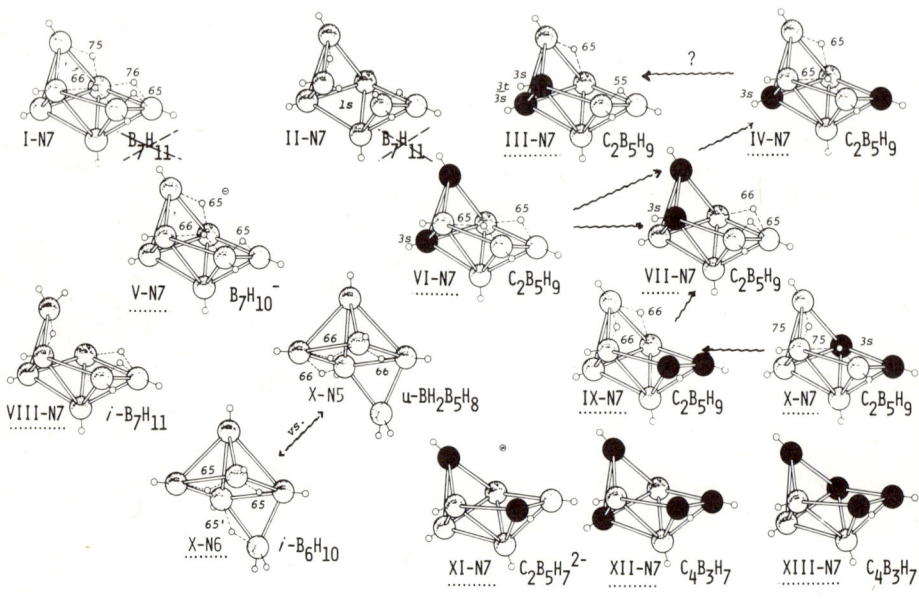

FIG. 7. The $[B_7H_{11}]$ family of nido-carboranes.

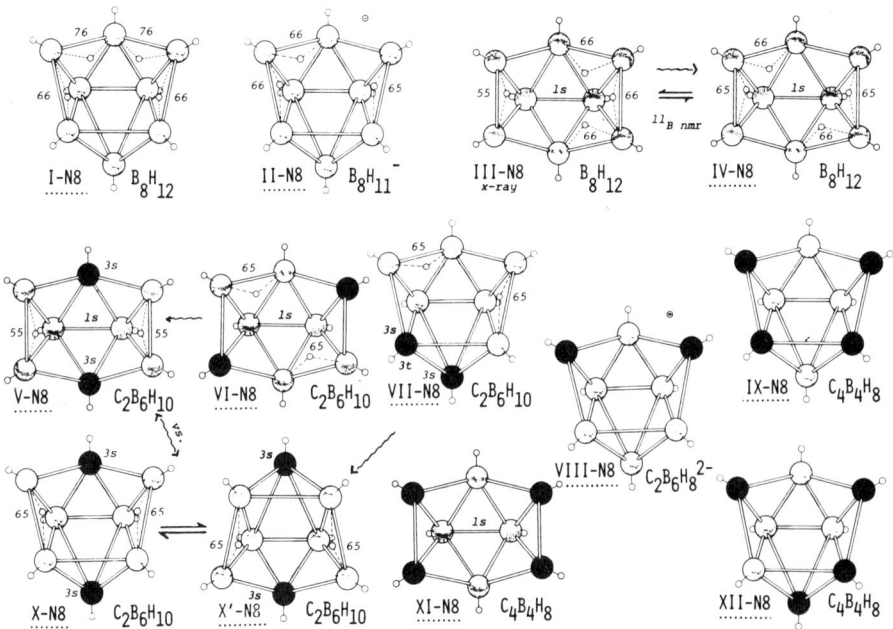

FIG. 8. The B_8H_{12} family of *nido*-carboranes.

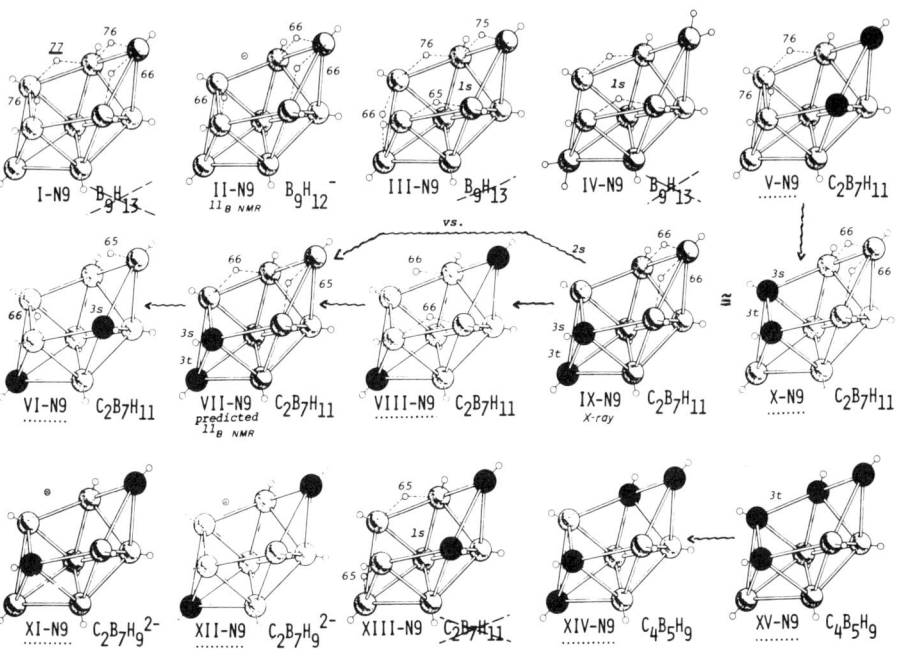

FIG. 9. The $[B_9H_{13}]$ family of *nido*-carboranes.

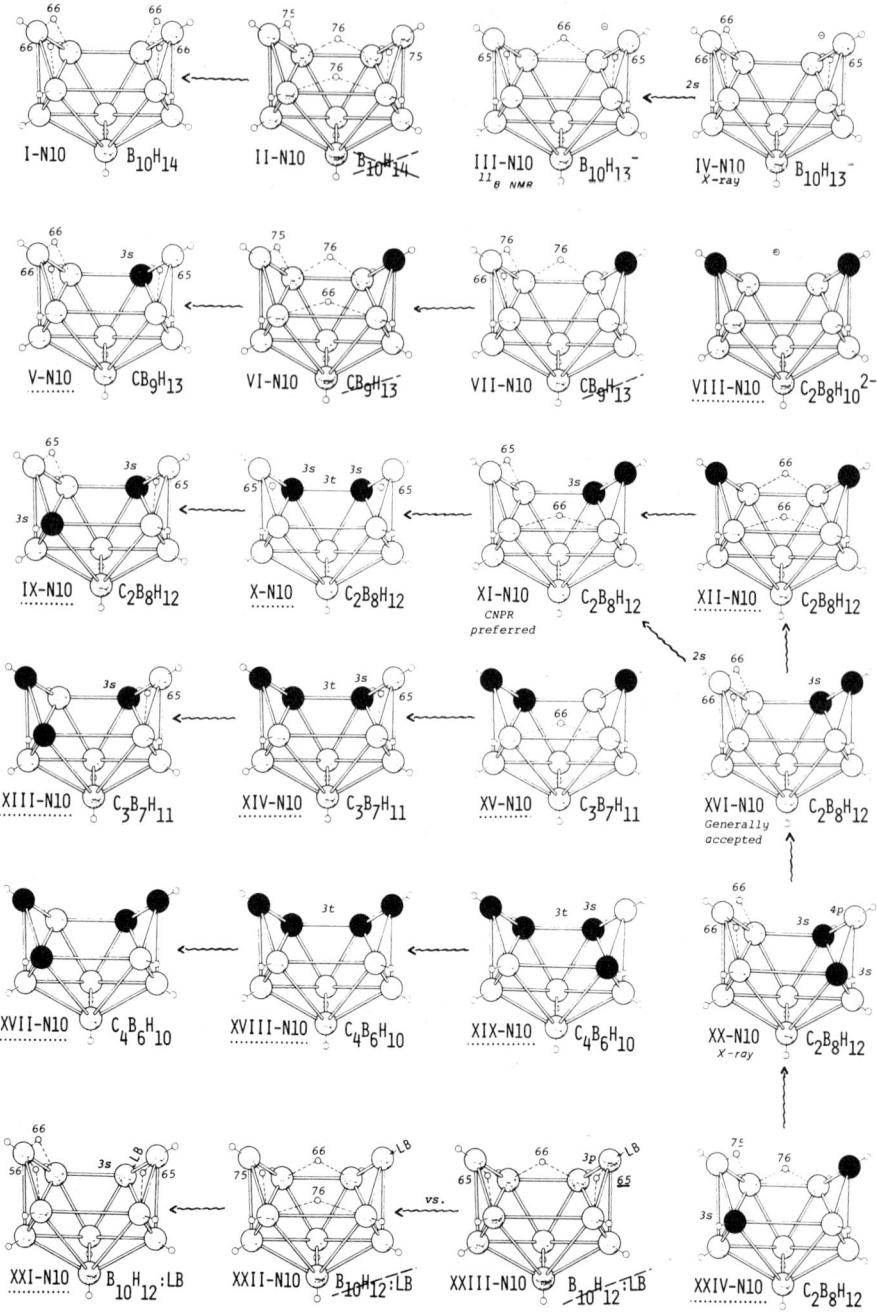

FIG. 10. The $B_{10}H_{14}$ family of *nido*-carboranes. (LB, Lewis base.)

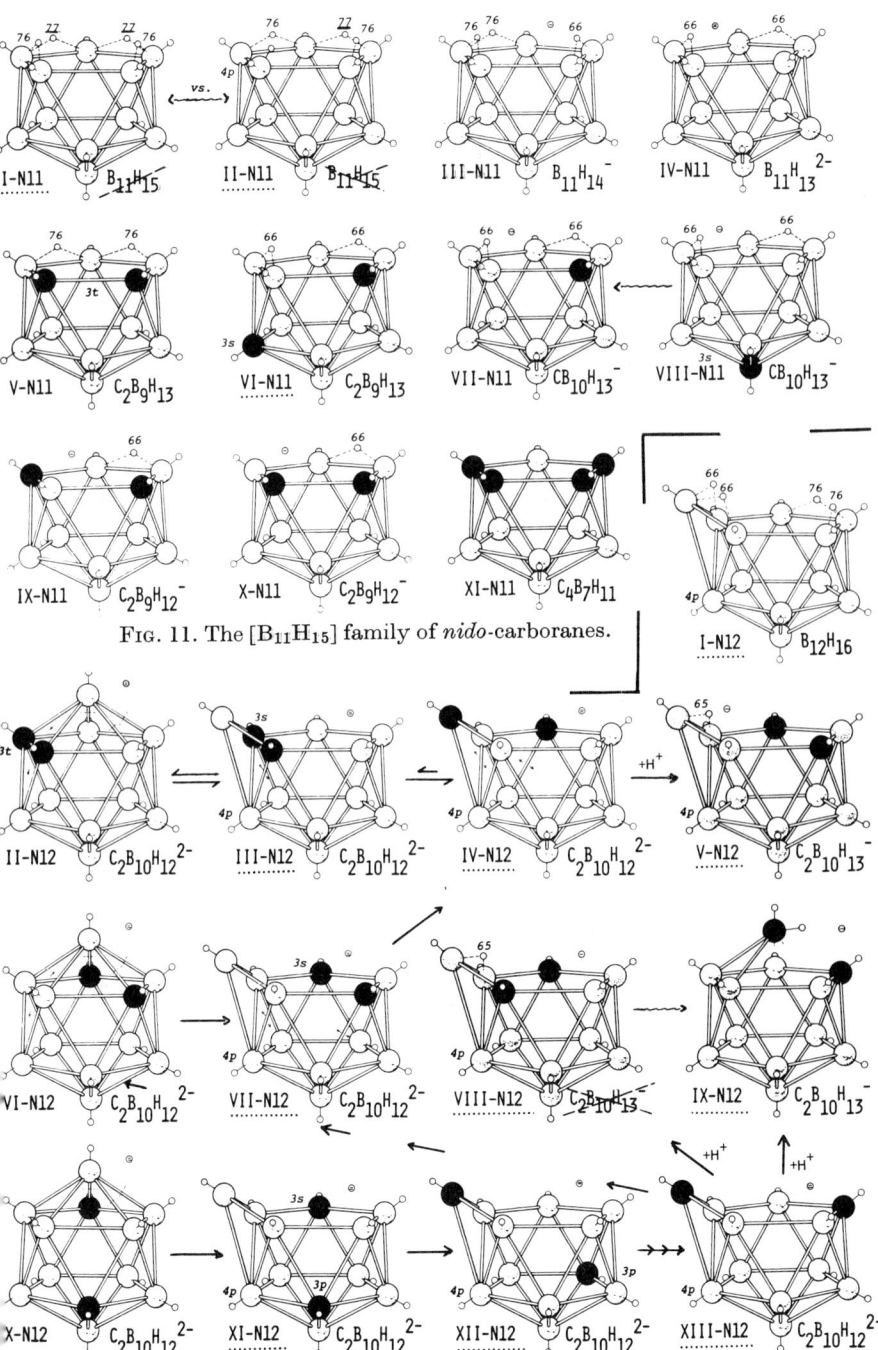

FIG. 11. The [$B_{11}H_{15}$] family of *nido*-carboranes.

FIG. 12. The [$B_{12}H_{16}$] family of *nido*-carboranes.

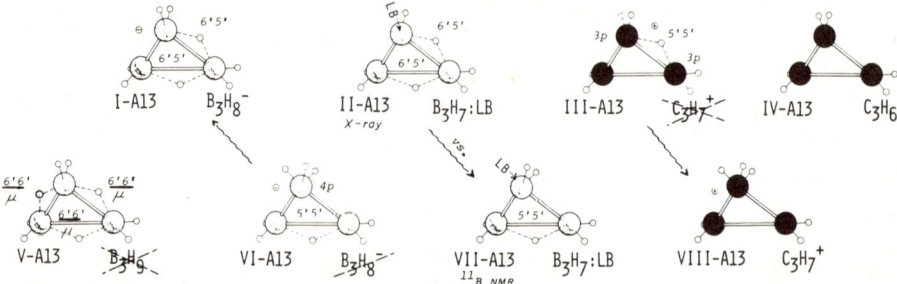

Fig. 13. The [B_3H_9] family of *arachno*-carboranes. (LB, Lewis base.)

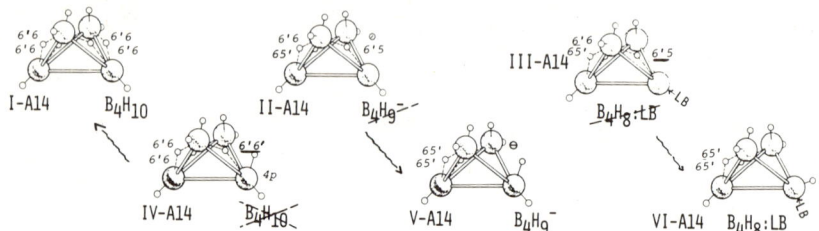

Fig. 14. B_4H_{10} family of *arachno*-carboranes. (LB, Lewis base.)

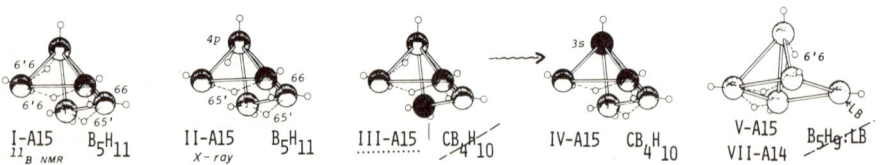

Fig. 15. The B_5H_{11} family of *arachno*-carboranes. (LB, Lewis base.)

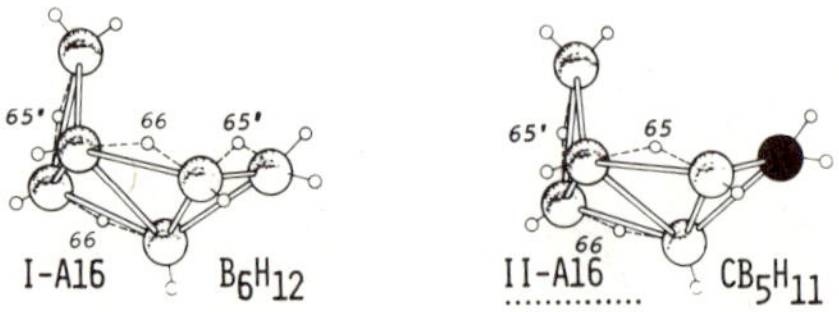

Fig. 16. The B_6H_{12} family of *arachno*-carboranes.

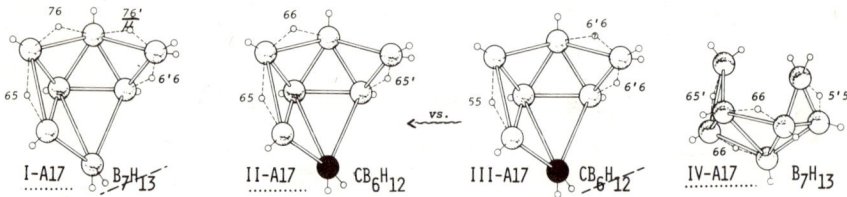

Fig. 17. The [B_7H_{13}] family of *arachno*-carboranes.

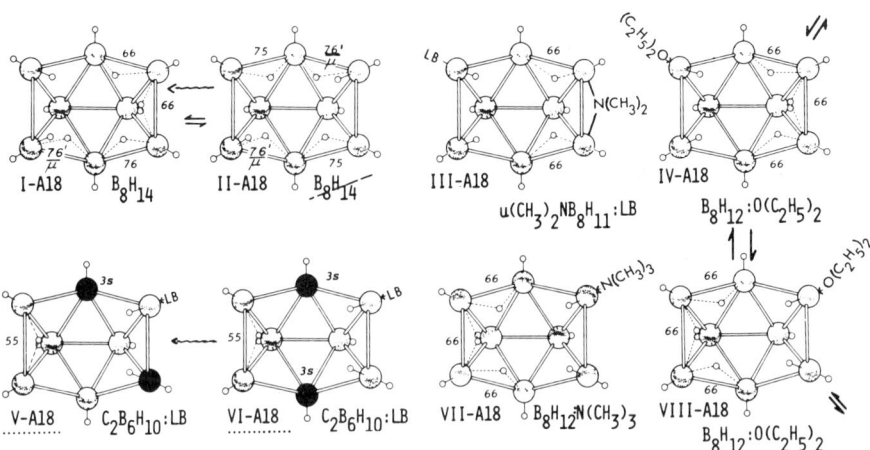

FIG. 18. The B_8H_{14} family of *arachno*-carboranes. (LB, Lewis base.)

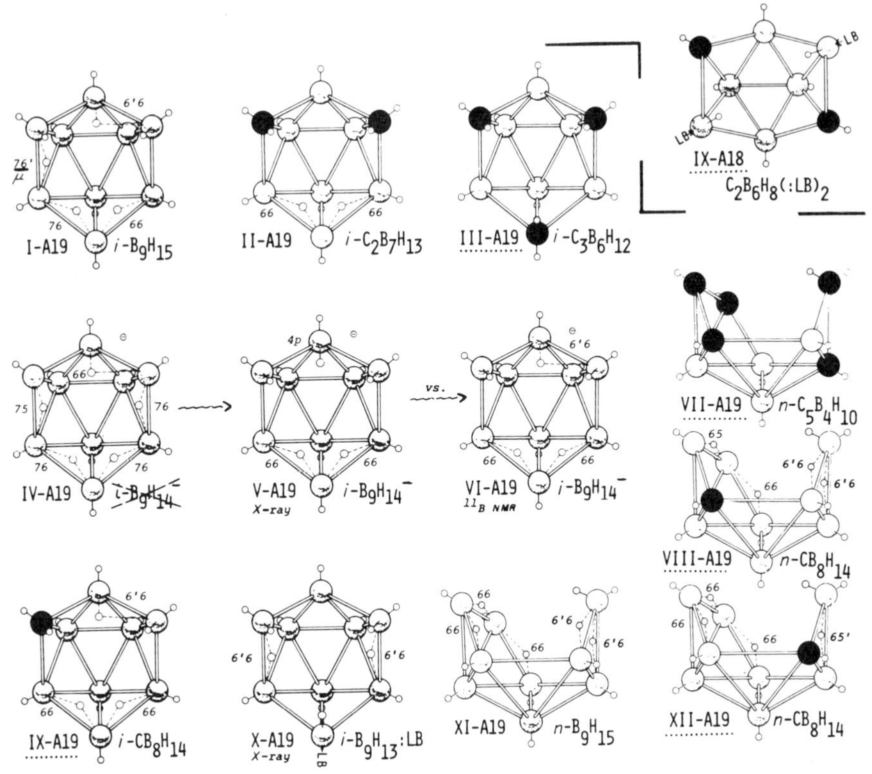

FIG. 19. The B_9H_{15} family of *arachno*-carboranes. (LB, Lewis base.)

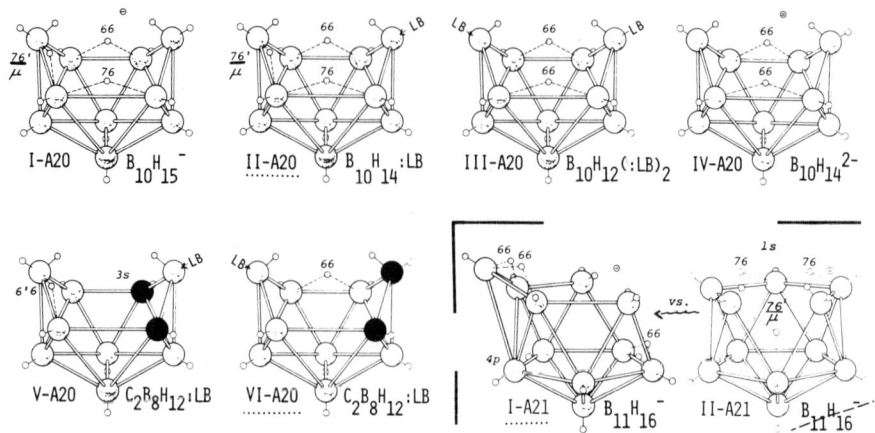

FIG. 20. The [$B_{10}H_{16}$] family of *arachno*-carboranes. (LB, Lewis base.)

FIG. 21. The [$B_{11}H_{17}$] family of *arachno*-carboranes.

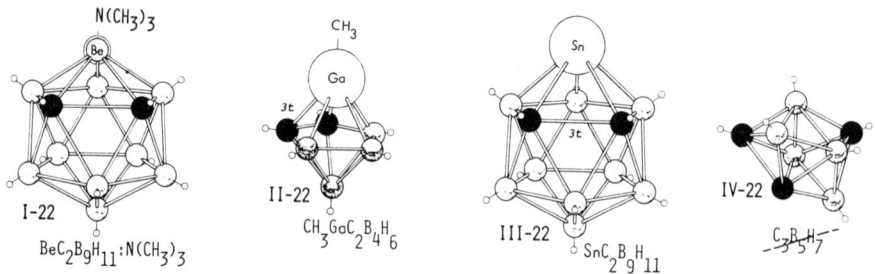

FIG. 22. The BH group-substituted carboranes.

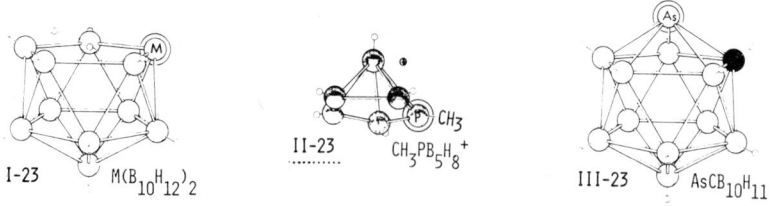

FIG. 23. The CH group-substituted carboranes.

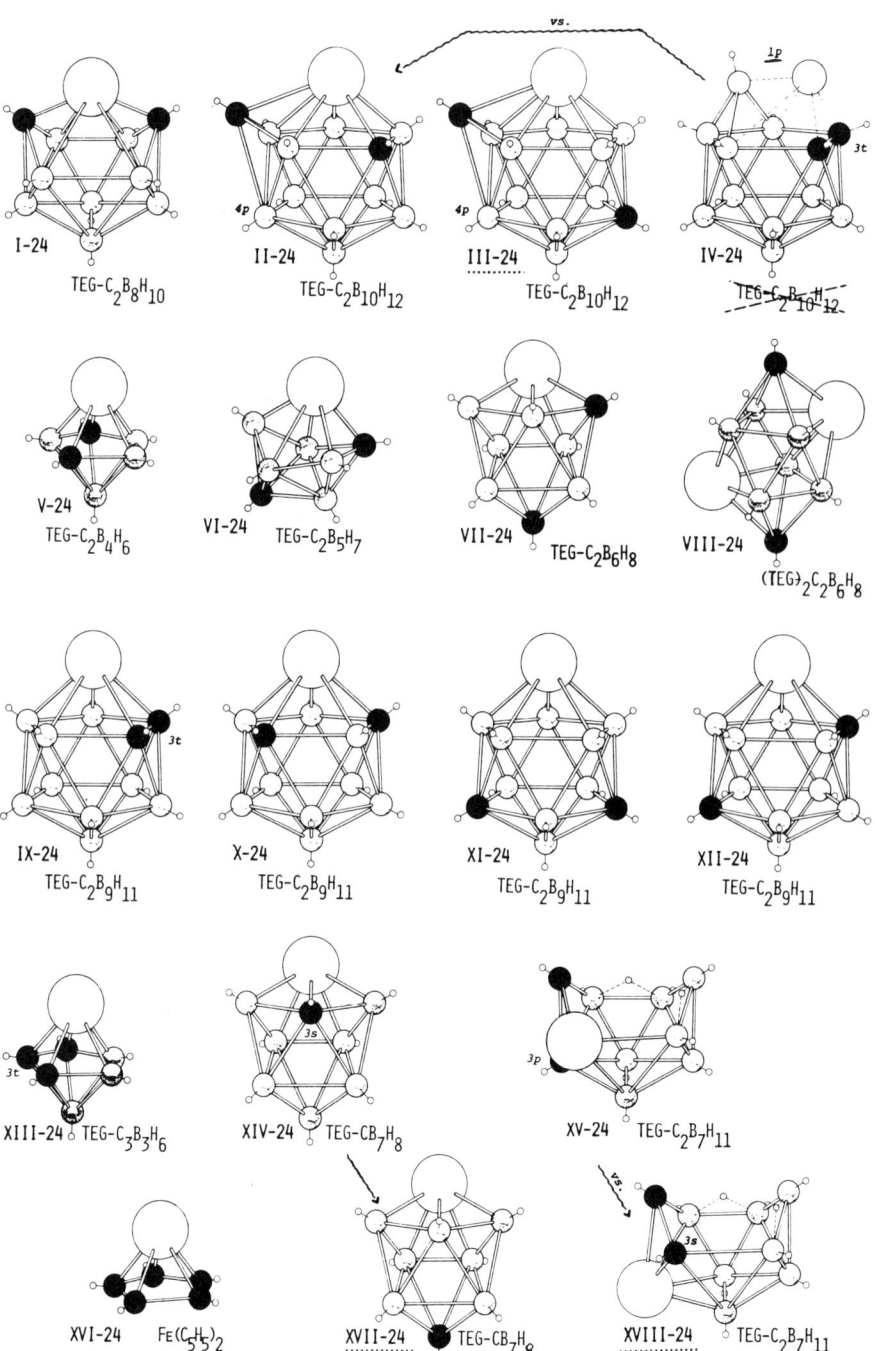

FIG. 24. The transition element group (TEG)-substituted carboranes.

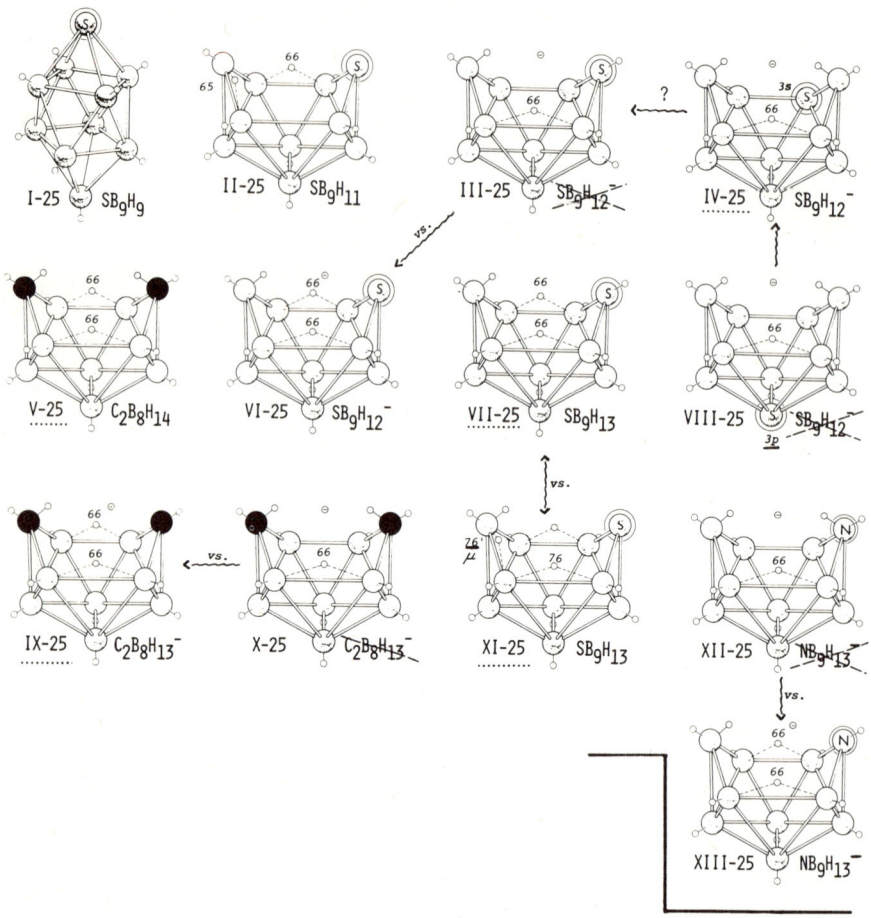

FIG. 25. Comparison of carbathia and carbaaza-substituted carboranes.

Fig. 1) within which to place four bridge hydrogens, may not exist because of steric crowding of the bridge hydrogens. The present author was unable to account for the predilection of the nido eight-vertex B_8H_{12} (III-N8) to assume the more open arachno-structural fragment, at least in the crystal phase (82), although to test this exception the isoelectronic $C_2B_6H_{10}$ (V-N8) was isolated.

3. It was also suggested (164) on empirical grounds that the "all-carbon" carborane $C_5H_5^+$ (V-N5) would have the skeletal configuration of its parent borane B_5H_9. The following year Hoffman et al. (134) published their conclusions, based on theoretical grounds, that $C_5H_5^+$ would have a "B_5H_9-like "structure.

Within months, both Wade *(147–152)* and Rudolph et al. *(115, 123)* pointed out that our empirical formula categories for closo, nido, and arachno species (i.e., $C_{0-2}B_nH_{n+2}$ and $C_{0-4}B_nH_{n+4}$ and $C_{0-6}B_nH_{n+6}$) which correlate with the structural rules are similar to Hückel electron-counting rules; i.e., $n+1$, $n+2$, and $n+3$ pairs of skeletal electrons characterize the closo, nido, and arachno classes, respectively *(158)*. Others have subsequently paraphrased these same thoughts *(67, 93)*. The antecedent B_nH_{n+4} and B_nH_{n+6} borane categorizations date back through the "systematics" *(110)* of Parry and Edwards (1959) to the original works *(133)* of Stock (1933). The first positively identified *hypho*-borane Lewis base adduct $[(CH_3)_3P)_2B_5H_9]$ has been structurally characterized by Shore et al. *(33a)* and has the structure anticipated by Lipscomb *(82)* for $B_5H_{11}^{2-}$. (*Hypho*, from the Greek word meaning "net", is applied to those compounds that are isoelectronic with $C_{0-8}B_nH_{n+8}$ or that have $n+4$ framework electron pairs.) Other hypho species, e.g., B_6H_{10}: $P(CH_3)_3$ *(88a)* and $B_5H_{12}^-$ *(117a)*, have also been characterized.

Although many investigators had long been aware that the closo, nido, and arachno *(164)* categories (Table 1) monotonically differed by electron pairs (or their isoelectronic equivalents), it was not generally recognized that electrons associated with endohydrogens could legitimately be considered as skeletal electrons, and, thus, empirical formulas and related isoelectronic relationships *(158)* were focused upon rather than electron count. Both approaches result in the same conclusions and are complementary but the more straightforward skeletal electron counting *(115, 123, 147, 148–152)* seems to be gaining greater acceptance.

Wade expanded the 1971 hypothesis to incorporate metal hydrocarbon π complexes, electron-rich aromatic ring systems, and aspects of transition metal cluster compounds [a parallel that had previously been noted by Corbett *(19)* for cationic bismuth clusters]. Rudolph and Pretzer chose to emphasize the redox nature of the closo, nido, and arachno interconversions within a given size framework, and based the attendant opening of the deltahedron after reduction (diagonally downward from left to right in Fig. 1) on first- and second-order Jahn–Teller distortions *(115, 123)*. Rudolph and Pretzer have also successfully utilized the author's approach to predict the most stable configuration of SB_9H_9 (I-25) *(115)* and other thiaboranes.

A number of studies published since 1971 confirm several of the initial predictions. Two groups at Indiana University [Siedle et al. *(128)* and Streib et al. *(61)*] have revealed that the nine-vertex nido structures $B_9H_{12}^-$ (II-N9) and the isoelectronic C,C-dimethyl derivative of $C_2B_7H_{11}$ (IX-N9) *do have nonicosahedral fragment structures* (see Figs. 1 and 9;

TABLE I
Correlation Carboranes by Empirical Formula

	closo-		nido-	
	Carboranes	Boranes	Carboranes	Boranes
4				
5	$C_2B_3H_5$		$C_5H_5^+$	B_5H_9
			$B_5H_8^-$	
6	$C_2B_4H_6$	$C_1B_5H_7$	$C_4B_2H_6$ $C_3B_3H_7$ $C_2B_4H_8$ $C_1B_5H_9$	B_6H_{10}
	$[C_1B_5H_6^-]$		$C_2B_4H_7^-$	
	$B_6H_6^{2-}$			
7	$C_2B_5H_7$			
	$B_7H_7^{2-}$			
8	$C_2B_6H_8$			B_8H_{12}
	$B_8H_8^{2-}$			
9	$C_2B_7H_9$			
	$B_9H_9^{2-}$			$B_9H_{12}^-$
10	$C_2B_8H_{10}$		$[C_4B_6H_{10}]$ $[C_3B_7H_{11}]$ $C_2B_8H_{12}$	$B_{10}H_{14}$
	$C_1B_9H_{10}^-$ $[B_{10}H_{11}^-]$		$B_{10}H_{12}^{2-}$ $B_{10}H_{13}^-$	
	$B_{10}H_{10}^{2-}$			
11	$C_2B_9H_{11}$		$C_2B_9H_{12}^-$ $C_2B_9H_{13}$	$B_{11}H_{15}$
	$C_1B_{10}H_{11}^-$		$C_1B_{10}H_{13}^-$ $B_{11}H_{14}^-$	
	$B_{11}H_{11}^{2-}$		$C_2B_9H_{11}^{2-}$	
12	$C_2B_{10}H_{12}$		$C_1B_{10}H_{11}^{3-}$ $B_{11}H_{13}^{2-}$	
	$C_1B_{11}H_{12}^-$			
	$B_{12}H_{12}^{2-}$			
	C_2 $\quad$ C_1 $\quad$ B_nH_{n+2}		C_4 $\quad$ C_3 $\quad$ C_2 $\quad$ C_1 $\quad$ B_nH_{n+4}	
	$C_{0-2}B_nH_{n+2}$		$C_{0-4}B_nH_{n+4}$	
	$n+1$ skeletal electron pairs		$n+2$ skeletal electron pairs	

	arachno-		
Carboranes		$B_3H_8^-$	Boranes
			B_4H_{10} 4
			B_5H_{11} 5
C_6H_6		$[C_2B_4H_{10}]$	B_6H_{12} 6
$C_7H_7^+$			7
			B_8H_{14} 8
	$C_2B_7H_{13}$		B_9H_{15} 9
		$B_9H_{14}^-$	10
	$B_{10}H_{14}^{2-}$	$B_{10}H_{15}^-$	11
			12

| C_6 | C_5 | C_4 | C_3 | C_2 | C_1 | B_nH_{n+6} |

$C_{0-6}B_nH_{n+6}$

$n + 3$ skeletal electron pairs

see also prediction 1 in the preceding). Moreover, Masamune (*89*) recently demonstrated that $C_5H_5^+$ (V-N5) apparently has the B_5H_9 (I-N5) configuration (see prediction 3). In addition, Hogeveen and Kwant (*59*) have isolated a permethyl derivative of $C_6H_6^{2+}$ that is isoelectronic and isostructural with $C_4B_2H_6$ (IX-N6) and B_6H_{10} (I-N6) and which reinforces to an even greater extent the structural parallels in borane, carborane, and carbonium ion chemistries.

An investigation of $C_2B_6H_{10}$ (V-N8; isoelectronic with B_8H_{12} and possessing even fewer bridge hydrogens) did not resolve the dilemma regarding eight-vertex nido compounds. The possible assumption of an arachno eight vertex structure by *nido*-$C_2B_6H_{10}$ suggests that something may be uniquely different about the nido eight-vertex polyhedral fragment.

If the bonding in B_8H_{12} and in $C_2B_6H_{10}$ is described in terms of 2- and 3-center bonds, then four 3-center bonds should reside within the skeletal framework and, in spite of the more open arachno structure, the required four 3-center bonds are successfully accommodated, at least in the crystalline phase. The possibility that the true nido structures (I-N8 for B_8H_{12} and X-N8 for $C_2B_6H_{10}$) may exist in the fluid phases is under investigation.

There have been complementary reports by several investigators encouraging an expansion of the original deltahedron (*164*) deltahedral fragment postulate. Among these reports are the revelations by Siedle *et al.* (*129*) and Schaeffer *et al.* (*132*) of the two different tautomeric structures (involving bridge hydrogen placement) of $B_{10}H_{13}^-$ in the fluid phase (III-N10; unencumbered tautomer) as opposed to the crystal phase (IV-N10; encumbered tautomer) where crystal packing forces evidently play a role (*78, 144*). These results herald the demise of a second belief that bridge and endohydrogens occupy the same locations in both the unencumbered fluid phase as well as in the encumbered crystalline phase. Other kinds of bridge hydrogens are also known to be distorted by the crystal environment, e.g., F_2H^- (*156*).

Shore's investigations (*65*) of the structure of $B_4H_9^-$ (V-A14), to be compared with Parry and Paine's studies (*39, 109*) of B_3H_7:LB (VII-A13), as a function of temperature, coupled with Onak's *et al.* study (*88, 106*) of Lewis base migration in the isomerization of *nido*-$C_2B_4H_6$:$N(CH_3)_3$ (i.e., VI-N6 is produced from VII-N6) expose certain patterns that relate to the predictable equivalence in some cases and to the predictable nonequivalence in other cases of a Lewis base (:LB) and/or a hydride (H^-) in such structures.

Beaudet *et al.* (*3*) have demonstrated that the bridge hydrogen of CB_5H_7 (III-C4) is on the cage (i.e., located over a triangular facet) and is thus involved in a 4-center bond.

Hawthorne and his associates (*60*, *155*) have reassuringly shown that the previous suggestions (*160*) are correct regarding bridge hydrogen placement in $CB_{10}H_{13}^-$ (VIII-N11) and the two isomers of $C_2B_9H_{12}^-$ (IX- and X-N11), as well as the anticipated (*159*) placement of the carbon in *closo*-$CB_{10}H_{11}^-$ [this in spite of a previously published alternative structure (*62*)].

The following sections expand on the deltahedron–deltahedral fragment hypothesis (structural rule 1), and in the order of decreasing importance add three additional rules, one involving the placement of BE hydrogens (rule 2), next the placement of the various heteroelements (with emphasis on carbon) (rule 3) (*172*), and finally the structural accommodations of boron, including the influences of endohydrogens (rule 4). All are shown to have their roots *firmly and simply* embedded in CNPR considerations.

What is thus offered is in effect a "back of the envelope" systematization of carborane–borane structures of predictive and teaching utility.

II. Structural Rules

A. Reappraisal of Deltahedron–Deltahedral Fragment Hypothesis (Rule 1)

In its abridged form the deltahedron–deltahedral fragment hypothesis (*164*) itemizes the various closo deltahedra in the left vertical column in Fig. 1. Deletion of one highest coordination vertex site from each of the closo deltahedra produces the nido deltahedral fragments displayed in the middle vertical column of Fig. 1. Removal of one additional highest coordination vertex from the nido-deltahedral fragments (necessarily adjacent to the open faces in the case of boranes and carboranes) produces the skeletal configurations characteristic of the arachno series in the right vertical column in Fig. 1.

There is an exception: in addition to the expected normal arachno nine-vertex fragment characteristic of n-B_9H_{15} (XI-A19) (*78*, *130*), there is a fragment generated by the removal of a low coordination vertex which is reflected in the structures of both i-B_9H_{15} (I-A19) (*21*) and its isoelectronic analog i-$C_2B_7H_{13}$ (II-A19) (*136*).

The nido ten-vertex and arachno ten-vertex deltahedral fragments are coincidently the same. *Such ten-vertex nido and arachno species are not isoelectronic*, although they are frequently misjudged as being isoelectronic, and a number of investigators have assigned incorrect structures based on comparison of ^{11}B NMR spectra of compounds that were mistakenly thought to be isoelectronic.

Figure 1 was initially generated by removing high-coordination vertices horizontally (*164*) (shown as process H in Fig. 1), but the consequences of removing low-coordination vertices (shown as process L, L', or L" in Fig. 1) are also illuminating. (The primes reflect the number of "bonds" necessarily reintroduced following the vertex removal.)

If one vertex and its attendant bonds are removed from a ball-and-stick model of the closo twelve-vertex icosahedron and two bonds are subsequently inserted into the open face, the eleven-vertex deltahedron results. If from each resulting smaller deltahedron any one of the lowest-coordination vertices, and its attendant bonds, are monotonically removed and one bond is inserted, the next smaller deltahedron results in all cases, from the icosahedron to the trigonal bipyramid. It was the exact reverse of this primitive ball-and-stick degradation concept (process L) which allowed the correct bisdisphenoid (*154*) structure for $C_2B_6H_8$ (VI-C2) to be anticipated (*172*) prior to its production.

A similar appraisal of the vertical middle row in Fig. 1 (the nido fragments) shows that basic process L also applies in all cases excepting that, in following low coordination vertex removal, bonds must be inserted in the ten- to nine-vertex and six- to five-vertex cases (process L').

The vertical relationships between adjacent arachno fragments are illustrated in the right vertical row in Fig. 1. The ten- to nine-vertex arachno transformation (via process L) yields the unpredicted (by process H) *i*-nine-vertex arachno fragment instead of the *n*-nine-vertex arachno fragment. And, to convert the hypothetical seven-vertex arachno fragment to the known six-vertex arachno fragment, removal of a high-coordination edge vertex is required; removal of the low-coordination vertex produces an arachno fragment that is identical to the six-vertex nido fragment. The alternative six-vertex arachno fragment is very unlikely to be observed for any *arachno*-carborane.

B. STRUCTURAL PREFERENCES OF VARIOUS MOIETIES: PRELUDE TO THE EXPOSITION OF RULES 2, 3, AND 4

In order to recognize the patterns that pervade carborane chemistry, it is instructive to conduct a "molecular census," to ascertain where and in what coordination-number environments the various groups, e.g., BH, BH_2, CH, CH_2, CH_3, and Lewis base equivalents (B:LB, BH:LB, and BH_2:LB, where LB = Lewis base) are most frequently found. Table II itemizes the various groups; obvious trends may be observed and examples of the extreme cases are given.

The term *coordination number* of a given atom is used herein to define the total number of other atoms with which the given atom is

TABLE II

COORDINATION NUMBER RANGE OF VARIOUS GROUPS

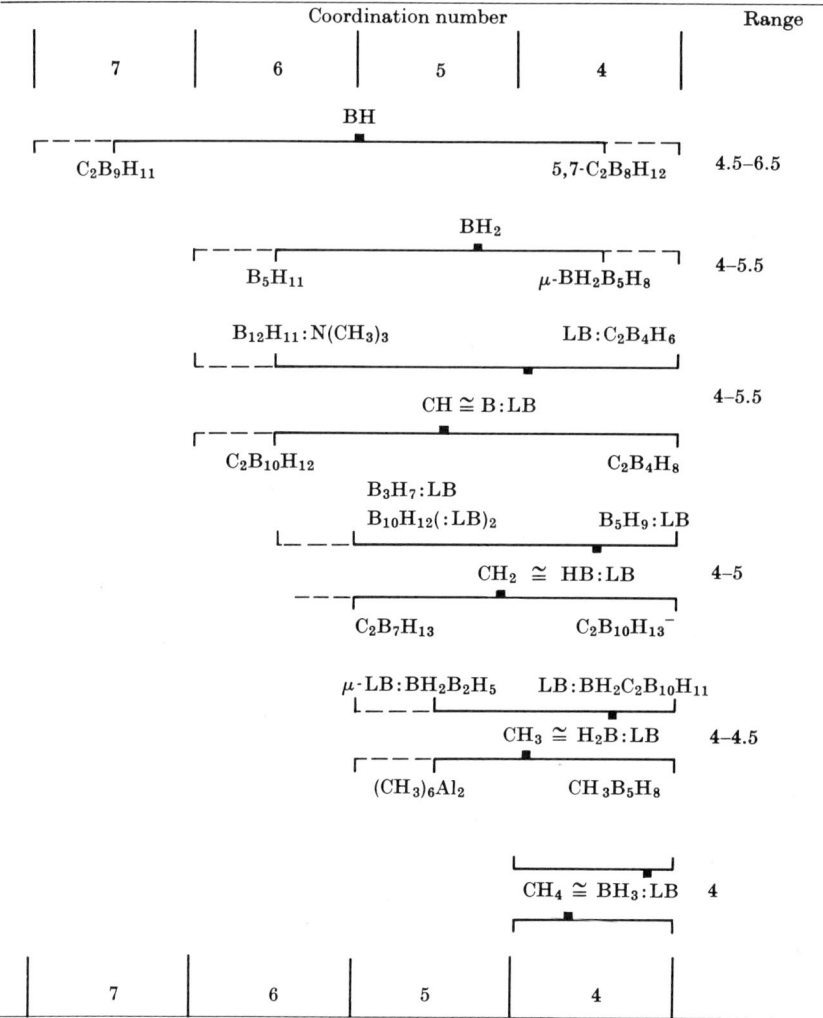

associated. The borons in B_2H_6 are, therefore, each five-coordinate as they are associated with two terminal hydrogens, two bridge hydrogens, and one other boron.

Note that BH groups are most amenable to varying coordination-number environments (5-7), whereas BH_2 groups (less electrons to delocalize) are less amenable to high-coordination number environments

TABLE III

VIOLATIONS OF VARIOUS RULES

Rule 1	Rule 2	Rule 3	Rule 4
Primary Skeletal configurations that are not deltahedra or deltahedral fragments displayed in Fig. 1	**Primary** Bridge hydrogens associated with more than two borons (i.e., XXX-bridge hydrogens)	**Primary** Carbons found in sites two coordination numbers higher than otherwise present, or adjacent to bridge hydrogens	**Primary** BH groups found in seven- or four-coordinate sites, or BH$_2$ groups found in six-coordinate sites, or BH$_3$ groups found in five-coordinate sites
←—?——	**Secondary** Two-coordinate bridge hydrogens occupying higher-coordination sites when lower-coordination alternatives are present, or the presence of 77-, 76-, and 6'6-bridge hydrogens	**Secondary** Carbons found in sites one coordination number higher than otherwise present	——?—→
Secondary Skeletal configurations displayed in Fig. 1 but in incorrect series ←—?——		**Tertiary** Carbons adjacent when both are over four-coordinate	

TABLE IV

EXAMPLES OF VARIOUS RULE VIOLATIONS

Rule 1	Rule 2	Rule 3	Rule 4
1p None or maybe $C_2B_{10}H_{13}^-$ (IX-N12)	2p CB_5H_7 (III-C4) $B_{10}H_{11}^-$ (VIII-C4)	3p None or maybe [$C_2B_4H_6$:LB^a(VII-N6)] TEG-$C_2B_7H_{11}$(XV-24)	←—?—→
←—?—→	2s $B_{10}H_{13}^-$ (IV-N10) $C_2B_7H_{11}$ (IX-N9) $C_2B_8H_{12}$ (XI,XVI,XX-N10) $B_{11}H_{15}$ (I-N11) i-B_9H_{15} (I-A19) B_7H_{13} (I-A17)	3s 1,6-$C_2B_8H_{10}$ (I-C3) $CB_{10}H_{13}^-$ (VIII-N11)	4p $C_2B_9H_{11}$ (II-C4) $B_{11}H_{15}$ (II-N11) $B_3H_8^-$ (VI-A13)
1s B_8H_{12} (III-N8) $C_2B_6H_{10}$ (V-N8) B_7H_{13} (IV-A17)		3t 1,2-$C_2B_4H_6$ (III-C3) 1,2-$C_2B_{10}H_{12}$ (II-C3)	←—?—→
←—?—→			

^{a} LB, Lewis base.

and probably more amenable to lower-coordination number environments. Carbon is less adaptable than boron, thus B:LB groups (isoelectronic with CH groups) are found to be only slightly more tractable than carbon to the various structural environments (in one or two cases B:LB groups have been found adjacent to bridge hydrogens).

The CH_2 groups are less tractable than CH groups. The seeming predilection of carbon to be found in a five-coordinate CH_2 environment with the *arachno*-carboranes rather than in a five-coordinate CH environment, e.g., $C_2B_7H_{13}$ (II-A19) is thought to be the result of the endohydrogen's preference rather than that of the carbon. (Both are "electron-sufficient" and prefer to engage in localized bonding if possible.) However, in the absence of the endohydrogen requirements, a five-coordinate CH group would be preferred over a five-coordinate CH_2 group.

The definition here of electron-deficient as compared to electron-sufficient is keyed to the number of electrons lacking in the boron species as compared to the equivalent carbon species; thus, B_4H_{10} is four electrons deficient compared to a neutral C_4H_{10}. In almost all cases the number of electrons deficient coincides exactly with the number of simplistic 3-center bonds [the $s + t$ of Lipscomb's styx terminology (*82*)] necessary to describe the structure. Again, B_5H_9 (I-N5) requires five 3-center bonds whereas $B_5H_8^-$ (II-N5) would require four 3-center bonds and $C_5H_5^+$ (V-N5) require one 3-center bond. All three species have one skeletal 3-center BBB or CCC bond, whereas the other 3-center bonds are expressed as 3-center bridge hydrogens. An alternative and more general definition of electron deficiency has been used by Wade (*147a*).

Rule 1 and the additional rules 2, 3, and 4 are displayed (in abridged form) in Tables III and IV (during the exposition of the rules, the reader may benefit by referring to these tables).

C. Bridge and Endohydrogen Considerations (Rule 2)

The recognition of the significance of the BE hydrogens in carborane structures and function is the feature that made possible the formulation of CNPR theory. The relative importance of bridge hydrogen placement as opposed to carbon placement was settled in favor of the bridge hydrogens when Grimes deduced the structure of $C_2B_3H_7$ to be IV-N5 (*32, 33, 45, 168*) instead of VIII-N5 (*162*). Both bridge hydrogens and carbons favor sites of lowest-coordination number; however, in the case of $C_2B_3H_7$ the bridge hydrogen preempted the lowest-coordination environment at the expense of carbon (*168*).

Were the isoelectronic anion (i.e., $C_2B_3H_6^-$) to be prepared, the most stable configuration should have adjacent basal carbons to accommodate the basal bridge hydrogen as well as the carbon. Either cis or trans basal carbons should be anticipated in the isoelectronic dianion ($C_2B_3H_5^{2-}$) which would have no bridge hydrogens. Rudolphs's Extended Hückel Molecular Orbital calculations would favor cis basal carbons in the latter case.

Although not realized until recently, in all probability a bridge hydrogen, if forced to assume a position upon the cage [as is the case with CB_5H_7 (III-C4)], would locate over a triangular face (3) and would hence be associated with three borons via a 4-center bond as illustrated in IX-N5.

In contrast to the convention of counting the adjacent atoms alone when assigning rudimentary coordination numbers to either the carbon or boron, *the bridge hydrogen convention applied here counts the coordination numbers of the two (or rarely three) borons with which the bridge hydrogen is associated*. For example, all of the two-coordinate bridge hydrogens in $B_{10}H_{14}$ (I-N10) are referred to as 66-bridge hydrogens since all of the borons associated with the bridge hydrogens are six-coordinate (including the bridge hydrogens in the count). The four bridge hydrogens in B_6H_{10} (I-N6) are similarly labeled as two 66-bridge hydrogens and two 65-bridge hydrogens. Since, as can be seen from Table II, BH_2 groups are less amenable to higher coordination situations, a prime is added when one (or both) of the borons associated with a bridge hydrogen is a BH_2 group; thus the bridge hydrogens in B_2H_6 are 5'5'-bridge hydrogens whereas those in B_4H_{10} (I-A14) are 6'6-bridge hydrogens.

Bridge hydrogens prefer the lowest-coordination sites, avoiding if possible borons whose coordination numbers are 7 or 6'. There are patterns beyond the scope of the present manuscript suggesting that a bridge hydrogen associated with a seven-coordinate BH group is less content (i.e., more labile or more acidic) than with a 6'-coordinate BH_2 group. Nevertheless, all known compounds with 77-, 76'- and 6'6'-bridge hydrogens are unstable and are considered as equivalently undesirable throughout this manuscript.

Disregarding steric congestion, it is much more than coincidence that the "stable" *nido*-boranes B_5H_9 (I-N5), B_6H_{10} (I-N6), and $B_{10}H_{14}$ (I-N10) incorporate only 66- and 65-bridge hydrogens, whereas the unobserved structures for the unstable *nido*-boranes B_7H_{11} (I-N7), B_8H_{12} (I-N8), and B_9H_{13} (I-N9) would contain 76-, 75-; two 76-; and two 76- and 77-bridge hydrogens, respectively.

The latter two species, B_8H_{12} (I-N8) and B_9H_{13} (I-N9), may be considered to "eliminate" the overly coordinated 7X-bridge hydrogens by either opening into the arachno configuration B_8H_{12} (*121*) (III- and IV-N8) which incorporates 66-bridge hydrogens or by the elimination of one proton from B_9H_{13} to form $B_9H_{12}^-$ (II-N9) which also incorporates only 66-bridge hydrogens!

Bridge hydrogens are thought to seek sites between lowest coordinated borons because of the bridge hydrogen's attraction to the most "electron available" environment. Noted throughout the text are the many cases where electronegative anions (more electrons available) stabilize more highly coordinated bridge hydrogens than are stabilized on isoelectronic neutral molecules. When heteroatoms furnish the requisite skeletal electrons within neutral molecules, the heteroatoms become positively charged. Such positively charged locations are shunned by bridge hydrogens and, thus, we find no bridge hydrogens neighboring carbon, sulfur or nitrogen (or rarely borons associated with Lewis bases as they are isoelectronic with carbon). Indeed, as discussed in Section III, E, even terminal hydrogens avoid sulfur and nitrogen. The positive charge centered about the heteroatom that repels adjacent bridge hydrogens, might also affect more remote locations. Such heteroatoms could possibly account for bridge hydrogens in selected compounds being found in slightly higher coordinated locations more distant from the heteroatom when lower coordinated locations were available closer to the heteroatom (see XI-N10 v. XVI-N10 and VII-N9 v. IX-N9). This possible longer range effect of heteroatoms is currently neglected but should be considered seriously in future investigations.

It is safe to assume that any XXX-bridge hydrogen on a cage, e.g., as in CB_5H_7 (III-C4) (*3*), and associated with three 6-coordinate borons (a 666-bridge hydrogen) would be in a very undesirable situation. Similarly, three coordinate bridge hydrogens in the hypothetical closo species ($B_{12}H_{14}$) and ($B_{10}H_{12}$) [or $B_{10}H_{11}^-$ (VIII-C4)] would be 777- and 776-bridge hydrogens, respectively. This could account for the fact that such species have heretofore only been known as $B_{12}H_{12}^{2-}$ and $B_{10}H_{10}^{2-}$ ions (*95*) and that the related acids ((H_3O^+)$_2B_nH_n^{2-}$) are very strong acids.

Under rule 2, the instability of a three-coordinate bridge hydrogen occupying any cage site (e.g., 666) is pointed out as a primary violation, whereas a two-coordinate bridge hydrogen occupying a higher coordination site when a lower coordination site is available is defined as a secondary violation. The various violations are noted about the molecules within the figures by the symbols (2s), etc. For example, (2s) denotes that the molecule depicted involves a secondary violation of rule 2 (see Tables III and IV).

Compound B_3H_9 (V-A13), which would have three very undesirable 6'6'-bridge hydrogens, has never been isolated, and both $B_{11}H_{15}$ [I-N11 (28) with perhaps two 77-bridge hydrogens] and $B_{10}H_{15}^-$ [I-A20 (122) with a 76'-bridge hydrogen] are known to be very unstable. The closely related B_2H_6 (two 5'5'-bridge hydrogens) and $B_{11}H_{14}^-$ [III-N11 (1), two 76-bridge hydrogens] are more stable, and as would be expected, the related $B_{11}H_{13}^{2-}$ (IV-N11) with two compliant 66-bridge hydrogens is much more stable (34).

In those instances where packing forces in the crystalline state evidently induce the bridge hydrogens in molecules, such as $B_{10}H_{13}^-$ [(III-N10 (129) vs. IV-N10 (132)] and probably $C_2B_7H_{11}$ [IX-N9 vs. VII-N9 (61)], to assume less desirable (i.e., higher-coordination) situations, to date, only conflicts between structurally preferable XX-bridge hydrogens (where X = 6 or 5 rather than 6' or 7) appear to be involved.

The 6'6-bridge hydrogens in B_4H_{10} (I-A14) are biased toward the more desirable 6-boron and away from the 6'-boron (82).

D. Carbon and Other Heteroelement Atom Considerations (Rule 3)

From the discovery of the first carboranes it was noted that the carbons were in the apex positions in $C_2B_3H_5$ (I-C4) (40, 90, 126, 175) but were in the equatorial positions in $C_2B_5H_7$ (V-C2) (104); i.e., *they were in the lowest-coordination positions possible in each case.* Furthermore, it was observed that the carbons, if one or both were over four-coordinate, tended to be separate or to separate at least in the *closo*-carboranes; e.g., the 1,6-$C_2B_4H_6$ (IV-C2) (40, 90, 126, 175) isomer was produced by heating the 1,2-$C_2B_4H_6$ (III-C3) isomer (100). These CNPR considerations made it possible to publish the correct (and only the correct) carbon position isomers for both 2,4-$C_2B_5H_7$ (V-C2) (4, 5, 104) and 1,7-$C_2B_6H_8$ (VI-C2) (172), in spite of ambiguous ^{11}B NMR data.

Upon closer scrutiny it becomes apparent that these considerations revealed by carbon's predilections are general and apply to the other heteroelement carboranes as well.

It is now recognized that the predisposition of most, if not all, two-coordinate bridge hydrogens are more important (rule 2s) than the positioning of carbon when carbon's coordination number choices differ by 1 (rule 3s). Isomers in violation of the foregoing rules are known (i.e., carbons are found in ideal situations at the expense of the ideal placement of the bridge hydrogens), but in such cases the ideal carbon isomers are less stable than their ideal bridge hydrogen analogs. A carbon would probably be least likely to be found two coordination numbers higher (i.e., in a six- rather than a four-coordinate site) to accommodate the desire of bridge hydrogens to migrate to slightly lower-coordination sites. When such a situation is eventually encountered it will be defined as a primary violation of rule 3 (see Tables III and IV and XI-N12 and XII-N12).

The bridge hydrogen vs. carbon competition for low-coordination sites is further complicated by the capability of the bridge hydrogens to migrate (*157*) into ideal sites even under ambient conditions if not in the encumbered crystalline phase. Carbon migration (in the absence of seven-coordinate borons), in contrast, usually requires elevated temperatures for migration to other sites; e.g., the 1,2-$C_2B_4H_6$ (III-C3) rearrangement (*100*) into 1,6-$C_2B_4H_6$ (IV-C2). Furthermore, under pyrolytic conditions, carbons never (to date) migrate in such a fashion as to become adjacent (violating rule 3t) in the final products although such migration might be necessary to attain the more stable, i.e., ideal bridge hydrogen structure.

Thus, although thermodynamically best isomers may be predicted (with bridge hydrogens in optimal positions and carbons in next best positions), frequently the chemical precursors kinetically dictate that the carbons will be placed in the wrong positions for producing the best isomer. In these cases the bridge hydrogens (mobile at ambient conditions) migrate to the best positions available, whereas the carbons are immobile unless elevated temperatures are involved. It is these same mobile bridge hydrogens which, on occasion, succumb to crystal packing forces and assume different bridge hydrogen configurations (*129, 132*) in the encumbered crystal phase (e.g., $B_{10}H_{13}^-$ has structure IV-N10 rather than III-N10) and, on occasion, even become endohydrogens rather than bridge hydrogens [e.g., $B_9H_{14}^-$, (V-A19) (*43*); B_9H_{13}:LB (X-A19) (*153*); and B_5H_{11} (II-A15) (*79*)] to accommodate such crystal packing forces.

E. BORON CONSIDERATIONS (RULE 4)

As perceived in Table II, BH groups are found to have the greatest coordination number range (5–7) while BH_2 groups have a more restricted range (4–6) probably due to the localizing effect of the additional endo-terminal hydrogen that militates against the high-coordination predispositions of the electron-deficient boron.

In order to catalog specific undesirable architectural loci under as few rule violations as possible, "content or malcontent" endohydrogens have been included (when not adjacent to bridge hydrogens) under rule 4.

For cataloging purposes the borons are considered as speciously more important than the endohydrogens when comparing content or malcontent BH_2 groups (rule 4). By contrast, if bridge hydrogens neighbor a BH_2 group such architectural conditions are considered as content or malcontent bridge hydrogens (rule 2) rather than content or malcontent BH_2 groups under rule 4.

Although avoiding organizational difficulties, this approach generates a separate problem because, as will become apparent, there is a continuum of BE hydrogens, some pure bridge (covered under rule 2) and some pure endo (covered under rule 4), with several specific examples defined as somewhere in between pure bridge and pure endo [e.g., B_5H_{11} (I- and II-A15)].

Seven-coordinate BH groups [e.g., $C_2B_9H_{11}$ (II-C4) (*2, 135, 145*)] as well as six-coordinate BH_2 groups not associated with bridge hydrogens [e.g., B_5H_{11} (II-A15)] are both associated with structural instability; both cases are considered to be violations of rule 4p. In the former case it may be assumed that it is malcontent boron that finds itself in an overly coordinated situation, whereas in the latter case it is understood to be a combination of malcontent boron and malcontent endohydrogen.

When bridge hydrogens are involved as neighbors of seven-coordinate BH or six-coordinate BH_2 groups, then the architectural indiscretions are considered to be secondary violations of rule 2 rather than involving rule 4. In this contrived fashion, tabulating a single undesirable locus as violating more than one rule is avoided.

The suggested rules having been itemized in the previous sections, a chemical rationalization of their origins follows, after which the various types and families of boranes and carboranes are discussed in detail.

F. PRIMARY, SECONDARY, AND TERTIARY EXPRESSIONS OF RULES 1, 2, 3, AND 4

Had the patterns become apparent in the numerical order of importance rather than in the order rule 3(1956), 1(1970), 2(1971), and 4(1972), their organization would have been simpler.

The rules in overall decreasing order of importance essentially state that the ideal structures for carboranes will be based on most spherical deltahedra (rule 1); the BE hydrogens will tend to be placed in the lowest possible coordination environments (rule 2); when elements to the right of boron in the periodic table are incorporated into the deltahedron or deltahedral fragment, they will tend to preempt low-coordination sites (e.g., carbon) or, if electron-deficient, high coordination sites (rule 3); and, lastly, boron will eschew seven-coordinate BH or six-coordinate

BH$_2$ environments (rule 4). In actual fact, however, selected compounds either comply marginally or violate the rules to varying degrees. In the sense that these few exceptions define the limits of the rules, it is useful to break the rules down into subsets of varying importance. Thus a primary violation of rule three (3p) may be a more serious offense than a secondary violation of the more important rule 2 (2s).

The rules are displayed in abridged form in Tables III and IV. They are deliberately described in the negative sense; i.e., rather than describing or listing the innumerable architectural features which do conform to the various rules, Tables III and IV emphasize the limited numbers of architectural features that are considered to be either on the borderline of acceptability or verging on violating the various rules.

In an oversimplified manner, the relative importance of the various rules (i.e., 1, skeletal; 2, bridge hydrogen; 3, carbon; 4 boron) have been rationalized on CNPR considerations as outlined in the following.

It has long been recognized that when there are fewer electrons available for bonding, more bonds are produced, e.g., in the series B_2H_6, C_2H_6, and N_2H_6 (i.e., $2NH_3$), 12, 14, and 16 electrons are available, respectively, for bonding (although in inverse order 9, 7, and 6 bonds are observed). It follows that the select series of most spherical deltahedra, in accordance with rule 1, maximizes the geometrical opportunities for multiple bonding and that the most electron-deficient *closo*-boranes ($B_nH_n^{2-}$) and *closo*-carboranes ($C_{0-2}B_nH_{n+2}$) will be found in the closed, most spherical, deltahedral configurations.

In countering the icosahedronism belief (*164*) the series of most spherical deltahedra (left-hand column in Fig. 1) may have been overemphasized. The most spherical deltahedra were correctly preferred because of coordination number smoothing; i.e., it follows that the eight-vertex bisdisphenoid incorporating five- and six-coordinate vertices is preferable when all of the atoms are boron ($B_8H_8^{2-}$) or boron and carbon [$C_2B_6H_8$ (VI-C2)] as opposed to the hexagonal bipyramid (seven- and five-coordinate vertices) and perhaps the bicapped trigonal antiprism (*154*) (four- and six-coordinate vertices). The ideal four-coordinate sites for carbon in the latter deltahedron involve overly acute angles as in $C_2B_3H_5$ (I-C4).

As the initial paper (*164*) was concerned only with boron and carbon, those cases wherein the skeletal atoms involved would be more content in different coordination sites and where, therefore, different deltahedra (with differently coordinated vertices) would be preferred were not considered. However, Wade has deduced that $(CO)_3CrC_6H_6$ is a nido compound (four-coordinate carbon and nine-coordinate chromium) derived from the preferred, in this case, eight-vertex (*154*) hexagonal bipyramid by the removal of one high-coordination vertex.

Wade also incorporates electron-rich species such as $C_5H_5^-$ and $C_4H_4^{2-}$ as arachno species derived by removal of two nonadjacent high-coordination vertices from the pentagonal and tetragonal bipyramids, respectively, and has, moreover, correlated many aspects of seemingly unrelated classes of compounds beyond the recognized borders of carborane chemistry.

When the systems under consideration have additional electrons available for skeletal bonding, fewer bonds are required and the more open, nido and arachno deltahedra fragments describe their structures instead of closed deltahedra.

Of the atoms within the carboranes, the bridge hydrogens (rule 2) are in the most alien environments (as compared to the common one-coordinate terminal hydrogens). Most bridge hydrogens are 100% over their normal coordination environment (two-coordinate), whereas some are 200% over normal (three-coordinate).

The carbons (rule 3) within the carboranes are found in four-, five-, and six-coordinate situations, never over 50% above the normal four-coordinate carbon.

Boron (rule 4) is regarded as most tractable and, therefore, least important; boron has been found to be from three- to seven-coordinate in various molecular configurations.

In discussing the various types and classes of carboranes (*45, 78, 123, 131, 147, 147a, 158, 164*) in the following sections, the relationship of structure and rule are brought into focus.

III. Carboranes, Their Analogs and Derivatives

A. *closo*-CARBORANES

All of the *closo*-carboranes discovered to date conform to most of the aforementioned rules and were the models from which rules, 1, 3, and 4 were first inferred.

The most stable *closo*-carboranes violate no rules (Fig. 2), whereas a group of slightly less stable *closo*-carborane isomers that will rearrange into the more stable isomers (Fig. 3) violate only rules (3s and 3t). A less stable group of *closo*-carboranes (Fig. 4) is restricted to unfavorable deltahedra, and the least stable known *closo*-carborane (III-C4) violates the very important bridge hydrogen rule (2p) by having a bridge hydrogen of necessity on the cage. Included in Fig. 4 are also selected, and as yet undiscovered, deltahedra.

1. Stable closo-Carboranes

Before its discovery (and based on CNPR considerations), it was suggested (*172*) that 1,10-$C_2B_8H_{10}$ (I-C2) was destined to be the most stable dicarba-*closo*-carborane (two ideal sites for carbon; eight ideal sites for boron) as has since been confirmed (*137, 138*). Although the same considerations might apply to an even greater extent to 1,5-$C_2B_3H_5$ (I-C4) (*40, 126, 175*), the steric arrangement about carbon (small angles) evidently works against stability in the latter case. The isomers of $C_2B_{10}H_{12}$ (II-C2, III-C2, and II-C3) (*45, 158*) are less stable than 1,10-$C_2B_8H_{10}$ (I-C2); their carbons would evidently prefer to be less highly coordinated, but within the icosahedral isomers of $C_2B_{10}H_{12}$ that choice does not exist.

Bigger cages, up to but not including seven-coordinate vertices (i.e., less steric strain), tend to accompany increased stability; thus 1,6-$C_2B_4H_6$ (IV-C2) and 2,4-$C_2B_5H_7$ (V-C2) are progressively more stable. Compound $C_2B_6H_8$ (VI-C2) is apparently even more stable and, incidentally, is the only parent *closo*-carborane that is comprised of a *dl*-pair (*172*).

2. Rearrangement-Prone closo-Carboranes

closo-Carborane isomers that are known to exist and have been isolated but which, upon heating, rearrange into their more stable relatives (Fig. 2), are shown in Fig. 3.

These are not the only rearrangement-prone isomers known to exist. Short-time high-temperature pyrolysis of $C_2B_4H_8$ (III-N6) (*107*) and B_2H_6 produces over 400 compounds, as deduced from gas chromatograph–mass spectroscopic studies (*171*). Although many of the 400 compounds are comprised of the innumerable methyl and polymethyl derivatives of the known carboranes, other compounds that (based on chromatographic retention time and monoisotopic mass spectra) reveal the virtually certain existence of four isomers of $C_2B_5H_7$, six isomers of $C_2B_8H_{10}$, and at least two isomers each of $C_2B_7H_9$ and $C_2B_6H_8$ in addition to myriad others are observed.

Stibr *et al.* (*132a*) have reported the 1,2-isomer of $C_2B_8H_{10}$.

3. Unstable or Unknown closo-Carboranes

A third group of unstable or unknown *closo*-carboranes is displayed in Fig. 4.

Compounds $C_2B_3H_5$ (I-C4) and 2,3-$C_2B_9H_{11}$ (II-C4) are sterically unfavored closo species that uniquely react with B_2H_6 (41) and CH_3Li (49) to produce more stable nido counterparts:

$$closo\text{-}C_2B_3H_5 + B_2H_6 \longrightarrow nido\text{-}C_2B_6H_{10} + H_2$$
$$closo\text{-}C_2B_9H_{11} + LiCH_3 \longrightarrow nido\text{-}LiCH_3C_2B_9H_{11}$$

The more stable closo-carboranes, by contrast (displayed in Figs. 2 and 3), react with these same reagents at higher temperatures to form other closo-carboranes or to form dilithio-closo derivatives such as $Li_2C_2B_5H_5$ (169).

Those species related to $C_2B_9H_{11}$ (II-C4) (2, 135, 145), i.e., $CB_{10}H_{11}^-$ (62, 72, 155) and $B_{11}H_{11}^{2-}$ (70), due to the presence of the seven-coordinate boron (violation of rule 4p) tend to scramble their borons on an NMR time scale via a valence bond tautomerism mechanism (142, 155) which possibly involves the transient and reversible opening of the eleven-vertex closo structure to the eleven-vertex nido structure (55, 96).

When no carbons are present, i.e., $B_{11}H_{11}^{2-}$, all borons become equivalent on an NMR time scale (142, 155). In the case of $CB_{10}H_{11}^-$ (155) more restricted scrambling is noted, as the carbon remains in the lowest-coordination position possible, whereas for $C_2B_9H_{11}$ (II-C4) no scrambling is noted as both carbons occupy and remain in the lowest-coordination positions.

It has been implied that open 3-center bonds involving carbon (142) inhibit scrambling; more likely, the carbon's predisposition to avoid higher-coordination vertices would inhibit the scrambling mechanism partially in $CB_{10}H_{11}^-$ and completely in $C_2B_9H_{11}$ (II-C4). Should the isoelectronic species, $CH_3B_{11}H_{10}^{2-}$ and $LB:B_{11}H_{10}^-$, be formed, less fluxional species might be produced with the methyl group in the former stabilizing a seven-coordinate boron (Table II), whereas the Lewis base would be expected to stabilize one of the five-coordinate sites.

As anticipated (159), the structure for $CB_{10}H_{11}^-$, based on rules 3p and 3s, is correct in spite of published ^{11}B NMR data that (62) place the carbon in the seven-coordinate position instead of a five-coordinate position.

Onak's CB_5H_7 (III-C4) (101) is under investigation by Beaudet (3), who has determined that the lone $\overline{666}$-bridge hydrogen is almost certainly above a triangular face. This undesirable situation (a violation of rule 2p) probably accounts for its instability, i.e., in the presence of a Lewis base (proton acceptor) the related closo-anion $CB_5H_6^-$ is produced (116). More recently, R. R. Rietz has reported (private communication, 1974) a second CB_nH_{n+1} closo-carborane, CB_6H_8.

In like manner the presumed intermediate $B_{10}H_{11}^-$ (in brackets in the equation below) would, in the absence of appropriate Lewis bases, probably have the structure VIII-C4; under our reaction conditions, however, it evidently reacts further to produce $B_{10}H_{14}$ (I-N10) and an iodo derivative of I-N10 *(170)*:

$$closo\text{-}B_{10}H_{10}{}^{2-} \xrightarrow{+H^+} closo\text{-}[B_{10}H_{11}{}^-] \xrightarrow{+HI} nido\text{-}B_{10}H_{12}I^- \xrightarrow{+H^+} nido\text{-}B_{10}H_{13}I$$

$$nido\text{-}B_{10}H_{13}I + HI \rightleftharpoons nido\text{-}B_{10}H_{14} + I_2$$

A. B. Burg suggested that HI would (if it worked) be safer than mixing the borane $B_{10}H_{10}{}^{2-}$ and a comparable superacid. In a recent communication, Shore *(127)* suggests that a $B_{10}H_{11}{}^-$ (VIII-C4) salt has been isolated.

Structure IV-C4 is the logical thirteen-vertex polyhedron that would be expected if the L' process illustrated in the left vertical column in Fig. 1 were simply reversed to generate the next larger polyhedron from the icosahedron. The parent IV-C4 has never been observed, but Dunks *et al.* *(17, 23, 26)* have produced derivatives wherein a transition element group occupies one of the two undesirably high-coordinated boron positions. Large polyhedra such as V-C4 or VI-C4 with even more seven-coordinate borons and, thus, susceptible to valence bond tautomerism would also appear to be possible. Perhaps such species will one day be observed as their transition metal-substituted analogs wherein the higher-coordinated vertices are replaced with atoms of transition elements that are more tolerant of seven-coordinate sites than boron. Structure VI-C4 is an alternate choice for V-C4. A unique problem arises when the structure of $C_2B_2H_4$ is considered. If $C_2B_2H_4$ (VII-C4) were to be constructed about the vertices of that deltahedron one vertex smaller than the $C_2B_3H_5$ (I-C4) deltahedron, it should be much more unstable. The degeneracy of the bonding molecular orbitals for tetrahedral $B_4H_4{}^0$ is such that $B_4H_4{}^{2-}$ (or presumably $C_2B_2H_4$) is not favored *(48, 82)*.

Just as methyl or alkyl groups tend to stabilize small-ring organic compounds and also $C_2B_3H_5$ [e.g., the $(CH_3)_2C_2B_3(C_2H_5)_3$ *(75)* derivative of $C_2B_3H_5$ is much more stable], perhaps a $H_2C_2B_2(CH_3)_2$ or a $(CH_3)_2C_2B_2(C_2H_5)_2$ derivative of VII-C4 should be sought.

B. *nido*-CARBORANES

Almost all of the *nido*-carboranes have bridge hydrogens, and the organization of their structures is much more complicated than *closo*-carborane structures because the bridge hydrogens (rule 2) apparently outrank the carbons (rule 3) and come into conflict as to low-coordination site preference (168).

Among the nido-carboranes there are isomers wherein the carbons speciously outrank bridge hydrogens, but they appear to be less stable than isomers that allow the bridge hydrogens to occupy the optimal sites (see rules 2s and 3s in Tables III and IV).

By contrast, within the previously discussed closo-species, CB_nH_{n+2}, only one example [CB_5H_7 (III-C4)] is known; both the lone 666-bridge hydrogen and the carbon could preempt lowest coordination without compromising one another in all members of the series.

Diborane B_2H_6 (with 5'5'-bridge hydrogens) may be considered as the smallest nido-borane. Although larger nido-boranes such as B_3H_7 and B_4H_8 have been postulated as intermediates, neither has been isolated. The related $B_4H_7^-$ anion (with three 66-bridge hydrogens) has been investigated by Kodama et al. (74); their suggested most probable tetrahedral structure would seemingly have no obvious site for the additional proton that would be necessary to produce the parent B_4H_8 from the $B_4H_7^-$ anion.

Matteson and Mattschei (91) have reported evidence for a CB_3H_7 that should be isoelectronic with $B_4H_7^-$; structural studies should be of great interest.

Plesek and Hermanek have prepared a unique di-nido-carborane $C_4B_{18}H_{22}$ (113e).

The major nido-carborane families are discussed below in the order of increasing complexity.

1. B_6H_{10} Family of nido-Carboranes

The only complete series of $C_{0-4}B_nH_{n+4}$ nido-carboranes is the I-, II-, III-, IV-, IX-N6 series (158) displayed in Fig. 6. In this series, the four-coordinate carbons monotonically replace borons about the edge, without compromising the optimal bridge hydrogen positions whatsoever. In the B_6H_{10} family, there need be no competition between bridge hydrogens and carbon for favored positions. Ignoring the presence of bridge hydrogens in Fig. 6, there is, in each structure, 1 six-coordinate apex associated with 5 four-coordinate edge positions. The carbons are ideally four-coordinate in all cases, while the bridge hydrogens (sixty-six- and sixty-five-coordinate in the parent B_6H_{10}) improve (as does the stability) to fifty-five-coordinate in $C_3B_3H_7$ (9, 45). Compound $C_4B_2H_6$ (IX-N6) (108, 111) is extremely stable (no bridge hydrogens); the peralkyl derivative of IX-N6 tolerates hot sulfuric acid without effect (7).

Two $C_2B_4H_7^-$ anions, V-N6 and VIII-N6 (102), with ideal four-coordinate carbons and 55-bridge hydrogens, have been prepared by Onak et al. (*88, 106*). Two nido Lewis base analogs ($C_2B_4H_6$:LB), VI-N6 and VII-N6 (*88, 106*) have also been identified. An elegant study by Onak reveals an isomerization wherein the Lewis base in an architecturally compromising situation (i.e., in VII-N6 the Lewis base and a bridge hydrogen are affiliated with the same boron) rearranges or migrates under mild conditions in such a fashion that the undesirable feature is eliminated (VI-N6).

A boron to which is attached a Lewis base is isoelectronic with carbon and thus the Lewis base–bridge hydrogen feature in VII-N6 is undesirable. After the rearrangement the Lewis base is not on the apex (VI-N6), i.e., since it is isoelectronic with carbon it prefers the four-coordinate base site and thus becomes isoelectronic with IV-N6.

Shore (the most prolific producer of new boranes in recent years) has added $B_6H_9^-$ and $B_6H_{11}^+$ to the pantheon of new (N6) boranes (*63, 64*); the former will be found to have two adjacent 65-bridge hydrogens and a nonadjacent 55-bridge hydrogen whereas in the latter, five 66-bridge hydrogens should be present.

The vacant edge-base bond in B_6H_{10} (I-N6) involves 2 five-coordinate borons; probably the low coordination of the B—B bond (unique among neutral boranes) accounts for both the capacity of B_6H_{10} to act as a Lewis base in accepting a variety of Lewis acids (*63*) as well as the short bond length.

At the other extreme, Hogeveen and Kwant (*59*) have produced the permethyl derivative of $C_6H_6^{2+}$ which is isoelectronic with B_6H_{10}; the cation $C_5BH_6^+$, yet to be described, would be expected to have a similar structure with the boron in the apex position.

2. $B_{10}H_{14}$ Family of nido-Carboranes

In contrast to the nido-B_6H_{10} family, which has only four-coordinate edge positions, the edge positions in the $B_{10}H_{14}$ family are both four- and five-coordinate (disregarding bridge hydrogens), whereas the cage positions are six-coordinate. Thus, in the $B_{10}H_{14}$ family the ideal placement of bridge hydrogens and carbons about the edge may be in conflict.

The known structure of $B_{10}H_{14}$ in both the fluid and crystal phases (*82*) is the structure displayed as I-N10, which incorporates four content 66-bridge hydrogens. For illustrative purposes only, the rotation of the four bridge hydrogens, as displayed in II-N10, would produce two, much less stable, 76-bridge hydrogens and two 75-bridge hydrogens. Not surprisingly, evidence of $B_{10}H_{14}$ in the II-N10 configuration does not exist.

By contrast, consider the two known structures of $B_{10}H_{13}^-$. The III-N10 structure (129) optimally minimizes the coordination of the bridge hydrogens and, accordingly, is precisely the structure observed in the unencumbered fluid phase. The 66-bridge hydrogen in $B_{10}H_{13}^-$ (III-N10) occupies a site that is not desirable in the parent $B_{10}H_{14}$ (I-N10) but is desirable in the related anion $B_{10}H_{13}^-$ (III-N10). As in many cases in which hydrogen tautomerism is possible (even probable), a slightly less preferred arrangement is observed in the crystal (132) as opposed to the fluid phase. In the encumbered crystal phase, $B_{10}H_{13}^-$ evidently exists as IV-N10, which coincidently reflects the bridge hydrogen placement in the parent $B_{10}H_{14}$ (I-N10). Perhaps a crystal involving another cation would allow the more desirable III-N10 configuration of $B_{10}H_{13}^-$ to be observed even in the crystal phase.

At variance with our view that coordination number considerations are of first-order importance (which favors III-N10), others (83, 86) favor both structures for $B_{10}H_{13}^-$ (III-N10, based on smoother charge distribution, and the IV-N10 structure, based on 3-center resonance considerations). Probably charge and resonance considerations are of importance, and possibly all three considerations are subtly interrelated.

Accenting the premise that bridge hydrogen placement is more important (32) than carbon placement leads to the conclusion that the structure V-N10 for the isomers of CB_9H_{13} will be more stable than either VI- or VII-N10. In the latter two cases, a carbon would be in an optimal four-coordination position rather than a five-coordination position (avoiding violating rule 3s), but the bridge hydrogens would suffer high-coordination penalties (violating the more important rule 2s) as a result. The CB_9H_{13} isomer (VI-N10) should be less stable than the parent $B_{10}H_{14}$ (I-N10), whereas the V-N10 isomer of CB_9H_{13} should be of equivalent or greater stability than $B_{10}H_{14}$.

As illustrated in the following sections, there are many examples wherein carbon occupies sites 1 coordination number higher than is otherwise available; but never (to date) has an isomer been observed in which carbon occupies a site that is 2 coordination numbers higher than is otherwise available. Thus, no candidate structures in Fig. 10 are illustrated wherein carbons might be placed in the six-coordination situations when four-coordination alternatives are available. When such isomers are produced they should be less stable than those illustrated and will be cataloged as structures that violate rule 3p.

In contemplating dicarba-*nido*-decaborane species, it is easy to predict the most stable dianion, $C_2B_8H_{10}^{2-}$ (VIII-N10). With no bridge hydrogens to complicate the issue, the most stable isomer should be the isomer illustrated; less stable isomers with carbons in five-coordinate sites should also be observed.

Two isomers of $C_2B_8H_{12}$ have been identified, XX-N10 (*38*) or one of the two bridge hydrogen tautomers, XI- or XVI-N10 (*118*). Because of rule 2s the XI-N10 tautomer should be preferred, but crystal packing forces may favor the XVI-N10 tautomer in the encumbered crystalline phase. Alternatively, in XI-N10 or XVI-N10 [as opposed to $B_{10}H_{13}^-$ (III-N10)] there would be a polarization of the molecule such that the carbon rich end of the molecule would become positively charged and perhaps the bridge hydrogen would tend to migrate toward the relatively negatively charged boron rich zone. XI-N10 and XVI-N10 differ only by the presence of a 65-bridge hydrogen in the former and a 66-bridge hydrogen in the latter.

The known isomers of $C_2B_8H_{12}$ are but two members of a spectrum of six or seven probable isomers yet to be discovered. All are displayed horizontally in Fig. 10 as IX-, X-, XI-, XII-N10 and vertically down the right-hand side of Fig. 10 as XVI-, XX-, and XXIV-N10. Accepting the premise that bridge hydrogen placement is more important than carbon placement suggests that IX- and X-N10 will be the more stable isomers, whereas those at the other end of the spectrum will be less stable.

Identical considerations suggest that the stabilities of the three isomers of $C_3B_7H_{11}$ will decrease in the order XIII- > XIV- > XV-N10. So also the $C_4B_6H_{10}$ isomers should be less stable in the order XVII- (*20*, *165*), XVIII-, XIX-N10 (*14*). Brown *et al.* (*14a*), at the University of Liverpool, have shown unequivocally that $H_4C_4B_6(CH_3)_6$ has an adamantane-type structure with sp^2 boron rather than the carborane structure with sp^3 boron.

The *nido*-decaborane and *arachno*-decaborane Lewis base adducts should be discussed in the light of the foregoing considerations. It is well known that the very stable decaborane (I-N10) reacts in excess Lewis base to produce the very stable $B_{10}H_{12}$ $(:LB)_2$ (III-A20); both species violate no rules. It may be presumed that the first step in this sequential transformation is the production of the transient intermediate *arachno*-$B_{10}H_{14}:LB$ (II-A20). Such an intermediate, possessing both $\overline{76'}$- and 76-bridge hydrogens should be quite unstable encouraging perhaps the loss of hydrogen which, in turn, would produce *nido*- $B_{10}H_{12}:LB$ in either the XXII-N10 (most probable) or XXIII-N10 configurations. Both of these candidate structures have structural handicaps, i..e, 76- and 75-bridge hydrogens are seen in XXII-N10, whereas a bridge hydrogen neighbors a surrogate carbon (a boron to which is coupled a Lewis base) in XXIII-N10. Fortuitously for such species, they have been prepared in the presence of excess Lewis base, and the simple addition of the second Lewis base to either XXII- or XXIII-N10 produces the very stable *arachno*-$B_{10}H_{12}$ $(:LB)_2$ (III-A20) which has content 66-bridge hydrogens and violates no rules whatsoever.

Were this reaction carried out slowly at the lowest possible temperature and in an inert solvent with decaborane in excess rather than with the Lewis base in excess, there might be sufficient time for a Lewis base rearrangement to take place (see rearrangement VII-N6 to VI-N6). The Lewis base could effectively migrate to a higher coordination position and the predicted to-be-stable $B_{10}H_{12}$:LB isomer (XXI-N10) should be produced from the intermediate XXII-N10. Only a 3s violation occasioned by the placement of the surrogate carbon would compromise the XXI-N10 structure for $B_{10}H_{12}$:LB.

An excellent review including research on "the intermediate dicarba-*nido*-boranes" has been compiled by Plesek and Hermanek (*113b*). Many other new compounds are revealed. Previously, they had reported (*113*) the high yield synthesis of 5,6-$C_2B_8H_{12}$ (XI- and XVI-N10).

Stibr *et al.* (*132a*) have reported the nido isomers of $C_2B_8H_{11}^-$ and $C_2B_8H_{10}^{2-}$.

3. B_5H_9 Family of nido-Carboranes

In the B_5H_9 family (disregarding bridge hydrogens), there is a five-coordinate apex position associated with 4 four-coordinate edge positions. This limited number of edge sites complicates the structures of several isoelectronic analogs of B_5H_9.

A reappraisal of the previously discussed B_6H_{10} and $B_{10}H_{14}$ families reveals the following trends: there is no bridge hydrogen congestion and there are sufficient four-coordinate sites in the B_6H_{10} family (Fig. 6) for carbon substitution without compromising the ideal 66- and 65-sites for bridge hydrogen occupation. The $B_{10}H_{14}$ family (Fig. 10) differed from the B_6H_{10} family in that in accommodating bridge hydrogen coordination number considerations (rule 2s) the carbons, when present, are in cases forced to accept higher five-coordination sites in violation of rule 3s. However, in all of the $B_{10}H_{14}$ family cases the requisite five-coordination sites are available about the open face rather than on the cage or in a cage position.

The B_5H_9 family, by contrast (Fig. 5), is the first case discussed wherein the overriding importance of bridge hydrogen accommodation (i.e., the preempting of edge sites by the bridge hydrogens) forces the carbon into the apex or lone five-coordinate cage position [$C_2B_3H_7$ (IV-N5)]. As far as carbon is concerned, the apex and base positions only differ by 1 in coordination number. At the other extreme, i.e., in the absence of any bridge hydrogens, the occupation of four-coordinate edge positions by all carbons present [e.g., C_4BH_5 (VI-N5)] should be expected.

The pentaborane-type species $C_5H_5^+$ (V-N5) (*89, 134, 164*) was discussed in Section I, B.

4. B_9H_{13} Family of nido-Carboranes

The next family to be discussed, i.e., the nine-vertex *nido*-carboranes, includes the complications of the $B_{10}H_{14}$ family (i.e., differingly coordinated edge positions) as well as one additional complication not encountered in the B_6H_{10}, $B_{10}H_{14}$, and B_5H_9 families. The additional perturbation arises when a selected cage position and certain edge positions are of equivalent coordination number (disregarding bridge hydrogens).

In the previous paper (*164*) the difficulties of placing four bridge hydrogens about the unfavorably puckered five-membered face of the preferred nine-vertex nido fragment was pointed out; thus the polyhedral fragment displayed in Fig. 9 for B_9H_{13} (I-N9) should be unstable due to extensive bridge hydrogen congestion. Three of the four bridge hydrogens (77-, 76-, 76-) in I-N9 are also overly coordinated. The simple removal of the acutely offensive 77-bridge hydrogen to produce $B_9H_{12}^-$ (II-N9) removes all objections simultaneously, i.e., removes bridge hydrogen crowding and generates three compliant 66-bridge hydrogens. During the preparation of this manuscript, Siedle *et al.* (*128*) furnished precisely the ^{11}B and ^1H NMR spectra that are required by the numbers and kinds of hydrogens and borons (six types of boron in the ratio of 2:2:2:1:1:1) for the projected structure for $B_9H_{12}^-$ (II-N9) (*164*). In other words, the B_9H_{13} structure (I-N9) should not be stable (considering bridge hydrogen congestion and coordination number); however, the anion, $B_9H_{12}^-$ (II-N9), differing by only one bridge hydrogen, should be stable. Were $B_9H_{12}^-$ related to III-N9, bridge hydrogen tautomerism would have resulted in three kinds of boron in the ratio of 3:3:3.

In Fig. 9, there is an alternative III-N9 structure for the parent borane, B_9H_{13}. Such a structure, formed by the breaking of one high coordination bond, produces in I-N9 the skeletal arrangement characteristic of i-B_9H_{15} (I-A19), i.e., an arachno structure rather than the nido structure (I-N9). The III-N9 structure may be more probable than the I-N9 structure and has the precedent of B_8H_{12} (I-N8) which incorporates 76-bridge hydrogens preferring, by the breaking of one high-coordination bond, the III-N8 structure (involving only 66-bridge hydrogens) characteristic of the *arachno*-borane B_8H_{14} (I-A18), at least in the crystal phase. However, in the case of B_9H_{13} (III-N9), undesirable 76- and 75-bridge hydrogens remain, even following the "nido to arachno" transformation (suggestive of instability). It seems that structure III-N9 for B_9H_{13} has "one strike against it" but may be more promising than I-N9.

Amplifying the difficulties in selecting a satisfactory structure for B_9H_{13}, consider the following: for every electron that is deficient there should be one 3-center bond (82) involving either three borons, two borons and a carbon, or two borons and a hydrogen (i.e., a bridge hydrogen).

In a given molecular system the total number of 3-center bonds will be dictated by the empirical formula with an ideal number expressed within the cage, whereas if extra hydrogens are present they may be expressed as bridge hydrogens.

What would happen if the "idealized" structure for B_9H_{13} (I-N9), which should have a total of nine 3-center bonds, opened to produce the III-N9 structure to avoid bridge hydrogen congestion? We suggest that, since nine 3-center bonds are present in any case, the open III-N9 structure would be less hospitable to the five, required, skeletal 3-center bonds than would be I-N9. The four extra 3-center bridge hydrogens would, in both cases, account for the other four electrons that are deficient. Since there are no other extra hydrogens, i.e., endohydrogens, to be impressed into service as bridge hydrogens, the structure III-N9 has no recourse but to account for the requisite five electrons deficient by somehow including five 3-center bonds within the more open skeleton.

Consider the further iniquity of converting two of the four 3-center bridge hydrogens (in III-N9) into endohydrogens. With only two 3-center bonds consequently allowed to be expressed as 3-center bridge hydrogens, "conservation of 3-center bonds" would require that seven instead of five 3-center bonds would have to be expressed within the already sundered skeleton.

The probability of four-center or five-center bonds completely accounting for the skeletal electron deficiencies in the B_9H_{13} structure (IV-N9) is a probability that must be rejected. Structure IV-N9 for B_9H_{13} is extremely unlikely. By contrast, others have described the same B_9H_{13} structure (IV-N9) as "eminently satisfactory" (83, 86). This is but one of several examples that illustrate the diametrically opposing predictions resulting from other theories as compared to CNPR theory.

When contemplating potential $C_2B_7H_{11}$ carborane isomers, as displayed horizontally in the middle row of Fig. 9, and noting that in contrast to the B_6H_{10}, $B_{10}H_{14}$, and B_5H_9 families (Figs. 6, 10, and 5), one cage position in the nine-vertex nido species has a coordination number as low as four of the five positions around the open face, the five-coordinate cage position and edge positions should be considered as equivalently satisfactory for carbon's occupancy. Again, based on the greater demand of bridge hydrogens than that of carbons for lower-coordination environments, we predict that the various isomers displayed for $C_2B_7H_{11}$ would decrease in stability in the order VI- > VII- > VIII- > IX- > X-N9, whereas isomer V-N9 would be predicted to be the least stable (one unstable isomer containing 76- and 75-bridge hydrogens is not shown). Following the patterns and parallels discussed above involving the $C_2B_8H_{12}$ isomers in Fig. 10, it is easy to follow the reasons for the predicted order of stability.

A C,C-dimethyl isomer of $C_2B_7H_{11}$ has been identified (*118*) and is displayed as IX-N9 (X-ray crystal structure) (*61*). The bridge hydrogen tautomer of IX-N9 (i.e., VII-N9) should be observed in the unencumbered fluid phase to minimize the bridge hydrogen coordination numbers.

Near the completion of this manuscript, Rietz and Schaeffer revealed (*119*) that the parent compound $C_2B_7H_{11}$ almost certainly does not have the same structure as the C,C-dimethyl derivative which has the structure IX-N9. It is interesting to note that a BH_2 group is present in the parent $C_2B_7H_{11}$ that is not present in the C,C-dimethyl derivative (IX-N9). The authors suggest, as one candidate, a structure similar to the XIII-N9 configuration with one of the 65-bridge hydrogens in XIII-N9 biasing completely to the five-coordinate boron thus producing a five-coordinate BH_2 group. This structure may be correct yet there are several alternatives.

The presence of BH_2 groups in nido species, e.g., $B_{11}H_{15}$ (II-N11) and the unstable isomer of $C_2B_9H_{13}$ (not shown) is a possibility that has not been rejected. And as will be seen below under the discussion of the *arachno*-boranes B_5H_{11} (I-A15 versus II-A15) and $B_9H_{14}^-$ (V-A19 versus VI-A19), bridge hydrogen versus endohydrogen status is not considered sacred; in fact, a continuum of BE hydrogens is preferred. Yet both of the two hydrogens in the BH_2 group (in the parent $C_2B_7H_{11}$) spin couple to the boron with exactly the same coupling constant ($J = 125$ cps), which should not be expected in any "XIII-N9-like" structure. As mentioned earlier, the concept of conservation of 3-center bonds with regard to the opening of a nido structure into an arachno structure and the concurrent conversion of a bridge hydrogen into an endohydrogen, is considered to be electronically improbable paralleling the arguments advanced against structure IV-N9 for B_9H_{13} (above).

The acceptance of an N9 structure (necessarily incorporating vicinal carbons) in accordance with rules 1–4 would favor a VII-N9 structure for the parent $C_2B_7H_{11}$ with the 65-bridge hydrogen converting into an endohydrogen (to produce a five-coordinate BH_2 group). Such a structure is unlikely because the presence of two methyl groups (on remote carbons) would not be expected to cause the "extra hydrogens" to occupy different sites.

The following thoughts suggest an alternate structure. The highly symmetrical triplet in the ^{11}B NMR spectrum is reminiscent of the BH_2 groups in B_5H_{11} (I-A15) and B_4H_{10} (I-A14). In those cases the two "different" terminal hydrogens of the BH_2's neighbor one or two bridge hydrogens and are presumed to become NMR equivalent by rapid fluxional behavior on an NMR time scale. Consider also that $\mu\text{-}(CH_3)_2BB_5H_8$ exists (*36, 37*) and may be considered as either a derivative of B_5H_9 or B_6H_{10} (see X-N5 or X-N6, both in Fig. 7); i.e., a precedent for a dialkyl-substituted BH_2-containing *nido*-borane exists.

A $\mu\text{-}BH_2$ or a $\mu\text{-}BH_2H$ derivative of a $C_2B_6H_{10}$ isomer (i.e., $\mu\text{-}BH_2HC_2B_6H_8 \cong C_2B_7H_{11}$) might account for the data. Such a compound (not shown) could resemble one of the $C_2B_6H_{10}$ structures in Fig. 8 with an added BH_3 group (two terminal hydrogens and one neighboring bridge hydrogen) replacing both bridge hydrogens and resembling VII-N7. The carbons could be vicinal in any number of ways; the fluxional BH_2 group would account for the precisely equivalent coupling, and, since the parent $C_2B_7H_{11}$ and the dimethyl derivative of $C_2B_7H_{11}$ (with the known structure IX-N9) would not be isoelectronic, the radically different ^{11}B and ^{1}H NMR spectra would be explained.

The BH_3 groups may be removed from similar species such as B_2H_6, B_4H_{10}, and B_5H_{11}; thus the parent *nido*-$C_2B_7H_{11}$ might provide an easy route to a new closo isomer of $C_2B_6H_8$ with adjacent carbons by similar loss of a BH_3 group. Subsequently, R. R. Rietz reported (private communication, 1974) the isolation and identification of 1,2-$C_2B_6H_8$ from the parent $C_2B_7H_{11}$ in about 12% yield.

Of the several $C_2B_7H_9{}^{2-}$ isomers, XI- and XII-N9 are predicted to be most stable; carbon in these cases would be expected to occupy the lone four-coordinate position. In a similar vein, the $C_4B_5H_9$ isomers XIV- and XV-N9 would be expected to be stable although other stable $C_4B_5H_9$ isomers are possible.

5. B_7H_{11} *Family of* nido-*Carboranes*

Only a few comments need be expended relative to this group of *nido*-carboranes since, as far as is known, *no species such as B_7H_{11} or any species isoelectronic with B_7H_{11} (I-N7) has ever been reported*. On first principles we would select I-N7 as the structure for B_7H_{11}; however, partially paralleling the situation for B_9H_{13} (I-N9), there are undesirable 75- and 76-bridge hydrogens in addition to the probable bridge hydrogen congestion. The removal of one bridge hydrogen to generate $B_7H_{10}{}^-$ (V-N7) seemingly could alleviate both of these problems; perhaps the V-N7 structure may one day be observed. It has been predicted (*84*) that B_7H_{11} would have the structure II-N7 produced by simultaneously opening up the I-N7 structure and converting two bridge hydrogens into endohydrogens. Paralleling the arguments advanced against the B_9H_{13} structure IV-N9, the II-N7 configuration (84) for B_7H_{11} is extremely improbable.

A more likely structure for B_7H_{11} is displayed as VIII-N7 and is related to either X-N5 (36, 37) (μ-$BH_2B_5H_8$) or X-N6 (both shown in Fig. 7). There is no doubt that a dimethyl derivative of either X-N5 or X-N6 (both shown in Fig. 7) has been produced which, upon heating, reverts to the dimethyl derivative of B_6H_{10} (I-N6) (37). The suggested VIII-N7 structure for i-B_7H_{11} would bear the same relationship as X-N6 to i-B_6H_{10} (i.e., minimize bridge hydrogen coordination numbers) or as X-N5 would bear to B_5H_9 (I-N5).

Reasonable 3-center bond descriptions of both i-B_6H_{10} (X-N6) and i-B_7H_{11} (VIII-N7), however, are not obvious; thus μ-$BH_2B_5H_8$ (X-N5) is favored as well as a similar μ-$BH_2B_6H_9$ structure for B_7H_{11}.

Six hypothetical isomers of $C_2B_5H_9$ are displayed and, for reasons similar to those invoked in the cases of $C_2B_7H_{11}$ and $C_2B_8H_{12}$ (in Figs. 9 and 10), the isomer III-M7 should be anticipated to be most stable, whereas isomer X-N7 would be the least stable. The isomer of $C_2B_5H_7{}^{2-}$ predicted to be the most stable and the two isomers of $C_4B_3H_7$ expected to be most stable are displayed as XI-, XII-, and XIII-N7, respectively.

6. [B_8H_{12}] Family of nido-Carboranes

The known structure of the *nido*-borane B_8H_{12} in the crystalline state (82) has been determined to be III-N8, which is in contrast to the preferred B_8H_{12} nonicosahedral structure (I-N8) (164) and is similar to the eight-vertex arachno species displayed in Fig. 18. Perhaps the two 76-bridge hydrogens in I-N8 militate against stability (since there would be no bridge hydrogen crowding). It was thought that perhaps an isoelectronic carborane derivative such as $C_2B_6H_{10}$ with two fewer hydrogens of lower-coordination number might assume the literal nido structure, e.g., X-N8.

Since a rather unstable $C_2B_6H_{10}$ *nido*-carborane had been known for several years to exist, this isomer of $C_2B_6H_{10}$ was structurally identified (41) as having (probably) the arachno structure of V-N8 instead of the nido structure X-N8, but this is not yet sure.

Structure X-N8 differs from V-N8 by only one high-coordination edge bond and the presence or absence of such a bond would not affect the coordination numbers of either carbon. There is no precedent to indicate that the more desirable 55-bridge hydrogens in V-N8, as opposed to the 65-bridge hydrogens in X-N8, would constitute a sufficient driving force for opening into the arachno configuration. If the X-N8 structures were actually a pair of valence bond tautomers, i.e., the unique edge bond alternating positions between the two borons as shown in X-N8 and X'-N8 (rapidly on an NMR time scale), then the data could also be rationalized in terms of a valence bond tautomeric pair.

Perhaps the B_8H_{12} structure I-N8 and the $C_2B_6H_{10}$ structure X-N8 will be found to be preferred in the fluid phases. Microwave and/or electron diffraction studies of B_8H_{12} and $C_2B_6H_{10}$ in the unencumbered vapor phases are in progress and should adjudicate this structural dilemma. In any case the assumption of either the V-N8 or X-N8 structures (both ideal for bridge hydrogens) rather than alternatives that would be ideal for carbons is again indicative that accommodating the bridge hydrogens (rule 2s) is more important than accommodating the carbons (rule 3s).

It would appear that rule 1 correctly predicts the structures of all nido species (see Fig. 1) except perhaps the structures of the eight-vertex nido species that, at least in the two cases known, tentatively seem to assume the arachno configuration. Candidate structures for $C_4B_4H_8$ are XI-, XII-, and/or IX-N8; the most probable candidate for a true nido structure is VIII-N8 ($C_2B_6H_8^{2-}$).

When the conservation of 3-center bondedness relating to B_9H_{13} (III- or IV-N9) was discussed, the penalty for opening up the skeletal framework by the breaking of an edge bond was pointed out to be that of greater difficulty in harboring the required number of 3-center bonds. Possibly the eight-vertex nido compounds have a singular electronic solution to this dilemma and can uniquely accept the requisite four 3-center bonds within an arachno-skeletal framework. As a possible precedent, the eleven-vertex closo compound $CB_{10}H_{11}^-$ seems (at least transiently) to open reversibly into an eleven-vertex nido structure (155). In any event, none of the bridge hydrogens in B_8H_{12} (III-N8) are converted into endohydrogens which is the greatest criticism of IV-N9 for B_9H_{13} and II-N7 for B_7H_{11}.

7. [$B_{11}H_{15}$] Family of nido-Carboranes

Compound $B_{11}H_{15}$ has had a checkered history; it may have been prepared (28), but probably not. The straightforward structure would be I-N11 (incorporating four bridge hydrogens and seven skeletal 3-center bonds) but the presence of two $\overline{77}$- and two $\overline{76}$-bridge hydrogens (also very congested) should render such a species very unstable. Conversion of one bridge hydrogen into an endohydrogen (II-N11) would produce a six-coordinate BH_2 group (violation of rule 4p) and would require an eighth 3-center bond to be expressed within the skeletal framework (more acceptable in larger frameworks), but would still leave one very undesirable 77- and two 76-bridge hydrogens; these are very unsatisfactory solutions. Perhaps an edge bond breaks (not shown, but similar to II-A21 and paralleling the situation with B_8H_{12}; i.e., the preferred CNPR I-N8 structure opens into the III-N8 structure) which would produce a six-membered, favorably puckered open face with additional room for the four bridge hydrogens. Although the problem ought to be less serious with larger molecules, the opening should make a less favorable environment for the requisite, seven, skeletal 3-center bonds and three 7X-bridge hydrogens would remain. Removal of a single bridge hydrogen from I-N11 produces $B_{11}H_{14}^-$ (III-N11) (1) which is stable in spite of its compromising 76-bridge hydrogens, but then the negative charge probably furnishes the attraction to keep the two 76-bridge hydrogens content. An alternative (not shown) with a six-coordinate (violates rule 4p) BH_2 group and two 66-bridge hydrogens must also be considered for $B_{11}H_{14}^-$. Moreover nido-$B_{11}H_{14}^-$ may always be associated with at least one Lewis base [e.g., $B_{11}H_{14}$ (:LB)] which would relate to it arachno-$B_{11}H_{16}^-$ in Fig. 21. The seven-coordinated boron would facilitate fluxional behavior and account for the single kind of boron seen in the ^{11}B NMR spectrum of $B_{11}H_{14}^-$ in etherial solution. When two bridge hydrogens are removed from $B_{11}H_{15}$ to produce $B_{11}H_{13}^{2-}$ (IV-N11) (1, 34), a very stable species that incorporates two content 66-bridge hydrogens is obtained.

One of the two isomers of $C_2B_9H_{13}$ is also in question; the known marginally stable isomer of $C_2B_9H_{13}$ (V-N11) (60, 160) has adjacent carbons and adjacent 76-bridge hydrogens. Attempts to isolate the less stable isomer (presumably with nonadjacent carbons) have failed.

A structure for the less stable isomer of $C_2B_9H_{13}$ with nonadjacent carbons about the open face incorporating one 66-bridge hydrogen and one endohydrogen on the boron between the carbons (a violation of rule 4p) might possibly be correct or, perhaps, one edge bond breaks between the carbons (see IX-N11) which would allow the introduction of a 75-bridge hydrogen into a "II-A21-like" structure.

A likely candidate structure for an even more stable isomer of $C_2B_9H_{13}$ is VI-NII (161), which only violates rule 3s and has content 66-bridge hydrogens; other variants with carbon in cage positions are probable. Another isomer of $C_2B_9H_{13}$ has been prepared from 1,12-$C_2B_{10}H_{12}$ (II-C2). The structure (113a) resembles VI-N11; however, the cage carbon is para to the edge carbon rather than meta. Furthermore, Rietz (private communication, 1974) has also detected a second, more stable isomer of $C_2B_9H_{13}$ from the pyrolysis of $C_2B_7H_{11}$. A structure with one carbon in a six-coordinate cage position (violation of 3s) but with content 66-bridge hydrogens (resembling VI-N11 or VIII-N11) should be suspected.

Two isomers of $CB_{10}H_{13}^-$ are known: VII-N11 (*140, 160*) and the less stable isomer VIII-N11 (*6*), which incorporates one 3s violation.

If VIII-N11 ($CB_{10}H_{13}^-$) were neutralized to $CB_{10}H_{14}$, then it might be more acidic than $B_{11}H_{14}^-$ (III-N11) since the negative charge in the latter should result in more content 76-bridge hydrogens. Lewis base analogs of the $C_2B_9H_{12}^-$ isomer (IX-N11) have been prepared from $C_2B_9H_{11}$ (II-C4) by Chowdhry *et al.*; the bridge hydrogens are found about the open face but the Lewis base group in *nido*-LB:7,9-$C_2B_9H_{11}$ is found upon the cage (violation of rule 3s) (*16*).

The bridge hydrogens in the isomers of $C_2B_9H_{12}^-$ (IX-N11 and X-N11) are positioned (*60*) exactly as previously suggested (*160*). In the absence of complicating bridge hydrogens, the related species such as the two obvious isomers of $C_2B_9H_{11}^{2-}$ and $C_4B_7H_{11}$ (XI-N11) (*20*) are very stable.

Plesek *et al.* (*113d*) have prepared the nido compound $NC_2B_8H_{11}$ which is isoelectronic with $C_4B_7H_{11}$ (XI-N11). They suggest a CHBH-NHBHCH pentagonal open face, and, since the nitrogen donates two electrons (see Section III, E) and, consequently, becomes quite positively charged, we would anticipate their reported anion $NC_2B_8H_{10}^-$ would differ from the parent by simply lacking the most acidic nitrogen-attached proton. They also reported an $NC_2B_8H_{13}$ which they show is a derivative of 5,6-$C_2B_8H_{12}$ with a bridging NH_2 group.

8. [$B_{12}H_{16}$] Family of nido-Carboranes

Were *nido*-$B_{12}H_{16}$ (I-N12) to be discovered, it would be expected to have the basic skeletal structure of the most probable thirteen-vertex closo polyhedron as typified by $C_2B_{11}H_{13}$ (IV-C4), with one of the 2 seven-coordinate vertex positions removed and embracing four bridge hydrogens (I-N12) about the perimeter in the lowest possible coordination situations. As a consequence of the presence of a seven-coordinate BH group, wholesale bridge hydrogen and valence bond tautomerism should be expected; quite possibly the ^{11}B NMR spectra of $B_{12}H_{16}$ or of related anions such as $B_{12}H_{15}^-$ or $B_{12}H_{14}^{2-}$ would be simple doublets at ambient temperatures.

It has been noted by workers in both the USSR (*178–181*) and United States (*24, 31, 42*) that all closo isomers of $C_2B_{10}H_{12}$ (II-C2, III-C2, and

II-C3) add two electrons to become the corresponding nido-$C_2B_{10}H_{12}^{2-}$ dianions which would be isoelectronic with $B_{12}H_{16}$ and which should have skeletal structures resembling I-N12.

An attempt to relate the known closo-icosahedral "parents" to their nido twelve-vertex progeny follows.

Since the two carbons in 1,2-$C_2B_{10}H_{12}$ (II-C3) are furnishing the two extra electrons (as compared to $B_{12}H_{12}^{2-}$), any overall dipole should have its positive center "between" the carbons. Moreover, since the closo structure, upon opening into a nido structure, should preferably have its carbons about the resulting open face, the most likely electron-deficient recipients of any additional electrons would be those borons neighboring the carbons, which would become surrogate carbons upon receipt of electrons. Compound 1,2-$C_2B_{10}H_{12}$ (II-C3) possibly accepts two electrons to produce the hypothetical intermediate (II-N12) which by one dsd rearrangement (involving the carbon–carbon bond) would produce the transient intermediate (III-N12).

There is a special and very important feature of the anticipated open nido twelve-vertex structures in Fig. 12: repetition of single Lipscomb dsd rearrangements (denoted by the two-headed arrows) monotonically allows the six skeletal atoms about the open face to rotate about the second tier of five skeletal atoms (two-tier dsd rotation). Each dsd rearrangement (*85, 163*) (valence bond tautomerism) recreates the same configuration and involves only the motion of two skeletal atoms (in the ball-and-stick representation) and would allow carbons, if located in different tiers, to migrate apart. Such wholesale valence bond tautomerism is known to accompany the presence of seven-coordinate BH groups, e.g., $B_{11}H_{11}^{2-}$ and $CB_{10}H_{11}^{-}$ (*142, 155*).

In order to place at least one carbon in the lowest-coordination edge position, configuration III-N12 would tend to rearrange into the preferred IV-N12 by a single dsd rearrangement. Rapid fluxional behavior, engendered by the presence of the seven-coordinate boron which, utilizing the intermediate (III-N12), would make the CH groups in IV-N12 equivalent on an NMR time scale, should be anticipated. Upon reoxidation the reverse process, i.e., IV-N12 → III-N12 → II-N12 → II-C3, would take place and thus the original 1,2-$C_2B_{10}H_{12}$ would be produced.

Turning to 1,7-$C_2B_{10}H_{12}$ (III-C2), the addition of electrons, again between the carbons, should produce the hypothetical intermediate (VI-N12) which would rearrange into the intermediate VII-N12 via a single dsd permutation and finally rearrange into IV-N12 (to accommodate carbon preference); Structure IV-N12 is thus exactly the same species as is produced from 1,2-$C_2B_{10}H_{12}$. In this fashion the experimental observations that both 1,2- and 1,7-$C_2B_{10}H_{12}$ produce, upon reduction, the same $C_2B_{10}H_{12}{}^{2-}$ anion and produce only 1,2-$C_2B_{10}H_{12}$ upon reoxidation may be rationalized.

Compound 1,12-$C_2B_{10}H_{12}$ (II-C2) does not have the same options available as the 1,2- and 1,7-isomers (all borons are equivalent), and the configurational sequence, 11-C2 → [X-N12] → [XI-N12] → [XII-N12] → [XIII-N12], is envisaged as most probable. In these fluxional species, two-tier dsd rearrangement would account for the production of the intermediate XII-N12 structure from XI-N12 were it necessary. Were the second carbon trapped within or restricted to the second tier (a violation of rule 3p), then the position farthest away from the seven-coordinate boron would be the best position for the second carbon (i.e., XII-N12). Actually, carbon would prefer to be about the open face, which suggests that the ideal carbon positions upon the twelve-vertex nido skeleton would be those displayed in XIII-N12 and, given the fluxional characteristics that accompany seven-coordinate BH groups, intermediate XII-N12 should be able to rearrange into the more stable XIII-N12 isomer. Reoxidation of XIII-N12 would move any one of the four edge borons (probably all four are equivalent on an NMR time scale) into an apex position thus producing 1,7-$C_2B_{10}H_{12}$ (III-C2) from 1,12-$C_2B_{10}H_{12}$ (II-C2), exactly as is experimentally known to be the case.

In summary, the reduction–oxidation of both 1,2- and 1,7-$C_2B_{10}H_{12}$ to produce 1,2-$C_2B_{10}H_{12}$ and the reduction–oxidation of 1,12-$C_2B_{10}H_{12}$ to produce only 1,7-$C_2B_{10}H_{12}$ can tentatively be accounted for on a structural basis.

Were a proton added to the $C_2B_{10}H_{12}{}^{2-}$ dianion IV-N12 (derived from either 1,2-$C_2B_{10}H_{12}$ or 1,7-$C_2B_{10}H_{12}$), the $C_2B_{10}H_{13}{}^{-}$ anion V-N12 should be anticipated, and one carbon should be expected to defer to the bridge hydrogen preference. Whether the added proton becomes a tautomerizing bridge hydrogen or an endohydrogen is probably unimportant; similar protons are observed in $B_9H_{14}{}^{-}$ (V-A19 vs. VI-A19) and in B_5H_{11} (I-A15 vs. II-A15). But since there are no triplets in the ^{11}B NMR spectrum, the V-N12 structure incorporating a bridge hydrogen is favored in the fluid phase.

If a proton is added to the dianion XIII-N12 (derived from 1,12-$C_2B_{10}H_{12}$), CNPR rules suggest that a $C_2B_{10}H_{13}{}^{-}$ anion with the structure VIII-N12 should be expected. However, since NMR data indicated that the "extra" proton is on the carbon, the pseudoicosahedral fragment structure (IX-N12), known to be correct in the crystalline phase, must be accepted as correct even in the fluid phase (*24, 141*).

A previously disturbing feature of the published ^{11}B NMR spectra

(24) of the two $C_2B_{10}H_{13}^-$ anions was that they appeared sufficiently different not to be representative of isoelectronic species. Thus, the finding (141) that IX-N12 has a "bridging" CH_2 group and is more of an N11 derivative than an N12 compound eliminates this dilemma: the two ions are not isostructural and, therefore not isoelectronic. That is not to say that $C_2B_{10}H_{13}^-$ (IX-N12) is an alkyl derivative of $CB_{10}H_{13}^-$ (VII-N11), however, because evidently no bridge hydrogens are generated.

Were electron exchange possible (i.e., $C_2B_{10}H_{12} + C_2H_{10}B_{12}^{2-} \rightleftharpoons C_2B_{10}H_{12}^{2-} + C_2H_{10}H_{12}$) then, in the presence of any unreacted $C_2B_{10}H_{12}$, structure XIII-N12 would be expected to slowly convert into IV-N12 as has been observed (24) experimentally.

Wade points out that Wing's "slipped" dicarbollide species (176) such as $Cu(C_2B_9H_{11})_2$ could be considered as a dinido species, each $CuC_2B_9H_{11}$ unit filling twelve of the vertices of the thirteen-vertex polyhedron typified by IV-24.

C. arachno-CARBORANES

The *arachno*-carboranes and their analogs may be discussed in somewhat less complicated terms than their nido relatives. The trend is as follows: in the *closo*-species there are no endohydrogens and there are a sufficient number of low-coordination sites so that the bridge hydrogen and the carbon need not be in competition. In the nido species, endohydrogens are also generally absent but there are more bridge hydrogens or carbons, and complicated competition between bridge hydrogens and carbon for low-coordination sites is pervasive. By contrast, within the arachno species, endohydrogens are usually present and carbons (reluctantly), when also present, frequently collaborate with the endohydrogens in occupying the lowest-coordination sites together. Thus, instead of bridge hydrogens and carbons competing for lowest-coordination sites, as in the nido species, frequently the endohydrogens and carbons appear to occupy jointly the lowest-coordination sites within the arachno species to accommodate the endohydrogen preferences.

Among the *arachno*-boranes, it would appear to date that the bridge hydrogens are more acidic than the endohydrogens of five-coordinate BH_2 groups, but that the endohydrogens of five-coordinate CH_2 groups are more acidic than the bridge hydrogens. As is discussed in more detail in Section IV, the proton acidity among the *arachno*-carboranes and related species is, in the order of increasing acidity, BHB bridge hydrogen < endo CH (above four-coordinate) < endo NH $\cong$ SH. Thus anion formation involves the loss of bridge hydrogens in the *arachno*-boranes, e.g., $B_{10}H_{15}^-$ (*122*) to form $B_{10}H_{14}^{2-}$ and B_4H_{10} to form $B_4H_9^-$ (*65*), as opposed to loss of endo heteroatom hydrogens to produce $C_2B_7H_{12}^-$ (*136*) and $SB_9H_{12}^-$ (*57*) etc., when *arachno*-carboranes or heteroatom *arachno*-boranes are involved.

Inasmuch as bridge hydrogens and endohydrogens are considered to constitute a continuum (e.g., some "extra" hydrogens are pure endo, some pure bridge, and some in between with characteristics of both), it should not be surprising that among the *arachno*-boranes there are several conflicts as to whether selected extra hydrogens are endo or bridge and that, in the encumbered crystal phase packing forces may sometimes cause a bridge hydrogen to become an endohydrogen or vice versa.

The smallest *arachno*-borane is apparently $B_2H_7^-$ (*13, 35, 58*) with a 4″4″- or a 5″5″-bridge hydrogen, stabilized by the presence of the negative charge.

1. [B_3H_9] Family of arachno-*Carboranes*

The parent B_3H_9 (V-A13) has apparently never been observed. This probably results from the presence of three 6′6′-bridge hydrogens accompanied by the 3 six-coordinate BH_2 groups. Disproportionation of B_3H_9 into B_2H_6 is favored as more content 5′5′-bridge hydrogens and five-coordinate BH_2 groups are produced.

The removal of one bridge hydrogen from B_3H_9 (V-A13) produces $B_3H_8^-$ (I-A13) (*82*), incorporating marginally acceptable 6′5′-bridge hydrogens. The acidic nature of the 6′5′-bridge hydrogens is probably sated by the compensating presence of the negative charge.

For illustrative purposes an alternative structure for $B_3H_8^-$ should be compared. For example, could not one bridge hydrogen in I-A13 become a terminal hydrogen as depicted in VI-A13? Such a configuration would violate rule 4p but the remaining 5′5′-bridge hydrogens should be more content. At least in the encumbered crystal phase, the $B_3H_8^-$ structure (I-A13) is observed; perhaps the "almost-as-satisfactory" alternative VI-A13 facilitates the capacity for tautomerism, known to be rampant in this ion in the liquid phase.

In the isoelectronic B_3H_7:LB species, the liquid phase structure (VII-A13) is definitely indicated to be correct by Parry and Paine (*39, 109*), rather than a structure with elements of both VII-A13 and II-A13 which may be more stable in the crystal phase (*97*).

At this juncture it is useful to consider the situation that would obtain were a bridge hydrogen neighboring a carbon (CH group) or a surrogate carbon (i.e., a B:LB group). A carbon donating electrons into an electron-deficient sink (as in a carborane) would, as a result, become positively charged as shown in structure a; this situation would be partially alleviated if the carbon preempted the neighboring bridge hydrogen and the electrons involved structure b and even more so if the carbon preempted the neighboring bridge hydrogen and the electrons involved and then released the proton while retaining the two electrons (structure c). Such a proton could perhaps relocate elsewhere on the molecule as is illustrated in Section III, E.

```
      H                         H                                H:LB⁺
      |                         |
      ⟋ δ⁺                      |                         δ⁻
H—B      C—H      H—B         C—H       H—B          C̈—H
 /|\   /|\         /|\    ⎯⎯   /|\         /|\   ⎯⎯    /|\
    ⎯⎯
    (a)                    (b)                      (c)


      H                         H                                H⁺
      |                         |
      ⟋  +                      | δ⁺
H—B      N—H      H—B         N—H       H—B          N̈—H
 /|\   /|\         /|\    ⎯⎯   /|\         /|\   ⎯⎯    /|\

    (d)                    (e)                      (f)
```

This sequence parallels, first, the reasons bridge hydrogens tend not to neighbor carbons (or borons that are attached to Lewis bases), second, the reasons CH_2 (or BH:LB) groups are preferentially found among the *arachno*-carboranes, and, third, the reasons that when such species produce anions, the endohydrogens of CH_2 groups are the most acidic. In seeming support of this, were a nitrogen in a similar environment and, thus, required to donate twice as many electrons as carbon, the nitrogen involved should reflect to a greater extent the trends observed with carbon. Indeed, as is seen in Section III, E and in Fig. 25, not only has a bridge hydrogen neighboring a nitrogen never been observed (structure d) but the hypothetical endohydrogen of an NH_2 group (structure e) is evidently too acidic to remain on nitrogen and only arachno species with NH groups (structure f) are to be anticipated under normal circumstances. Superacids or liquid hydrogen iodide might produce species incorporating NH_2 or isoelectronic SH groups from their related anions.

The picture becomes somewhat more complete when the structural choices available for the $C_3H_7^+$ cation are considered. This species is known as its trialkyl derivative, norbornyl cation (*99*). It has been unequivocally demonstrated by NMR spectra that the extra hydrogen in norbornyl cation assumes an endo position (VIII-A13). A good way to view the $C_3H_7^+$ alternatives is to recognize that one 3-center bond is mandatory and that the 3-center bond could be expressed as (a) and CHC-bridge hydrogen (III-A13) or, more likely, as either (b) a 3-center bond involving the three carbons (CCC) or (c) a CHH 3-center bond involving two of the three terminal hydrogens in the CH_3 group of VIII-A13. The CCC 3-center bond alternative (b) is preferable since it distributes the necessarily present positive charge among the three carbons rather than involving less tolerant hydrogens. And just as the negative charge could stabilize the 6'5'-bridge hydrogens in $B_3H_8^-$ (I-A13), it is likely that the positive charge in $C_3H_7^+$ would militate against stability of a 5'5'-bridge hydrogen in III-A13.

Probably the CH_3 group in $C_3H_7^+$ (VIII-A13), the bridging CH_3 groups in $(CH_3)_6Al_2$, and the bridging $LB:BH_2$ groups in $LB:B_3H_7$ (VII-A13) should be considered as isoelectronic. The CH_3 group as depicted in VIII-A13 is, of course, not actually a "CH_3 group" (it was excised for illustrative purposes from the nonbornyl cation); it is a trialkyl CH_2 group and, therefore, more amenable to the five-coordination situation (see Table II).

2. B_4H_{10} Family of arachno-Carboranes

The parent B_4H_{10} (I-A14) is not a very stable borane (82), perhaps due to the four marginal 6'6-bridge hydrogens. For illustrative purposes, if a bridge hydrogen were to be shifted into an endo position (IV-A14), then an even worse situation incorporating a very undesirable 6'6'-bridge hydrogen would be produced.

By contrast, if one bridge proton was removed from B_4H_{10} (I-A14) to produce $B_4H_9^-$, then the two structural choices would be II-A14, related to the parent B_4H_{10} (I-A14), and V-A14. Shore et al. (65) have carried out an elegant series of experiments showing that the V-A14 structure for $B_4H_9^-$ is preferred at ambient temperatures. Based on hydrogen and boron considerations (rules 2 and 4), the structure V-A14 for $B_4H_9^-$ should have been anticipated as most satisfactory.

Species $B_4H_8:LB$ is known to have structure VI-A14 (15, 25, 77, 98) which again should have been expected. The unfavored structure (III-A14) for $B_4H_8:LB$ would have the Lewis base attached to a boron (surrogate carbon) neighboring a bridge hydrogen in an unsatisfactory conflict of interests. The Lewis base in this case would not have the option of migrating to another boron in order to eliminate this structural indiscretion, as was the case in Onak's (88, 106) rearrangement of structure VII- to VI-N6.

A CB_3H_9 compound would be expected to have a VI-A14-type structure.

3. B_5H_{11} Family of arachno-Carboranes

The structure of B_5H_{11} is apparently I-A15 in the liquid phase (173, 174), although it is reported (94) to have a different structure (II-A15) in the crystal phase. The difference in position (i.e., bridge hydrogen vs. endohydrogen) in B_5H_{11} would be indistinguishable between I-A15 and II-A15 and could not be differentiated in the vapor phase by present microwave techniques. In any case, B_5H_{11} is an unstable compound and in the liquid phase structure (I-A15), the two 6'6-bridge hydrogens, although undesirable, are probably less undesirable than would be the overly coordinated BH_2 group (violation of rule 4p) in the crystal phase structure (II-A15). Evidently packing forces favor II-A15 in the latter case.

A large research group at Indiana University (18, 80, 105, 120) has obtained splendid line-narrowed ^{11}B NMR spectra which, however, should also favor I-A15. Yet if a continuum of BE hydrogens (especially in the *arachno*-boranes) is simply accepted, then polemics as to whether the bridge hydrogen or endohydrogen structures are correct are of little substance.

When carborane analogs of B_5H_{11} are discovered, it would be expected that an isomer of CB_4H_{10} such as IV-A15 (violating only rule 3s) would be more stable than the isomer III-A15 which contains two undesirable 6'6-bridge hydrogens. Near the completion of this manuscript, Matteson and Mattschei reported (91) a CB_4H_{10} isomer for which they suggest the IV-A15 structure.

The Lewis base adduct of B_5H_9 (B_5H_9:LB) may be considered to be isoelectronic with CB_4H_{10} with a LB:B group substituting for the CH group as in IV-A15 or in III-A15. Williams et al. (103), by contrast, once suggested the structure V-A15 for B_5H_9:LB, which would be isoelectronic with a methylene derivative of B_4H_{10} (I-A14).

In the light of the currently recognized importance of BE hydrogens, a comparison of all three structures (III-, IV-, and V-A15) reveals that the bridge hydrogen environments improve in the order V-A15 (four 6'6-bridge hydrogens), III-A15 (two 6'6- and one 65-bridge hydrogens), to VI-A15 which has three uncompromised 65- and 66-bridge hydrogens.

The V-A15 structure was first suggested for B_5H_9:LB to account for the apex-to-base methyl migration of 1-$CH_3B_5H_8$ in the presence of a Lewis base (i.e., two sets of borons become equivalent in a V-A15 intermediate), but the Lewis base equivalents of IV-A15 or III-A15 would also account for this result. Kodama (*73*) has verified that the rearrangement involves an adduct rather than the ion pair (LB:H^+ and $CH_3B_5H_7^-$).

The IV-A15- or III-A15-type structures for B_5H_9:LB would also be ideal precursors for a di-Lewis base adduct since the structure of $B_5H_9(P(CH_3)_3)_2$ has one of the Lewis base units on the apex (*127*) and one on the base.

4. B_6H_{12} Family of arachno-*Carboranes*

The structure of B_6H_{12} is I-A16 (*18, 82*). The more stable isomer of a possible monocarba derivative (CB_5H_{11}) should have the structure II-A16, with three 66-, 65'- ,and 65-bridge hydrogens rather than the structure of a methylene derivative of B_5H_{11} (I-A15) (e.g., $CH_2B_5H_9$ with the methylene bridging the two BH_2 groups and encompassing four 6'6-, 6'6-, 66-, and 65-bridge hydrogens).

In contrast, Shore *et al.* (*117a*) have shown that the preferred structure of the isoelectronic $B_6H_{11}^-$ is best described as related to B_5H_{11} (I-A15), i.e., μ-$BH_3B_5H_8^-$ with four 65'-, 66-, and 65-bridge hydrogens.

5. B_7H_{13} Family of arachno-*Carboranes*

The parent B_7H_{13} might be expected to have a structure resembling a μ-$BH_3B_6H_{10}$ with 76- and 75'- and two 66-bridge hydrogens or, less likely, I-A17. However, B_7H_{13} has never been isolated, perhaps because a BH_3 group could be easily lost to produce the structurally related B_6H_{10} (I-N6). Removal of one proton from μ-$BH_3B_6H_{10}$ could produce μ-$BH_3B_6H_9^-$ (or $B_7H_{12}^-$) with no overly coordinated bridge hydrogens. The II-A17 isomer of CB_6H_{12} should be more stable than its bridge hydrogen tautomer (III-A17) because of bridge hydrogen considerations. Possibly a methylene derivative of B_6H_{12} ($CH_2B_6H_{10}$) should be anticipated.

There is an alternative structural pattern that suggests another configuration for B_7H_{13}: the parent *arachno*-boranes are built up by the monotonic additions of borons in the series $B_3H_8^-$, B_4H_{10}, B_5H_{11}, and B_6H_{12} (I-A13, I-A14, I-A15, and I-A16, respectively) in Figs. 13 through 16.

One additional boron (i.e., one additional triangle) is added in each case after which the two endohydrogens (first) and the bridge hydrogens (second) are placed in lowest possible coordination sites. Such a candidate structure for B_7H_{13} may be easily envisioned and is displayed as IV-A17. Inasmuch as the aforementioned series also becomes progressively less stable with increasing molecular weight, it is anticipated that such a structure as IV-A17 for B_7H_{13} would be extremely unstable.

6. B_8H_{14} Family of arachno-Carboranes

The parent B_8H_{14} (22) is unstable, as might be anticipated from the compromising presence of at least one 76′-bridge hydrogen in either of the possible candidate structures I- or $\overline{\text{II}}$-A18. The ^{11}B NMR spectrum, reflecting the 4:2:2 kinds of boron, implies rapid tautomerism involving species such as I- and II-A18. An anion, $B_8H_{13}^-$, with the 76′-bridge hydrogen missing from I-A18, should be more stable.

Removal of the least stable or most acidic 76′-bridge hydrogen from the most favored isomer of B_8H_{14} (I-A18) and the substitution of one of the terminal hydrogens with a Lewis base and one of the bridge hydrogens with a —$(CH_3)_2N$— group produces the known crystal phase structure for μ-$(CH_3)_2NB_8H_{11}$:LB (III-A18) (81, 82), which violates no rules.

Species B_8H_{12} (III-N8) is stabilized in the presence of ethers; indeed, its ^{11}B NMR spectrum changes dramatically in ether (22) revealing adduct formation and, thus, probably an isoelectronic relationship with B_8H_{14} (I-A18). Bridge hydrogen tautomerism plus ether exchange (see IV- and VIII-A18, etc.) probably accounts for the relatively simple ^{11}B NMR spectrum of B_8H_{12} in ether. A static adduct is evidently produced from B_8H_{12} in the presence of CH_3CN, as opposed to $O(C_2H_5)_2$, which probably has the structure VII-A18, compatible with the much more complex ^{11}B NMR spectrum observed (22).

Studies are under way to determine the structures of the $C_2B_6H_{10}$ (V-N8 or X-N8) Lewis base adducts, i.e., $C_2B_6H_{10}$:LB. The structural antecedent (V-N8) would favor VI-A18, but V-A18 should be more stable, i.e., satisfying one carbon and one endohydrogen more completely without compromising the 55-bridge hydrogen.

If hydrogen was lost and an additional Lewis base incorporated, i.e., if $C_2B_6H_8$ (:LB)$_2$ was produced, the most stable isomer should have structure IX-A18. It is assumed that the two remaining endohydrogens in IX-A18 would be almost but not quite equally content on either the isoelectronic LB:B or CH groups. However, a residual capacity of boron to accommodate the higher-coordination numbers (Table II) should bias the endohydrogens towards association with the LB:B moieties rather than with the CH groups.

7. B_9H_{15} Families of arachno-Carboranes

The parent species i-B_9H_{15} (I-A19) (21) and n-B_9H_{15} (XI-A19) (76, 125, 130) are displayed in Fig. 19. The latter is more stable, probably because it has only two marginally satisfactory 6'6-bridge hydrogens compared to the very unsatisfactory 76'-bridge hydrogen in I-A19. Compound n-B_9H_{15} (XI-A19) is known to be capable of ejecting BH_3 groups to form B_8H_{12} (III-N8). No carborane isomers of n-B_9H_{15} (XI-A19) are known; a few are suggested such as VII-, VIII-, and XII-A19.

The less stable i-B_9H_{15} isomer (I-A19) incorporates at least two (perhaps three) endohydrogens and four (perhaps three) bridge hydrogens as well as one very unfavorable 76'-, one 76-, and two 66-bridge hydrogens; BE hydrogen tautomerism accounts for the ^{11}B NMR spectrum.

One *arachno*-carborane isomer of i-B_9H_{15}, namely $C_2B_7H_{13}$, is known; its structure is II-A19 (*136*). The carbons and endohydrogens collaborate in occupying five-coordinate CH_2 sites primarily because the "extra" hydrogens' desire to occupy an endo (terminal hydrogen) site rather than a bridge hydrogen site. This is in contrast to *nido*-carboranes wherein the extra hydrogens were almost always bridge hydrogens, and carbons and bridge hydrogens avoided adjacent structural sites in all cases. It follows that the most stable isomer of i-$C_3B_6H_{12}$ should have the structure III-A19.

In these cases, I-, II-, and III-A19, conservation of the 3-center bonds would require five (or six) 3-center bonds to be expressed within the skeleton matched with four (or three) 3-center bridge hydrogens, respectively. There are several examples demonstrating that the nine-vertex arachno system is flexible enough to adapt to both electronic alternatives. The structures for i-B_9H_{13}:LB (X-A19) and i-$B_9H_{14}^-$ (A-V19) are both known in the encumbered crystalline phases and both must accommodate six 3-center bonds within the skeletal framework. By contrast, i-$C_2B_7H_{13}$ (II-A19) absorbs only five 3-center bonds within the skeletal framework.

Others have noted that the encumbered X-ray-determined structures of i-$B_9H_{14}^-$ (V-A19) (*43*) and i-B_9H_{13}:LB (X-A19) (*82*) both involve two bridge hydrogens and three endohydrogens. However, the relative positions of these five extra hydrogens are not the same in both cases. The differences have been attributed to the presumed influences of charge smoothing and/or resonance stabilization.

Our prejudices favor either crystal packing considerations or, more likely, the fact that X-A19 has fewer desirable locations for bridge hydrogen placement (i.e., locations adjacent to the positively charged LB:BH group are shunned).

In the unencumbered liquid phase, i-$B_9H_{14}^-$ should have structure VI-A19 and i-CB_8H_{14} should have a similar structure (IX-A19). The conflicting structures for $i\text{-}B_9H_{14}^-$ (V- and VI-A19) actually constitute just another case wherein it makes little difference whether the unique hydrogen is endo or bridge; the absolute position of the unique hydrogen in space would be essentially the same (whether it were bridge or endo) and crystal packing forces would have little difficulty in centering the unique hydrogen as an endohydrogen even if it were a bridge hydrogen in the unencumbered fluid phase. This situation parallels the B_5H_{11} case (I- and II-A15) in which it makes little difference whether the reference is to bridge or to endohydrogens. Evidently the V-A19 or VI-A19 structure for $i\text{-}B_9H_{14}^-$ is favored over $n\text{-}B_9H_{14}^-$ since removal of a proton (probably a 6'6-bridge hydrogen) from $n\text{-}B_9H_{15}$ produces $i\text{-}B_9H_{14}^-$.

8. $[B_{10}H_{16}]$ Family of arachno-Carboranes

The parent *arachno*-borane $B_{10}H_{16}$ has never been observed (it would have two very undesirable 76'- and two 76-bridge hydrogens); however, the marginally stable derivative $B_{10}H_{15}^-$ has been studied by Rietz et al. (*122*) and the data as interpreted by Siedle (*128*) seemingly indicate that the anticipated $B_{10}H_{15}^-$ structure (I-A20) is correct even though it contains one 76'-bridge hydrogen (probably stabilized by the negative charge); tautomerism should be rampant in the liquid phase and no doubt accounts for the ^{11}B NMR data. The formal removal of one bridge hydrogen (as an H^+) from $B_{10}H_{15}^-$ (I-A20) or the addition of two electrons to $B_{10}H_{14}$ (I-N10) produces the very stable $B_{10}H_{14}^{2-}$ (IV-A20) with two rather stable 66-bridge hydrogens (*69*). The structure of $B_{10}H_{12}(:LB)_2$ (III-A20), with which $B_{10}H_{14}^{2-}$ is isoelectronic and isostructural has been accepted for many years (*82*). The most stable structure for $C_2B_8H_{14}$ analog of $B_{10}H_{16}$ should also have the III-A20 structure with CH_2 groups replacing LB:BH group (see also V-25). Stibr et al. (*132a*) isolated and identified the most stable isomer of $C_2B_8H_{14}$ (V-25) as well as arachno isomers of $C_2B_8H_{12}^{2-}$.

The Lewis base adduct of the $C_2B_8H_{12}$ isomer (XX-N10) (*38*) has been isolated and should have structure V-A20 ($C_2B_8H_{12}$:LB) which incorporates an undesirable 6'6-bridge hydrogen. By contrast, the Lewis base adduct of the other known isomer of $C_2B_8H_{12}$ (XI- or XVI-N10) (*118*) should have structure VI-A20 ($C_2B_8H_{12}$:LB) and should be more stable due to the uncompromised 66-bridge hydrogen.

9. $[B_{11}H_{17}]$ Family of arachno-Carboranes

Just as the removal of one highest-coordination vertex from the most likely, closo, thirteen-vertex polyhedron (e.g., IV-C4) produces the most probable candidate, namely, nido twelve-vertex polyhedral fragment (see Fig. 12), the removal of one additional high-coordination edge position (adjacent to the open face) should produce the eleven-vertex arachno fragment (marred by a seven-coordinate boron) as typified by I-A21 with a heptagonal open face for occupation by the six extra hydrogens.

No eleven-vertex arachno species is known, but $B_{10}H_{13}^-$ (III- and IV-N10) (129, 132) in the presence of diborane indulges in total boron and hydrogen exchange, and a transient arachno intermediate $B_{11}H_{16}^-$ has been suggested (124).

$$B_{10}H_{13}^- + [BH_3] \leftrightharpoons [B_{11}H_{16}^-]$$

The expected configuration for $B_{11}H_{16}^-$ (I-A21) allows both endo-hydrogens to occupy lowest possible coordination sites and allows for satisfactory 66-bridge hydrogen sites; the seven-coordinate boron (violation of rule 4p) is unfavorable, but, in light of the predisposition of seven-coordinate borons to facilitate wholesale valence bond tautomerism (e.g., $B_{11}H_{11}^{2-}$), this feature would account for the complete scrambling of all borons. An alternative and less probable structure for $B_{11}H_{16}^-$ is represented as II-A21, which is produced by breaking one bond in the I-N11 skeletal configuration; a very undesirable $\overline{76'}$-bridge hydrogen would be present (possibly stabilized by the negative charge), but the seven-coordinate boron (violation of 4p) would be avoided.

D. HETEROATOM CARBORANE ANALOGS

There are a number of heteroatoms that can substitute for either boron or carbon in the carboranes. The groups that are as electron-deficient as BH groups are listed vertically to the left of the center line in Table V, whereas those that are as capable as carbon in donating electrons are listed to the right of the center line. The transition elements for the most part electronically substitute for boron and occupy high-coordination sites, but upon electron demand the transition element may also substitute for carbon and concomitantly occupy low-coordination sites. Several transition element moieties, by contrast, are one more electron deficient than boron and occupy, as would be anticipated, high-coordination positions and require additional electron donors (CH groups) to counter the electronic deficit (XIII-24).

In abridged form, we shall discuss in order (a) BH-substituted heteroatom carborane analogs (Fig. 22), (b) CH-substituted heteroatom carborane analogs (Fig. 23), (c) transition element group (TEG)-substi-

TABLE V

Isoelectronic Equivalents of Various Groups

			High			"CH"	Low		
				"BH"					
			−3e	−2e	−1e	0	+1e	+2e	+3e
II	a		Be$^+$	Be/BeH$^+$	BeH	BeH$^-$			
				Mg/		(BeNR$_3$)			
	b			Zn/		ZnH$^-$	ZnH$_2^{2-}$		
							$\overline{\text{ZnB}_{10}\text{H}_{12}}^{2-}$		
				Cd/		CdH$^-$	$\overline{\text{CdH}_2}^{2-}$		
							$\overline{\text{CdB}_{10}\text{H}_{12}}^{2-}$		
							$\overline{\text{HgB}_{10}\text{H}_{12}}^{2-}$		
III	a		B$^+$	B/BH$^+$	BH	BH$^-$			
					$\underline{}$	B(NR$_3$)			
			Al$^+$	Al/AlH$^+$	AlH	AlH$^-$			
	b		Ga$^+$	Ga/GaH$^+$	$\overline{\text{GaH}}$	GaH$^-$			
			In$^+$	In/GH$^+$	$\overline{\text{InH}}$	InH$^-$			
IV	a			C$^+$	C/CH$^+$	CH	CH$^-$		
				$\underline{}$		$\underline{}$	(CNR$_3$)		
				Si$^+$	Si	SiH	SiH$^-$		
	b			Ge$^+$	Ge	GeH	GeH$^-$		
				Sn$^+$	Sn	$\overline{\text{SnH}}$	SnH$^-$		
V	a				N$^+$	N	NH	NH$^-$	
					P$^+$	$\overline{\text{P}}$/PH$^+$	PH	PH$^-$	
	b				As$^+$	As	AsH	AsH$^-$	
					Sb$^+$	$\overline{\text{Sb}}$	SbH	SbH$^-$	
VI	a					S$^+$	S/SH$^+$	SH	
	b					Se$^+$	Se	SeH	
						Te$^+$	Te	TeH	

tuted carborane analogs (Fig. 24), and (d) the coherent interesting consequences that arise in those cases in which the two extra electrons required for stability are donated by one heteroatom, such as S or NH (Fig. 25), rather than carbon.

1. BH-Substituted Heteroatom Carborane Analogs

Groups from Table V such as $BeN(CH_3)_3$ (I-22) *(114)*, AlR *(92, 177)* GaC (II-22) *(46)*, Sn (III-22) *(146)*, and perhaps a bare carbon (IV-22) *(139)* are found substituting isoelectronically for BH groups.

These heteroatom moieties should opt, in the most stable isomers, for the higher-coordination-numbered positions when choices are present.

2. CH-Substituted Heteroatom Carborane Analogs

Several groups are known that are evidently isoelectronic with CH groups and as such, when choices exist, preempt lowest-coordination sites.

For example, the Zn, Cd, Ni, and Hg moieties in $M(B_{10}H_{12})_2^{2-}$ *(44, 47)* substitute for CH groups (or BH^- groups) in two molecules of $B_{11}H_{13}^{2-}$ (IV-N11), and, thus, I-23 becomes isoelectronic with $B_{11}H_{13}^{2-}$ (IV-N11), with Zn occupying an edge site as would a CH group *(71)*.

3. Transition Element Heteroatom Carborane Analogs *(50, 155)*

For simplicity, TEG in Fig. 24 stands for transition element group including the transition element atom and attached moieties (e.g., $C_5H_5^-$, $(CO)_n$, and $C_2B_9H_{11}^{2-}$). For the purposes of this discussion the interest is in the transition element atom insofar as it substitutes for BH groups and not in what other flotsam is attached. In general, the transition elements substitute for BH groups and when the coordination number is unusually high (even for boron) the transition element atom preempts the highest position. Thus, in $C_5H_5\ CoC_2B_8H_{10}$ (I-24) *(52)* and in $C_5H_5\ CoC_2B_{10}H_{12}$ (II-24) *(17, 23, 26)* the transition element atoms are found in the seven-coordinate positions; structure III-24 should be a slightly more stable isomer than II-24 when both the transition atom and carbon are in optimal positions. The second most desirable location for carbon should be the six-coordinate site farthest removed from the seven-coordinate positions if the choice is between six-coordinate sites. Reassuringly, exactly as predicted, the most stable isomer *closo*-TEG-$C_2B_{10}H_{12}$ (III-24) is produced upon heating II-24 [Dustin *et al.* *(26a)*]. These authors also subscribe to the old *(172)* carbon placement considerations (rule 3) but point out the "mysterious" predilection of the TEG and the carbon to become separated in the most stable isomers.

That the initially suggested structure for TEG-$C_2B_{10}H_{12}$ (IV-24) *(53)* has been shown to be incorrect augurs well for the CNPR approach, since IV-24 was based on an illogical thirteen-vertex deltahedron (violation of rule 1p); see IV-C4 for the more reasonable deltahedral alternative.

Many other transition atoms simply substitute for the standard six-coordinate BH groups. Representative examples are in the seven-, eight-, nine- (45), ten-, eleven-, and twelve-vertex (50) closo species displayed as V-, VI-, VII- VIII-, and I-24 and the group IX-, X-, XI-, and XII-24, which illustrate the diversity possible.

In the icosahedral group of compounds (IX- to XII-24) there are no obviously preferred vertices for carbon occupation and/or transition element occupation, and it is interesting to note that, upon heating one isomer of TEG-$C_2B_9H_{11}$ (IX-24), a number of other isomers are formed (51). It would appear that there is no overriding first-order driving force that tends to place carbons adjacent to or apart from the transition atom in the icosahedron. The carbons in such species tend to become nonadjacent upon heating, just as occurs when the parent 1,2-$C_2B_{10}H_{12}$ isomer (II-C3) is treated to produce the 1,7- and 1,12-isomers (II- and III-C2).

There are groups incorporating a transition metal element atom that contribute fewer skeletal electrons than a BH group; these include the $(CO)_3Mn$ unit of $(CO)_3MnC_3B_3H_6$ (XIII-24) (45) [to be compared with $C_2B_5H_7$ (V-C2)] and the C_5H_5Fe unit of ferrocene $Fe(C_5H_5)_2$ (XVI-24) [to be compared with $C_4B_2H_6$ (IX-N6)], wherein the transition element occupies the highest-coordination position available (substituting for a boron); but since the transition element contributes one less electron than boron, the inclusion of an additional carbon to make up the difference is necessitated. Alternatively, the iron group in $C_2B_3H_7Fe(CO)_3$ is only as electron deficient as a BH group (10) (not shown). Both compounds are isoelectronic with B_6H_{10} (I-N6).

It has been suggested (27) that the XIV-24 structure accounts for the various types and abundances of boron in the ^{11}B NMR spectrum (2:2:2:1) of $(CO)_3CoCB_7H_8$. Although XIV-24 may be correct, XVII-24 should have the same numbers of types and abundance ratios and should be more stable, as the carbon would be in the more desirable, low-coordination number position.

More disturbing is the proposed (68) structure XV-24 (to account for boron type and abundance in the ^{11}B NMR spectrum) rather than XVIII-24. It is important not to lose sight of the fact that the thermodynamically less stable isomers may frequently be produced during a gentle synthesis; XV-24 and XVIII-24, however, present a striking exception to the thoughts thus far presented in this manuscript. Structure XV-24 is quite possibly correct but it should be much less stable than XVIII-24. In the present author's choice (XV-24), a Co that is isoelectronic (in this case with boron) is found in a lower-coordination edge site (violating rule 4p) while one carbon is found in the highest possible coordination sites (a violation of rule 3p). The preferred XVIII-24 isomer would also have exactly the boron types and abundances required to account for the observed ^{11}B NMR spectrum.

As one illustrative example, since Co is seen to inhabit BH or highest-coordination sites in general, it is noteworthy that a cobalt in its more reduced form (i.e., furnishing an additional electron into the cage) substitutes, in such cases, for a CH group and is, therefore, found, as would be expected, in a low-coordination site (I-23).

E. ONE HETEROATOM DONATING TWO ELECTRONS

In the two series enumerated in the following, it would be expected that the extra electrons would be progressively more difficult to delocalize smoothly: first series—$B_{10}H_{10}^{2-} < CB_9H_{10}^{-} < 1,10\text{-}C_2B_8H_{10} < 1,6\text{-}C_2B_8H_{10} < B_{10}H_8(LB)$, $< SB_9H_9$; second series—$B_{12}H_{12}^{2-} < CB_{11}H_{12}^{-} < 1,12\text{-}C_2B_{10}H_{12} < 1,2\text{-}C_2B_{10}H_{12} < SB_{11}H_{11} < (CH_3)_3\text{-}NCB_{10}H_{10}$. The order may be explained as follows: B_nH_n species are not predicted to be stable; two additional electrons are required and thus $B_nH_n^{2-}$ species are known. For illustrative and bookkeeping purposes these two additional electrons may be defined as "requisite" electrons. In isoelectronic cases such as the closo species $(C_2B_nH_{n+2})$, each carbon donates one of the two requisite electrons.

In the $B_nH_n^{2-}$ species the skeletal electrons and the concomitant negative charges would be as evenly distributed throughout the molecular skeleton as possible. When carbon (or carbons) donate the requisite electrons, the carbons would tend to become positively charged and, thus, tend to distort the donated electrons back toward the carbons. *Thus, the most electron-rich environments are at the same time the most positively charged.*

When the requisite electron that is donated comes from a Lewis base, passes through a boron and on into the skeletal system, then even less complete donation and even more polarization of positive and negative charges should be anticipated. When the two requisite electrons are donated by one heteroelement atom, such as S or NH, the polarization of positive and negative charge should increase substantially over the previous cases. This effect should be maximized in the case where a Lewis base donates electrons to carbon [e.g., the $(CH_3)_3NC$ group in the last example cited] which, in turn, passes the requisite two electrons on into the cage system. In the latter case the carbon donates the two requisite electrons that had been "borrowed" from the Lewis base.

Consider the ten skeletal atom series of thiaboranes in Fig. 25, *closo*-SB_9H_9 (*115*), *nido*-SB_9H_{11}, and *arachno*-$SB_9H_{12}^-$ (*57*). Based on first principles, elucidated in the previous sections and as displayed in Figs. 2, 10, and 20, the sulfur atoms would be placed in the carbon preferred positions as represented by I- and probably II- and III-25, respectively. The presence of bare sulfurs in structures I- and II-25 could be expected. Logically, it would have been expected (incorrectly, as it turns out) that a terminal hydrogen would be placed on the sulfur atom in $SB_9H_{12}^-$ as displayed in III-25. The hypothetical parent of structure III-25 (with one more bridge hydrogen) is represented as VII-25.

Refocusing attention on $SB_9H_{12}^-$ (III-25), the sulfur atom, doing the job of two carbons (i.e., contributing two electrons into the boron skeletal framework), should become much more positively charged than would be a carbon in a comparable situation. Structure III-25 should then be highly polar with a positive charge on sulfur and a negative charge distributed among the borons. This polarization in $SB_9H_{12}^-$ (III-25), if unsatisfactory, could be alleviated in at least three ways, e.g., the migration of a positively charged proton from the sulfur onto the cage to produce the zwitterion tautomer VI-25 or, alternatively, by placing the sulfur in alien, more highly coordinated positions as displayed in structure IV- or VIII-25. The proton migration to produce the zwitterion [dipolar ion or inner salt; $SB_{12}H_{12}^-$ (VI-25)] is completely compatible with the ^{11}B NMR spectrum, and the absence of an SH group in the ^1H NMR spectrum demonstrates that the author's choice (*57*) of VI-25, as the structure for $SB_9H_{12}^-$, is correct (*128*).

Two effects become apparent in the above cases and they should work in opposition: electron-sufficient carbon donates electrons into the skeletal framework but (*a*) wants and frequently gets the least alien, lowest available coordination site (rule 3) [see, for example, $2,4-C_2B_5H_7$ (V-C2)]; alternatively (*b*), could not the carbons more efficiently donate their extra skeletal electrons if they were in the most highly coordinate apex sites of the pentagonal bipyramid (i.e., a *closo*-$1,7-C_2B_5H_7$)? Although effect (*a*) seems most important in the case of carbon, it may not be in the case of sulfur which accounts for the inclusion here of candidate structures IV- or VIII-25 for $SB_9H_{12}^-$ for illustrative purposes, even though they are known to be incorrect for the structures thus far discovered. Perhaps effect (*b*) will be found to be the dominant theme in some subsequent investigations.

Although the suggested rearrangement to produce the zwitterion is a satisfying way to visualize the preference of structure VI-25 over III-25 in the case of $SB_9H_{12}^-$, in actuality the sulfur attached proton in the hypothetical parent SB_9H_{13} (VII-25) can simply be recognized as most acidic, and, therefore, the anion VI-25 would simply be derived from the parent VII-25 by the removal of that sulfur-attached proton.

This rationalization of the data, with an extreme case of charge smoothing as the driving force, is made more believable as a result of Shore's (63) placement of a proton in the one available vacancy about the base of B_6H_{10} (I-N6) to produce $B_6H_{11}{}^+$ which has no such driving force [a compound predicted to be stable by Lipscomb (82)].

If one accepts the explanation for the correctness of structures VI-25 and XIII-25 as based on dipolarization considerations due to the necessity of S and N to donate two electrons, would there not be at least a similar, smaller but noticeable effect in the case of carbon analogs wherein each carbon donates only one electron into the cage?

Only one example has been published [i.e., $C_2B_7H_{13}$ (II-A19)] and it confirms these suspicions (136). The two bridge hydrogens and the two endohydrogens of the CH_2 groups in $C_2B_7H_{13}$ are acidic, and although all four hydrogens exchange to produce $C_2B_7H_9D_4$ when $C_2B_7H_{13}$ (II-A19) is mixed with excess D_2O and K_2CO_3, the salt Cs^+ $C_2B_7H_{12}{}^-$, upon neutralization by DCl, produces $C_2B_7H_{12}D$, with the deuterium on the carbon. This reveals that the endohydrogens of the CH_2 groups are more acidic than the bridge hydrogens and that the unencumbered fluid phase structure of $C_2B_7H_{12}{}^-$ should be that of $C_2B_7H_{13}$ (II-A19) minus an endohydrogen from one of the two CH_2 groups.

Thus, within the arachno species, the endohydrogens of BH_2, CH_2, and $NH_2(SH)$ groups are progressively more acidic; the latter three are more acidic than most two-coordinate bridge hydrogens.

When the *arachno*-carborane $C_2B_8H_{14}$ (V-25) is eventually discovered its anion $C_2B_8H_{13}{}^-$ should have structure IX-25 rather than X-25 structure based on the greater acidity of an endohydrogen attached to carbon as compared to a bridge hydrogen; stated another way, the zwitterionic configuration (IX-25) will be preferred.* Were it possible to produce $B_{10}H_{11}(:LB)_2{}^-$ from $B_{10}H_{12}(:LB)_2$ (III-A20), it should be isoelectronic with $C_2B_8H_{13}{}^-$ (IX-25).

Before leaving structures II-, III-, and IV-25, although III-25 is unstable with respect to VI-25 for an isomer of $SB_9H_{12}{}^-$, there is reason to suspect that the IV-25 isomer of $SB_9H_{12}{}^-$ may one day be discovered unless sulfur (as compared to carbon) has a greater predilection than carbon for low-coordination sites (which has not yet been ascertained). This suggests that the known *nido*-SB_9H_{11} (II-25) should be reconsidered.

* Recently, Plesek and Hermanek (113b) compiled a review that includes their research and "the intermediate dicarba-*nido*-boranes"; many other new compounds are revealed. Stibr et al. (132a) had previously isolated the most stable isomer of $C_2C_8H_{14}$ (V-25).

An alternative nido-SB_9H_{11} (not shown) would place the sulfur in the higher-coordinate five position (as in IV-25) but result in a more ideal placement of the bridge hydrogens (i.e., two 65-bridge hydrogens as opposed to one 66- and one 65-bridge hydrogen in II-25).

Returning to Fig. 25 and focusing on the hypothetical parent compound SB_9H_{13} (VII-25) that contains two 66-bridge hydrogens, one may ask: Would not that compound also tend to transfer its sulfur-attached proton onto a boron–boron bond to form a zwitterionic tautomer (XI-25) resembling $B_{10}H_{15}^-$ (I-A20) (*122*)? Since $\overline{76'}$- and 76-bridge hydrogens are present in XI-25, the tendency to lose that very acidic bridge proton from XI-25 to become the anion should be considerable. Thus both VII-25 or its tautomeric isomer (XI-25) should be very acidic. In seeming confirmation of this supposition, it has been impossible to date to add a proton to $SB_9H_{12}^-$ (VI-25) to produce either the parent SB_9H_{13} (VII-25) or its zwitterionic counterpart XI-25. Indeed (*57*), compound $H_3O^+SB_9H_{12}^-$ is as strong an acid as $(H_3O^+)_2B_{12}H_{12}^{2-}$ or sulfuric acid for that matter. Perhaps the parent SB_9H_{13} (VII- or XI-25) could be produced from $SB_9H_{12}^-$ (VI-25) in superacid, liquid HI, or in Shore's HCl plus BCl_3 mixture (*127*):

$$SB_9H_{12}^- + BCl_3 + HCl \rightarrow SB_9H_{13} + BCl_4^-$$

Rudolph's SB_9H_9 (I-25) (*115*), where S substitutes for two CH groups, is stable (no extra hydrogens complicate matters); but would not the terminal NH hydrogen in an isoelectronic analog such as NB_9H_{10} be very acidic? Such a compound when synthesized should tend to lose the acidic NH to form $NB_9H_9^-$. A question remains. In the event that this latter anion could be protonated via Shore's techniques or in liquid HI or in superacid (carefully), would the proton locate terminally upon the nitrogen or over a triangular face reminiscent of VIII-C4?

IV. Bridge and Endohydrogens and Relative Lowry-Brønsted Acidity

In its current form, CNPR theory has resulted from the recognition that the second most important consideration (rule 2) within carborane structures is the coordination numbers involving the BE hydrogens.

The relative Lowry-Brønsted acidities of the various groups within the carboranes and related species are tentatively listed in Table VI. Further experiments to determine the equilibrium positions involving species such as

$$SB_9H_{12}^- + B_{10}H_{11}^- \rightleftharpoons SB_9H_{13} + B_{10}H_{10}^{2-}$$

TABLE VI

Tentatively Suggested Proton Acidity of Various Groups as a Function of Coordination Number

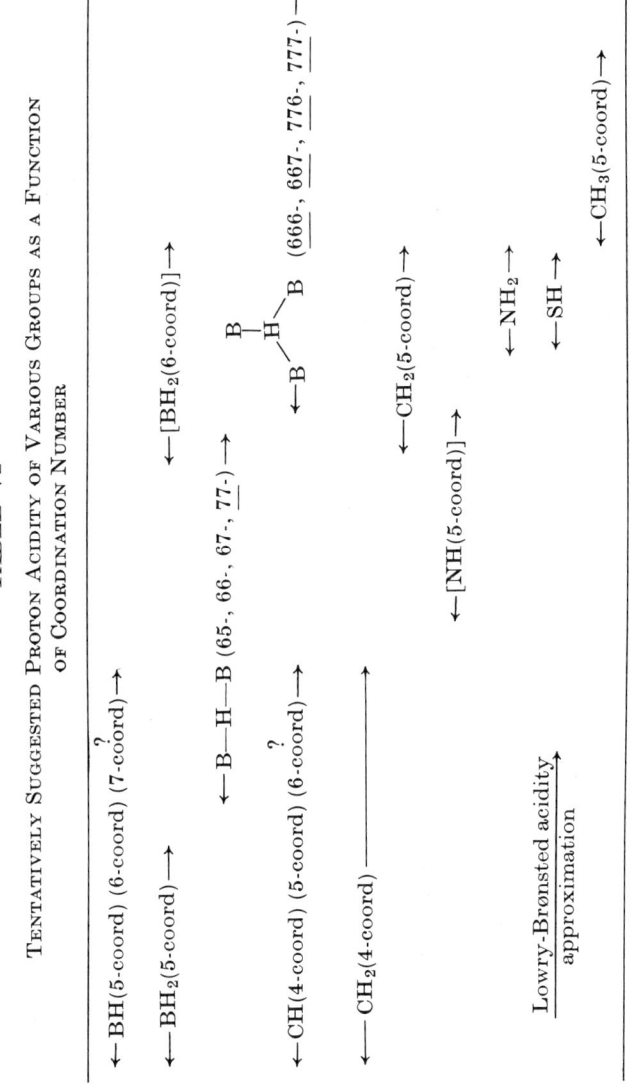

may reveal relative SH and B_3H (protonated face) acidities complicated by the presence of negative charges.

There are, of course, several kinds of BH and CH groups. Depending on the coordination numbers of the borons and carbons involved, some are more acidic than others. The most interesting acidity order that can now be tentatively handled involves the B_2H (protonated edge) group or the 3-center bridge hydrogens which have been shown [as predicted by Parry and Edwards (*110*)] to differ in degree, i.e., to be more acidic when larger electron-deficient sinks are involved. Hermanek, and coworkers have added the fused *nido*-boranes (*56, 113c*). In the order of increasing acidity are $B_5H_9 < B_6H_{10} < B_{10}H_{14}$ (*8, 64, 66*) $< B_{16}H_{20} < B_{18}H_{22}$ (isomers).

Shore has pointed out that in acidity B_4H_{10} (I-A14) (*65*) lies between $B_{10}H_{14}$ (I-N10) and B_6H_{10} (I-N6). When the various "kinds" of bridge hydrogens are considered, as well as the degree within which they differ, then B_4H_{10} is not an anomalous example. Within the series $B_{18}H_{22}$, $B_{16}H_{20}$, $B_{10}H_{14}$, B_6H_{10}, and B_5H_9 the most highly coordinated bridge hydrogens (and thus the most acidic) are what have been described above as relatively content 66-bridge hydrogens. By contrast, much less content 6'6-bridge hydrogens are present in B_4H_{10}.

When one takes into account then, both the largeness of the electron-deficient delocalized electron sink (*56, 110, 113c*) (degree) and the larger effect of the kind of bridge hydrogen (see underlined species in Table VII), then B_4H_{10} falls into line with $B_{18}H_{22}$, $B_{16}H_{20}$, $B_{10}H_{14}$, B_6H_{10}, and B_5H_9. Many other bits and pieces of information also seem to fit.

Shore *et al.* (*12*) have confirmed that the presence of electron-withdrawing groups (i.e., ClB_5H_8) promote greater bridge hydrogen acidity, whereas the presence of electron-donating groups (i.e., $CH_3B_5H_8$) lessen bridge hydrogen acidity when compared to the parent B_5H_9. Bridge hydrogen locations are also influenced by such groups (e.g., as in $CH_3B_6H_9$ and BrB_6H_9) (*11*).

In Table VII are illustrated, in greatly oversimplified fashion, the relationships of the known boranes and some carboranes. The lack of acidity of B_2H_6 (*8*) may be attributed to both its small size and 5'5'-bridge hydrogens.

Why is $C_2B_4H_8$ (III-N6) so much more acidic than $C_2B_4H_7^-$ (VIII-N6) and less acidic than B_6H_{10} (I-N6) although all three species are isoelectronic? The B_6H_{10} has a 66-bridge hydrogen to lose and is six electrons deficient, whereas the isoelectronic $C_2B_4H_8$ is only four electrons deficient and has a less acidic 65-bridge hydrogen to lose. Moreover, once $C_2B_4H_7^-$ is formed, only a 55-bridge hydrogen remains, and it is only three electrons deficient and negatively charged besides! These

TABLE VII

OVERSIMPLIFIED LOWRY-BRØNSTED ACIDITY AS A FUNCTION OF "KIND AND DEGREE" OF VARIOUS BRIDGE HYDROGENS

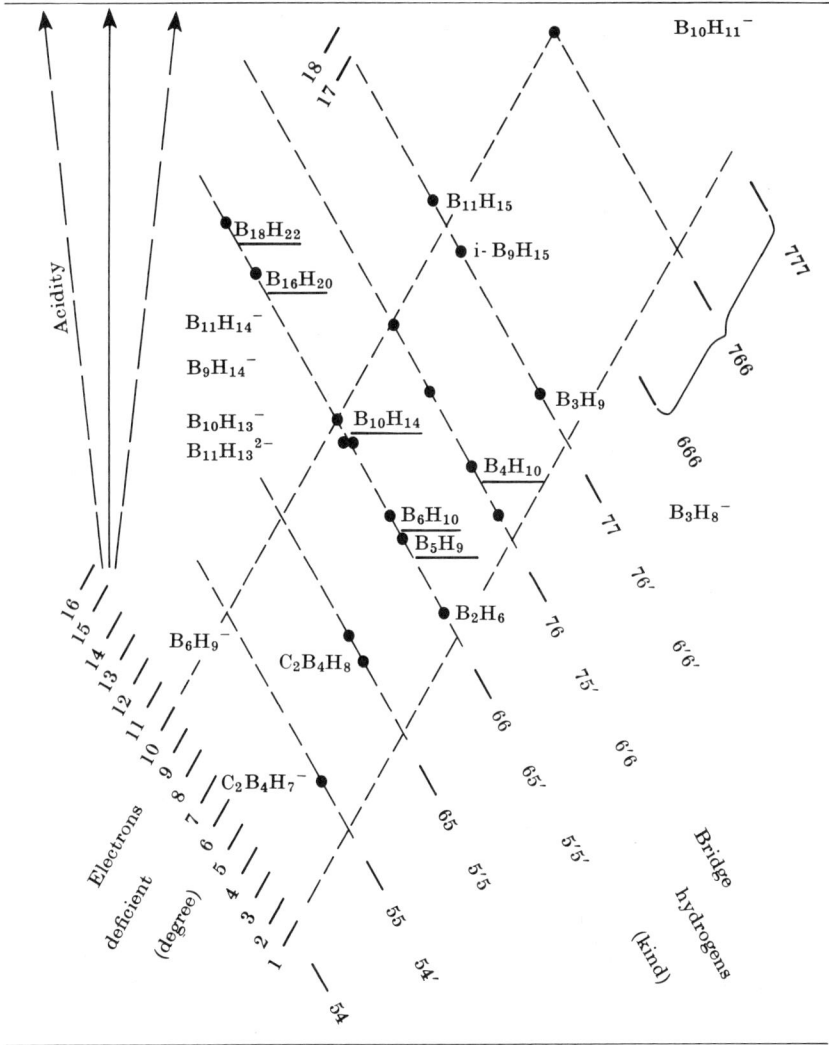

observations give rise to the question: Could bridge hydrogen acidities be utilized as a tool to infer structure?

Why are the i-$B_9H_{14}^-$ (VI-A19) and $B_{11}H_{14}^-$ (III-N11) anions stable, but their parents i-B_9H_{15} (I-A19) and $B_{11}H_{15}$ (I-N11) are unstable?

The latter have $\overline{76'}$- and $\overline{77}$-bridge hydrogens and are more electron-deficient (see Table VII); it is also reasonable to suggest here that the stability of any of the overly coordinated bridge hydrogens in the anions is greatly enhanced by the presence of the negative charge. Many problems suggest themselves for investigation; among the various atoms within the carboranes, the multivalent hydrogens are evidently the most sensitive to coordination number environment.

In Table VII it has been assumed for illustrative purposes that both differences in degree and kind are linearly related, that the members of the various groups (e.g., $\overline{6'6'}$-, $\overline{76'}$- and $\overline{77}$-) are equal, that many other isoelectronic differences such as the presence of carbons as opposed to negative charges play no role, and that the total number of bridge hydrogens is not relevant, etc. Of course, these assumptions are only partially correct; for example, the presence of negative charge, although extremely important, is neglected (e.g., a 65-bridge hydrogen $B_6H_9^-$ would be much less acidic than a 65-bridge hydrogen on the isoelectronic $C_2B_4H_8$). Of course, $B_5H_8-B_5H_8$ should be considered as a B_5H_9 derivative, not a 10-boron compound in Table VII.

When more data are accumulated, the various borane anions, carboranes, and neutral boranes should only be compared with each other, rather than all together as in Table VII.

V. Conclusions

There are a number of additional points of view (*29, 30, 54, 87, 112, 143*) which the discerning reader will wish to consider; some are complementary, some compatible, some competitive, and some present alternatives to the CNPR theory.

On the preceding pages there has been an attempt to present carborane structural chemistry as clear and orderly. In fact, however, many bits of information are missing, perhaps have been misplaced, or misinterpreted; additional research is needed to fill in the blanks and to correct the errors.

The discriminating reader will note, for instance, that only the neighboring atoms are considered in the assignment of coordination numbers to the carbons (rule 3) and the borons (rule 4). The identities (whether hydrogen, carbon, or boron, etc.) of the neighboring atoms have been neglected. The elements that were "bonded" to the borons involved in the bridge hydrogen count (rule 2) have also been neglected. Advances in CNPR theory may be anticipated when these second-order coordination number considerations are appraised.

For a more general overview, based in part on CNPR theory, which encompasses not only electron-deficient carboranes but also includes metal hydrocarbon π-complexes, cluster and ring compounds, see the preceding article in the present volume, by K. Wade.

ACKNOWLEDGMENTS

It is a pleasure to acknowledge the financial support of the Office of Naval Research. Special credit is due Dr. Alan J. Gotcher for his organizational adjuvancy, critical review, and many helpful discussions and suggestions. We are also grateful to Drs. R. E. Kesting, J. F. Ditter, P. A. Miller, and Mr. F. McGee for their editorial aid; to Drs. K. Wade (*152*) and R. W. Rudolph for apprising us of the much more general chemical consequences of the CNPR classification system when electron counting is emphasized; and to Drs. T. P. Onak, K. Harmon, and A. R. Siedle for bringing to the writer's attention many corrobative examples.

REFERENCES

1. Aftandilian, V. O., Miller, H. C., Parshall, G. W., and Muetterties, E. L., *Inorg. Chem.* **1**, 734 (1962).
2. Barry, T. E., Tebbe, F. N., and Hawthorne, M. F., *Tetrahedron Lett.* p. 715 (1965).
3. Beaudet, R. A., McKown, G. L., Don, B. P., Vergamini, P. J., and Jones, L. H., *Chem. Commun.* p. 765 (1974).
4. Beaudet, R. A., and Poynter, R. L., *J. Amer. Chem. Soc.* **86**, 1258 (1964).
5. Beaudet, R. A., and Poynter, R. L., *J. Chem. Phys.* **43**, 2166 (1965).
6. Beer, D. C., Burke, A. R., Englemann, T. R., Storhoff, G. N., and Todd, L. J., *Chem. Commun.* p. 1611 (1971).
7. Binger, P., *Tetrahedron Lett.* No. 24, p. 2675 (1966).
8. Bond, A. C., and Pinsky, M. L., *J. Amer. Chem. Soc.* **92**, 7685 (1970).
9. Bramlett, C. L., and Grimes, R. N., *J. Amer. Chem. Soc.* **88**, 4269 (1966).
10. Brennon, J. P., Grimes, R. N., Schaeffer, R., and Sneddon, L. G., *Inorg. Chem.* **12**, 2266 (1973).
11. Brice, V. T., Johnson, H. D., and Shore, S. G., *J. Amer. Chem. Soc.* **95**, 6629 (1973).
12. Brice, V. T., and Shore, S. G., *Inorg. Chem.* **12**, 309 (1973).
13. Brown, H. C., Sthehle, P. F., and Tierney, P. A., *J. Amer. Chem. Soc.* **79**, 2020 (1957).
14. Brown, M. P., Holliday, A. K., and Way, G. M., *Chem. Commun.* p. 532 (1973).
14a. Brown, M. P., Halliday, A. K., Way, G. M., and Whittle, R. B., *Chem. Commun.* p. 532 (1973).
15. Centofanti, L. F., Kodama, G., and Parry, R. W., *Inorg. Chem.* **8**, 2072 (1969).
16. Chowdhry, V., Pretzer, W. R., Rai, D. N., and Rudolph, R. W., *J. Amer. Chem. Soc.* **95**, 4560 (1973).
17. Churchill, M., see Dustin et al. (*26*, footnote 10).
18. Clouse, A. O., Moody, D. C. Rietz, R. R., Roseberry, T., and Schaeffer, R., *J. Amer. Chem. Soc.* **95**, 2496 (1973).
19. Corbett, J. D., *Inorg. Chem.* **7**, 198 (1968).
20. Ditter, J. F., unpublished identification.
21. Dobson, J., Keller, P. C., and Schaeffer, R., *J. Amer. Chem. Soc.* **87**, 3522 (1965); *Inorg. Chem.* **7**, 399 (1968).

22. Dobson, J., and Schaeffer, R., *Inorg. Chem.* **7**, 402 (1968).
23. Dunks, G. B., McKown, M. M., and Hawthorne, M. F., *J. Amer. Chem. Soc.* **93**, 2541 (1971).
24. Dunks, G. B., Wiersema, R. J., and Hawthorne, M. F., *J. Amer. Chem. Soc.* **95**, 3174 (1973).
25. DuPont, J. A., and Schaeffer, R., *J. Inorg. Nucl. Chem.* **15**, 310 (1960).
26. Dustin, D. F., Dunks, G. B., and Hawthorne, M. F., *J. Amer. Chem. Soc.* **95**, 1109 (1973).
26a. Dustin, D. F., Evans, W. J., Jones, C. J., Hawthorne, M. F., Wiersema, R. J., Gong, H., and Chan, S., *J. Amer. Chem. Soc.* **96**, 3085 (1974).
27. Dustin, D. F., and Hawthorne, M. F., *Inorg. Chem.* **12**, 1380 (1973).
28. Edwards, L. J., and Makhlouf, J. M., *J. Amer. Chem. Soc.* **88**, 4728 (1966).
29. Epstein, I. R., *Inorg. Chem.* **12**, 709 (1973).
30. Epstein, I. R., Marynick, D. S., and Lipscomb, W. N., *J. Amer. Chem. Soc.* **95**, 1760 (1973).
31. Fein, M. M., Bobinski, J., Mays, N., Schwartz, N., and Cohen, M. S., *Inorg. Chem.* **2**, 1111 (1963).
32. Franz, D. A., and Grimes, R. N., *J. Amer. Chem. Soc.* **92**, 1438 (1970).
33. Franz, D. A., Miller, V. R., and Grimes, R. N., *J. Amer. Chem. Soc.* **94**, 412 (1972).
33a. Fratini, A. V., Sullivan, G. W., Denniston, M. L., Hertz, R. K., and Shore, S. G., *J. Amer. Chem. Soc.* **96**, 3013 (1974).
34. Fritchie, C. J., *Inorg. Chem.* **6**, 1199 (1967).
35. Gaines, D. F., *Inorg. Chem.* **2**, 523 (1963).
36. Gaines, D. F., *Accounts Chem. Res.* **6**, 416 (1973).
37. Gaines, D. F., and Iorns, T. V., *J. Amer. Chem. Soc.* **92**, 4571 (1970).
38. Garrett, P. M., Ditta, G. S., and Hawthorne, M. F., *J. Amer. Chem. Soc.* **93**, 1265 (1971).
39. Glore, J. D., Rathke, J. W., and Schaeffer, R., *Inorg. Chem.* **12**, 2175 (1973).
40. Good, C. D., and Williams, R. E., U.S. Patent 3,030,289 (1959); *Chem. Abst.* **57**, 12534b (1962).
41. Gotcher, A. J., Ditter, J. F., and Williams, R. E., *J. Amer. Chem. Soc.* **95**, 7514 (1973); see also Reilly, T., and Burg, A. B., *Inorg. Chem.* **13**, 1250 (1974).
42. Grafstein, D., and Dvorak, J., *Inorg. Chem.* **2**, 1128 (1963).
43. Greenwood, N. N., Gysling, H. J., McGinnety, J. A., and Owen, J. D., *Chem. Commun.* p. 505 (1970).
44. Greenwood, N. N., McGinnety, J. A., and Owen, J. D., *J. Chem. Soc., A*, p. 809 (1971).
45. Grimes, R. N., "Carboranes," Academic Press, New York, 1970.
46. Grimes, R. N., and Rademaker, W. J., *J. Amer. Chem. Soc.* **91**, 6498 (1969).
47. Guggenberger, L. J., *J. Amer. Chem. Soc.* **94**, 114 (1972).
48. Hall, J. H., Epstein, I. R., and Lipscomb, W. N., *Inorg. Chem.* **12**, 915 (1973).
49. Hawthorne, M. F., in "The Chemistry of Boron and its Compounds" (E. L. Muetterties, ed.), Wiley, New York, 1967, Chapter 5.
50. Hawthorne, M. F., *Pure Appl. Chem.* **29**, 547 (1972).
51. Hawthorne, M. F., *Pure Appl. Chem.* **29**, 557–560 (1972).
52. Hawthorne, M. F., *Pure Appl. Chem.* **29**, 564 (1972).
53. Hawthorne, M. F., *Pure Appl. Chem.* **29**, 565 (1972).
54. Hawthorne, M. F., and Dunks, G. B., *Science* **178**, 462 (1972).

55. Hawthorne, M. F., Tebbe, F. N., and Garrett, P. M., private communication.
56. Hermanek, S., and Plotova, H., *Collect. Czech. Chem. Commun.* **36**, 1639 (1971).
57. Hertler, W. R., Klanberg, F., and Muetterties, E. L., *Inorg. Chem.* **6**, 1696 (1967).
58. Hertz, R. K., Johnson, H. D., and Shore, S. G., *Inorg. Chem.* **12**, 1875 (1973).
59. Hogeveen, H., and Kwant, P. W., *Tetrahedron Lett.* No. 19, p. 1665 (1973).
60. Howe, D. V., Jones, C. J., Wiersema, R. J., and Hawthorne, M. F., *Inorg. Chem.* **10**, 2516 (1971).
61. Huffman, J. C., and Streib, W. E., *Chem. Commun.* p. 665 (1972).
62. Hyatt, D. E., Sholer, F. R., Todd, L. J., and Warner, J. L., *Inorg. Chem.* **6**, 2229 (1967).
63. Johnson, H. D., Brice, V. T., Brubaker, G. L., and Shore, S. G., *J. Amer. Chem. Soc.* **94**, 6711 (1972).
64. Johnson, H. D., Geanangel, R. A., and Shore, S. G., *Inorg. Chem.* **9**, 908 (1970).
65. Johnson, H. D., and Shore, S. G., *J. Amer. Chem. Soc.* **92**, 7586 (1970).
66. Johnson, H. D., Shore, S. G., Mock, N. L., and Carter, J. C., *J. Amer. Chem. Soc.* **91**, 2131 (1969).
67. Jones, C. J., Evans, W. J., and Hawthorne, M. F., *Chem. Commun.* p. 543 (1973).
68. Jones, C. J., Francis, J. N., and Hawthorne, M. F., *J. Amer. Chem. Soc.* **94**, 8391 (1972).
69. Kendall, D. S., and Lipscomb, W. N., *Inorg. Chem.* **12**, 546 (1973).
70. Klanberg, F., and Muetterties, E. L., *Inorg. Chem.* **5**, 1955 (1966).
71. Klanberg, F., Wegner, P. A., Parshall, G. W., and Muetterties, E. L., *Inorg. Chem.* **7**, 2072 (1968).
72. Knoth, W. H., *J. Amer. Chem. Soc.* **89**, 1274 (1967).
73. Kodama, G., *J. Amer. Chem. Soc.* **94**, 5907 (1972).
74. Kodama, G., Engelhardt, U., Lafrenz, C., and Parry, R. W., *J. Amer. Chem. Soc.* **94**, 407 (1972).
75. Koster, R., *Abstr. Proc. Int. Symp. Organometal. Chem. 2nd, 1965*, p. 97 (1965).
76. Kotlensky, V. W., and Schaeffer, R., *J. Amer. Chem. Soc.* **80**, 4517 (1958).
77. LaPrade, M. D., and Nordman, C. E., *Inorg. Chem.* **8**, 1669 (1969).
78. M. F. Lappert, and H. J. Emeleus (eds.), *MTP Int. Rev. Sci., Inorg. Chem. Ser. 1* **1**, pp. 79 and 139 (1972).
79. Lavine, L. R., and Lipscomb, W. N., *J. Chem. Phys.* **22**, 614 (1954).
80. Leach, J. B., Onak, T. P., Spielman, J., Rietz, R. R., Schaeffer, R., and Sneddon, L. G., *Inorg. Chem.* **9**, 2170 (1970).
81. Lewin, R., Simpson, P. G., and Lipscomb, W. N., *J. Amer. Chem. Soc.* **85**, 478 (1963).
82. Lipscomb, W. N., "Boron Hybrides," Benjamin, New York, 1963.
83. Lipscomb, W. N., *Inorg. Chem.* **3**, 1683 and Fig. 12 (1964).
84. Lipscomb, W. N., *Inorg. Chem.* **3**, 1683 and Fig. 7 (1964).
85. Lipscomb, W. N., *Science* **153**, 373 (1966).
86. Lipscomb, W. N., *Pure Appl. Chem.* **29**, 493 and 509 (1972).
87. Lipscomb, W. N., *Accounts Chem. Res.* **6**, 257 (1973).
88. Lockman, B., and Onak, T. P., *J. Amer. Chem. Soc.* **94**, 7923 (1972).
88a. Mangion, M., Clayton, W. R., Long, J., and Shore, S. G., *J. Amer. Chem. Soc.* in press (1975).

89. Masamune, S., Sakai, M., Ona, H., and Jones, A. J., *J. Amer. Chem. Soc.* **94**, 8956 (1972).
90. Mastryukov, V. S., Dorofeeva, O. V., Vilkov, L. V., Zigach, A. F., Laptev, V. T., and Petrunin, A. B., *Chem. Commun.* p. 276 (1973).
91. Matteson, D. S., and Mattschei, P. K., *Inorg. Chem.* **12**, 2472 (1973).
92. Mikhailov, B. M., and Potapova, T. V., *Izv. Akad. Nauk SSSR, Ser. Khim* **5**, 1153 (1968).
93. Mingos, D. M. P., *Nature (London), Phys. Sci.* **236**, 99 (1972).
94. More, E. B., Dickerson, R. E., and Lipscomb, W. N., *J. Chem. Phys.* **27**, 209 (1957).
95. Muetterties, E. L., and Knoth, W. H., "Polyhedral Boranes," Dekker, New York, 1968.
96. Muetterties, E. L., and Knoth, W. H., "Polyhedral Boranes," p. 68, Dekker, New York, 1968, see also Williams, *163*, Fig. 22.
97. Nordman, C. E., and Reimann, C., *J. Amer. Chem. Soc.* **81**, 3538 (1959).
98. Norman, A. D., and Schaeffer, R., *J. Amer. Chem. Soc.* **88**, 1143 (1966).
99. Olah, G. A., White, A. M., DeMember, J. R., Commeyras, A., and Lui, C. Y., *J. Amer. Chem. Soc.* **92**, 4627 (1970).
100. Onak, T. P., Drake, R. P., and Dunks, G. B., *Inorg. Chem.* **3**, 1686 (1964).
101. Onak, T. P., Drake, R. P., and Dunks, G. B., *J. Amer. Chem. Soc.* **87**, 2505 (1965).
102. Onak, T. P., and Dunks, G. B., *Inorg. Chem.* **5**, 439 (1966).
103. Onak, T. P., Gerhart, F. J., and Williams, R. E., *J. Amer. Chem. Soc.* **85**, 1754 (1963).
104. Onak, T. P., Gerhart, F. J., and Williams, R. E., *J. Amer. Chem. Soc.* **85**, 3378 (1963).
105. Onak, T. P., and Leach, J. B., *J. Amer. Chem. Soc.* **92**, 3513 (1970).
106. Onak, T. P., Lockman, B., and Haran, G., *J. Chem. Soc., Dalton Trans.* p. 2115 (1973).
107. Onak, T. P., Williams, R. E., and Weiss, H. G., *J. Amer. Chem. Soc.* **84**, 2830 (1962).
108. Onak, T. P., and Wong, G. T. F., *J. Amer. Chem. Soc.* **92**, 5226 (1970).
109. Paine, R. T., and Parry, R. W., *Inorg. Chem.* **11**, 268 (1972).
110. Parry, R. W., and Edwards, J. L., *J. Amer. Chem. Soc.* **81**, 3554 (1959).
111. Pasinski, J. P., and Beaudet, R. A., *Chem. Commun.* p. 928 (1973).
112. Pauling, L., *J. Inorg. Nucl. Chem.* **32**, 3745 (1970).
113. Plesek, J., and Hermanek, S., *Chem. Ind. (London)* p. 1267 (1971).
113a. Plesek, J., and Hermanek, S., *Chem. Ind. (London)* p. 381 (1973).
113b. Plesek, J., and Hermanek, S., *Pure Appl. Chem.* **39**, 431 (1974).
113c. Plesek, J., Hermanek, S., and Hanousek, F., *Collect. Czech. Chem. Commun.* **33**, 699 (1968).
113d. Plesek, J., Stibr, B., and Hermanek, S., *Chem. Ind. (London)* p. 649 (1972).
113e. Plesek, J., and Hermanek, S., *Chem. Ind. (London)* p. 890 (1972).
114. Popp, G., and Hawthorne, M. F., *J. Amer. Chem. Soc.* **90**, 6553 (1968).
115. Pretzer, W. R., and Rudolph, R. W., *J. Amer. Chem. Soc.* **95**, 931 (1973).
116. Prince, S. R., and Schaeffer, R., *Chem. Commun.* p. 451 (1968).
117. Reddy, J. M., and Lipscomb, W. N., *J. Chem. Phys.* **31**, 610 (1959).
117a. Remmel, R. J., Johnson II, H. D., Jaworiwsky, I. S., and Shore, S. G., *J. Amer. Chem. Soc.* **97**, 5395 (1975).
118. Rietz, R. R., and Schaeffer, R., *J. Amer. Chem. Soc.* **93**, 1263 (1971).

119. Rietz, R. R., and Schaeffer, R., *J. Amer. Chem. Soc.* **95**, 6254 (1973).
120. Rietz, R. R., Schaeffer, R., and Sneddon, L. G., *J. Amer. Chem. Soc.* **92**, 3514 (1970).
121. Rietz, R. R., Schaeffer, R., and Sneddon, L. G., *Inorg. Chem.* **11**, 1242 (1972).
122. Rietz, R. R., Siedle, A. R., Schaeffer, R. O., and Todd, L. J., *Inorg. Chem.* **12**, 2100 (1973).
123. Rudolph, R. W., and Pretzer, W. R., *Inorg. Chem.* **11**, 1974 (1972).
124. Schaeffer, R., and Tebbe, F., *J. Amer. Chem. Soc.* **85**, 2020 (1963).
125. Schaeffer, R., and Walter, E., *Inorg. Chem.* **12**, 2209 (1973).
126. Shapiro, I., Good, C. D., and Williams, R. E., *J. Amer. Chem. Soc.* **84**, 3837 (1962).
127. Shore, S. G., personal communication.
128. Siedle, A. R., Ph.D. Thesis, p. 61. Indiana University, Bloomington (1973); $B_9H_{12}^-$ data to be published with Garber, A. R., Bodner, G. M., and Todd, L. J.
129. Siedle, A. R., Bodner, G. M., and Todd, L. J., *J. Inorg. Nucl. Chem.* **33**, 3671 (1971).
130. Simpson, P. G., and Lipscomb, W. N., *J. Chem. Phys.* **35**, 1340 (1961).
131. Snaith, R., and Wade, K., See Knoth (*72*, Chapter 4, p. 139).
132. Sneddon, L. G., Huffman, J. C., Schaeffer, R. O., and Streib, W. E., *Chem. Commun.* p. 474 (1972).
132a. Stibr, B., Plesek, J., and Hermanek, S., *Chem. Ind. (London)* p. 649 (1972).
133. Stock, A. E., "Hydrides of Boron and Silicon," Cornell Univ. Press, Ithaca, New York, 1933.
134. Stohrer, W. D., and Hoffmann, R., *J. Amer. Chem. Soc.* **94**, 1661 (1972).
135. Tebbe, F. N., Garrett, P. M., and Hawthorne, M. F., *J. Amer. Chem. Soc.* **86**, 4222 (1964).
136. Tebbe, F. N., Garrett, P. M., and Hawthorne, M. F., *J. Amer. Chem. Soc.* **88**, 607 (1966).
137. Tebbe, F. N., Garrett, P. M., and Hawthorne, M. F., *J. Amer. Chem. Soc.* **90**, 869 (1968).
138. Tebbe, F. N., Garrett, P. M., Young, D. C., and Hawthorne, M. F., *J. Amer. Chem. Soc.* **88**, 609 (1966).
139. Thompson, M. L., and Grimes, R. N., *J. Amer. Chem. Soc.* **93**, 6677 (1971).
140. Todd, L. J., Hyatt, D. E., and Scholer, F. R., *153rd Meet., Amer. Chem. Soc.* Paper 57 (1967).
141. Tolpin, E. I., and Lipscomb, W. N., *Chem. Commun.* p. 257 (1973).
142. Tolpin, E. I., and Lipscomb, W. N., *J. Amer. Chem. Soc.* **95**, 2384 (1973).
143. Travers, N. F., see Lappert and Emeleus (*78*, Chapter 3, p. 79).
144. Travers, N. F., see Lappert and Emeleus (*78*, Chapter 3, p. 109).
145. Tsai, C., and Streib, W. E., *J. Amer. Chem. Soc.* **88**, 4513 (1966).
146. Voorhees, R. L., and Rudolph, R. W., *J. Amer. Chem. Soc.* **91**, 2173 (1969).
147. Wade, K., *Chem. Commun.* p. 792 (1971).
147a. Wade, K., "Electron Deficient Compounds," Nelson, London, 1971.
148. Wade, K., *Inorg. Nucl. Chem. Lett.* **8**, 559 (1972).
149. Wade, K., *Inorg. Nucl. Chem. Lett.* **8**, 563 (1972).
150. Wade, K., *Inorg. Nucl. Chem. Lett.* **8**, 823 (1972).
151. Wade, K., *Nature (London) Phys. Sci.* **240**, 71 (1972).
152. Wade, K., *New Sci.* **62**, 615 (1974).

153. Wang, F. E., Simpson, P. G., and Lipscomb, W. N., *J. Chem. Phys.* **35**, 1335 (1961).
154. Wells, A. F., "Structural Inorganic Chemistry," 3rd ed., p. 100 Oxford Univ. Press (Clarendon), London and New York 1962.
155. Wiersema, R. J., and Hawthorne, M. F., *Inorg. Chem.* **12**, 785 (1973).
156. Williams, J. M., and Shneemeyer, L. F., *Chem. Eng. News* **51**, No. 39, p. 20 (1973).
157. Williams, R. E., *J. Inorg. Nucl. Chem.* **20**, 198 (1961).
158. Williams, R. E., *in* "Progress in Boron Chemistry" (R. J. Brotherton and H. Steinberg, eds.), Vol. 2, Chapter 2, p. 37. Pergamon, Oxford, 1970.
159. Williams, R. E., *in* "Progress in Boron Chemistry" (R. J. Brotherton and H. Steinberg, eds.), Vol. 2, p. 45. Pergamon, Oxford, 1970.
160. Williams, R. E., *in* "Progress in Boron Chemistry" (R. J. Brotherton and H. Steinberg, eds.), Vol. 2, pp. 48–49. Pergamon, Oxford, 1970.
161. Williams, R. E., *in* "Progress in Boron Chemistry" (R. J. Brotherton and H. Steinberg, eds.), Vol. 2, p. 60. Pergamon, Oxford, 1970.
162. Williams, R. E., *in* "Progress in Boron Chemistry" (R. J. Brotherton and H. Steinberg, eds.), Vol. 2, p. 61. Pergamon, Oxford, 1970.
163. Williams, R. E., *in* "Progress in Boron Chemistry" (R. J. Brotherton and H. Steinberg, eds.), Vol. 2, p. 67, Fig. 22. Pergamon, Oxford, 1970.
164. Williams, R. E., *Inorg. Chem.* **10**, 210 (1971).
165. Williams, R. E., *Inorg. Chem.* **10**, 213 (1971).
166. Williams, R. E., *Inorg. Chem.* **10**, 214 (1971).
167. Williams, R. E., *Inorg. Chem.* **10**, p. 212, footnote 7 (1971).
168. Williams, R. E., *Inorg. Chem.* **10**, p. 213, footnote 9b (1971).
169. Williams, R. E., *Pure Appl. Chem.* **29**, 569 (1972), and references therein.
170. Williams, R. E., unpublished information.
171. Williams, R. E., and Ditter, J. F. *156th Nat. Meet., Amer. Chem. Soc.* INOR 130 (1968); see also Tech. Rep. No. 17, Office of Navel Research Contract Nonr 4381(00). ONR, Washington, D.C., 1969.
172. Williams, R. E., and Gerhart, F. J., *J. Amer. Chem. Soc.* **87**, 3513 (1965).
173. Williams, R. E., Gerhart, F. J., and Pier, E., *Inorg. Chem.* **4**, 1239 (1965).
174. Williams, R. E., Gibbins, S. G., and Shapiro, I., *J. Chem. Phys.* **30**, 320 (1959).
175. Williams, R. E., Good, C. D., and Shapiro, I. *140th Meet., Amer. Chem. Soc.* Paper 14N, p. 36 (1961).
176. Wing, R. M., *J. Amer. Chem. Soc.* **89**, 5599 (1967).
177. Young, D. A. T., Wiley, G. R., Hawthorne, M. F., Churchill, M. R., and Reis, A. H., *J. Amer. Chem. Soc.* **92**, 6663 (1970).
178. Zakharkin, L., and Kalinin, V., *Izv. Akad. Nauk SSSR, Ser. Khim.* p. 194 (1967).
179. Zakharkin, L. I., and Kalinin, V. N., *Izv. Akad. Nauk SSSR, Ser. Khim.* **10**, 2310 (1967).
180. Zakharkin, L. I., Kalinin, V. N., Kvasov, B. A., and Synakin, A. P., *Zh. Obshch, Khim.* **41**, 1726 (1971).
181. Zakharkin, L. I., Kalinin, V. N., and Podvistotskaya L., *Izv. Akad. Nauk SSSR, Ser. Khim.* p. 2310 (1967).

PREPARATION AND REACTIONS OF PERFLUOROHALOGENOORGANOSULFENYL HALIDES*

A. HAAS and U. NIEMANN

Chair of Inorganic Chemistry II, Ruhr-University, Bochum, Germany

I. Introduction	. . . .	143
II. Perfluorohalogenoorganosulfenyl Fluorides	. . . . .	144
A. Preparation	. . . .	144
B. Properties	. . . .	145
III. Perfluorohalogenoorganosulfenyl Chlorides	. . . . .	146
A. Preparation	. . . .	146
B. Properties	. . . .	154
IV. Perfluorohalogenoorganosulfenyl Bromides	. . . . .	155
V. Reactions of Perfluorohalogenoorganosulfenyl Halides and Related Reactions	. . . .	157
A. With Pseudohalides	. . . .	157
B. With Silver Perfluorohalogenocarboxylates	. . . .	163
C. With Perfluorohalogenothioketones	. . . . . .	165
D. With Ammonia, Primary and Secondary Amines, and Amides	.	167
E. With Arsines, Alcohols, Thioalcohols, Sulfinates, and Carbonyl Compounds	. . . .	172
F. With Alkanes, Alkenes, Alkynes, and Nitriles	. . . .	175
G. With Aromatics and Heteroaromatics	. . . . .	177
H. Cyclizations, Conversions and Reactions of Sulfenyl Chlorides with Metal Carbonyls	. . . .	188
VI. Characteristics of Perfluorohalogenoorganomercapto Groups	. .	189
References .	. . . .	190

I. Introduction

Perfluorohalogenoorganosulfenyl halides are, like thiols, mercaptides, and thioketones, key compounds for the synthesis of new perfluorohalogenoorganomercapto derivatives. The high reactivity of the S—X bond (X = Cl, Br) toward electrophilic as well as nucleophilic reagents causes them to be highly valued starting materials for the synthesis of new derivatives. The chemistry of the not very stable sulfenyl fluorides has

* English translation by Dr. N. Welcman, Department of Chemistry, Technion-Israel, Institute of Technology, Haifa, Israel.

hardly been investigated. However, the easily accessible and relatively stable sulfenyl chlorides have been studied extensively (42, 105, 106, 109).

Particularly worth mentioning is the application of sulfenyl compounds in plant preservation. Of primary importance in this respect are the reactions of sulfenyl chlorides with secondary amines, which, with proper choice of the starting materials, lead to compounds with good fungicidal properties. The reactions of $(CH_3)_2NSO_2N(C_6H_5)H$, $(CH_3)_2NSO_2N(p\text{-}CH_3\text{---}C_6H_4)NH$ (107), phthalimide (10, 108), and other secondary amines with the sulfenyl chlorides, $CFCl_2SCl$ (107) and $(CF_3S)_2\,CClSCl$ (108), furnish sulfenyl amides of great effectiveness.

II. Perfluorohalogenoorganosulfenyl Fluorides

A. Preparation

Evidence for the existence of this type of compound appeared only in 1967. Seel, Gombler, and Budenz (145) reacted trihalomethanesulfenyl chlorides with active potassium fluoride (prepared through degradation of potassium fluorosulfinate) at 150°C in the gas phase:

$$CF_nCl_{3-n}SCl + KF \longrightarrow CF_nCl_{3-n}SF \quad (n = 0, 1, 2, 3)$$

The products were identified by ^{19}F NMR spectroscopy. The chemical shifts δ (using $CFCl_3$ as an external standard) and spin-spin coupling constants J of liquid fluorides are listed in Table I. Compound CF_3SF

TABLE I

Chemical Shifts and Coupling Constants for Liquid Sulfenyl Fluorides

Compound	$\delta(R_f)$ (ppm)	$\delta(SF)$ (ppm)	J_{F-F}(Hz)
CCl_3SF	—	249	—
$CFCl_2SF$	31 (Doublet)	265 (Doublet)	4.85
CF_2ClSF	45 (Doublet)	297 (Triplet)	6.85
CF_3SF	58 (Doublet)	351 (Quartet)	27

could additionally be characterized by means of its infrared [$\nu(S\text{---}F) = 808\ cm^{-1}$] and mass spectra (among other fragments: CF_3SF^+, CF_3^+, SF^+) (144). The liquid product of the reaction between CCl_3SCl and HgF_2 or AgF, formulated in a previous publication (102) as CCl_3SF, proved later to be the isomeric $CFCl_2SCl$ (73, 99, 150).

The best method for the preparation of CF_3SF is the reaction of CF_3SCl with HgF_2 in a nickel or platinum apparatus at 130°C, followed by removal of the reaction products by condensation in liquid nitrogen (*144*). Analogous reactions with metal fluorides lead to other sulfenyl fluorides (*102, 189*):

$$CF_3CF_2CF_2SCl + AgF \xrightarrow[\text{autoclave}]{125°-160°C/6 \text{ hr}} CF_3CF_2CF_2SF$$

$$NF_2CCl_2SCl + AgF_2 \xrightarrow[\substack{\text{Monel metal} \\ \text{autoclave}}]{25°C/6-12 \text{ hr}} NF_2CCl_2SF$$

The compound isomeric with perfluoropropane-1-sulfenyl fluoride is formed by the pyrolysis of $[(CF_3)_2CF]_2SF_2$ (*136*):

$$(CF_3)_2CF\text{—}SF_2\text{—}CF(CF_3)_2 \xrightarrow{200°C} (CF_3)_2CFSF + SF_4 + (CF_3)_2CFCF(CF_3)_2$$

Difluorodifluoroaminomethanesulfenyl fluoride, NF_2CF_2SF, is formed among other products by the fluorination of AgSCN or KSCN (*32*). Perfluorooctanesulfenyl fluoride, $CF_3(CF_2)_7SF$, is obtained by electrolytic fluorination of the appropriate thiol (*152*).

A further reaction of general importance, the cleavage of disulfanes by means of a fluorinating agent, provides in the case of perfluoroaryl disulfanes the sulfenyl fluorides in good yield (*33*):

$$(4\text{-}R_fC_6F_4S)_2 + AgF_2 \xrightarrow[CFCl_2CF_2Cl]{20°C/10 \text{ hr}} 4\text{-}R_fC_6F_4SF \quad (R_f = F, CF_3)$$

B. Properties

All compounds, except the slightly green $(CF_3)_2CFSF$, are colorless liquids or colorless gases above their boiling points. The sulfenyl fluorides, $CF_nCl_{3-n}SF$ ($n = 0, 1, 2, 3$), are stable for some time in the liquid phase at $-50°C$ and in the gaseous state at low pressures (~10 torr, 20°C).

However, even at room temperature a fast isomerization takes place to sulfenyl chlorides containing the appropriately fluorinated trihalomethane:

$$CF_nCl_{3-n}SF \longrightarrow CF_{n+1}Cl_{2-n}SCl \quad (n = 0, 1, 2)$$

The speed of the fluorine–chlorine exchange at the sulfur atom decreases with increasing degree of fluorination at the carbon atom (*145*). Compound CF_3SF does not undergo exchange reactions at all; however, in the liquid phase it exists in equilibrium with the dimeric compound $CF_3SF_2SCF_3$. Both the monomer and the dimer are converted in a matter of a few hours into CF_3SSCF_3 and CF_3SF_3, particularly in the presence of potassium fluoride (*144*):

$$3CF_3SF \longrightarrow CF_3S(F_2)SCF_3 + CF_3SF \longrightarrow CF_3SSCF_3 + CF_3SF_3$$

The CF_3SF reacts with both base (e.g., Mg) and noble metals (e.g., Cu, Hg) to give the metal fluoride and CF_3SSCF_3. Only pure nickel made inert by treatment with SF_4 is somewhat stable toward CF_3SF (*144*).

The dihalodifluoroaminomethanesulfenyl fluorides, NF_2CX_2SF (X = Cl, F), are relatively stable substances. The dichloro compound is stable at 20°C toward isomerization and decomposition in metal containers; it reacts with oxygen to give $NF_2CCl_2S(O)F$ (*189*). On heating to 100°C in Pyrex containers, it decomposes to SO_2, $FN{=}CCl_2$, SiF_4 and minor quantities of $NF_2CCl_2S(O)F$. The difluoro compound hydrolyzes in water and bases and also shows oxidizing properties; in alcoholic solution iodine is liberated from an aqueous potassium iodide solution, sulfur being deposited at the same time (*32*).

III. Perfluorohalogenoorganosulfenyl Chlorides

A. Preparation

This class of compound has experienced considerable growth of interest in recent years through numerous new publications. Several common approaches for the preparation are discussed in the following.

1. *Fluorination with Metal Fluorides or HF*

The reaction of CCl_3SCl with an alkali metal fluoride (NaF) in a high-boiling polar solvent, e.g., tetramethylenesulfone or acetonitrile, leads in good yield to the important trifluoromethanesulfenyl chloride, CF_3SCl (*158, 159*):

$$CCl_3SCl + NaF \xrightarrow{170°-250°C} CF_3SCl \text{ (47\% yield)}$$

Among the by-products obtained in this reaction are minor quantities of CF_2ClSCl as well as the compounds CF_3SSCF_3 and $CF_3S(O)F$ (resulting from the reaction of CF_3SF_3 and SiO_2); this can be explained only by the intermediate formation of the acid fluoride, CF_3SF (*43*). This compound disproportionates into CF_3SF_3 and CF_3SSCF_3:

$$3CF_3SF \longrightarrow CF_3SF_3 + CF_3SSCF_3$$

This indirect proof of the appearance of CF_3SF leads to the conclusion that fluorination of sulfenyl chlorides of the series $CF_nCl_{3-n}SCl$ (n = 0, 1, 2) with alkali metal fluorides follows the mechanism observed in the formation of sulfenyl fluorides: the initial chlorine–fluorine exchange at the sulfur atom is followed by isomerization to the sulfenyl chloride containing the corresponding more highly fluorinated methyl group.

In accord with this mechanism, trichloromethanethiosulfenyl chloride, CCl_3SSCl, cannot be fluorinated under analogous reaction conditions (44). Plainly, the isomerization is here impossible on steric grounds because of the additional sulfur atom.

Further, the metal fluorides SbF_3 (with admixture of small quantities of $SbCl_5$), HgF_2, and AgF are also found suitable for the fluorination of sulfenyl halides, e.g.,

$$CF_2ClSCl + SbF_3 \longrightarrow CF_3SCl \text{ (11\% yield)} \quad (187)$$

$$CCl_3SCl, CFCl_2SCl + SbF_3/SbCl_5 \xrightarrow{150°C} CF_2ClSCl \quad (43)$$

$$CCl_3SCl + HgF_2 \text{ or } AgF \longrightarrow CFCl_2SCl \quad (150)$$

The bifunctional chlorocarbonylsulfenyl chloride, obtainable through partial hydrolysis of trichloromethanesulfenyl chloride (162), is fluorinated by SbF_3 at the carbonyl group (66, 67):

$$CCl_3SCl \xrightarrow[-2HCl]{H_2O/H_2SO_4} Cl-\overset{\overset{O}{\|}}{C}-SCl + SbF_3 \xrightarrow{85°-95°C} F-\overset{\overset{O}{\|}}{C}-SCl$$

On fluorinating $CCl_3SN(C_2H_5)_2$ (prepared from CCl_3SCl and diethylamine) (185), a mixture of the mono- and difluoro compounds is obtained. The following cleavage with HCl provides the corresponding sulfenyl chlorides (128, 185):

$$CCl_3SN(C_2H_5)_2 + SbF_3 \longrightarrow CF_nCl_{3-n}SN(C_2H_5)_2 \xrightarrow{HCl} CF_nCl_{3-n}SCl \; (n = 1, 2)$$

In a direct reaction of the amide with hydrogen fluoride only $CFCl_2SCl$ is obtained (128). This points again to an isomerization of the CCl_3SF formed initially. The reaction between CCl_3SCl and an excess of anhydrous hydrogen fluoride leads to the same product (128). This reaction is also feasible on a technical scale if the reagents are vigorously mixed and pressure as well as high temperature are applied. Under such conditions small quantities of CF_2ClSCl are also formed (109). According to a patented procedure (141), gas-phase fluorination with hydrogen fluoride at 180°C in the presence of a chromium oxide–fluoride catalyst [prepared by fluorination of chromium(III) hydroxide] furnishes CF_3SCl in 74% yield.

2. *Addition of Chlorine or Chlorine Monofluoride to the C=S Double Bond of Perhalogenothiocarbonyl Compounds*

An important and widely applicable method for the synthesis of perfluorohalogenosulfenyl halides is based on the ability of the halogens

(Cl_2, Br_2) and interhalogen compounds (ClF) to add to C=S double bonds.

By this method of addition of chlorine to fluorothiocarbonyl chloride, thiocarbonyl difluoride (109), or hexafluorothioacetone (120), the respective sulfenyl chlorides are obtained:

$$\underset{F}{\overset{X}{>}}C=S + Cl_2 \longrightarrow F-\underset{Cl}{\overset{X}{\underset{|}{C}}}-SCl \quad (X = Cl, F)$$

$$\underset{CF_3}{\overset{CF_3}{>}}C=S + Cl_2 \longrightarrow \underset{CF_3}{\overset{CF_3}{>}}\underset{}{\overset{Cl}{\underset{|}{C}}}-SCl$$

Analogously, it is possible to react the linear compounds $CF_3SC(X)S$ (X = F, SCF_3), obtainable through catalytic dimerization or trimerization, in order to produce new sulfenyl chlorides (51, 53):

$$\underset{F}{\overset{F}{>}}C=S \begin{cases} \xrightarrow[-40°C]{KF} CF_3S-\overset{F}{\underset{|}{C}}=S \xrightarrow[-78°C]{Cl_2} CF_3S-\overset{F}{\underset{\underset{Cl}{|}}{C}}-SCl \\ \xrightarrow[-78°C]{CsF} \underset{CF_3S}{\overset{CF_3S}{>}}C=S \xrightarrow[-78°C]{Cl_2} \underset{CF_3S}{\overset{CF_3S}{>}}C\underset{Cl}{\overset{SCl}{<}} \end{cases}$$

Trifluoromethylthiocarbonyl fluoride can be converted to the chloride by halogen exchange with aluminum trichloride. By chlorine addition this compound furnishes another sulfenyl chloride (55):

$$CF_3S-\overset{F}{\underset{|}{C}}=S \xrightarrow{AlCl_3} CF_3S-\overset{Cl}{\underset{|}{C}}=S \xrightarrow{Cl_2} CF_3S-\overset{Cl}{\underset{\underset{Cl}{|}}{C}}-SCl$$

It is worth mentioning that fluorothiocarbonyl isothiocyanate, produced by the reaction of CSFCl with metal thiocyanates (52), combines with chlorine quantitatively at low temperatures at the C=S double bond, without suffering an attack on the isothiocyanate group (21):

$$F-\overset{S}{\underset{}{\overset{\|}{C}}}-Cl + MSCN \xrightarrow{-MCl} F-\overset{S}{\underset{}{\overset{\|}{C}}}-NCS \quad (M = Ag, K, NH_4)$$

$$F-\overset{S}{\underset{}{\overset{\|}{C}}}-NCS \xrightarrow[-78°C]{Cl_2} F-\overset{SCl}{\underset{\underset{Cl}{|}}{\underset{|}{C}}}-NCS$$

Only at elevated temperatures is the isothiocyanate group also attacked by further addition of chlorine with simultaneous elimination of sulfur dichloride. It is thus converted into the isocyanide dichloride group, without isolation of an intermediate:

$$\text{F}-\underset{\underset{\text{Cl}}{|}}{\overset{\overset{\text{SCl}}{|}}{\text{C}}}-\text{NCS} + \text{Cl}_2 \xrightarrow{40°C} \text{FCl}_2\text{C}-\text{NCS} + \text{SCl}_2$$

$$\text{FCl}_2\text{C}-\text{NCS} + \text{Cl}_2 \xrightarrow{70°C} \text{FCl}_2\text{C}-\text{N}=\text{CCl}_2 + \text{SCl}_2$$

The more highly fluorinated isothiocyanates, $\text{F}_2\text{ClC}-\text{NCS}$ and $\text{F}_3\text{C}-\text{NCS}$, even at 80°C with chlorine yield only the iminochloromethanesulfenyl chlorides (21):

$$\text{ClF}_2\text{C}-\text{NCS} + \text{Cl}_2 \longrightarrow \text{ClF}_2\text{C}-\text{N}=\overset{\overset{\text{SCl}}{|}}{\text{C}}\text{Cl}$$

$$\text{F}_3\text{C}-\text{NCS} + \text{Cl}_2 \longrightarrow \text{F}_3\text{C}-\text{N}=\overset{\overset{\text{SCl}}{|}}{\text{C}}\text{Cl}$$

It may be assumed, therefore, that highly electronegative groups, such as CF_2Cl or CF_3, stabilize the imino compounds. Similar observations were made with aliphatic and aromatic iminochloromethanesulfenyl chlorides; the former are very unstable, but the latter are stable to some degree (190).

In the presence of catalytic quantities of iodine, chlorination again proceeds further to the isocyanide dichlorides (21):

$$\text{ClF}_2\text{C}-\text{N}=\overset{\overset{\text{SCl}}{|}}{\text{C}}\text{Cl} + \text{Cl}_2 \xrightarrow{\text{I}_2} \text{ClF}_2\text{C}-\text{N}=\text{CCl}_2 + \text{SCl}_2$$

$$\text{F}_3\text{C}-\text{N}=\overset{\overset{\text{SCl}}{|}}{\text{C}}\text{Cl} + \text{Cl}_2 \xrightarrow{\text{I}_2} \text{F}_3\text{C}-\text{N}=\text{CCl}_2 + \text{SCl}_2$$

Chlorine monofluoride combines smoothly with thiocarbonyl compounds. For instance, the sulfenyl chlorides of the series $\text{CF}_n\text{Cl}_{3-n}$-SCl ($n = 1, 2, 3$), which were already prepared by other methods, are obtained in good yield:

$$\text{Cl}_n\text{F}_{2-n}\text{C}=\text{S} + \text{ClF} \longrightarrow \text{Cl}_n\text{F}_{3-n}\text{CSCl} \quad (n = 0, 1, 2) \; (20)$$

Addition of ClF to the dimeric and trimeric thiocarbonyl difluoride furnishes sulfenyl chlorides which were not accessible for a long time (20):

$$CF_3S-\underset{F}{C}=S + ClF \longrightarrow CF_3S-CF_2-SCl$$

$$\underset{CF_3S}{\overset{CF_3S}{>}}C=S + ClF \longrightarrow \underset{CF_3S}{\overset{CF_3S}{>}}C\underset{F}{\overset{SCl}{<}}$$

In the latter reaction, rather large quantities of the disulfane, $(CF_3S)_2$-$CFSSCF_3$, are formed, as well as the decomposition products $CF_3SC(S)F$, CF_3SCl, and CF_3SCF_2SCl. Formation of these compounds is probably due to decomposition of the $(CF_3S)_2CFSCl$ initially formed to $CF_3SC(S)F$ and to CF_3SCl and the subsequent reactions with excess of ClF:

$$\underset{CF_3S}{\overset{CF_3S}{>}}C\underset{F}{\overset{SCl}{<}} \longrightarrow CF_3S-\underset{F}{C}=S + CF_3SCl$$

$$\downarrow +ClF \qquad\qquad \downarrow +ClF$$

$$CF_3S-CF_2-SCl \qquad CF_3SF$$

The relatively stable sulfenyl fluoride eventually combines in a competing reaction with unreacted $(CF_3S)_2C=S$ to give $(CF_3S)_2C(F)SSCF_3$.

3. Chlorolysis of Perfluorohalogenoorganodisulfanes

Chlorolysis of the corresponding disulfanes is a favorable procedure for the preparation of aliphatic, aromatic, and heterocyclic sulfenyl chlorides under not too demanding conditions [low temperature; mild chlorinating agents, such as SO_2Cl_2 (14) or CH_3SCl_3 (15)]. Perfluorinated sulfenyl chlorides can also be prepared by this procedure in special cases: a mixture of CF_3SSCF_3 and chlorine reacts in a Pyrex Carius tube under UV irradiation to form sulfenyl chloride in an equilibrium reaction (84):

$$CF_3SSCF_3 + Cl_2 \underset{}{\overset{h\nu}{\rightleftarrows}} 2CF_3SCl \quad \text{(about 50\% yield)}$$

The chlorination of $C_6F_5SSC_6F_5$ in an inert solvent furnishes the perfluorobenzene sulfenyl chloride (137):

$$C_6F_5SSC_6F_5 + Cl_2 \xrightarrow[CCl_4]{20°C/2\ hr} 2C_6F_5SCl$$

By passing an HCl-free stream of chlorine into a cooled solution of bis(2,3,5,6-tetrafluoropyridyl)disulfane, the sulfenyl chloride is obtained in 73% yield (6):

[Structural reaction: perfluoropyridyl disulfide + Cl₂ → perfluoropyridyl-SCl]

$$\text{(F}_4\text{C}_5\text{N)-SS-(C}_5\text{NF}_4\text{)} + \text{Cl}_2 \xrightarrow[\text{CCl}_4]{0°\text{C}/2\text{hr}} 2\,(\text{F}_4\text{C}_5\text{N})\text{-SCl}$$

By the use of higher temperatures as well as by partial UV irradiation, it is possible to cleave the longer-chain disulfanes, e.g.,

$$\text{CF}_3\text{CF}_2\text{CF}_2\text{SSCF}_2\text{CF}_2\text{CF}_3 + \text{Cl}_2 \xrightarrow[25°-130°\text{C}]{h\nu} 2\,\text{CF}_3\text{CF}_2\text{CF}_2\text{SCl} \quad (102)$$

$$\text{CF}_2\text{ClCF}_2\text{SSCF}_2\text{CF}_2\text{Cl} + \text{Cl}_2 \xrightarrow{80°-90°\text{C}} 2\,\text{CF}_2\text{ClCF}_2\text{SCl} \quad (100)$$

$$(\text{CF}_3)_2\text{CFSSCF}(\text{CF}_3)_2 + \text{Cl}_2 \xrightarrow[\text{reflux}]{h\nu} 2\,(\text{CF}_3)_2\text{CFSCl} \quad (3)$$

$$(\text{CF}_2\text{ClCCl}_2\text{CF}_2\text{CFCl})_2\text{S}_2 + \text{Cl}_2 \xrightarrow{150°-200°\text{C}} 2\,\text{CF}_2\text{ClCCl}_2\text{CF}_2\text{CFClSCl} \quad (88)$$

$$[(\text{CF}_3)_2\text{CF}(\text{CF}_2)_4]_2\text{S}_2 + \text{Cl}_2 \xrightarrow{150°\text{C}} 2\,(\text{CF}_3)_2\text{CF}(\text{CF}_2)_4\text{SCl} \quad (74)$$

$$(\text{CF}_2\text{ClCFCl})_2\text{S}_x + \text{Cl}_2 \xrightarrow{120°\text{C}/7\text{ hr}} 2\,\text{CF}_2\text{ClCFClSCl} \quad (x = 2, 3) \quad (5)$$

On heating the sulfane mixture $[\text{C}_4\text{F}_9\text{CF}(\text{CF}_3)_2]_2\text{S}_n$ (average sulfur chain length $n = 2.5$) with chlorine for 125 hr at 105°C, the perfluorohexane-2-sulfenyl chloride is formed in appreciable yield (86, 127). The patent literature (5, 19, 87–89, 127) abounds with long-chain sulfenyl chlorides prepared according to this latter procedure, but physical data are lacking.

Cyclic disulfanes also undergo cleavage reactions. 5,5-Difluoro-3-chloro-1,2,4-dithiazole reacts with chlorine to give various sulfenyl chlorides depending on temperature:

$$\text{F}_2\text{C}(\text{S-S})\text{C}(\text{Cl})\text{=N} + \text{Cl}_2 \xrightarrow{-50°\text{C}/2\text{hr}} \text{CF}_2\text{ClN=C(Cl)-SSCl} \quad (45)$$

$$+ \text{Cl}_2 \xrightarrow{50°\text{C}/12\text{hr}} \text{CF}_2\text{ClN=C(Cl)-SCl} \quad (21)$$

Bifunctional sulfenyl chlorides are obtained in the reaction of 5,5,6,6-tetrafluoro-1,2,3,4-tetrathiane or perfluoro-1,2,5-trithiepane (104):

$$\text{F}_2\underset{\text{S}}{\overset{\text{S}}{\bigcirc}}\text{S} + \text{Cl}_2 \xrightarrow{\text{CHCl}_3} \text{ClSCF}_2\text{CF}_2\text{SCl}$$

$$\text{F}_2\underset{\text{S}}{\overset{\text{S}}{\bigcirc}}\text{S} + \text{Cl}_2 \xrightarrow[\text{HCl}]{0°\text{C}} \text{ClSSCF}_2\text{CF}_2\text{SSCl}$$

$$\text{F}_2\underset{\text{S—S}}{\overset{\text{S}}{\bigcirc}}\overset{\text{F}_2}{\text{F}_2} + \text{Cl}_2 \xrightarrow[\text{17 days}]{h\nu} \text{ClSCF}_2\text{CF}_2\text{SCF}_2\text{CF}_2\text{SCl}$$

Cleavage with chlorine can also lead to the formation of fragments of different size, e.g.,

$$\text{CF}_3\text{SN}{=}\overset{\overset{\text{Cl}}{|}}{\text{C}}{-}\text{SSCF}_3 + \text{Cl}_2 \xrightarrow{100°\text{C}/4.5\text{ hr}} \text{CF}_3\text{SN}{=}\overset{\overset{\text{Cl}}{|}}{\text{C}}{-}\text{SCl} + \text{CF}_3\text{SCl} \quad (35,\ 36)$$

To some extent this reaction is followed by one yielding an isocyanide dichloride:

$$\text{CF}_3\text{SN}{=}\overset{\overset{\text{Cl}}{|}}{\text{C}}\text{SCl} + \text{Cl}_2 \longrightarrow \text{CF}_3\text{SN}{=}\text{CCl}_2 + \text{SCl}_2$$

4. Reactions of Perfluorohalogenoorganothiols or -mercaptides with Chlorine

Chlorination of thiols serves for the preparation of temperature-sensitive sulfenyl halides, since the main reaction products are easily separated from the side products by fractional condensation, e.g.,

$$\text{C}_6\text{F}_5\text{SH} + \text{Cl}_2 \xrightarrow[\text{CCl}_4]{-10°\text{C}} \text{C}_6\text{F}_5\text{SCl} + \text{HCl} \quad (124,\ 137)$$

$$\text{CF}_3\overset{\overset{\text{O}}{\|}}{\text{C}}{-}\text{SH} + \text{Cl}_2 \xrightarrow{-78°\text{C}} \text{CF}_3\overset{\overset{\text{O}}{\|}}{\text{C}}{-}\text{SCl} + \text{HCl} \quad (135)$$

The reactions take place in several steps (115). First, the thiol is chlorinated to the appropriate sulfenyl chloride; then the latter reacts further with additional thiol to give the disulfane; finally, chlorolysis of the disulfane takes place.

A similar reaction occurs when heavy metal mercaptides are cleaved with chlorine, e.g.,

$$\left.\begin{array}{c}\text{AgSCF}_3\\ \\ \text{Hg(SCF}_3)_2\end{array}\right\} + 2\text{Cl}_2 \xrightarrow{-22°\text{C}} 2\text{CF}_3\text{SCl} + \left\{\begin{array}{c}\text{AgCl}\\ \\ \text{HgCl}_2\end{array}\right. \quad (29,\ 84)$$

$$\left.\begin{array}{c}\text{Pb(SC}_6\text{F}_5)_2\\ \\ \text{Hg(SC}_6\text{F}_5)_2\end{array}\right\} + 2\text{Cl}_2 \xrightarrow[\text{CCl}_4]{0°\text{C}} 2\text{C}_6\text{F}_5\text{SCl} + \left\{\begin{array}{c}\text{PbCl}_2\\ \\ \text{HgCl}_2\end{array}\right. \quad (124,\ 137)$$

5. Photolytically Initiated Addition of Sulfur Chlorides to Perfluorohalogenoolefins

Addition of sulfur chlorides to perfluoroolefins furnishes a series of interesting sulfenyl chlorides (100):

$$CF_2{=}CF_2 + SCl_2 \xrightarrow{100°-110°C/5\ hr} CF_2ClCF_2SCl + CF_2ClCF_2SSCl$$

$$CF_2{=}CF_2 + S_2Cl_2 \xrightarrow{100°-120°C/6\ hr}$$

$$CF_2ClCF_2SCl + (ClCF_2CF_2)_2S + (ClCF_2CF_2S)_2 + (ClCF_2CF_2S)_2S$$

The reaction with SCl_2 can be effected advantageously under UV irradiation in the presence of phosphorus trichloride (127), e.g.,

$$CF_3CF{=}CF_2 + SCl_2 \xrightarrow[5\%\ PCl_3]{h\nu/45°-50°C} CF_3CFClCF_2SCl\ \text{or}\ CF_3C(CF_2Cl)FSCl$$

$$CFCl{=}CF_2 + SCl_2 \xrightarrow[5\%\ PCl_3]{h\nu/45-50°C} CF_2ClCFClSCl\ \text{or}\ CFCl_2CF_2SCl$$

The mixture of isomers formed in any of these reactions could not be separated.

6. Other Reactions

Photolysis of a mixture of N_2F_4 and thiophosgene results in a remarkable addition; from a mixture of several compounds dichlorodifluoroaminomethanesulfenyl chloride can be separated (188):

$$\begin{array}{c}F\\F\end{array}\!\!>\!N{-}N\!<\!\!\begin{array}{c}F\\F\end{array} + \begin{array}{c}Cl\\Cl\end{array}\!\!>\!C{=}S \xrightarrow{h\nu} \begin{array}{c}F\\F\end{array}\!\!>\!N{-}\underset{\underset{Cl}{|}}{\overset{\overset{Cl}{|}}{C}}{-}SCl$$

Finally, it should be mentioned that in the meantime the series of the chlorodithioperfluorohalogenomethanes, $CF_nCl_{3-n}SSCl$ ($n = 1, 2, 3$), has been completed, although by very different means: $CFCl_2SSCl$, together with $CFCl_2SCl$ and S_2Cl_2, can be obtained by chlorination of bis(fluorodichloro)methane polysulfides, $CFCl_2S_nCFCl_2$ ($n > 2$) (119). The polysulfides are produced by warming $CFCl_2SCl$ with sulfur, or, alternatively, as side products in the industrial synthesis of $CFCl_2SCl$ from CCl_3SCl and HF (109). Compound $CF_2ClSSCl$ is formed in an attempt to fluorinate the isocyanide dichloride $F_2ClCSSN{=}CCl_2$; the isocyanide dichloride is an addition product of thiocarbonyl difluoride and $ClSN{=}CCl_2$ (35, 36):

$$F_2C{=}S + ClSN{=}CCl_2 \xrightarrow[30\ hr]{h\nu} F_2ClCSSN{=}CCl_2 \xrightarrow[110°C]{SbF_3/TMS} CF_2ClSSCl + ClCN$$

The mechanism of this reaction is unknown. Compound CF_3SSCl can be prepared in 93% yield by splitting the disulfanes, R_2NSSCF_3 ($R = CH_3$, C_2H_5), that are obtained from N,N-dialkylaminosulfenyl chloride and $Hg(SCF_3)_2$, with hydrogen chloride (16):

$$R_2NSSCF_3 + 2HCl \xrightarrow{20°C} CF_3SSCl + R_2NH \cdot HCl$$

B. PROPERTIES

The sulfenyl chlorides are yellow or slightly yellow colored liquids, with the exception of C_3F_7SCl, which is orange (102) and $ClSCF_2CF_2SCF_2CF_2SCl$, which is colorless (104). They are air-stable at room temperature. The photolysis by irradiation for 14 days of CF_3SCl in a quartz tube leads to the formation of sulfur, CF_3Cl, S_2Cl_2, SCl_2, and CF_3SSCF_3 (84). These compounds are water-stable for several hours at room temperature, but they are decomposed by bases (85). They are freely soluble in most organic solvents, but immiscible with water.

Hydrolysis, which has been extensively studied in the case of CF_3SCl (85), is more complicated than implied by the equation

$$CF_3SCl + H_2O \longrightarrow CF_3SOH + HCl$$

On shaking CF_3SCl with an excess of water for 3 hr at 20°C, the yellow coloration disappears. After additional shaking for 9 hr, 55–60% CF_3SSCF_3 and COS are obtained as well as CF_3SO_2H. The solution, when made alkaline, also contains fluoride ions, the concentration of which corresponds to a 6% hydrolysis. The formation of the compounds isolated is best interpreted by the equation:

$$3CF_3SCl + 2H_2O \longrightarrow CF_3SSCF_3 + CF_3SO_2H + 3HCl$$

However, on shaking CF_3SCl with water for 12 to 24 hr, starting with too little and gradually increasing the amount of water until a fourfold excess is present, $CF_3SSO_2CF_3$ can also be isolated in an amount equivalent to that of CF_3SSCF_3:

$$4CF_3SCl + 2H_2O \longrightarrow CF_3SSCF_3 + CF_3SO_2SCF_3 + 4HCl$$

The alkaline hydrolysis with a 15% NaOH solution generally proceeds as follows:

$$3CF_3SCl + 4NaOH \longrightarrow CF_3SSCF_3 + CF_3SO_2Na + 3NaCl + 2H_2O$$

If the reaction is run at 70°C, then CF_3SSCF_3 hydrolyzes to sulfur, S^{2-}, F^-, and CO_3^{2-}. At 95°C a complete hydrolysis of CF_3SO_2Na takes place producing CHF_3.

The hydrolysis of $CF_3—C(O)SCl$ proceeds as follows (159):

$$3CF_3\overset{O}{\underset{\|}{C}}-SCl + 8OH^- \longrightarrow 3CF_3COO^- + 3Cl^- + 4H_2O + (S^{2-}, S, S_2O_3^{2-})$$

Sulfenyl chlorides can be oxidized by means of chlorine water or hydrogen peroxide to sulfonyl chlorides:

$$RSCl + 2Cl_2 + 2H_2O \longrightarrow RSO_2Cl + 4HCl \quad [R = CF_3(85), CF_2ClCFCl(5), C_6F_5(134)]$$

$$ClSCF_2CF_2SCl + 4Cl_2 + 4H_2O \longrightarrow ClSO_2CF_2CF_2SO_2Cl + 8HCl \quad (104)$$

On shaking sulfenyl chlorides with mercury or a solution of potassium iodide, disulfanes are produced:

$$2RSCl + 2KI \longrightarrow RSSR + I_2 + 2KCl \quad [R = CFCl_2 (150), CF_2ClCF_2 (100)]$$

$$xClSCF_2CF_2SCF_2CF_2SCl + 2xKI \longrightarrow (SCF_2CF_2SCF_2CF_2S)_x + 2xKCl + xI_2 \quad (104)$$

$$xClSCF_2CF_2SCl + 2xKI \longrightarrow (SCF_2CF_2S)_x + 2xKCl + xI_2 \quad (104)$$

$$2CF_3SCl + 2Hg \longrightarrow CF_3SSCF_3 + Hg_2Cl_2 \quad (84)$$

$$xClSCF_2CF_2SCl + 2xHg \longrightarrow (SCF_2CF_2S)_x + xHg_2Cl_2 \quad (104)$$

$$4CF_3\overset{O}{\underset{\|}{C}}-SCl + 5Hg \longrightarrow CF_3\overset{O}{\underset{\|}{C}}-SS-\overset{O}{\underset{\|}{C}}-CF_3 + Hg(SC\overset{O}{\underset{\|}{-}}CF_3)_2 + 2Hg_2Cl_2 \quad (135)$$

No reaction occurs between CF_3SCl and CO within 8 hr at 100°C. However, on irradiation of a mixture of CF_3SCl and CO with Pyrex-filtered light ($\lambda > 300$ nm), predominantly CF_3SSCF_3 and $COCl_2$ are formed, as well as minor amounts of $CF_3S-C(O)Cl$ (157).

Investigations of the halogen exchange between solid AgCl and liquid sulfenyl chlorides of the type $R_1R_2R_3CSCl$ (R_1, R_2, R_3 = F, Cl, CF_3, CF_3S) with the aid of radioactive ^{36}Cl (12) indicate that highly electronegative substituents induce polarization of the S—Cl bond and thereby increase the heterolytic reactivity of this bond.

IV. Perfluorohalogenoorganosulfenyl Bromides

Relatively few perfluorohalogenoalkanesulfenyl bromides are presently known. These are orange-to-red colored substances which are prepared with difficulty and are too unstable to arouse interest. Apart from a few specific syntheses they may be obtained by the methods used for the preparation of sulfenyl chlorides.

Thiocarbonyl compounds add bromine to the C=S double bond, e.g.,

$$\underset{F}{\overset{X}{>}}C{=}S + Br_2 \longrightarrow \underset{F}{\overset{X}{>}}\underset{Br}{\overset{|}{C}}{-}SBr \quad (X = Cl, F) \quad (187)$$

$$\underset{CF_3}{\overset{CF_3}{>}}C{=}S + Br_2 \longrightarrow \underset{CF_3}{\overset{CF_3}{>}}\underset{|}{\overset{Br}{\underset{}{C}}}{-}SBr \quad (120)$$

Dimeric and trimeric thiocarbonyl fluorides react in the same manner (53):

$$CF_3S{-}\overset{X}{\underset{|}{C}}{=}S + Br_2 \xrightarrow{20°C} CF_3S{-}\underset{Br}{\overset{X}{\underset{|}{C}}}{-}SBr \quad (X = F, SCF_3)$$

Compound $(CF_3S)_2CBrSBr$ could not be obtained in the pure state: ^{19}F NMR spectra have shown that in this case an equilibrium reaction takes place, which moves to the right only to the extent of 90%.

A chlorine–bromine exchange in fluorodichloromethanesulfenyl chloride by means of hydrogen bromide provides a route to additional sulfenyl bromides (109). After the initial halogen exchange at the sulfur, further addition of hydrogen bromide causes stepwise substitution at the methyl group:

$$CFCl_2{-}SCl \xrightarrow{48\% \; HBr} CFCl_2{-}SBr \xrightarrow{HBr}$$
$$CFClBr{-}SBr \xrightarrow{HBr} CFBr_2{-}SBr$$

Several methods are available for the synthesis of CF_3SBr. Either CF_2BrSBr is fluorinated at 100°C with antimony trifluoride (186) or CF_3SCl is reacted with bromine cyanide over activated carbon at 70°C (36). Apart from that it is obtained always contaminated with CF_3SSCF_3 in the reaction between $Hg(SCF_3)_2$ and bromine (31) these compounds react at 0°C to give a mixture composed of 55% CF_3SBr and 45% CF_3SSCF_3.

The complete disproportionation of $CF_3S(O)Br$ to CF_3SBr and CF_3SO_2Br can also be regarded as a method for synthesizing CF_3SBr, since it can easily be separated from the sulfonyl bromide (131).

Trifluoroacetylsulfenyl bromide is obtained in a way analogous to that for the corresponding sulfenyl chloride, by bromination of the thiol (135):

$$CF_3{-}C(O)SH + Br_2 \longrightarrow CF_3{-}C(O)SBr + HBr$$

Bromine, too, causes the fission of cyclic sulfanes (104), e.g.,

$$\underset{F_2}{\overset{F_2}{\diagup}}\overset{S-S}{\underset{S-S}{\diagdown}}\diagup + Br_2 \longrightarrow BrSCF_2CF_2SBr$$

$$\underset{F_2}{\overset{F_2}{\diagup}}\overset{S}{\underset{S-S}{\diagdown}}\overset{F_2}{\diagup}_{F_2} + Br_2 \longrightarrow BrSCF_2CF_2SCF_2CF_2SBr$$

Like CF_3SSCl, although in lower yield, CF_3SSBr can be prepared, by fission of the S—N bond in N,N-dialkylaminotrifluoromethyl disulfanes with HBr (16):

$$R_2NSSCF_3 + 2BHr \longrightarrow CF_3SSBr + R_2NH \cdot HBr \quad (R = CH_3, C_2H_5)$$

The decrease in yield can be accounted for by partial decomposition of the quite unstable CF_3SSBr:

$$2CF_3SSBr \longrightarrow CF_3SSCF_3 + S_2Br_2 \quad (\longrightarrow Br_2 + 2S)$$

Bromination of CF_3SSCl with boron tribromide proceeds considerably better:

$$3CF_3SSCl + BBr_3 \longrightarrow 3CF_3SSBr + BCl_3$$

From this reaction CF_3SSBr can be isolated in 65% yield.

V. Reactions of Perfluorohalogenoorganosulfenyl Halides and Related Reactions

A. WITH PSEUDOHALIDES

The reaction of perfluorohalogenoalkanesulfenyl halides with silver pseudohalides leads to a group of compounds capable of undergoing a host of chemical changes. The first substances prepared in this way were the derivatives of trifluoromethanesulfenyl chloride (28):

$$CF_3SCl + AgX \longrightarrow CF_3SX + AgCl \quad (X = CN, SCN, SeCN, OCN)$$

The thiocyanate is stable toward water for some period of time. However, the alkaline hydrolysis, in analogy with normal thiocyanates (156), causes rapid decomposition. Intermediates CF_3SH and HNCO are formed through cleavage of the CF_3S—C bond, but whereas the cyanate remains intact, the trifluoromethanethiol hydrolyzes further (85):

$$CF_3SCN \xrightarrow{OH^-(H_2O)} CF_3SH + NCO^-$$
$$\downarrow OH^-(H_2O)$$
$$F^- + S^{2-} + CO_3^{2-}$$

The thermally unstable CF_3SSCN decomposes readily at room temperature within a few minutes to CF_3SSCF_3 and polythiocyanate (*85, 116, 117*). The selenocyanate, on the other hand, is stable up to 300°C and its decomposition at 500°C yields selenium quantitatively. By contrast, phenylsulfenylselenocyanate decomposes readily at 175°C according to the equation:

$$C_6H_5SSeCN \longrightarrow (C_6H_5S)_2 + (CN)_2 + Se_3(CN)_2 + Se$$

The isocyanate has been investigated more intensively than the other pseudohalides (*39*). In its preparation from CF_3SCl and $AgOCN$, the linear dimeric $(CF_3S)_2NC(O)NCO$ is also formed in 25% yield, besides CF_3SNCO (75%). A cyclic dimer is obtained by heating the monomer to 100°C for several hours. As has been shown in spectroscopic investigations (*24, 40*), it has the structure of a planar uretidine-1,3-dione ring with *trans*-CF_3S groups. Hydrolysis of CF_3SNCO and of uretidine-1,3-dione furnishes a symmetrically disubstituted urea and carbon dioxide:

$$\left. \begin{array}{c} CF_3SNCO \\ F_3CS \underset{N}{\overset{O}{\underset{\|}{C}}} \underset{\underset{O}{\overset{\|}{C}}}{N} SCF_3 \end{array} \right\} \xrightarrow[-CO_2]{H_2O} CF_3S-NH-\overset{O}{\underset{\|}{C}}-NH-SCF_3$$

By contrast, the linear dimer gives an unsymmetrically substituted urea:

$$(CF_3S)_2NC(O)NCO + H_2O \longrightarrow (CF_3S)_2NC(O)NH_2 + CO_2$$

Hydrolysis of the monomer also furnishes symmetrically disubstituted urea, in constrast to that of CF_3NCO (*7*):

$$CF_3SNCO + H_2O \longrightarrow \langle CF_3S-NH-COOH \rangle \xrightarrow{-CO_2}$$
$$\langle CF_3S-NH_2 \rangle \xrightarrow{CF_3SNCO} CF_3S-NH-C(O)-NH-SCF_3$$

The IR spectra of the CF_3S derivatives of urea, as well as of the deuterated compound, have been discussed at great length (*71*).

In the presence of catalytic quantities of anhydrous sodium acetate, the cyclic trimer, tris(trifluoromethylmercapto) isocyanurate, is obtained at 100°C (*28, 40*).

Compound CF_3SNCO undergoes further reactions typical of organic isocyanates with hydrogen halides (*41*), amines (*28*), and other species (*39, 69*), as is shown in Table II.

TABLE II
REACTION OF CF_3SNCO WITH VARIOUS COMPOUNDS

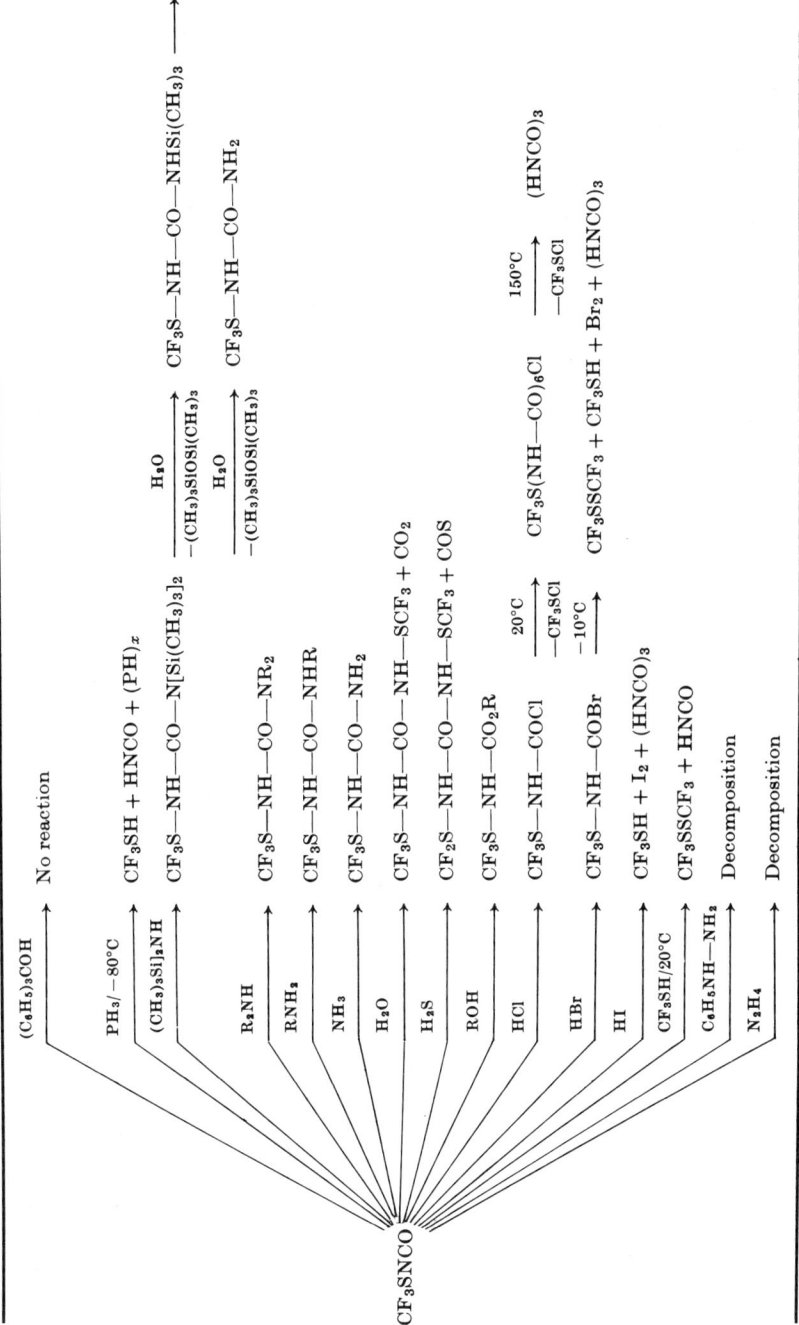

TABLE III

Reaction of 2,4-Bis(trifluoromethylmercapto)uretidine-1,3-dione with Various Compounds

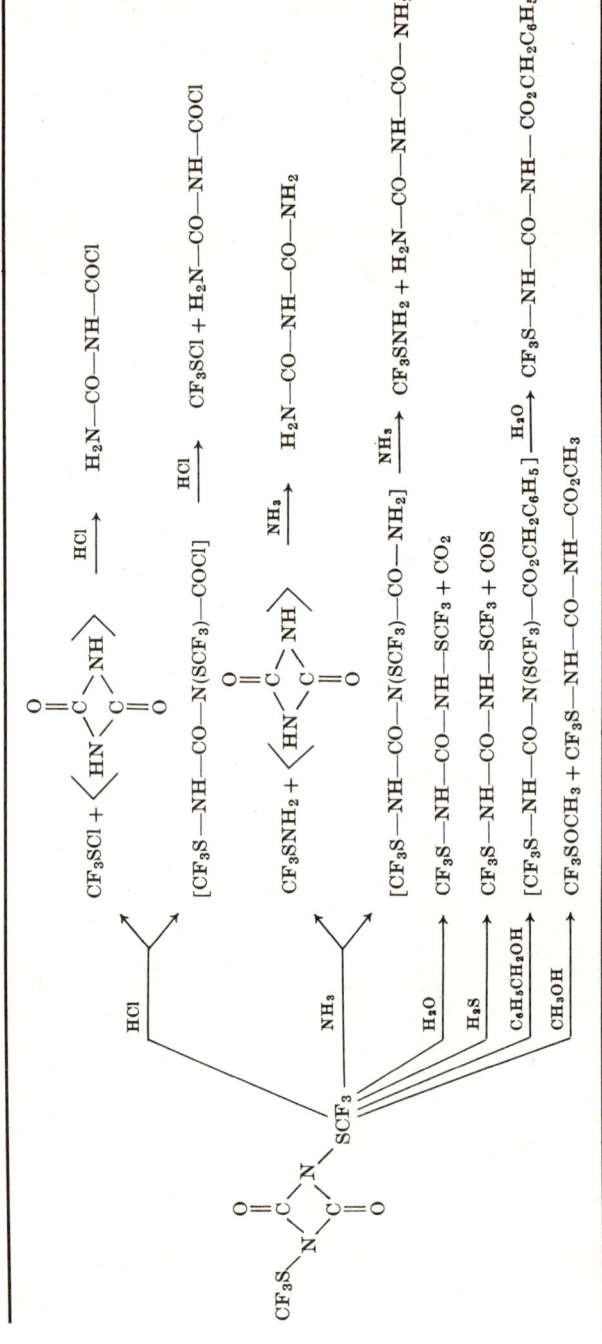

The chemical behavior of the cyclic dimer is quite analogous to that of other uretidine diones (*69*). Some characteristic reactions are summarized in Table III. As can be seen in Table III the mechanism of the reactions with ammonia and hydrogen chloride is not quite clear. The isolated products can be obtained either via a uretidine-dione or via a linear intermediate.

In the reaction of trifluoromethylmercaptosulfenyl chloride with AgNCO the formation of only the monomeric isocyanate is observed (in 87% yield) (*16*):

$$CF_3SSCl + AgNCO \longrightarrow CF_3SSNCO + AgCl$$

The additional sulfur atom seems to reduce the reactivity of the S—Cl bond to nucleophilic attack only to a slight extent. Hydrolysis yields, here too, the symmetrically disubstituted N,N'-bis(trifluoromethylmercaptosulfenyl) urea:

$$2CF_3SSNCO + H_2O \longrightarrow CF_3SS-NH-CO(O)-NH-SSCF_3 + CO_2$$

Difluorochloro- and fluorodichloromethanesulfenyl chlorides react directly with AgOCN to give the monomeric isocyanates, $CF_2ClSNCO$ and $CFCl_2SNCO$, whereas in benzene solution the corresponding isocyanurates are formed (*61*). Freshly prepared $CF_2ClSNCO$ either trimerizes or dimerizes to a uretidine dione within a month when kept at 20°C.

The sulfenylthiocyanates $CF_2ClSSCN$ and $CFCl_2SSCN$ are formed as monomers in benzene but are very unstable and decompose readily to the disulfane and polythiocyanate:

$$2CF_nCl_{3-n}SSCN \longrightarrow CF_nCl_{3-n}SSCF_nCl_{3-n} + (SCN)_x \quad (n = 1, 2)$$

Compound CF_3S—$CFCl$—SCl (*53*) reacts with AgCN, AgSCN, and AgOCN to give the corresponding pseudohalides, which differ only very slightly in their chemical properties from the respective $CFCl_2S$ derivatives. It is noteworthy that $(CF_3S)_2CCl$—SCl does not react under any circumstances with metal pseudohalides; by contrast, $(CF_3)_2CCl$—SCl (*120*) yields the compounds $(CF_3)_2CCl$—SCN and $(CF_3)_2CCl$—$SSCN$. The sulfenylthiocyanate is remarkably stable: even at 135°C no polythiocyanate separates out.

Pentafluorobenzene sulfenyl chloride, C_6F_5SCl, forms a very stable thiocyanate, as well as the compounds C_6F_5SSCN (a yellow polymeric substance of variable composition) and C_6F_5SSeCN (decomposes quantitatively at 50°C), neither of which could be isolated in the pure state at room temperature (*124*).

The isocyanate prepared in benzene solution (*123, 124*) consists, according to osmometric measurements, of a mixture of the dimer and trimer. It does not react with compounds containing active hydrogen through addition to the N=C double bond, but, instead, through the formation of a sulfenyl halide and cyanuric acid:

$$3C_6F_5SNCO + 3HX \longrightarrow 3C_6F_5SX + (HNCO)_3 \quad (X = Cl, Br)$$

The mass spectrum shows the molecular ion of the monomer only; the $C_6F_5S^+$ ion is also present in relatively high intensity. These findings suggest that the S—N bond in the polymeric C_6F_5SNCO is very weak and consequently the compound depolymerizes quite readily.

When reacting fluorocarbonyl sulfenyl chloride with silver pseudo-halides (66, 67), it is observed that it reacts only monofunctionally in contrast to the bifunctional chloro compound, e.g.,

$$\underset{\substack{\|\\O}}{F-C}-SCl + AgSCN \longrightarrow \underset{\substack{\|\\O}}{F-C}-SSCN + AgCl$$

The sulfenyl thiocyanate is unstable and decomposes at 20°C within a few days in a remarkable way:

$$2F-\underset{\substack{\|\\O}}{C}-SSCN \longrightarrow COF_2 + CO_2 + S + (SCN)_x$$

Presumably COF_2 is split off initially from 2 moles of sulfenylthiocyanate and the remaining decomposition products are formed in additional intermediate steps:

$$2F-\underset{\substack{\|\\O}}{C}-SSCN \longrightarrow COF_2 + NCSS-\underset{\substack{\|\\O}}{C}-SSCN \longrightarrow$$

$$COS + S(SCN)_2 \longrightarrow S + (SCN)_x \quad (118)$$

Of interest is the reaction with silver cyanide: at low temperatures (−80°C) the thiocyanate forms initially and then, as shown by ^{19}F NMR measurements, it undergoes a slow transmutation at 30°C into the isothiocyanate (67):

$$\underset{\substack{\|\\O}}{FC}-SCN \longrightarrow \underset{\substack{\|\\O}}{FC}-NCS$$

Compound FC(O)SNCO shows the well-known chemical properties of an isocyanate; it also dimerizes to a uretidine dione:

$$2F-\underset{\substack{\|\\O}}{C}-SNCO \xrightarrow[\text{2-3 days}]{60°-65°C} F-\underset{\substack{\|\\O}}{C}S-N\underset{\substack{\|\\O}}{\overset{\overset{\substack{O\\\|}}{C}}{\underset{C}{\big|}}}N-SCF$$

B. With Silver Perfluorohalogenocarboxylates

In the reaction of sulfenyl chlorides with silver carboxylates mixed anhydrides are formed; they are known as sulfenyl carboxylates (90, 130):

$$R-SCl + AgO-\overset{O}{\underset{\|}{C}}-R' \longrightarrow R'-\overset{O}{\underset{\|}{C}}-O-S-R + AgCl$$

The reaction of sulfenyl chlorides, $CF_nCl_{3-n}SCl$ ($n = 1, 2, 3$) with silver trifluoroacetate furnishes stable halogenated sulfenyl carboxylates of the general formula

$CF_nCl_{3-n}SOC(O)CF_3$ (62):

$$CF_nCl_{3-n}SCl + AgO-\overset{O}{\underset{\|}{C}}-CF_3 \xrightarrow{-30°C/0.5\ hr} CF_nCl_{3-n}SO\overset{O}{\underset{\|}{C}}-CF_3 + AgCl$$

These colorless liquids are stable indefinitely at room temperature.

Trifluoromethanesulfenyl trifluoroacetate decomposes after a short time on irradiation with UV light to the symmetrical anhydrides $[CF_3C(O)]_2O$ and CF_3SOSCF_3; the latter is unstable and disproportionates:

$$2CF_3SOSCF_3 \longrightarrow CF_3SSCF_3 + CF_3SO_2SCF_3 \quad (63)$$

By contrast, $CF_2ClSOC(O)CF_3$, and $CFCl_2SOC(O)CF_3$ decarboxylate after a short UV irradiation with formation of the sulfanes, CF_2ClSCF_3 (154) and $CFCl_2SCF_3$ (62), respectively.

Thermal decomposition at 170°C (2 days) of $CF_3SOC(O)CF_3$ proceeds also with splitting off of CO_2:

$$CF_3SO-\overset{O}{\underset{\|}{C}}-CF_3 \longrightarrow CF_3SCF_3 + CO_2$$

If a chlorine atom is substituted for a fluorine atom in the acetate, the corresponding sulfenyl carboxylate cannot be isolated; instead, only decomposition products are obtained:

$$CF_3SCl + AgO-\overset{O}{\underset{\|}{C}}-CF_2Cl \xrightarrow{-AgCl}$$

$$CF_3SO-\overset{O}{\underset{\|}{C}}-CF_2Cl \longrightarrow [CF_2ClC(O)]_2O + CF_3SSCF_3 + CF_3SO_2SCF_3$$

The reaction of CF_3SCl with silver salts of higher perfluorocarboxyacids likewise does not lead to sulfenyl carboxylates, but to the carboxyacid anhydride, CF_3SSCF_3 and $CF_3SO_2SCF_3$ (62). However, upon irradiation of the reaction mixture with UV light, without attempting to isolate the sulfenyl carboxylate, the corresponding sulfanes are obtained (139):

$$CF_3SCl + CF_3CF_2C(O)OAg \xrightarrow[2\ hr]{h\nu} CF_3SCF_2CF_3 + CO_2$$
$$(70\%)$$

$$CF_3SCl + CF_3CF_2CF_2C(O)OAg \xrightarrow[4\ hr]{h\nu} CF_3SCF_2CF_2CF_3 + CO_2$$
$$(50\%)$$

These examples show unmistakably that the substituent at the carboxyl group also influences the stability of the sulfenyl carboxylate.

Fluorocarbonylsulfenyl chloride reacts in the following manner with silver trifluoroacetate:

$$\underset{\text{F-C-SCl}}{\overset{O}{\|}} + AgO\underset{\|}{\overset{O}{\text{-C-CF}_3}} \xrightarrow{-10°C/12\ hr} \underset{\text{F-C-SO-C-CF}_3}{\overset{O\quad\quad O}{\|\quad\quad\|}} \quad (48, 68)$$

The sulfenyl carboxylate is, on complete exclusion of moisture, a stable, colorless liquid, which on irradiation with UV light decomposes to $FC(O)SCF_3$ and CO_2. The sulfane is a good starting material for the synthesis of additional perfluoro compounds: By means of fluorine–chlorine substitution with BCl_3 (20°C, 2 days), it is possible to obtain $ClC(O)SCF_3$, whose existence previously could be shown only by mass spectrometric study of the products of irradiation of a mixture of CF_3SCl and CO (157). As an acid chloride, $ClC(O)SCF_3$ can be esterified with alcohols, but hydrolysis to the free acid fails.

Treatment of $FC(O)SCF_3$ with CsF affords bis(trifluoromethyl)-dithiocarbonate, $(CF_3S)_2CO$, with the evolution of COF_2; $(CF_3S)_2CS$ is also formed as a by-product. The following reaction mechanism explains the formation of the products isolated:

$$CF_3S\overset{O}{\overset{\|}{-C}}-F \underset{}{\overset{CsF}{\rightleftharpoons}} CF_3S-\underset{F}{\overset{|\overset{\ominus}{O}|\ \ Cs^\oplus}{\underset{|}{-C-}}}-F \xrightarrow{-F_2C=O} CF_3S^\ominus Cs^\oplus \xrightarrow[-CsF]{+CF_3S-\overset{O}{\overset{\|}{C}}-F} \underset{CF_3S}{\overset{CF_3S}{\diagdown}}C=O$$

$$\Big\updownarrow$$

$$F_2C=S + CsF \xrightarrow{+2F_2C=S} \underset{CF_3S}{\overset{CF_3S}{\diagdown}}C=S$$

The primary step in this reaction scheme—the addition of alkali metal fluorides to C=O double bonds—was first observed with perfluorocarbonyl fluorides and hexafluoroacetone (132).

C. WITH PERFLUOROHALOGENOTHIOKETONES

Addition of halogens or interhalogens to perfluorohalogenothioketones, as already described, represents an important method for the preparation of complex sulfenyl halides.

In an attempt to add sulfenyl halides to the C=S double bond, only the unsymmetrical disulfanes are formed, instead of mercapto-substituted sulfenyl halides (54, 56, 72), e.g.,

$$\underset{F}{\overset{F}{>}}C=S + CF_3SCl \xrightarrow{h\nu} \begin{cases} CF_3S-\underset{F}{\overset{F}{C}}-SCl \quad \text{(×)} \\ CF_3S-S-\underset{F}{\overset{F}{C}}-Cl \end{cases}$$

This reaction has been carefully studied since the well-known syntheses of perhalogenated disulfanes have always led to symmetric products.

Various factors suggest a free-radical mechanism:

1. Sulfenyl halides do not react with thiocarbonyls unless irradiated with UV light.
2. Compounds such as CF_3SCl decompose when irradiated (84):

$$2CF_3SCl \longrightarrow CF_3SSCF_3 + Cl_2 \quad (50\%)$$

3. There are always products present that are formed through the combination of two R_fS radicals. For example, the reaction of CSF_2 with CF_3SCl proceeds by the following photochemical mechanism (84):

$$CF_3S-Cl \longrightarrow CF_3S^\cdot + {}^\cdot Cl \quad \text{(a)}$$

$$CF_3S^\cdot + CSF_2 \longrightarrow CF_3S-SCF_2{}^\cdot \quad \text{(b)}$$

$$CF_3S-SCF_2{}^\cdot + CF_3S-Cl \longrightarrow CF_3S-SCF_2Cl + CF_3S^\cdot \quad \text{(c)}$$

$$Cl^\cdot + SCF_2 \longrightarrow ClF_2CS^\cdot \quad \text{(d)}$$

$$ClF_2CS^\cdot + CF_3S-Cl \longrightarrow CF_3S-SCF_2Cl + {}^\cdot Cl \quad \text{(e)}$$

According to the first reaction, the S—Cl bond is homolytically cleaved. The radicals thus formed react [Eqs. (b) and (d)] to give the precursor of the end product, which is produced according to Eqs. (c) and (e) together with the newly formed radicals of reaction (a). The appearance of the free radicals $CF_3S^\cdot$, $Cl^\cdot$, and $ClF_2CS^\cdot$ is supported by the presence of the by-products $CF_3S\!-\!SCF_3$, $CF_2ClS\!-\!SCF_2Cl$, and $CF_2ClS\!-\!Cl$, as proved unequivocally by the ^{19}F NMR spectrum. The following equations illustrate the method of their formation:

$$CF_3S\cdot + \cdot SCF_3 \longrightarrow CF_3S\!-\!SCF_3$$
$$CF_2ClS\cdot + \cdot SCF_2Cl \longrightarrow CF_2ClS\!-\!SCF_2Cl$$
$$Cl\cdot + \cdot Cl \longrightarrow Cl_2$$
$$CSF_2 + Cl_2 \longrightarrow CF_2ClS\!-\!Cl$$

Since no compound is found of the formula $CF_3S\!-\!SCF_2Cl$, it is unlikely that $CF_3S\cdot$ and $Cl\cdot$ react according to

$$CF_3S\cdot + F_2C\!=\!S \longrightarrow CF_3SCF_2\!-\!S\cdot$$
or
$$Cl\cdot + F_2C\!=\!S \longrightarrow \cdot CF_2SCl$$

Other disulfanes can be prepared similarly, as summarized in Table IV. The yields are substantially higher when sulfenyl halides of the formula $CF_nCl_{3-n}SX$ (X = Cl, Br) are reacted. Within a homologous series the

TABLE IV

PREPARATION OF DISULFANE COMPOUNDS

Disulfane	Starting materials		Reaction time (hr)	Yield (%)
	Thiocarbonyl compound	Sulfenyl halide		
$CF_3S\!-\!SCF_2Cl$	CSF_2	CF_3SCl	0.3	67
$CF_3S\!-\!SCFCl_2$	$CSFCl$	CF_3SCl	36	56
$CF_2ClS\!-\!SCFCl_2$	CSF_2	$CFCl_2SCl$	20	62
$CF_2BrS\!-\!SCF_2Br$	CSF_2	CF_2BrSBr	0.3	100
$CF_2BrS\!-\!SCBrClF$	CSF_2	$CFClBrSBr$	2	78
$CFClBrS\!-\!SCBrClF$	$CSFCl$	$CFClBrSBr$	6	75
$CF_3S\!-\!CClFS\!-\!SCF_3$	$CF_3SC(S)F$	CF_3SCl	24	29
$CF_3S\!-\!CCl_2S\!-\!SCF_3$	$CF_3SC(S)Cl$	CF_3SCl	24	24
$CF_3S\!-\!CClFS\!-\!SCF_2Cl$	CSF_2	$CF_3S\!-\!CClFSCl$	16	33
$CF_3S\!-\!CClFS\!-\!SCFCl_2$	$CSFCl$	$CF_3S\!-\!CClFSCl$	22	60.5
$CF_3S\!-\!CCl_2S\!-\!SCF_2Cl$	CSF_2	$CF_3S\!-\!CCl_2SCl$	71	32.6
$CF_3S\!-\!CCl_2S\!-\!SCFCl_2{}^a$	$CSFCl$	$CF_3S\!-\!CCl_2SCl$	62	—
$(CF_3)_2\!-\!CClS\!-\!SCF_2Cl$	CSF_2	$(CF_3)_2\!-\!CClSCl$	9	25
$(CF_3S)_2\!-\!CFS\!-\!SCF_2Cl$	CSF_2	$(CF_3S)_2\!-\!CFSCl$	4	30
$(CF_3S)_2\!-\!CFS\!-\!SCFCl_2$	$CSFCl$	$(CF_3S)_2\!-\!CFSCl$	150	25
$(CF_3S)_2\!-\!CFS\!-\!SCCl_3$	$CSCl_2$	$(CF_3S)_2\!-\!CFSCl$	250	15
$(CF_3S)_2\!-\!CFS\!-\!SCFCl\!-\!SCF_3$	$CF_3SC(S)F$	$(CF_3S)_2\!-\!CFSCl$	150	25
$(CF_3S)_2\!-\!CFS\!-\!SCCl_2\!-\!SCF_3$	$CF_3SC(S)Cl$	$(CF_3S)_2\!-\!CFSCl$	150	20

[a] This compound was not obtained in the pure state.

reactions become faster with increasing degree of fluorination of the starting compound.

The same reaction can also be applied to halogenocarbonylsulfenyl chlorides and affords perhalogenated disulfanes with interesting functional groups (68):

$$F-\underset{\underset{O}{\|}}{C}-SCl + S=CF_nCl_{2-n} \xrightarrow{h\nu} F-\underset{\underset{O}{\|}}{C}-SSCF_nCl_{3-n} \quad (n = 0, 1, 2)$$

$$F-\underset{\underset{O}{\|}}{C}-SCl + CF_3S-\underset{\underset{S}{\|}}{C}-F \xrightarrow{h\nu} CF_3S-CFCl S-\underset{\underset{O}{\|}}{S}C-F$$

$$Cl-\underset{\underset{O}{\|}}{C}-SCl + S=CF_2 \xrightarrow{h\nu} ClCS-SCF_2Cl$$

$$Cl-\underset{\underset{O}{\|}}{C}-SCl + S=CF_2 \xrightarrow{h\nu} Cl\underset{\underset{O}{\|}}{C}S-SCF_2Cl$$

Compound ClC(O)SSCCl₃ is obtained from FC(O)SSCCl₃ through fluorine–chlorine exchange with BCl₃; ClC(O)SSCF₃ and FC(O)SSCF₃ are formed as follows:

$$2X-\underset{\underset{O}{\|}}{C}-SCl + Hg(SCF_3)_2 \longrightarrow 2X-\underset{\underset{O}{\|}}{C}-SSCF_3 + HgCl_2 \quad (X = Cl, F)$$

In the homologous series $ClC(O)SSCF_nCl_{3-n}$ ($n = 0-3$), compound $ClC(O)SSCFCl_2$ is missing: it cannot be synthesized either directly from ClC(O)SCl and CSFCl nor through halogen exchange in $FC(O)SSCFCl_2$, since in this case the fluorine atom of the halomethyl group is also exchanged (48).

The fluorothiocarbonyl isothiocyanates may also be reacted successfully in place of the thiocarbonyl compounds (21), e.g.,

$$F-\underset{\underset{S}{\|}}{C}-NCS + CF_3S-Cl \xrightarrow{h\nu} CF_3S-\underset{\underset{NCS}{|}}{\overset{\overset{F}{|}}{S}C}-Cl$$

$$F-\underset{\underset{NCS}{|}}{\overset{\overset{S}{\|}}{C}}-NCS + F-\underset{\underset{}{}}{\overset{\overset{Cl}{|}}{C}}-SCl \xrightarrow[HCl]{h\nu} F-\underset{\underset{NCS}{|}}{\overset{\overset{Cl}{|}}{C}}-S-S-\underset{\underset{NCS}{|}}{\overset{\overset{Cl}{|}}{C}}-F$$

D. With Ammonia, Primary and Secondary Amines, and Amides

In the reaction between stoichiometric quantities of CF_3SCl and ammonia in a Carius tube at low temperature the trifluoromethylmercaptoamine is formed (30):

$$CF_3SCl + 2NH_3 \xrightarrow{-45°C} CF_3SNH_2 + NH_4Cl$$

This reaction can also be carried out in an autoclave at 10°C (*70*) or, in high yield, by introducing CF_3SCl into liquid ammonia at −80°C (*64, 70*).

The mixed amines, $CF_nCl_{3-n}SNH_2$ ($n = 1, 2$), are synthesized from the corresponding sulfenyl chlorides and ammonia at −60°C with the use of a perhalogenated solvent (*58*):

$$CF_nCl_{3-n}SCl + 2NH_3 \xrightarrow[CCl_3F]{-60°C} CF_nCl_{3-n}SNH_2 + NH_4Cl \quad (n = 1, 2)$$

The corresponding pentafluorobenzene derivative is obtained by adding the sulfenyl chloride dropwise to a saturated solution of ammonia in diethyl ether (*138*):

$$C_6F_5SCl + 2NH_3 \xrightarrow[ether]{0°C} C_6F_5SNH_2 + NH_4Cl$$

A more complete substitution of the trihalomethylmercaptoamines is achieved most readily at −60°C in *n*-pentane or $CFCl_3$ solution in the presence of pyridine as an HCl acceptor and with an additional mole of sulfenyl chloride (*58*):

$$CF_{3-n}Cl_nSNH_2 + CF_{3-n}Cl_nSCl + C_5H_5N \longrightarrow$$
$$(CF_{3-n}Cl_nS)_2NH + C_5H_5N \cdot HCl \quad (n = 0, 1, 2)$$

$$CF_3SNH_2 + CF_nCl_{3-n}SCl + C_5H_5N \longrightarrow$$
$$CF_3S(CF_nCl_{3-n}S)NH + C_5H_5N \cdot HCl \quad (n = 0, 1, 2)$$

$$CFCl_2SNH_2 + CF_2ClSCl + C_5H_5N \longrightarrow CFCl_2S(CF_2ClS)NH + C_5H_5N \cdot HCl$$

Compound $(C_6F_5S)_2NH$ too, like the monosubstituted amine, is obtained by adding C_6F_5SCl dropwise to a solution of NH_3 in ether (*138*). In this case the concentration of ammonia is kept slightly lower. This compound is also formed as the sole product of the direct reaction of C_6F_5SCl and ammonia in a Carius tube (*123*).

All bis(mercapto)amines can be converted into the corresponding tris(mercapto)amines in the presence of, for example, trimethylamine as HCl acceptor. In this way it is possible to prepare $(C_6F_5S)_3N$ from $(C_6F_5S)_2NH$ and C_6F_5SCl (*138*):

$$(C_6F_5S)_2NH + C_6F_5SCl + (CH_3)_3N \xrightarrow[ether]{34°C} (C_6F_5S)_3N + (CH_3)_3N \cdot HCl$$

A trihalomercaptoamine with three different substituents is formed as follows:

$CFCl_2S(CF_2ClS)NH + CF_3SCl + (CH_3)_3N \xrightarrow[n\text{-pentane}]{-60°C}$

$$\begin{array}{c} CF_3S \\ CF_2ClS \end{array}\!\!\!\!\!\!>\!\!N\!\!-\!\!SCFCl_2 + (CH_3)_3N \cdot HCl \quad (58)$$

For the synthesis of other triply sulfenylated amines, it is best to start with the primary ones:

$$\left.\begin{array}{c} CF_3SNH_2 \\ CF_2ClSNH_2 \\ CFCl_2SNH_2 \end{array}\right\} + 2CF_nCl_{3-n}SCl + 2(CH_3)_3N \xrightarrow[n\text{-pentane}]{-60°C} \left\{\begin{array}{l} CF_3SN(SCF_nCl_{3-n})_2 + \\ \quad 2(CH_3)_3N \cdot HCl \\ CF_2ClSN(SCCl_{3-n}F_n)_2 + \\ \quad 2(CH_3)_3N \cdot HCl \\ CFCl_2SN(SCCl_{3-n}F_n)_2 + \\ \quad 2(CH_3)_3 N \cdot HCl \end{array}\right.$$

$(n = 1, 2)$

In an attempt to prepare tris(mercapto)amines of the type $(CF_3S)_2NSCF_nCl_{3-n}$ $(n = 0, 1, 2)$ from $(CF_3S)_2NH$ and the appropriate sulfenyl chloride, it was observed that, besides the expected bis(trifluoromethylmercapto)derivatives, other products were also formed (57):

$(CF_3S)_2NH + CF_2ClSCl \longrightarrow (CF_3S)_2NSCF_2Cl + CF_3SN(SCF_2Cl)_2 + (CF_3S)_3N$

$(CF_3S)_2NH + CFCl_2SCl \longrightarrow (CF_3S_2)NSCFCl_2 + CF_3SN(SCFCl_2)_2 + (CF_3S)_3N$

$(CF_3S)_2NH + CCl_3SCl \longrightarrow (CF_3S)_2NSCCl_3 + (CF_3S)_3N$

A comparison of yields showed that formation of $(CF_3S)_3N$ is favored with growing degree of chlorination of the sulfenyl chlorides, whereas the nucleophilic substitution at the sulfenyl sulfur is reduced by the imide nitrogen. It is, therefore, to be assumed that with increasing content of chlorine, the elimination of CF_3SCl rather than HCl from the intermediate is favored on energetic and steric grounds:

$$\left[\begin{array}{c} \quad\quad H \\ \quad\quad | \quad\quad \ominus \\ CF_3S\!-\!\overset{\oplus}{N}\ldots|S\!-\!CF_nCl_{3-n} \\ \quad\quad | \quad\quad | \\ \quad\quad SCF_3 \quad Cl \end{array}\right] \quad (n = 0, 1, 2)$$

The CF_3SCl thus formed affords in the competing reaction

$(CF_3S)_2NH + CF_3SCl \longrightarrow (CF_3S)_3N + HCl$

tris(trifluoromethylmercapto)amine, whereas the intermediate $(CF_3S)NH(SCF_nCl_{3-n})$ reacts either with $CF_nCl_{3-n}SCl$ to give $CF_3SN(SCF_nCl_{3-n})_2$ or with CF_3SCl to give $(CF_3S)_2NSCF_nCl_{3-n}$.

All amines so far prepared are stable toward humid air; they are also rather insensitive toward hydrolysis, as can be demonstrated by the reaction of CF_3SNH_2 with water (*30*). However, alkaline and acid aqueous solutions completely destroy these molecules. In hydrochloric acid, CF_3SCl and NH_4Cl are formed initially from CF_3SNH_2; CF_3SCl then reacts further:

$$3CF_3SNH_2 + 6HCl \longrightarrow 3CF_3SCl + 3NH_4Cl$$
$$\downarrow +2H_2O$$
$$CF_3SO_2^- + CF_3SSCF_3 + 3Cl^- + 4H^+$$

The R_fSN compounds react quite readily with hydrogen chloride through fission of the S—N bond; in this reaction the electronegative chlorine moves to the electropositive sulfur and the proton joins the negative nitrogen:

$$R_fSN{<} + HCl \longrightarrow R_fSCl + HN{<} \quad (69)$$

Compound CF_3SNH_2 is incompletely decomposed by bases at 20°C to give F^-, CO_3^{2-}, S^{2-}, NH_3, and CHF_3; at 75°C, however, hydrolysis is complete, and the products contain additionally sulfur, but no CHF_3 (*30*). Compound CF_3SNH_2 is not stable towards UV light and the products of irradiation include CF_3SSCF_3, NH_4F, NH_4SCN, and hydrazinium fluoride. It is to be assumed that the initial step of the photolysis is the homolytic fission of CF_3SNH_2 to $CF_3S^{\cdot}$ and $^{\cdot}NH_2$.

Compound $C_6F_5SNH_2$ is not very stable and decomposes in a short time:

$$2C_6F_5SNH_2 \longrightarrow (C_6F_5S)_2NH + NH_3 \quad (138)$$

Compound $(CF_3S)_3N$ even if it contains only minor impurities decomposes quantitatively when heated under reflux:

$$2(CF_3S)_3N \longrightarrow 3CF_3SSCF_3 + N_2 \quad (70)$$

Condensation reactions of $(CF_3S)_2NH$ have been thoroughly studied (*57*): with SCl_2 compound $(CF_3S)_2NSN(SCF_3)_2$ is not formed—in analogy with the formation of $(CH_3)_2NSN(CH_3)_2$ from dimethylamine (*11*)—but rather CF_3SCl, among other compounds, is obtained:

$$3(CF_3S)_2NH + 6SCl_2 \longrightarrow 6CF_3SCl + S_4N_3Cl + HCl$$

The reaction with benzoyl chloride proceeds like a substitution reaction:

$$(CF_3S)_2NH + C_6H_5-\underset{\underset{O}{\|}}{C}-Cl \longrightarrow C_6H_5-\underset{\underset{O}{\|}}{C}-NH-SCF_3 + CF_3SCl$$

In both cases it may be assumed that the electron-withdrawing effect of both CF_3S groups lowers the nucleophilic character of the imide nitrogen to such an extent that it is no longer able to attack—as is the rule with amines—the electrophilic reaction partner (sulfur or carbonyl carbon), but instead forms a weak bond with it. From this intermediate, as is the case with the reactions of $CF_nCl_{3-n}SCl$ ($n = 0, 1, 2$), compound CF_3SCl and the substitution product can be formed as shown in Scheme 1.

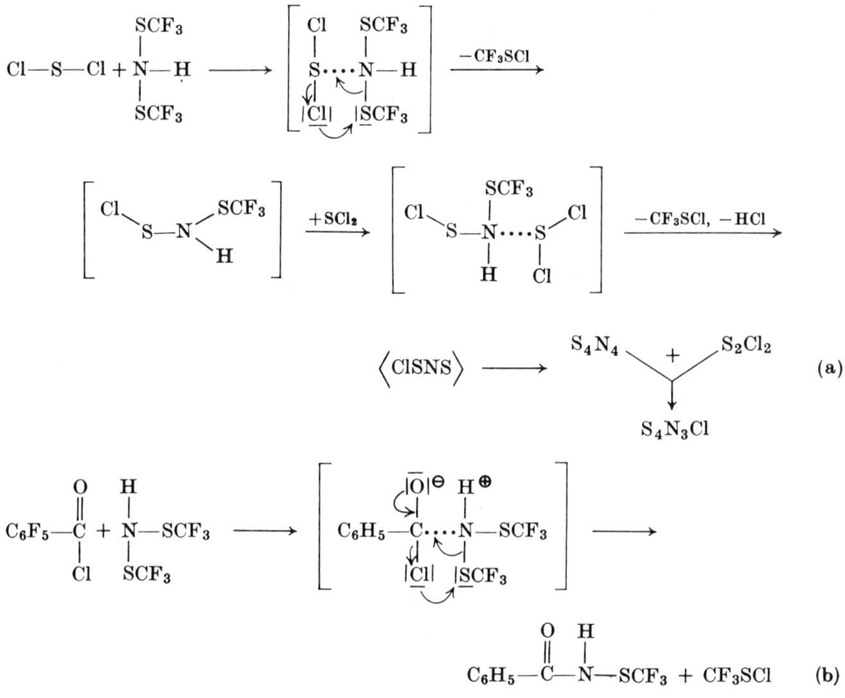

SCHEME 1

These same considerations apply to the reaction of $(CF_3S)_3N$ with SCl_2:

$$3(CF_3S)_3N + 6SCl_2 \longrightarrow 9CF_3SCl + S_4N_3Cl + S_2Cl_2$$

The disubstituted amines form with pyridine and trimethylamine 1:1 adducts, the stability of which increases with the degree of fluorination (59).

Thus, for example, $(CF_3S)_2NH \cdot N(CH_3)_3$ and $(CF_3S)_2NH \cdot NC_5H_5$ are stable and can be distilled without decomposition; similarly $CF_3S(CF_2ClS)_2NH \cdot N(CH_3)_3$ and $(CF_2ClS)_2NH \cdot NC_5H_5$ are also stable. For the higher chlorinated $CF_3S(CFCl_2S)NH$, the pyridine adduct only is stable up to 0°C.

Condensation reactions between perfluorohalogenosulfenyl chlorides and amides as well as primary and secondary amines have been thoroughly investigated. Compounds RNH_2 and $RR'NH$ react with sulfenyl chlorides in the presence of an excess of the amine to give $RN(H)SR_f$ or $RR'NSR_f$, respectively. Amides condense only in the presence of a tertiary amine, such as $(CH_3)_3N$, $(C_2H_5)_3N$, or pyridine. These reactions can be performed in inert organic solvents but afford just as satisfactory yields without any solvent. Substances synthesized up to the present, as well as their physical data and biological activity, have been summarized (42).

E. With Arsines, Alcohols, Thioalcohols, Sulfinates, and Carbonyl Compounds

Perfluorohalogenoalkanesulfenyl halides readily undergo condensation reactions with numerous types of compounds; this is due to their polarization $R_fS^{\delta+}$—$Hal^{\delta-}$ caused by the strong electron-withdrawing action of the fluorine atoms. Thus, the reaction of dimethylarsine with trifluoromethanesulfenyl chloride affords $(CH_3)_2AsSCF_3$, $(CH_3)_2AsCl$, and trifluoromethanethiol:

$$2(CH_3)_2AsH + 2CF_3SCl \longrightarrow (CH_3)_2AsSCF_3 + (CH_3)_2AsCl + CF_3SH + HCl \quad (19)$$

Tetramethyldiarsine is cleaved by CF_3SCl at the As—As bond:

$$(CH_3)_2As\text{—}As(CH_3)_2 + CF_3SCl \longrightarrow (CH_3)_2AsSCF_3 + (CH_3)_2AsCl$$

Compounds containing As—S bonds react similarly:

$$(CH_3)_2As\text{—}SR + CF_3SCl \longrightarrow RS\text{—}SCF_3 + (CH_3)_2AsCl \quad (R = C_2H_5, C_6H_5)$$

The reaction with hydrogen sulfide leads to the formation of trisulfanes (84). Condensations with alcohols proceed very smoothly to thioperoxides (3, 110):

$$R_fSCl + ROH \longrightarrow R_fSOR + HCl$$

$[R_f = CF_3;\ R = CH_3,\ C_2H_5,\ (CH_3)_2CH,\ CF_3CH_2$

$R_f = CFCl_2;\ R = CH_3OCH_2,\ CH_3,\ ClCH_2CH_2,\ C_4H_9,\ C_{12}H_{25},$
$\quad C_6H_5,\ (CH_3)_3C,\ 4\text{-}C_6H_4Cl,\ 2.5\text{-}C_6H_3Cl_2$

$R_f = (CF_3)_2CF;\ R = CH_3]$

$$2CF_3SCl + HO\text{—}CH_2\text{—}CH_2\text{—}OH \longrightarrow CF_3SO\text{—}CH_2\text{—}OSCF_3 + 2HCl$$

and those with thiols to disulfanes (4, 19, 84, 110, 123):

$$R_fSCl + RSH \longrightarrow R_fSSR + HCl$$

$[R_f = R = CF_3$

$R_f = CF_3;\ R = CH_3,\ C_2H_5,\ C_6H_5$

$R_f = C_6F_5;\ R = CH_3,\ n\text{-}C_4H_9,\ C_6F_5]$

In all these reactions, HCl acceptors, such as pyridine (3) or triethylamine (110), are used. In this respect fluorocarbonylsulfenyl chloride behaves quite extraordinarily (48): with alcohols it reacts at the fluorocarbonyl group with formation of alkoxycarbonylsulfenyl chlorides.

$$F-\overset{O}{\underset{\|}{C}}-SCl + ROH \xrightarrow[-HCl]{30°-60°C} RO-\overset{O}{\underset{\|}{C}}-SCl \quad (R = CH_3, C_2H_5)$$

By contrast, thiols, even in excess, react to furnish exclusively alkyl- or arylhalocarbonyl disulfanes:

$$F-\overset{O}{\underset{\|}{C}}-SCl + RSH \xrightarrow[-HCl]{20°C} F-\overset{O}{\underset{\|}{C}}-S-SR \quad (R = C_2H_5, C_6H_5)$$

Sulfenyl chlorides react with zincalkyl (or -aryl) sulfinates and also with sodium benzene sulfinate to yield the corresponding esters of thiosulfonic acid (13, 111, 161):

$$(RSO_2)_2Zn + 2CF_3SCl \longrightarrow 2RSO_2SCF_3 + ZnCl_2$$
$$C_6H_5SO_2Na + CF_nCl_{3-n}SCl \longrightarrow C_6H_5SO_2SCF_nCl_{3-n} \quad (n = 0, 1, 2)$$

On thermolysis the latter are reduced to thiocarbonyldihalides:

$$C_6H_5SO_2SCF_nCl_{3-n} \xrightarrow{170°-250°C/1.5 \text{ hr}} S=CCl_{2-n}F_n \quad (n = 0, 1, 2)$$

The products of the reactions of perfluorohalogenomethanesulfenyl chlorides with ketones, diketones, and ketoesters are monosubstituted trihalomethylmercapto derivatives, as shown in the following examples (8):

$$RCH_2-\overset{O}{\underset{\|}{C}}-CH_3 + CF_3SCl \xrightarrow{-HCl} RCH(SCF_3)-\overset{O}{\underset{\|}{C}}-CH_3 \quad (R = H, CH_3)$$

$$CH_3-\overset{O}{\underset{\|}{C}}-CH_2-\overset{O}{\underset{\|}{C}}-CH_3 + CF_nCl_{3-n}SCl \xrightarrow{-HCl} CH_3-\overset{O}{\underset{\|}{C}}-\overset{SCF_nCl_{3-n}}{\underset{|}{CH}}-\overset{O}{\underset{\|}{C}}-CH_3 \quad (n = 1, 2, 3)$$

$$\underset{H_5C_2O}{\overset{O}{\diagdown}}C-CH_2-\overset{O}{\underset{\|}{C}}-CH_3 + CF_nCl_{3-n}SCl \xrightarrow{-HCl} \underset{H_5C_2O}{\overset{O}{\diagdown}}C-\overset{SCF_nCl_{3-n}}{\underset{|}{CH}}-\overset{O}{\underset{\|}{C}}-CH_3 \quad (n = 1, 2, 3)$$

The reactivity of the keto group is not influenced by the substitution: the reaction of 3-(trifluoromethylmercapto)-2-butanone with hydroxylamine hydrochloride gives the corresponding oxime in good yield. Cyclization of diketones and keto esters to pyrazole derivatives succeeds by the well-known method using phenylhydrazine:

$$CH_3-\underset{O}{\overset{\|}{C}}-\underset{SCF_nCl_{3-n}}{\underset{|}{CH}}-\underset{O}{\overset{\|}{C}}-CH_3 + C_6H_5-NH-NH_2 \xrightarrow{-2H_2O}$$

pyrazole with $Cl_{3-n}F_nCS$ and CH_3 substituents, N-N, C_6H_5 ($n = 1, 2, 3$)

$$H_5C_2O\underset{O}{\overset{\diagdown}{\diagup}}\underset{SCF_nCl_{3-n}}{\underset{|}{C-CH}}-\underset{O}{\overset{\|}{C}}-CH_3 + C_6H_5-NH-NH_2 \xrightarrow{-H_2O, -C_2H_5OH}$$

pyrazolone with $Cl_{3-n}F_nCS$ and CH_3 substituents, O=, N-NH, C_6H_5 ($n = 2, 3$)

The 3-oxopropionic acid ethyl ester can be made to react with CF_3SCl by way of metallation; mono- and disubstituted products are formed in this reaction in equal amounts (49):

$$\underset{H}{\overset{O=}{\diagdown}}C-\underset{|}{\overset{Na}{CH}}-C\underset{OC_2H_5}{\overset{\diagup O}{\diagdown}} + CF_3SCl \xrightarrow{-25°C/2\ hr}$$

$$\underset{H}{\overset{O=}{\diagdown}}C-\underset{|}{\overset{SCF_3}{CH}}-C\underset{OC_2H_5}{\overset{\diagup O}{\diagdown}} + \underset{H}{\overset{O=}{\diagdown}}C-\underset{\underset{SCF_3}{|}}{\overset{SCF_3}{\overset{|}{C}}}-C\underset{OC_2H_5}{\overset{\diagup O}{\diagdown}}$$

Condensation with urea does not lead in this case, as would be expected, to the uracil derivative with ring closure, because the reaction takes place only at the more reactive aldehyde function:

$$\underset{H}{\overset{O}{\underset{\|}{C}}}-\underset{\underset{SCF_3}{|}}{CH}-\underset{\underset{OC_2H_5}{\diagdown}}{\overset{O}{\overset{\|}{C}}} + H_2N-\overset{O}{\overset{\|}{C}}-NH_2 \quad \underset{160°C/24\ hr}{\overset{(HCOOH)}{\longrightarrow}} \quad \text{[uracil with CF}_3\text{S substituent]}$$

$$\downarrow -H_2O$$

$$H_2N-\overset{O}{\overset{\|}{C}}-NH-CH=\underset{\underset{SCF_3}{|}}{C}-\underset{\underset{OC_2H_5}{\diagdown}}{\overset{O}{\overset{\nearrow}{C}}}$$

F. WITH ALKANES, ALKENES, ALKYNES, AND NITRILES

Addition reactions of CF_3SCl to olefins and nitriles were intensively studied. It has been shown that these reactions have to be either initiated with UV light or carried out in a strongly polar solvent.

The UV-initiated addition of CF_3SCl to $CHF=CF_2$ leads to $CF_3SCFHCF_2Cl$ and CF_3SCF_2CHFCl, the latter being the main product (76). The additions of $CFCl=CF_2$ (25, 76), $CH_2=CHCl$ (76), and $CH_2=CHC(O)OCH_3$ (80) proceed analogously.

Irradiation of a mixture of $CF_2=CFX$ (X = CF_3, OCH_3) and CF_3SCl affords the main products $CF_3SCFXCF_2Cl$ and CF_3SCF_2CFClX, together with CF_3SSCF_3 and Cl_2. The chlorine thus liberated competes with CF_3SCl in combining with the C=C double bond of the starting materials.

Compound CF_3SCl combines in a polar solvent, e.g., tetramethylenesulfone, at 20° to 25°C with $CH_2=CHCl$ to produce $CF_3SCH_2CHCl_2$ in 67% yield (2). Analogously, $CF_3SCHClCHCl_2$ is formed in 80% yield from CF_3SCl and $CHCl=CHCl$ (1). At 80° to 100°C, CF_3SCl reacts with $CH_2=CH_2$ in tetramethylenesulfone to give $CF_3SCH_2CH_2Cl$ in 80% yield (101).

Under the influence of UV light CF_3SCl can be made to combine with $CF_3SCH=CH_2$ to yield $(CF_3S)_2CHCH_2Cl$ and $(CF_3S)_2CHCH_2SCF_3$ (75, 79). Similarly, other sulfenyl chlorides can also be added to C=C double bonds, e.g.,

$$R_fSCl + CH_2=CH_2 \longrightarrow R_fSCH_2CH_2Cl$$
$$[R_f = C_2F_5, \quad (CF_3)_2CF(CF_2)_4 \ (89), \ C_4F_9 \ (123)]$$

Sulfenyl chlorides combine also with C≡N triple bonds. The reaction of CF_3SCl with $(CN)_2C=C(CN)_2$ in CH_2Cl_2 in the presence of $[(C_2H_5)_4N]Cl$ results in $(CN)_2C=C(CN)CCl=NSCF_3$ (82, 83). Reactions between R_2NCN and CF_3SCl take a similar course (65):

$$R_2NC{\equiv}N + CF_3SCl \longrightarrow R_2N-CCl{=}NSCF_3 \quad (R = CH_3, C_2H_5)$$

Addition as well as condensation reactions occur with some monosubstituted derivatives:

$$3RNHCN + 4CF_3SCl \longrightarrow 2R(CF_3S)NCCl{=}NSCF_3 + RNHCN \cdot 2HCl$$
$$[R = CH_3, (CH_3)_2CH]$$

With $(CH_3)_3CN(H)CN$ only the condensation product, $(CH_3)_3CN(SCF_3)$-CN, is formed.

Both addition and substitution are also observed in the reaction of H_2NCN with CF_3SCl. If the reaction is carried out in ether at 0°C, $CF_3SN(H)CCl{=}NSCF_3$, results; however without a solvent and at 20°C, 23% $(CF_3S)_2NCCl{=}NSCF_3$ and 1% $CF_3SN(H)CN$ are additionally formed. The latter can be prepared in better yield (17%) from H_2HCN and $CF_3SN(H)CCl{=}NSCF_3$.

The acidic proton of the amine can be substituted in aqueous acetone by reaction with $AgNO_3$ to give $CF_3SN(Ag)CN$. The silver salt reacts with CF_3SCl to yield hexabis(trifluoromethylmercapto)melamine, which can also be synthesized from $(CF_3S)_2NCCl{=}NSCF_3$ or $CF_3SN(H)CCl{=}NSCF_3$ and amines [$(CH_3)_3N$, pyridine]. Melamine itself reacts with CF_3SCl in acetonitrile and in the presence of pyridine to give

only. Here, too, the hydrogen atoms are acidic and can be substituted with silver nitrate in aqueous acetone to give the trisilver salt. In the reaction with CF_3SCl it affords likewise the compound

Alkynes react with CF_3SCl only via prior metallation with a Grignard reagent or C_6H_5Li (*81*):

$$C_6H_5C{\equiv}CH \xrightarrow{C_2H_5MgBr} C_6H_5C{\equiv}CMgBr \xrightarrow{CF_3SCl} C_6H_5C{\equiv}CSCF_3$$
$$C_6H_5C{\equiv}CH \xrightarrow{C_6H_5Li} C_6H_5C{\equiv}CLi$$

Under the influence of UV light, condensations between CF_3SCl and hydrocarbons are also possible (77, 78), e.g.,

$$CF_3SCl + RH \xrightarrow{h\nu} CF_3S-R + ClR + CF_3SSCF_3 + HCl$$

$$\left(R = \text{H}\underset{\text{H}}{\diagup}\!\!\bigcirc,\ -CH_2-\bigcirc\right)$$

$$CF_3SCl + n\text{-}C_4H_{10} \xrightarrow{h\nu} CF_3SCH_2CH_2CH_2CH_3 + CF_3SCH(CH_3)C_2H_5 +$$
$$CH_3CHClC_2H_5 + ClCH_2CH_2C_2H_5 + CF_3SSCF_3 + HCl$$

Other sulfenyl chlorides react analogously (77).

G. With Aromatics and Heteroaromatics

The introduction of the perfluorohalogenosulfenyl group into aromatic compounds was accomplished long before the preparation of sulfenyl halides. The CF_3S-substituted compounds have been synthesized largely by chlorination of the side chain in arylmethylthioethers followed by chlorine–fluorine exchange. This method was applied for the first time in the synthesis of trifluoromethylmercaptobenzene (92, 140, 172):

$$\underset{SCH_3}{\bigcirc} \xrightarrow{Cl_2} \underset{SCCl_3}{\bigcirc} \xrightarrow{SbF_3 \text{ or } HF} \underset{SCF_3}{\bigcirc}$$

This reaction is also feasible if the benzene ring contains substituents, such as halogens or methyl, carboxyl, and nitro groups.

These compounds can be oxidized without difficulty to the corresponding sulfoxides or sulfones (92, 95, 140, 142, 167, 172, 176). On the other hand, the trihalomethylmercapto group is noticeably inert toward chemical changes in the aromatic ring, such as halogenation (92), nitration (174), reduction of nitro groups, diazotation of amino groups, and hydrolysis of nitrile groups (172).

Numerous derivatives play a considerable role as fungicides, insecticides, and pharmaceuticals (93, 95, 125, 153) as well as serving as intermediates in the syntheses of dyes (22, 94, 95, 165, 166, 168, 169, 171, 173–175, 177–182).

The reaction of aryl magnesium halides with perfluorohalogenosulfenyl chlorides takes place under considerably milder conditions (47, 148):

$$\text{Ar}-\text{MgX} + \text{CF}_n\text{Cl}_{3-n}\text{SCl} \xrightarrow[\text{ether or THF*}]{0°C} \text{Ar}-\text{SCF}_n\text{Cl}_{3-n} (+\text{ArCl} + \text{ArX})$$

$$\left(\text{Ar} = \text{C}_6\text{H}_5, \; \text{H}_3\text{C}-\!\!\!\bigcirc\!\!\!-, \; \text{H}_3\text{C}-\!\!\!\bigcirc\!\!\!; \; n=2,3; \; X = \text{Cl, Br, I}\right)$$

The perfluorohalogenomethylmercapto compound is formed in about 50% yield, the by-products being aryl halides from the Grignard reagent in 5–15% yield. It is assumed that the reaction follows an S_N2 mechanism with a cyclic intermediate state in analogy with the reaction of Grignard reagents with alkyl halides (98, 155), e.g.,

$$\begin{array}{c}\text{CF}_3-\overset{\delta+}{\text{S}}\cdots\text{Cl}\\ \quad\;\;\;\;\vdots\;\;\;\;\;\;\;\;\;\text{MgCl}\\ \text{H}_5\underset{\delta-}{\text{C}_6}\;\;\;\;\text{C}_6\text{H}_5\\ \quad\;\;\;\text{Mg}\\ \quad\;\;\;\;\;\;\text{Cl}\end{array} \longrightarrow \begin{array}{c}\text{CF}_3\text{S} + \text{MgCl}_2\\ |\\ \text{H}_5\text{C}_6 + \text{C}_6\text{H}_5\\ |\\ \text{MgCl}\end{array}$$

Aromatic compounds with electron-donor substituents, such as —N(CH$_3$)$_2$ or —OH, also react with CF$_3$SCl via direct condensation (4, 133):

$$2\,\text{C}_6\text{H}_5\text{N(CH}_3)_2 + \text{CF}_3\text{SCl} \xrightarrow[\text{ether}]{0°C} \text{CF}_3\text{S}-\!\!\!\bigcirc\!\!\!-\text{N(CH}_3)_2 + \text{C}_6\text{H}_5\text{N(CH}_3)_2 \cdot \text{HCl}$$

$$\text{R}-\text{C}_6\text{H}_4-\text{OH} + \text{C}_5\text{H}_5\text{N} + \text{CF}_3\text{SCl} \xrightarrow[\text{CHCl}_3]{0°C} \text{CF}_3\text{S}-\text{C}_6\text{H}_3(\text{R})-\text{OH} + \text{C}_5\text{H}_5\text{N}\cdot\text{HCl}$$

($R = H, o$-CH_3, m-CH_3, o-OH, m-Cl)

In the case of phenol derivatives, the intermediate formation of sulfenate-esters ArOSCF$_3$, with subsequent rapid conversion to the cyclic substituted product is discussed elsewhere (3).

In the presence of Friedel–Crafts catalysts (BF$_3$, FeCl$_3$, and others), comparatively less reactive compounds, such as benzene (50°C, 2 hr) and toluene (100°C, 4 hr), can be reacted in an autoclave with CF$_3$SCl to give C$_6$H$_5$—SCF$_3$ or a mixture of m- and p-CF$_3$S—C$_6$H$_4$CH$_3$, respectively. Chloro- and bromobenzene react under more vigorous conditions (200°C, 2 hr) with the formation of a mixture of ortho-, meta-, and para-

* THF, tetrahydrofuran.

isomers, the catalyst being anhydrous hydrogen fluoride (4). Compound C_6F_5SCl reacts with pentafluorobenzene in the presence of SbF_5 to give $C_6F_5SC_6F_5$ in 95% yield (184). Trifluoromethane sulfonic acid, introduced by Effenberger and Epple (27) for Friedel–Crafts acylations of aromatic compounds, seems to act as a particularly suitable catalyst. It is assumed that the perfluorosulfonic acid–carbonic acid anhydrides initially formed are responsible for the catalytic influence. On addition of 0.1 mole CF_3SO_3H the reaction of benzene with CF_3SCl may be carried out under mild conditions in high yield (47):

$$C_6H_6 + CF_3SCl \xrightarrow[CF_3SO_3H]{0°C/5\text{-}hr} C_6H_5\text{-}SCF_3 + HCl$$

(70%)

In the reaction of trifluoromethylmercaptobenzene with CF_3SCl in the presence of CF_3SO_3H, chlorine-substituted products are obtained primarily beside small quantities of bis(trifluoromethylmercapto)-substituted compounds:

$$C_6H_5\text{-}SCF_3 + CF_3SCl \xrightarrow[CF_3SO_3H]{20°C/5\ hr} CF_3S\text{-}C_6H_4\text{-}SCF_3 + \text{(ortho isomer)}$$

(4%) (1%)

$$+ Cl\text{-}C_6H_4\text{-}SCF_3 + \text{(ortho Cl isomer)}$$

(50%) (10%)

No investigations have as yet been undertaken to elucidate the mechanism of these reactions, although ordinary electrophilic aromatic substitutions are the most likely. The attacking species is apparently R_fS^+ or an "activated complex" formed through coordination of the perfluorohalogenosulfenyl chloride with a Lewis or a Brønsted acid (4).

Hydroquinone reacts with CF_3SCl to afford, not as expected CF_3S-substituted compounds, but chlorohydroquinones instead (143). However, carrying out the reaction with 4-methoxyphenol in the presence of a threefold excess of pyridine and an excess of CF_3SCl, 2,6-bis(trifluoromethylmercapto)-4-methoxyphenol is formed in good yield. With concentrated nitric acid it can be oxidatively cleaved to 2,6-bis(trifluoromethylmercapto)-1,4-benzoquinone. This, in turn, reacts with CF_3SH on addition of pyridine to give 2,3,5-tris(trifluoromethylmercapto)hydroquinone, which is converted to the respective quinone by oxidation with N_2O_4 in dichloromethane in the presence of $MgSO_4$.

Addition of CF_3SH to this quinone affords the tetra-substituted hydroquinone (*143*) (Scheme 2).

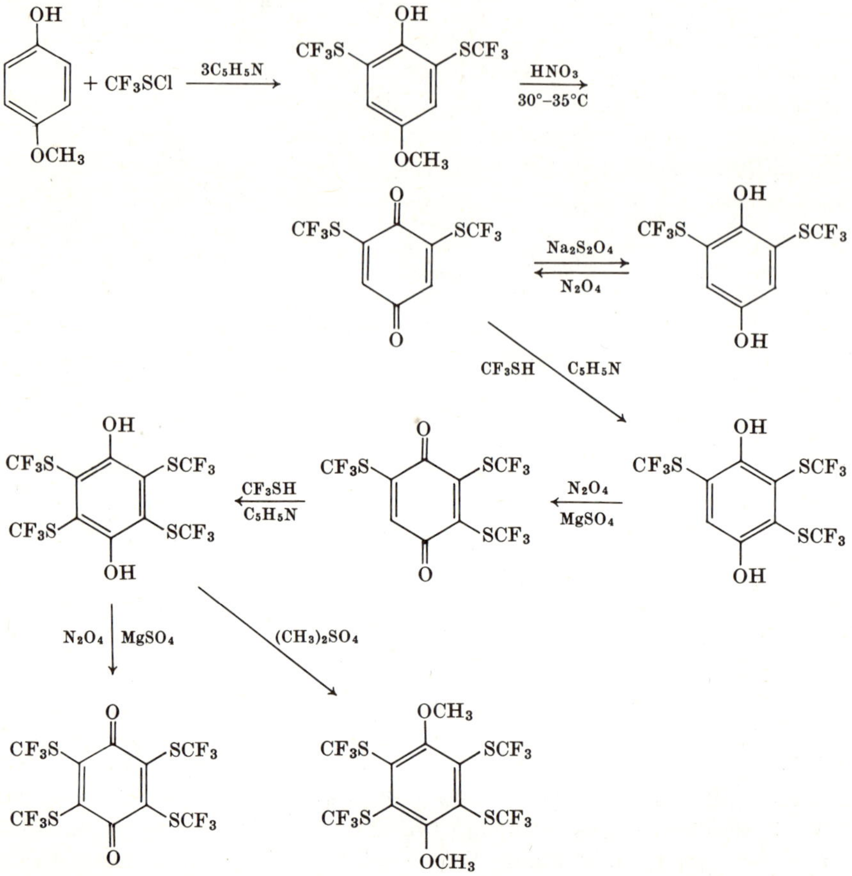

SCHEME 2

Heteroaromatics are subdivided, according to the electron influence of the heteroatom, into π-electron-deficient compounds and compounds with an excess of π electrons on the ring carbon atoms. The typical π-electron-deficient compound pyridine has so far been made to react only in one case: the reaction of lithium tetrakis(N-dihydropyridyl)-aluminate (LDPA) (*112–114*), obtainable from pyridine and lithium aluminum hydride, with trifluoromethanesulfenyl chloride in an excess of pyridine affords 3-trifluoromethylmercaptopyridine in low yield (13%) (*60*). This reaction probably occurs through sulfenylation of the 1,2-dihydropyridyl moiety of the LDPA with the formation of a 2,5-

dihydropyridine; this is followed by oxidation to the 3-substituted pyridine (34):

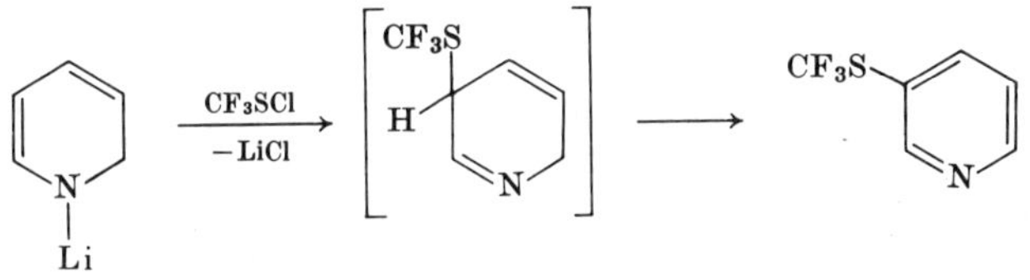

Compounds with excess of π electrons, as for example, pyrrole and thiophene, form a large number of substitution products in their reactions with perfluorohalogenosulfenyl halides (47, 60). Thus pyrrole reacts with an equimolar quantity of a sulfenyl chloride of the series CF_nCl_{3-n}-SCl ($n = 1, 2, 3$) with the formation of a mixture of isomers of monosubstituted compounds:

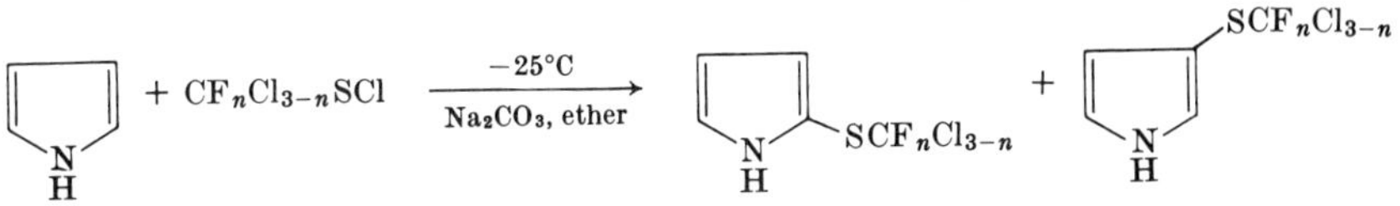

($n = 1, 2, 3$)

The yields decrease with decreasing degree of fluorination and at the same time the proportion of 2-substituted compounds increases.

An excess of sulfenyl chloride leads to disubstituted products only in the case of CF_3SCl:

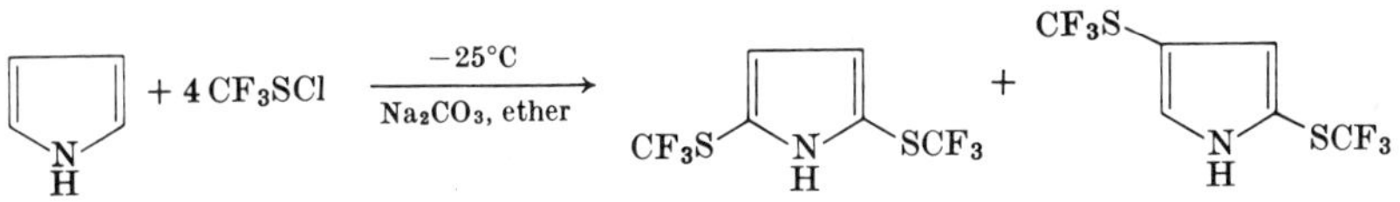

The reaction proceeds quantitatively in the direction shown only with a 1:4 excess of the sulfenyl chloride; the reaction in the stoichiometric ratio of 1:2 affords a mixture of mono- and disubstituted products.

With halogenocarbonylsulfenyl chlorides, 2-substituted compounds are formed exclusively:

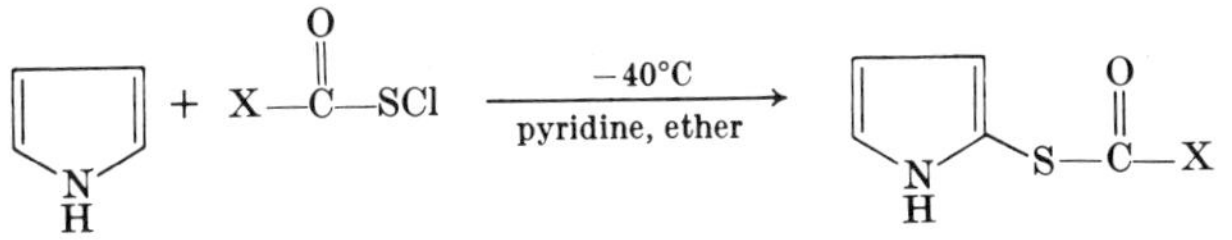

(X = F, Cl)

Attempts to N-substitute the pyrrole result exclusively in C-substituted products, e.g., [pyrrole-K] + CF₃SCl ⟶ [pyrrole-NH with SCF₃] + [pyrrole-NH with CF₃S and SCF₃]

Numerous N-substituted pyrroles undergo at elevated temperatures a conversion to C-substituted compounds. However, an analogous reaction course is not feasible since N-methylpyrrole also reacts with CF_3SCl to give C-substituted derivatives:

[N-methylpyrrole] + 4 CF₃SCl ⟶

[N-CH₃ pyrrole-SCF₃] + [CF₃S-N-CH₃ pyrrole-SCF₃] + [CF₃S-N-CH₃ pyrrole-SCF₃]

Owing to the extreme sensitivity of the pyrrole to acids, all reactions have to be carried out in high dilution and in presence of an HCl acceptor. The products can be kept for a prolonged period of time only in an extremely purified state.

Oxidation to the sulfoxide or sulfone proceeds selectively with the aid of m-chloroperbenzoic acid:

[pyrrole-SCF₃] →(m-Cl-C₆H₄-COOOH)→ [pyrrole-S(O)CF₃]
           →(2 m-Cl-C₆H₄-COOOH)→ [pyrrole-SO₂CF₃]

Indole and carbazole, which can be regarded theoretically as derivatives of pyrrole through its anellation with one or two benzene rings, show variable behavior toward sulfenyl halides:

[indole-Na] + $CF_nCl_{3-n}SCl$ →(benzene)→ [indole-SCF$_n$Cl$_{3-n}$]   (91)

($n = 1, 2$)

[indole] + CF₃SCl →(ether) [3-SCF₃-indole] (60)

[carbazole-N-MgI] + CF₃SCl →(ether, −MgClI) [carbazole-N-SCF₃] (60)

Electrophilic substitutions on thiophene, like those on benzene, can be carried out only in the presence of catalysts. In reactions with sulfenyl halides, SnCl₄ proved to be particularly suitable; in the case of the less reactive sulfenyl halides, Grignard reactions lead to the desired products:

[thiophene] + CF$_n$Cl$_{3-n}$SCl →(SnCl₄, −HCl) [2-SCF$_n$Cl$_{3-n}$-thiophene]
($n = 2, 3$)

[thiophene] + Cl—C(=O)—SCl →(SnCl₄, −HCl) [2-(S—C(=O)—Cl)-thiophene]

[2-MgBr-thiophene] + CFCl₂SCl →(ether, −MgBrCl) [2-SCFCl₂-thiophene]

[2-MgBr-thiophene] + F—C(=O)—SCl →(ether, −MgBrCl) [2-(S—C(=O)—F)-thiophene]

The reaction with fluorocarbonylsulfenyl chloride has to be carried out via the Grignard reagent, since, if SnCl₄ is used, a fluorine–chlorine exchange occurs.

The thiophene derivatives are relatively stable in the pure state. The fluorine atoms of the trifluoromethylmercapto group can be quantitatively substituted with chlorine by boron trichloride; with H_2O_2, oxidation to the sulfone takes place. Further substitution is achieved in both cases in the presence of perfluorosulfonic acids:

$$\text{CF}_3\text{S}-\underset{\text{H}}{\underset{|}{\text{N}}}-\text{SCF}_3 + \text{CF}_3\text{SCl} \xrightarrow{\text{CF}_3\text{SO}_3\text{H}} \text{CF}_3\text{S}-\underset{\text{H}}{\underset{|}{\text{N}}}\text{(SCF}_3\text{)}-\text{SCF}_3$$

$$\text{CF}_3\text{S}-\underset{\text{H}}{\underset{|}{\text{N}}}(\text{SCF}_3)-\text{SCF}_3 + \text{CF}_3\text{SCl} \xrightarrow{\text{CF}_3\text{SO}_3\text{H}} \text{CF}_3\text{S}-\underset{\text{H}}{\underset{|}{\text{N}}}(\text{SCF}_3)_2-\text{SCF}_3$$

$$\underset{\text{H}}{\underset{|}{\text{N}}}-\text{SCF}_n\text{Cl}_{3-n} + \text{CF}_n\text{Cl}_{3-n}\text{SCl} \xrightarrow{\text{C}_4\text{F}_9\text{SO}_3\text{H}}$$

$$\text{Cl}_{3-n}\text{F}_n\text{CS}-\underset{\text{H}}{\underset{|}{\text{N}}}-\text{SCF}_n\text{Cl}_{3-n} \;+\; \text{Cl}_{3-n}\text{F}_n\text{CS}-\underset{\text{H}}{\underset{|}{\text{N}}}(\cdot)-\text{SCF}_n\text{Cl}_{3-n}$$
$$(n = 1, 2, 3)$$

$$\underset{\text{S}}{}-\text{SCF}_3 + \text{CF}_n\text{Cl}_{3-n}\text{SCl} \xrightarrow{\text{CF}_3\text{SO}_3\text{H}} \text{Cl}_{3-n}\text{F}_n\text{CS}-\underset{\text{S}}{}-\text{SCF}_3$$
$$(n = 1, 2, 3)$$

$$\underset{\text{S}}{}-\text{SCF}_2\text{Cl} + \text{CF}_2\text{ClSCl} \xrightarrow{\text{CF}_3\text{SO}_3\text{H}} \text{ClF}_2\text{CS}-\underset{\text{S}}{}-\text{SCF}_2\text{Cl}$$

An interesting effect can be observed in the further reaction of 2,5-bis(trifluoromethylmercapto)thiophene with CF_3SCl: beside the small quantities of 2,3,5-tris(trifluoromethylmercapto)thiophene (1%), the main reaction product is 2,5-bis(trifluoromethylmercapto)-3-chlorothiophene:

$$\text{CF}_3\text{S}-\underset{\text{S}}{}-\text{SCF}_3 + \text{CF}_3\text{SCl} \xrightarrow{\text{CF}_3\text{SO}_3\text{H}} \begin{array}{c} \text{CF}_3\text{S}-\underset{\text{S}}{}(\text{SCF}_3)-\text{SCF}_3 \\ \\ \text{CF}_3\text{S}-\underset{\text{S}}{}(\text{Cl})-\text{SCF}_3 \end{array}$$

The latter reaction proceeds presumably through protonation of the starting material as follows:

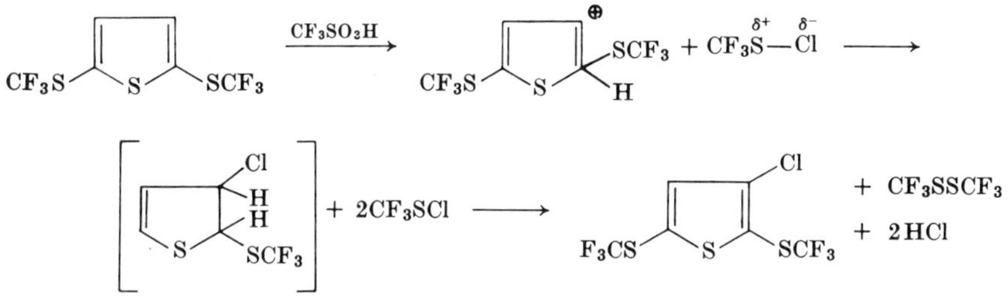

Reactions of uracil and structurally analogous compounds have been thoroughly studied (49, 50). Uracil (2,4-dihydroxypyrimidine), a component of the nucleic acids, can be classified as a heteroaromatic compound only with reserve, since the keto form dominates in the tautomeric equilibrium.

With CF_3SCl and in the presence of pyridine as HCl acceptor, a reaction takes place to give 5-trifluoromethylmercaptouracil. Under identical reaction conditions the sulfenyl halides $CF_nCl_{3-n}SCl$ ($n = 0, 1$) afford dinitrogen-substituted products:

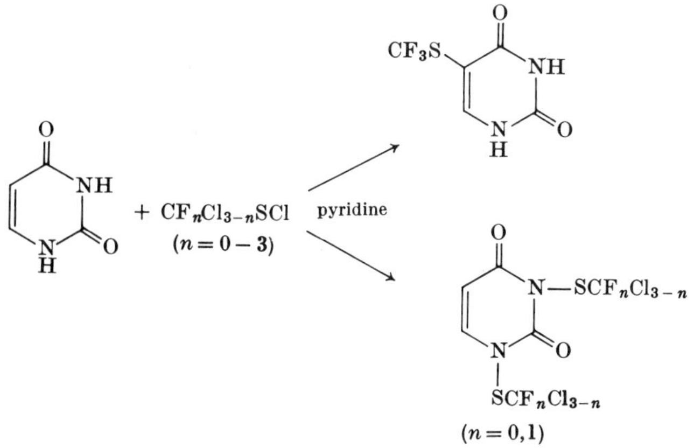

A theoretical interpretation of this different behavior is not yet available. In any case, the most obvious possibility that all sulfenyl halides attack initially at the nitrogen as the most electron abundant site (160), and that the CF_3S compound is then converted to the more stable 5-substituted product, is to be disregarded, since 1,3-dimethyluracil, with both nitrogen atoms blocked, gives with CF_3SCl the 5-substituted compounds in high yield.

Mononitrogen-substituted derivatives are obtained from sodium uracil and sulfenyl halides $CF_nCl_{3-n}SCl$ ($n = 0, 1, 2$) (49, 96):

$$\text{[uracil-Na]} + CF_nCl_{3-n}SCl \xrightarrow[\text{THF}]{-NaCl} \text{[N-SCF}_nCl_{3-n}\text{ uracil]}$$

The yield decreases with increasing degree of fluorination of the reactant sulfenyl halides. In the reaction with CF_3SCl, no appreciable quantity of the desired product could be isolated. Although in the mass spectrum of the reaction mixture a peak was observed corresponding to the molecular ion $M^+ = 212$, the position of the substituent could not be unequivocally determined.

All derivatives, sulfenylated at the nitrogen atom, are stable for a prolonged period of time only at low temperature. 5-Trifluoromethylmercaptouracil is an extraordinarily stable compound. In contrast to 5-bromouracil (129), the substituent cannot be exchanged by amines. Nevertheless, a complete fluorine–chlorine exchange can be brought about with the aid of boron trichloride, just as with other aromatic CF_3S compounds (170):

$$\text{[5-CF}_3\text{S-uracil]} + BCl_3 \xrightarrow[-BF_3]{\text{3 days/100°C}} \text{[5-CCl}_3\text{S-uracil]}$$

Oxidation to sulfoxide occurs with fuming nitric acid, whereas the sulfone is formed in sulfochromic acid:

$$\text{[5-CF}_3\text{S-uracil]} \xrightarrow[20°C/24\,hr]{\text{fuming HNO}_3} \text{[5-CF}_3\text{S(O)-uracil]}$$

$$\text{[5-CF}_3\text{S-uracil]} \xrightarrow[140°C/2\,hr]{CrO_3/H_2SO_4} \text{[5-CF}_3\text{SO}_2\text{-uracil]}$$

Perhalogenated sulfanes can be obtained by condensation of 5-mercaptouracil with sulfenyl halides, $CF_nCl_{3-n}SCl$ ($n = 1, 2, 3$):

[Reaction scheme: 5-mercaptouracil + $CF_nCl_{3-n}SCl \xrightarrow[THF]{-HCl}$ 5-($Cl_{3-n}F_nCSS$)-uracil]

Orotic acid (uracil-6-carboxylic acid), an intermediate in the biosynthesis of uracil, also reacts smoothly with CF_3SCl in pyridine to give the 5-substituted compound. The pyridinium salt initially formed can be easily cleaved with dilute hydrochloric acid:

[Reaction scheme: $C_5H_5N \cdot HOOC$-uracil + $CF_3SCl \xrightarrow{-C_5H_5N \cdot HCl}$ 5-CF_3S-6-HOOC-uracil]

The nucleic acid building stone, cytosine (4-amino-2-hydroxypyrimidine), again affords, under the same conditions, a 5-substituted product:

[Reaction scheme: cytosine + $CF_3SCl \xrightarrow{pyridine}$ 5-CF_3S-cytosine]

The result is surprising, as with the structurally quite similar adenine (6-aminopurine), N-substitution occurs only at the amino group (46):

[Reaction scheme: adenine + $CF_nCl_{3-n}SCl \xrightarrow{pyridine}$ $Cl_{3-n}F_nSC-NH$-purine]

($n = 1, 2, 3$)

However, NMR studies provide an explanation for the course of this reaction. They show that cytosine exists, at least in polar solvents, in the form of a zwitterion (103) and the electrophilic CF_3S group can no longer attack at the positively charged nitrogen.

Barbituric acid, a 2,4,5-trihydroxypyrimidine, which exists in the triketo form, is also able to react with CF_3SCl. Owing to the strong acidity of the methylene group, the reaction can take place in this case without an HCl acceptor as well in a suspension in THF. The 2:1 etherate formed initially splits off the solvent quantitatively on heating to 100°C:

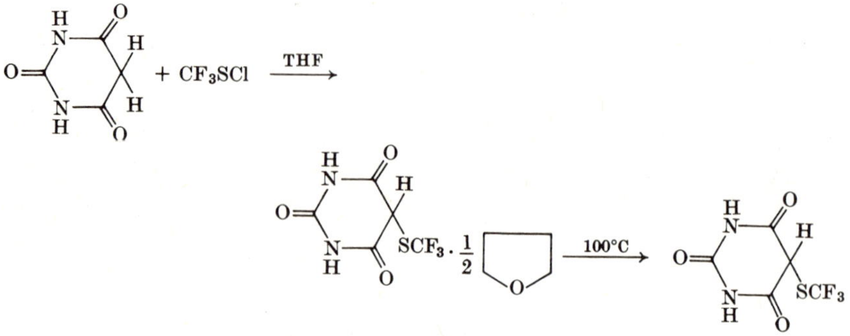

[barbituric acid] + CF$_3$SCl $\xrightarrow{\text{THF}}$ [5-SCF$_3$ barbituric acid · ½ THF] $\xrightarrow{100°C}$ [5-SCF$_3$ barbituric acid]

The IR spectrum suggests that the compound exists partly in the enolic form, just like the anhydrous 5-nitrobarbituric acid (*121*).

The proton remaining in the 5-position is evidently very unreactive. Metallation attempts with AgNO$_3$ lead only to the substitution of a nitrogen proton, as can be shown by the following ethylation:

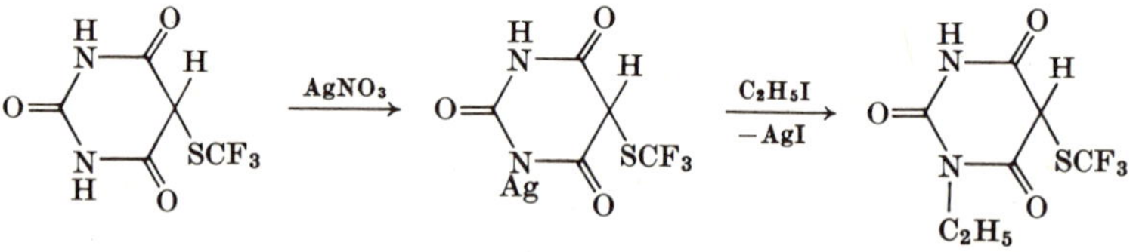

[5-SCF$_3$ barbituric acid] $\xrightarrow{\text{AgNO}_3}$ [N-Ag derivative] $\xrightarrow[-\text{AgI}]{\text{C}_2\text{H}_5\text{I}}$ [N-C$_2$H$_5$ derivative]

The latter compound differs considerably in its physical properties from the product resulting from a direct reaction of 5-ethylbarbituric acid and CF$_3$SCl:

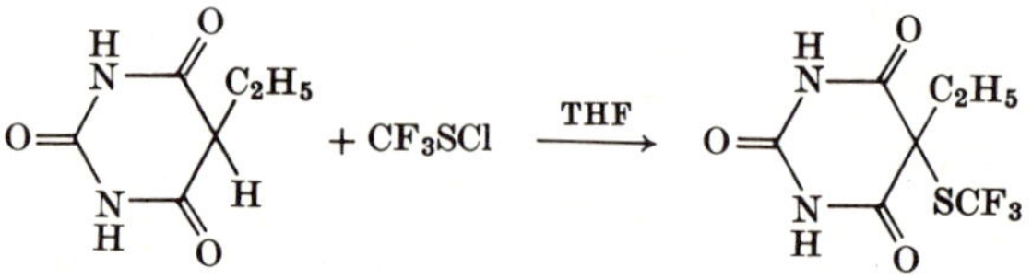

[5-ethylbarbituric acid] + CF$_3$SCl $\xrightarrow{\text{THF}}$ [5-ethyl-5-SCF$_3$ barbituric acid]

H. Cyclizations, Conversions and Reactions of Sulfenyl Chlorides with Metal Carbonyls

Heating FClC(NCS)SCl with exclusion of moisture (6–8 hr at 70°C) leads to the cyclic compounds 3-chloro-5,5-difluoro- and 3,5,5-trichloro-1,2,4-dithiazole. Presumably this ring-closure reaction takes place through the nonisolated 3,5-dichloro-5-fluoro-1,2,3-dithiazole (*20, 21*):

[Scheme showing: SCl, S, FC—N=C, Cl reacting to form a ring intermediate with F, Cl—C, N, CCl, S, S, which then gives F₂C—N, S, S, CCl (with CCl doubly bonded to N) and Cl₂C—N, S, S, CCl]

Attempts to react the difluoro-substituted ring with Hg(SCF₃)₂ did not lead to displacement of the chlorine by the CF₃S group, but instead CF₃SSCF₂NCS was formed quantitatively (21):

[Reaction scheme: F₂C—N ring with CCl + Hg(SCF₃)₂ → F₂C—N ring with CSCF₃ → CF₃SSCF₂NCS]

Similar conversions were also observed when an attempt was made to fluorinate the compounds CF₃SCFClSCl or (CF₃S)₂CClSCl with HgF₂ or HgCl₂. Here, too, only disulfanes resulted (20):

$$CF_3SC(X)ClSCl \xrightarrow{HgF_2} CF_3SSCCl_2X \quad (X=F, CF_3S)$$

Reactions of CF₃SCl with metal carbonyls have been investigated only to a minor extent. A reaction between Fe(CO)₅ and CF₃SCl takes place to give in poor yield the following binuclear complex, that exists in two isomeric forms (37):

[Structure of binuclear Fe complex with bridging S-CF₃ groups and CO ligands]

Reactions of Mn₂(CO)₁₀ with CF₃SCl take the following course (38):

$$2Mn_2(CO)_{10} + 4CF_3SCl \longrightarrow Mn_2(CO)_8(SCF_3)_2 + CF_3SSCF_3 + 2MnCl_2 + 12CO$$

VI. Characteristics of Perfluorohalogenoorganomercapto Groups

Characteristic physical data can be attributed to some perfluorohalogenomercapto groups in the large number of compounds synthesized. The most comprehensive data material applies to the following groups: CF₃S—, CF₂ClS—, CFCl₂S—, and C₆F₅S—.

Systematic IR spectroscopic studies furnish the following characteristic frequencies for the $CF_nCl_{3-n}S$— (n = 3, 2, 1) and C_6F_5S groups:

1. CF_3S group (122): $\nu_{as}(C-F) = 1205$–1155 cm^{-1}; $\nu_s(C-F) = 1135$–1095 cm^{-1}; $\delta_s(CF_3) = 765$–750 cm^{-1}; $\delta_{as}(CF_3) = 540$–510 cm^{-1}; $\nu(C-S) = 495$–445 cm^{-1}.

2. CF_2ClS group (61): $\nu_{as}(C-F) = 1200$–1090 cm^{-1}; $\nu_s(C-F) = 1090$–1050 cm^{-1}; $\nu(C-Cl) = 900$–850 cm^{-1}; $\delta_s(CF_2) = 680$–600 cm^{-1}; $\delta_{as}(CF_2) = 550$–500 cm^{-1}; $\nu(C-S) = 480$–400 cm^{-1}.

3. $CFCl_2S$ group (61): $\nu(C-F) = 1055$–1020 cm^{-1}; $\nu_{as}(C-Cl) = 850$–800 cm^{-1}; $\nu_s(C-Cl) = 760$–670 cm^{-1}; $\delta_s(C-F) = 570$–520 cm^{-1}; $\nu(C-S) = 450$–430 cm^{-1}.

4. C_6F_5S group* (9, 10): 1630 (m) 1510 (vs) cm^{-1}; ν(ring) = 1475 (vs), 1395 (m–s), 1360 (w), 1350 (w), 1270 (w–m), 1125 (m), 1080 (vs), 1055 (w) cm^{-1}; $\nu(C-F) = 970$ (vs), 905 (w) cm^{-1}; $\nu(C-S) = 860$ (vs), 715 (w–m) cm^{-1}. Bands with the intensity (w–m) do not appear in all compounds containing the C_6F_5S group.

The ^{19}F NMR chemical shifts of the $CF_nCl_{3-n}SN$ compounds (61) are all observed to be within very narrow limits: for CF_3SN-, $\delta = 51 \pm 4$ ppm; for CF_2ClSN-, $\delta = 38 \pm 3$ ppm; and for $CFCl_2SN-$, $\delta = 26 \pm 3$ ppm (relative to $CFCl_3$ as standard).

The electronegativity of the CF_3S group derived by different methods is found almost always to be 2.7 on Pauling's scale (12, 23, 28, 97). The Hammett–Taft and the Dewar constants obtained from pK_a values or from ^{19}F NMR shifts for CF_3S-substituted benzoic acids, anilines, and phenols or fluorobenzenes show that the CF_3S group has the highest electron affinity of all substituents containing divalent sulfur (17, 18, 26, 126, 146, 147, 149, 151, 163, 164, 183).

Acknowledgment

We are very grateful to Dr. N. Welcman for his help and useful remarks in preparing the manuscript.

References

1. Aleksandrov, A. M., Samusenko, Yu. V., Bratolyubova, A. G., and Yagupol'skii, L. M., *Zh. Org. Khim.* **9**, 69 (1973); see *Chem. Abstr.* **78**, 84215 (1973).
2. Aleksandrov, A. M., and Yagupol'skii, L. M., *Zh. Org. Khim.* **6**, 249 (1970); *J. Org. Chem. USSR* **6**, 242 (1970).
3. Andreades, S., U.S. Patent 3,081,350 (1961/1963); see *Chem. Abstr.* **59**, 5024 (1963).
4. Andreades, S., Harris, J. F., Jr., and Sheppard, W. A., *J. Org. Chem.* **29**, 898 (1964).
5. Banks, R. E., Haslam, G. M., Haszeldine, R. N., and Peppin, A., *J. Chem. Soc.*, C p. 1171 (1966).

* Intensities: weak (w); medium (m); strong (s); very strong (vs).

6. Banks, R. E., Haszeldine, R. N., Korsa, D. R., Rickett, F. E., and Young, I. M., *J. Chem. Soc.*, *C* p. 1660 (1969).
7. Barr, D. A., and Haszeldine, R. N., *J. Chem. Soc.*, *London* p. 3428 (1956).
8. Bayreuther, H., and Haas, A., *Chem. Ber.* **106**, 1418 (1973).
9. Beck, W., and Stetter, K. H., *Inorg. Nucl. Chem. Lett.* **2**, 383 (1966).
10. Beck, W., Stetter, K. H., Tadros, S., and Schwarzhans, K. E., *Chem. Ber.* **100**, 3944 (1967).
11. Blake, E. S., *J. Amer. Chem. Soc.* **65**, 1267 (1943).
12. Blaschette, A., Haas, A., and Klug, W., *Monatsh. Chem.* **101**, 1089 (1970).
13. Block, S. S., and Weidner, J. P., *Nature (London)* **214**, 478 (1967).
14. Brintzinger, H., Pfannstiel, K., Koddebusch, H., and Kling, K. E., *Chem. Ber.* **83**, 87 (1950).
15. Brower, K. R., and Douglas, I. B., *J. Amer. Chem. Soc.* **73**, 5787 (1951).
16. Bur-Bur, F., Haas, A., and Klug, W., *Chem. Ber.* **108**, 1365 (1975).
17. Bystrov, V. F., Utkanskaya, E. Z., and Yagupol'skii, L. M., *Opt. Spektrosk.* **10**, 138 (1961); *Opt. Spectrosc. (USSR)* **10**, 68 (1961); see *Chem. Abstr.* **55**, 10075 (1961).
18. Bystrov, V. F., Yagupol'skii, L. M., Stepanyants, A. U., and Fialkov, Yu. A., *Dokl. Akad. Nauk SSSR* **153**, 1321 (1963); *Proc. Acad. Sci. USSR, Chem. Sect.* **153**, 1019 (1963).
19. Cullen, W. R., and Dhaliwal, P. S., *Can. J. Chem.* **45**, 379 (1967).
20. Dahms, G., Diderrich, G., Haas, A., and Yazdanbakhsch, M., *Chem.-Ztg.* **98**, 109 (1974).
21. Dahms, G., Haas, A., and Klug, W., *Chem. Ber.* **104**, 2732 (1971).
22. Dickey, J. B., U.S. Patent 2,436,100 (1948); see *Chem. Abstr.* **42**, 3578 (1948).
23. Downs, A. J., Ph.D. Thesis, Cambridge University (1961).
24. Downs, A. J., and Haas, A., *Spectrochim Acta, Part A* **23**, 1023 (1967).
25. Durell, W. S., Stump, E. C., Westmorehand, G., and Padgett, C. D., *J. Polym. Sci., Part A* **3**, 4065 (1965).
26. Eaton, D. R., and Sheppard, W. A., *J. Amer. Chem. Soc.* **85**, 1310 (1963).
27. Effenberger, F., and Epple, G., *Angew. Chem. Int. Ed. Engl.* **11**, 299 and 300 (1972).
28. Emeléus, H. J., and Haas, A., *J. Chem. Soc.*, *London* p. 1272 (1963).
29. Emeléus, H. J., and MacDuffie, D. E., *J. Chem. Soc.*, *London* p. 2597 (1961).
30. Emeléus, H. J., and Nabi, S. N., *J. Chem. Soc.*, *London*. p. 1103 (1960).
31. Emeléus, H. J., and Pugh, H., *J. Chem. Soc.*, *London* p. 1108 (1960).
32. Englin, M. A., Makarov, S. P., Dubov, S. S., and Yakubovich, A. Ya., *Zh. Obshch. Khim.* **35**, 1412 (1965); *J. Gen. Chem. USSR* **35**, 1415 (1965).
33. Furin, G. G., Terent'eva, T. V., and Yakobson, G. G., *Izv. Sib. Otd. Akad. Nauk SSSR, Ser. Khim. Nauk* **14**, 78 (1972); see *Chem. Abstr.* **78**, 83964 (1973).
34. Giam, C. S., and Abbott, S. D., *J. Amer. Chem. Soc.* **93**, 1294 (1971).
35. Gielow, P., and Haas, A., *Chem.-Ztg.* **95**, 423 and 1010 (1971).
36. Gielow, P., and Haas, A., *Z. Anorg. Allg. Chem.* **394**, 53 (1972).
37. Grobe, J., and Kober, F., *Z. Naturforsch. B* **24**, 1346 (1969).
38. Grobe, J., and Kober, F., *J. Organometal. Chem.* **24**, 191 (1970).
39. Haas, A., *Chem. Ber.* **97**, 2189 (1964).
40. Haas, A., *Chem. Ber.* **98**, 111 (1965).
41. Haas, A., *Chem. Ber.* **98**, 1709 (1965).
42. Haas, A., *in* "Gmelin's Handbuch der anorganischen Chemie", 8th ed., Vol. 9, Suppl. Ser. Verlag Chemie, Weinheim, 1973.

43. Haas, A., unpublished results (1973).
44. Haas, A., and Bayreuther, H., unpublished results (1972).
45. Haas, A., and Diderrich, G., unpublished results (1973).
46. Haas, A., and Hellwig, V., *Chem. Ber.* **108**, 334 (1975).
47. Haas, A., and Hellwig, V., private communication.
48. Haas, A., Helmbrecht, J., Klug, W., Koch, B., Reinke, H., and Sommerhoff, J., *J. Fluorine Chem.* **3**, 1 (1973/1974).
49. Haas, A., and Hinsch, W., *Chem. Ber.* **104**, 1855 (1971).
50. Haas, A., and Hinsch, W., *Chem. Ber.* **105**, 1768 and 1887 (1972).
51. Haas, A., and Klug, W., *Angew. Chem., Int. Ed. Engl.* **5**, 845 (1966).
52. Haas, A., and Klug, W., *Angew. Chem., Int. Ed. Engl.* **6**, 940 (1967).
53. Haas, A., and Klug, W., *Chem. Ber.* **101**, 2609 (1968).
54. Haas, A., and Klug, W., *Chem. Ber.* **101**, 2617 (1968).
55. Haas, A., and Klug, W., unpublished results (1968).
56. Haas, A., Klug, W., and Marsmann, H., *Chem. Ber.* **105**, 820 (1972).
57. Haas, A., and Lorenz, R., *Z. Anorg. Allg. Chem.* **385**, 33 (1971).
58. Haas, A., and Lorenz, R., *Chem. Ber.* **105**, 273 (1972).
59. Haas, A., and Lorenz, R., *Chem. Ber.* **105**, 3161 (1972).
60. Haas, A., and Niemann, U., private communication.
61. Haas, A., and Oh, D. Y., *Chem. Ber.* **100**, 480 (1967).
62. Haas, A., and Oh, D. Y., *Chem. Ber.* **102**, 77 (1969).
63. Haas, A., and Peach, M. E., *Z. Anorg. Allg. Chem.* **338**, 299 (1965).
64. Haas, A., Peach, M. E., and Schott, P., *Angew. Chem., Int. Ed. Engl.* **4**, 440 (1965).
65. Haas, A., and Plaß, V., *Chem. Ber.* **105**, 2047 (1972).
66. Haas, A., and Reinke, H., *Angew. Chem., Int. Ed. Engl.* **6**, 705 (1967).
67. Haas, A., and Reinke, H., *Chem. Ber.* **102**, 2718 (1969).
68. Haas, A., Reinke, H., and Sommerhoff, J., *Angew. Chem., Int. Ed. Engl.* **9**, 466 (1970).
69. Haas, A., and Schott, P., *Chem. Ber.* **99**, 3103 (1966).
70. Haas, A., and Schott, P., *Chem. Ber.* **101**, 3407 (1968).
71. Haas, A., and Sheppard, N., *J. Chem. Soc., London* p. 6613 (1965).
72. Haas, A., and Yazdanbakhsch, M., unpublished results (1972).
73. Ham. N. S., *J. Amer. Chem. Soc.* **83**, 751 (1961).
74. Harris, J. F., Jr., U.S. Patent 3,048,569 (1960/1962); see *Chem. Abstr.* **57**, 16886 (1962).
75. Harris, J. F., Jr., U.S. Patent 3,062,894 (1961/1962); see *Chem. Abstr.* **58**, 8907 (1963).
76. Harris, J. F., Jr., *J. Amer. Chem. Soc.* **84**, 3148 (1962).
77. Harris, J. F., Jr., U.S. Patent 3,347,765 (1963/1967); see *Chem. Abstr.* **68**, 2581 (1968).
78. Harris, J. F., Jr., *J. Org. Chem.* **31**, 931 (1966).
79. Harris, J. F., Jr., *J. Org. Chem.* **32**, 2063 (1967).
80. Harris, J. F., Jr., *J. Org. Chem.* **37**, 1340 (1972).
81. Harris, J. F., Jr., and Joyce, R. M., U.S. Patent 3,062,893 (1960/1962); see *Chem. Abstr.* **58**, 7869 (1963).
82. Hartzler, H. D., U.S. Patent 3,201,449 (1962/1965); see *Chem. Abstr.* **63**, 16220 (1965).
83. Hartzler, H. D., *J. Org. Chem.* **29**, 1194 (1964).
84. Haszeldine, R. N., and Kidd, J. M., *J. Chem. Soc., London* p. 3219 (1953).

85. Haszeldine, R. N., and Kidd, J. M., *J. Chem. Soc., London* p. 2901 (1955).
86. Hauptschein, M., Braid, M., and Lawlor, F. E., *J. Amer. Chem. Soc.* **79**, 6248 (1957).
87. Hauptschein, M., Braid, M., and Lawlor, F. E., U.S. Patent 3,019,258 (1958/1962); see *Chem. Abstr.* **57**, 9666 (1962).
88. Hauptschein, M., Braid, M., and Lawlor, F. E., U.S. Patent 3,100,228 (1960/1963); see *Chem. Abstr.* **60**, 408 (1964).
89. Hauptschein, M., and Oesterling, R. E., U.S. Patent 3,209,036 (1963/1965); see *Chem. Abstr.* **63**, 17904 (1965).
90. Havlik, A. J., and Kharasch, N., *J. Amer. Chem. Soc.* **78**, 1207 (1956).
91. Hennart, C., French Patent 1,497,492 (1966/1967); see *Chem. Abstr.* **69**, 106555 (1968).
92. I. G. Farbenindustrie, AG., French Patent 820,796 (1937); see *Chem. Abstr.* **32**, 3422 (1938).
93. I. G. Farbenindustrie AG., British Patent 479,774 (1938); see *Chem. Abstr.* **32**, 5226 (1938).
94. I.G. Farbenindustrie AG., French Patent 828,572 (1938); see *Chem. Abstr.* **33**, 396 (1939).
95. I.G. Farbenindustrie AG., British Patent 503,920 (1939); see *Chem. Abstr.* **33**, 7586 (1939).
96. Jeney, E., and Zsolnai, T., *Zentralbl. Bakteriol., Parasitenk., Infektionskr. Hyg., Abt. 1: Orig.* **193**, 516 (1964); see *Chem. Abstr.* **62**, 3106 (1965).
97. Kagarise, R. E., *J. Amer. Chem. Soc.* **77**, 1377 (1955).
98. Kharasch, M. S., and Reinmuth, O., "Grignard Reactions of Nonmetallic Substances," p. 1048. Prentice-Hall, Englewood Cliffs, New Jersey, 1954.
99. Kloosterziel, H., *Rec. Trav. Chim. Pays-Bas* **80**, 1234 (1961).
100. Knunyants, I. L., and Fokin, A. V., *Izv. Akad. Nauk SSSR, Otd. Khim. Nauk* p. 705 (1955); *Bull. Acad. Sci. USSR, Div. Chem. Sci.* p. 627 (1955).
101. Knunyants, I. L., Rozhkov, I. N., Aleksandrov, A. M., and Yagupol'skii, L. M., *Zh. Obshch. Khim.* **37**, 1277 (1967); *J. Gen. Chem. USSR* **37**, 1210 (1967).
102. Kober, E., *J. Amer. Chem. Soc.* **81**, 4810 (1959).
103. Kokko, J. P., Goldstein, J. H., and Mardell, L., *J. Amer. Chem. Soc.* **83**, 2909 (1961).
104. Krespan, C. G., U.S. Patent 3,099,688 (1960/1963); see *Chem. Abstr.* **60**, 1597 (1964).
105. Kühle, E., *Synthesis* p. 561 (1970).
106. Kühle, E., *Synthesis* pp. 563 and 617 (1971).
107. Kühle, E., "The Chemistry of the Sulfenic Acids," Enke, Stuttgart, 1973.
108. Kühle, E., Haas, A., Klug, W., and Grewe, F., DOS. 1,908,680 (1969/1970); see *Chem. Abstr.* **73**, 109522 (1970).
109. Kühle, E., Klauke, E., and Grewe, F., *Angew. Chem.* **76**, 807 (1964).
110. Kühle, E., Klauke, E., and Unterstenhöfer, G., French Patent 1,339,765 (1961/1963); see *Chem. Abstr.* **60**, 5519 (1964).
111. Kühle, E., Klauke, E., and Weiss, W., German Patent 1,219,455 (1965/1966); see *Chem. Abstr.* **65**, 13553 (1966).
112. Lansbury, P. T., and Peterson, J. O., *J. Amer. Chem. Soc.* **83**, 3537 (1961).
113. Lansbury, P. T., and Peterson, J. O., *J. Amer. Chem. Soc.* **84**, 1759 (1962).
114. Lansbury, P. T., and Peterson, J. O., *J. Amer. Chem. Soc.* **85**, 2236 (1963).
115. Lecher, H., and Holschneider, F., *Ber. Deut. Chem. Ges.* **57**, 755 (1924).
116. Lecher, H., and Simon, K., *Ber. Deut. Chem. Ges.* **54**, 632 (1921).

117. Lecher, H., and Wittwer, M., *Ber. Deut. Chem. Ges.* **55**, 1474 (1922).
118. Lecher, H., and Wittwer, M., *Ber. Deut. Chem. Ges.* **55**, 1481 (1922).
119. Ludovici, W., private communication.
120. Middleton, W. J., and Sharkey, W. H., *J. Org. Chem.* **30**, 1384 (1965).
121. Mihai, F., and Nutin, R., *Rev. Roum. Chim.* **13**, 39 (1968); see *Chem. Abstr.* **69**, 105687 (1968).
122. Nabi, S. N., and Sheppard, N., *J. Chem. Soc., London* p. 3439 (1959).
123. Neil, R. J., and Peach, M. E., *J. Fluorine Chem.* **1**, 257 (1971/1972).
124. Neil, R. J., Peach, M. E., and Spinney, H. G., *Inorg. Nucl. Chem. Lett.* **6**, 509 (1970).
125. Nodiff, E. A., Lipschutz, S., Craig, P. N., and Gordon, M., *J. Org. Chem.* **25**, 60 (1960).
126. Orda, V. V., Yagupol'skii, L. M., Bystrov, V. F., and Stepanyants, A. U., *Zh. Obshch. Khim.* **35**, 1628 (1965); *J. Gen. Chem. USSR* **35**, 1631 (1965).
127. Pacini, H. A., Pavlath, A. E., and Foley, W. M., U.S. Patent 3,169,104 (1961/1965); see *Chem. Abstr.* **63**, 1702 (1965).
128. Petrov, K. A., and Neimysheva, A. A., *Zh. Obshch. Khim.* **29**, 3401 (1959); *J. Gen. Chem. USSR* **29**, 3362 (1959).
129. Phillips, A. P., *J. Amer. Chem. Soc.* **73**, 1061 (1951).
130. Putnam, R. E., and Sharkey, W., *J. Amer. Chem. Soc.* **79**, 6526 (1957).
131. Ratcliffe, C. T., and Shreeve, J. M., *J. Amer. Chem. Soc.* **90**, 5403 (1968).
132. Redwood, M. E., and Willis, C. J., *Can. J. Chem.* **45**, 389 (1967).
133. Richert, H., Belgian Patent 624,397 (1963); see *Chem. Abstr.* **59**, 9893 (1963).
134. Robson, P., Smith, T. A., Stephens, R., and Tatlow, J. C., *J. Chem. Soc., London* p. 3692 (1963).
135. Rochat, W. V., and Gard, G. L., *J. Org. Chem.* **34**, 4173 (1969).
136. Rosenberg, R. M., and Muetterties, E. L., *Inorg. Chem.* **1**, 756 (1962).
137. Sartori, P., and Golloch, A., *Chem. Ber.* **103**, 3936 (1970).
138. Sartori, P., and Golloch, A., *Chem. Ber.* **104**, 967 (1971).
139. Sauer, D. T., and Shreeve, J. M., *J. Fluorine Chem.* **1**, 1 (1971/1972).
140. Scherer, O., *Angew. Chem.* **52**, 457 (1939).
141. Scherer, O., Korinth, J., and Starck, D., German Patent 1,232,954 (1965/1967); see *Chem. Abstr.* **66**, 65093 (1967).
142. Schuhmacher, W., Scherer, O., and Müller, F., U.S. Patent 2,191,062 (1940); see *Chem. Abstr.* **34**, 4588 (1940).
143. Scribner, R. M., *J. Org. Chem.* **31**, 3671 (1966).
144. Seel, F., and Gombler, W., *Angew. Chem., Int. Ed. Engl.* **8**, 773 (1969).
145. Seel, F., Gombler, W., and Budenz, R., *Angew. Chem., Int. Ed. Engl.* **6**, 706 (1967).
146. Sheppard, W. A., *J. Amer. Chem. Soc.* **83**, 4860 (1961).
147. Sheppard, W. A., *J. Amer. Chem. Soc.* **85**, 1314 (1963).
148. Sheppard, W. A., *J. Org. Chem.* **29**, 895 (1964).
149. Sheppard, W. A., *Tetrahedron* **27**, 945 (1971).
150. Sheppard, W. A., and Harris, J. F., Jr., *J. Amer. Chem. Soc.* **82**, 5106 (1960).
151. Shepperd, W. A., and Taft, R. W., *J. Amer. Chem. Soc.* **94**, 1919 (1972).
152. Simons, J. H., U.S. Patent 2,519,983 (1950); see *Chem. Abstr.* **45**, 51 (1951).
153. Smith, Kline and French Laboratories, British Patents 851,951 and 851,952 (1960); see *Chem. Abstr.* **56**, 11444 (1962).
154. Sommerhoff, J., Dissertation, University of the Ruhr, Bochum (1972).
155. Swain, C. G., *J. Amer. Chem. Soc.* **70**, 1119 (1948).

156. Tarbell, D. S., and Harnish, D. P., *Chem. Rev.* **49**, 79 (1951).
157. Tattershall, B. W., and Cady, G. H., *J. Inorg. Nucl. Chem.* **29**, 2819 (1967).
158. Tullock, C. W., U.S. Patent 2,884,453 (1959); see *Chem. Abstr.* **53**, 16963 (1959).
159. Tullock, C. W., and Coffman, D. D., *J. Org. Chem.* **25**, 2016 (1960).
160. Veillard, A., and Pullmann, B., *J. Theor. Biol.* **4**, 37 (1963).
161. Weidner, J. P., and Block, S. S., *J. Med. Chem.* **10**, 1167 (1967).
162. Weiss, W., German Patent 1,224,720 (1964); see *Chem. Abstr.* **65**, 12112 (1966).
163. Yagupol'skii, L. M., Bystrov, V. F., Stepanyants, A. U., and Fialkov, Yu.A., *Zh. Obshch. Khim.* **34**, 3682 (1964); *J. Gen. Chem. USSR* **34**, 3731 (1964).
164. Yagupol'skii, L. M., Bystrov, V. F., and Utkanskaya, E. Z., *Dokl. Akad. Nauk SSSR* **135**, 377 (1960); *Proc. Acad. Sci. USSR, Phys. Chem. Sect.* **135**, 1059 (1960).
165. Yagupol'skii, L. M., and Gandel'sman, L. Z., *Zh. Obshch. Khim.* **35**, 1252 (1965); *J. Gen. Chem. USSR* **35**, 1259 (1965).
166. Yagupol'skii, L. M., and Gandel'sman, L. Z., *Zh. Obshch. Khim.* **37**, 2101 (1967); *J. Gen. Chem. USSR* **37**, 1992 (1967).
167. Yagupol'skii, L. M., and Gruz, B. E., *Zh. Obshch. Khim.* **31**, 1315 (1961); *J. Gen. Chem. USSR* **31**, 1219 (1961).
168. Yagupol'skii, L. M., and Kiprianov, A. J., *Zh. Obshch. Khim.* **22**, 2216 (1952); *J. Gen. Chem. USSR* **22**, 2273 (1952).
169. Yagupol'skii, L. M., Klyushnik, G. I., and Troitskaya, V. I., *Zh. Obshch. Khim.* **34**, 307 (1964); *J. Gen. Chem. USSR* **34**, 304 (1964).
170. Yagupol'skii, L. M., and Kondratenko, N. V., *Zh. Obshch. Khim.* **37**, 1770 (1967); *J. Gen. Chem. USSR* **37**, 1685 (1967).
171. Yagupol'skii, L. M., Krosovitskii, B. M., Blinov, V. A., Sidneva, K. M., and Pereyaslova, D. D., *Ukr. Khim. Zh.* **26**, 389 (1960); see *Chem. Abstr.* **55**, 3063 (1961).
172. Yagupol'skii, L. M., and Marenets, M. S., *Zh. Obshch. Khim.* **24**, 887 (1954); *J. Gen. Chem. USSR* **24**, 885 (1954).
173. Yagupol'skii, L. M., and Marenets, M. S., *Zh. Obshch. Khim.* **25**, 1771 (1955); *J. Gen. Chem. USSR* **25**, 1725 (1955).
174. Yagupol'skii, L. M., and Marenets, M. S., *Zh. Obshch. Khim.* **26**, 101 (1956); *J. Gen. Chem. USSR* **26**, 99 (1956).
175. Yagupol'skii, L. M., and Marenets, M. S., *Zh. Obshch. Khim.* **27**, 1395 (1957); *J. Gen. Chem. USSR* **27**, 1477 (1957).
176. Yagupol'skii, L. M., and Marenets, M. S., *Zh. Obshch. Khim.* **29**, 278 (1959); *J. Gen. Chem. USSR* **29**, 281 (1959).
177. Yagupol'skii, L. M., and Nazaretyan, V. P., *Ukr. Khim. Zh.* **33**, 617 (1967); see *Chem. Abstr.* **67**, 91670 (1967).
178. Yagupol'skii, L. M., and Troitskaya, V. I., *Zh. Obshch. Khim.* **29**, 2409 (1959); *J. Gen. Chem. USSR* **29**, 2374 (1959).
179. Yagupol'skii, L. M., and Troitskaya, V. I., *Zh. Obshch. Khim.* **29**, 2730 (1959); *J. Gen. Chem. USSR* **29**, 2697 (1959).
180. Yagupol'skii, L. M., Troitskaya, V. I., Gruz, B. E., and Kondratenko, N. V., *Zh. Obshch. Khim.* **35**, 1644 (1965); *J. Gen. Chem. USSR* **35**, 1645 (1965).
181. Yagupol'skii, L. M., Troitskaya, V. I., Levkow, I. I., Lifshits, E. B., Yufa, P. A., and Baroyn, N., *Zh. Obshch. Khim.* **37**, 166 (1967); *J. Gen. Chem. USSR* **37**, 174 (1967)

182. Yagupol'skii, L. M., Vishnevskaya, G. O., and Kaganovskaya, M. I., *Zh. Obshch. Khim.* **33**, 2721 (1963); *J. Gen. Chem. USSR* **33**, 2650 (1963).
183. Yagupol'skii, L. M., and Yagupol'skaya, L. N., *Dokl. Akad. Nauk SSSR* **134**, 1381 (1960); *Proc. Acad. Sci. USSR, Chem. Sect.* **134**, 1207 (1960).
184. Yakobson, G. G., Furin, G. G., and Terent'eva, T. V., *Izv. Akad. Nauk SSSR, Ser. Khim.* **9**, 2128 (1972); see *Chem. Abstr.* **78**, 3875 (1973).
185. Yarovenko, N. N., Motornyi, S. P., Vasil'eva, A. S., and Gershzon, T. P., *Zh. Obshch. Khim.* **29**, 2163 (1959); *J. Gen. Chem. USSR* **29**, 2129 (1959).
186. Yarovenko, N. N., and Vasil'eva, A. S., *Zh. Obshch. Khim.* **29**, 3786 (1959); *J. Gen. Chem. USSR* **29**, 3747 (1959).
187. Yarovenko, N. N., and Vasil'eva, A. S., *Zh. Obshch. Khim.* **29**, 3792 (1959); *J. Gen. Chem. USSR* **29**, 3754 (1959).
188. Zaborowski, L. M., and Shreeve, J. M., *J. Amer. Chem. Soc.* **92**, 3665 (1970).
189. Zaborowski, L. M., and Shreeve, J. M., *Inorg. Chim. Acta* **5**, 311 (1971).
190. Zumach, G., and Kühle, E., *Angew. Chem., Int. Ed. Engl.* **9**, 54 (1970).

CORRELATIONS IN NUCLEAR MAGNETIC SHIELDING. PART I

JOAN MASON

Open University, Milton Keynes, Buckinghamshire, England

I. Introduction	197
II. NMR Measurements and the Periodic Table	198
III. Theory and Physical Models of Nuclear Magnetic Shielding	202
A. Molecular Shielding Terms	202
B. The Cornwell Effect	206
C. The (Atomic) Local-Term Approximation	207
D. The Atom-plus-Ligand Local-Term Approximation	209
E. Additivity of Substituent Effects in Nuclear Magnetic Shielding	210
F. Ab Initio Calculations	214
G. Electronegativity Correlations	214
IV. Absolute Shielding	215
A. The Spin-Rotation Interaction	215
B. Other Methods for the Measurement of Absolute Shielding	217
C. Absolute Shielding Scales	218
V. Periodicity in Nuclear Magnetic Shielding	218
References	225

I. Introduction

A quarter of a century has elapsed since Proctor and Yu (74) found "an annoying ambiguity," in Bloch's phrase (6), in their determination of the magnetic moment of ^{14}N using an ammonium nitrate solution, and since Ramsey (75) proposed the theory of magnetic screening, based on Van Vleck's (91) theory of magnetic susceptibility. In the 1950s and early 1960s, useful approximations were developed, notably the local term approximation (58, 69, 81) which made it easier to interpret NMR shifts in chemical terms. But it has often been found difficult to make detailed interpretations and progress has been slow toward a physical model for nuclear magnetic screening comparable to the models with which the chemist understands other forms of spectroscopy, such as vibrational or electronic spectra. Perhaps as a result, the enthusiasm of the inorganic chemist for measuring shifts, particularly those of the less accessible nuclei, has diminished.

However, with the development of methods for the study of ^{13}C in natural abundance, the measurement of these shifts has become much easier. Pulse Fourier transform spectroscopy with decoupling and spectral accumulation has greatly increased sensitivity and speed of measurement and can produce simple spectra for difficult nuclei in complex molecules. Multiple pulsing can eliminate dipolar and quadrupolar effects in solids and give an accurate measure of shielding anisotropies (*94*). At the same time, other physical techniques, notably molecular beam, electric and magnetic resonance (*7*) and microwave spectroscopy (*28, 90*) are providing information on absolute shielding that can fill important gaps in our knowledge obtained from conventional NMR spectroscopy. In this article we collect evidence from the physical and chemical literature with which we can relate nuclear magnetic screening to molecular structure via the periodic table.

II. NMR Measurements and the Periodic Table

Figure 1 shows a periodic table of magnetic nuclei all of which have been observed by NMR except those shown in parentheses. Nuclei may be difficult to observe because of low sensitivity, which is proportional to the third power of the magnetic moment, because of low natural abundance, or because of quadrupolar or other broadening of the resonance lines. For the nuclei without electric quadrupole moments, Table I gives the order of diminishing *receptivity*, which is the product of the fractional natural abundance and the sensitivity relative to the proton at constant field. Table I shows that many nuclei of considerable interest, such as those of Sn, Pt, Pb, Hg, and Si, compare favorably in receptivity with ^{13}C, and shifts have been measured for a number of their compounds. Several complexes of Rh and W have been measured, and even $^{107, 109}$Ag (*8*) and ^{57}Fe have been observed; the measurement of the ^{57}Fe resonance in Fe(CO)$_5$ took 20 hr on a high-sensitivity pulse spectrometer (*83*).

For quadrupolar nuclei, rapid pulsing can improve the signal-to-noise ratio. They are often most readily measured by wide-line techniques however, as for example ^{17}O (*84*) and ^{14}N (*79*). Many more wide-line measurements have been made of ^{14}N, which is quadrupolar, than high-resolution measurements of ^{15}N, which has spin $\frac{1}{2}$ but is expensive. The broadening increases with the nuclear quadrupole moment, which is small for ^{2}H (2.8×10^{-3}), ^{6}Li (4.6×10^{-4}), and ^{17}O (-4×10^{-3}), and fairly small, between 10^{-2} and 10^{-1}, for ^{9}Be, ^{11}B, ^{14}N, ^{33}S, ^{35}Cl, and ^{37}Cl (the units are $e \times 10^{-24}$ cm^2, where e is the electronic charge) (*20, 70*). The broadening increases with asymmetry of the nuclear environment; thus, the line width increases a thousand-fold from the relatively sharp line

TABLE I

Order of Decreasing Receptivity of Nuclei of Spin ½

Nucleus	Natural abundance (%)	Sensitivity at constant field relative to an equal number of protons	Receptivity[a]
^{1}H	99.98	1	1
^{3}H	—	1.21	—
^{19}F	100	0.833	0.833
^{205}Tl	70.48	0.192	0.135
^{31}P	100	6.64×10^{-2}	6.64×10^{-2}
^{129}Xe	26.24	2.12×10^{-2}	5.56×10^{-3}
^{119}Sn	8.68	5.18×10^{-2}	4.50×10^{-3}
^{195}Pt	33.7	9.94×10^{-3}	3.35×10^{-3}
^{125}Te	7.03	3.16×10^{-2}	2.22×10^{-3}
^{207}Pb	21.11	9.13×10^{-3}	1.93×10^{-3}
^{113}Cd	12.34	1.09×10^{-2}	1.34×10^{-3}
^{199}Hg	16.86	5.72×10^{-3}	9.64×10^{-4}
^{171}Yb	14.27	5.50×10^{-3}	7.85×10^{-4}
^{169}Tm	100	5.51×10^{-4}	5.51×10^{-4}
^{77}Se	7.50	6.97×10^{-3}	5.23×10^{-4}
^{29}Si	4.70	7.85×10^{-3}	3.69×10^{-4}
→ ^{13}C	1.11	1.59×10^{-2}	1.76×10^{-4}
^{89}Y	100	1.17×10^{-4}	1.17×10^{-4}
^{109}Ag	48.65	1.01×10^{-4}	4.91×10^{-5}
^{103}Rh	100	3.12×10^{-5}	3.12×10^{-5}
^{183}W	14.28	6.98×10^{-5}	9.97×10^{-6}
^{15}N	0.365	1.04×10^{-3}	3.80×10^{-6}
^{57}Fe	2.245	3.38×10^{-5}	7.59×10^{-7}
^{3}He	10^{-6}	0.442	4.42×10^{-9}
^{239}Pu	—	2.9×10^{-3}	—

[a] The receptivity is the product of the natural abundance (expressed as a fraction) and the sensitivity at constant field (relative to an equal number of protons).

for aqueous Cl$^-$ to covalently bound chlorine, which may be difficult to observe. (This difference is turned to advantage in the "halogen probe" technique used in biological studies.) In many important fields of study, sharp lines are observed for quadrupolar nuclei because of symmetry about the nucleus, e.g., in complexes of ^{59}Co(III) or ^{105}Pt(II,IV). Broadening because of coupling to a quadrupolar nucleus can, of course, be removed by double resonance.

Paramagnetic broadening, due to unpaired electrons, is usually much greater than quadrupolar broadening, and the nuclear resonance

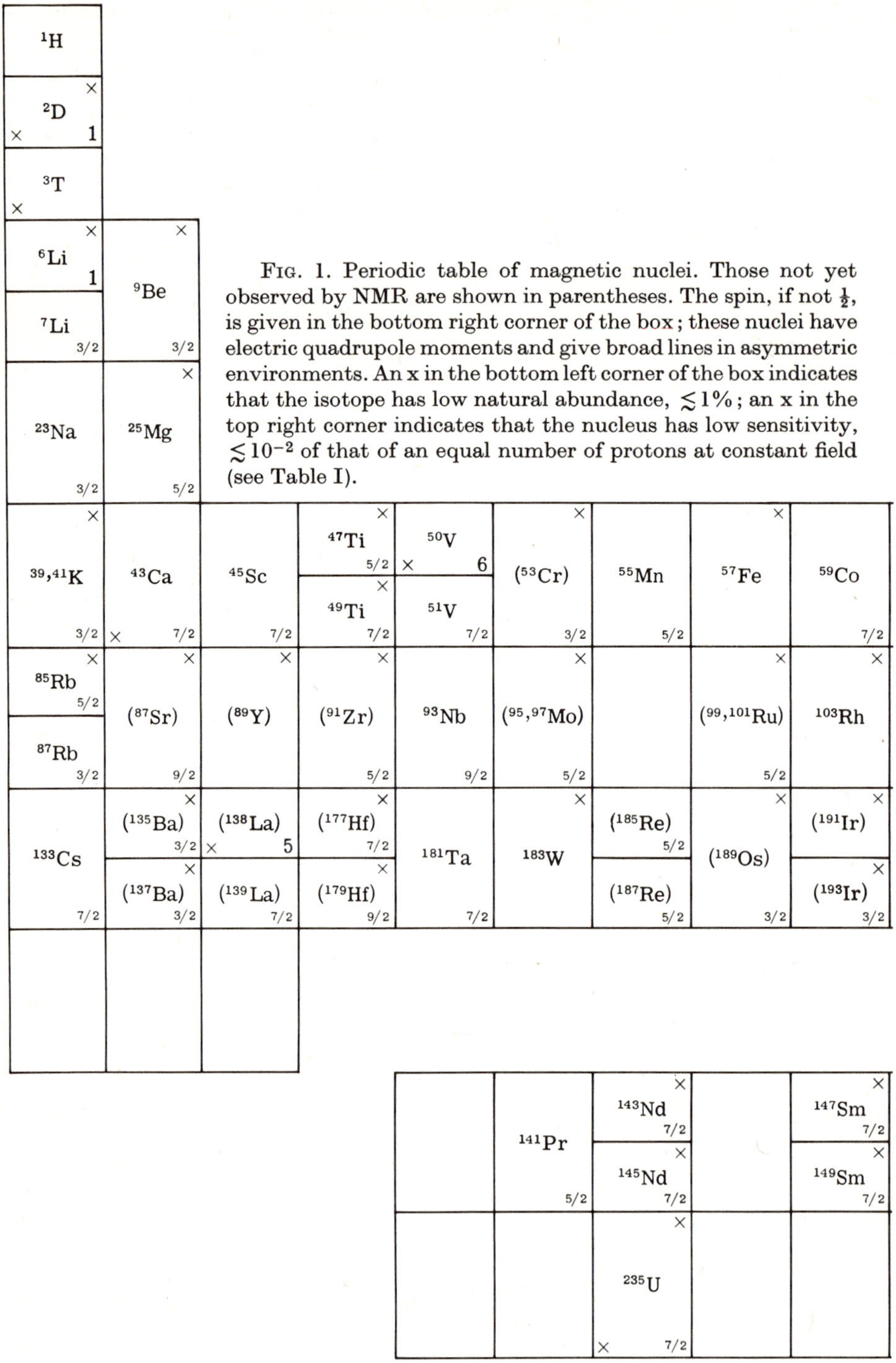

FIG. 1. Periodic table of magnetic nuclei. Those not yet observed by NMR are shown in parentheses. The spin, if not ½, is given in the bottom right corner of the box; these nuclei have electric quadrupole moments and give broad lines in asymmetric environments. An x in the bottom left corner of the box indicates that the isotope has low natural abundance, $\lesssim 1\%$; an x in the top right corner indicates that the nucleus has low sensitivity, $\lesssim 10^{-2}$ of that of an equal number of protons at constant field (see Table I).

								³He ×
		× ¹⁰B 3	× ¹³C	× ¹⁴N		¹⁷O	¹⁹F	× (²¹Ne)
		¹¹B 3/2 ×	×	× ¹⁵N ×		×	×	× 3/2
		²⁷Al 5/2	× ²⁹Si	× ³¹P	³³S × 3/2	× ³⁵,³⁷Cl 3/2		
× (⁶¹Ni) × 3/2	⁶³,⁶⁵Cu 3/2	× (⁶⁷Zn) 5/2	⁶⁹,⁷¹Ga 3/2	× ⁷³Ge 9/2	⁷⁵As 3/2	× ⁷⁷Se	⁷⁹,⁸¹Br 3/2	× (⁸³Kr) 9/2
× (¹⁰⁵Pd) 5/2	× ¹⁰⁷,¹⁰⁹Ag	× ¹¹¹,¹¹³Cd	¹¹³,¹¹⁵In 9/2	¹¹⁷,¹¹⁹Sn 7/2	¹²¹Sb 5/2 ¹²³Sb	(¹²³Te) × ¹²⁵Te	¹²⁷I 5/2	¹²⁹Xe × (¹³¹Xe) 3/2
× ¹⁹⁵Pt	× (¹⁹⁷Au)	¹⁹⁹Hg × ²⁰¹Hg 3/2	²⁰³Tl ²⁰⁵Tl 3/2	× ²⁰⁷Pb	(²⁰⁹Bi) 9/2			

¹⁵¹,¹⁵³Eu 5/2	× ¹⁵⁵,¹⁵⁷Gd 3/2	¹⁵⁹Tb 3/2	× ¹⁶¹,¹⁶³Dy 5/2	× ¹⁶⁵Ho 7/2	× ¹⁶⁷Er 7/2	× ¹⁶⁹Tm	× ¹⁷¹Yb × ¹⁷³Yb 5/2	¹⁷⁵Lu 7/2 ¹⁷⁶Lu 6

may be lost; or it may be detectable by wide-line techniques, as for example the ^{14}N resonance in liquid N_2O_4 at 10°C, at which temperature it is brown with NO_2, and exchanging with this as well (*1*).

III. Theory and Physical Models of Nuclear Magnetic Shielding

A. MOLECULAR SHIELDING TERMS

Ramsey gave the magnetic screening of a nucleus in a diamagnetic molecule as the resultant of a diamagnetic (σ_d) and a paramagnetic term σ_p (*75*). Lamb had shown (*45*) that a magnetic field induces electronic currents in a free atom that oppose the field and shield the nucleus (following Lenz's law, the Biot-Savart law, and Larmor's theorem). For an atom in a molecule, σ_d gives the shielding due to the rotation with the Larmor angular velocity of all the electrons of the molecule about the chosen origin, which we take at the nucleus of interest. This rotation is largely fictitious because of restraint by the other nuclei. Then σ_p represents similarly fictitious circulations of the valence electrons that reinforce the applied field. These arise from the slight unquenching in the magnetic field of the orbital angular momentum of the valence electrons which is quenched by directional bonding. The paramagnetic circulations are of $p_x \to p_y{}^*, d_x \to d_y{}^*$, etc., type ($x$ and y refer to different directions) allowed by the mixing in with the ground state by the magnetic field of excited states that have angular momentum, i.e., $n \to \pi^*$, $n \to \sigma^*$, $\sigma \to \pi^*$, $\pi \to \sigma^*$, $\pi^* \to \sigma^*$, $\sigma_x \to \sigma_y{}^*$, and so on. Although the diamagnetic and paramagnetic terms tend to cancel in conventional NMR experiments, σ_p can be measured directly for an isolated molecule in molecular beam or microwave experiments, as described below.

Ramsey obtained σ_d by first-order and σ_p by second-order perturbation theory (*76*); variational treatments give similar results (*14, 71, 73*). The term σ_p is sometimes called the second-order paramagnetic term and sometimes the high-frequency term (*14*), because of the dependence of the (temperature-independent) paramagnetism in molecules on the high-frequency matrix elements of the orbital moments (*91*).

The average values of the screening tensors for a particular nucleus can be written as

$$\sigma = \sigma_d + \sigma_p$$

with

$$\sigma_d = \frac{e^2}{3mc^2} \left\langle \psi^0 \left| \sum_i \mathbf{r}_i^{-1} \right| \psi^0 \right\rangle \tag{1}$$

and

$$\sigma_p = \frac{-e^2}{3mc^2} \sum_k \left[\frac{\left\langle \psi^0 \left| \sum_i \mathbf{L}_i \right| \psi^k \right\rangle \left\langle \psi^k \left| \sum_i \mathbf{L}_i \mathbf{r}_i^{-3} \right| \psi^0 \right\rangle + \text{c.c.}}{\Delta E_k} \right] \qquad (2)$$

where e and m are the electron charge and mass, respectively, c is the speed of light, ψ^0 is the wave function of the molecule in the ground state, ψ^k that in the kth excited state with energy ΔE_k above the ground state, and c.c. stands for complex conjugate. The factor 3 in the denominator arises from the averaging of the components of the shielding tensor about the three axes, i.e., $\sigma_{av} = \frac{1}{3}(\sigma_{xx} + \sigma_{yy} + \sigma_{zz})$. The $\mathbf{L}$ and $\mathbf{r}$ vectors are, respectively, the angular momentum and position of the ith electron relative to the origin, and sums are taken over all electrons i and excited states k. σ_d and σ_p thus increase indefinitely with molecular size. The magnitude of each depends on the choice of gauge origin but the sum is invariant, as for the two parts of the magnetic susceptibility (91).

Values of σ_d are not very difficult to calculate, for Eq. (1) involves only ground-state wave functions. Furthermore, Flygare and Goodisman (25) have shown that Eq. (3) gives values of σ_d that are reasonably accurate (within 1 to 2 ppm), using tabulated values of the free atom (Lamb) term (7, 55, 56) and a knowledge of the molecular geometry:

$$\sigma_d = \sigma_d \text{ (free atom)} + \frac{e^2}{3mc^2} \sum_\alpha \frac{Z_\alpha}{r_\alpha} \qquad (3)$$

where α runs over all nuclei except the nucleus of interest, Z_α is the atomic number of the αth nucleus, and r_α its distance from the origin. [Ramsey has given a corresponding relationship (78). The error is small, being given by the Hellman–Feynman theorem as the ratio of the bonding energy of an atom to the total electronic energy of the atom.] The term σ_d can, therefore, be treated as a physical correction applied to the resultant shielding to obtain the paramagnetic term from conventional NMR experiments; or applied to the paramagnetic term from spin-rotation experiments to obtain the total shielding.

Because of the $\langle r^{-1} \rangle$-dependence of σ_d, the innermost electron shells make the largest contribution, but the fall-off is fairly gradual, as we shall see in examples below. For free atoms, σ_d increases steadily with the number of electrons, as shown in Fig. 2, but the increase is periodic in molecules, as shown in Figs. 6–8 for the binary hydrides and fluorides.

Equation (2) shows that σ_p vanishes for electrons without angular momentum (symmetrically distributed), e.g., in s orbitals, closed shells, free atoms, or monatomic ions; similarly, $\sigma_p{}^\parallel$ vanishes in linear molecules. The value of σ_p is larger the more asymmetric the distribution of

p, d, or f electrons, the closer they are to the origin ($\langle r^{-3} \rangle$ large), and the lower the energy of the excited states involving a rotation of charge [$(\Delta E)^{-1}$ large]. When there are low-lying n → π* excited states, for example, σ_p correlates with the visible or ultraviolet absorption, the line shifting upfield as the absorption goes to shorter wavelengths (2, 23, 32).

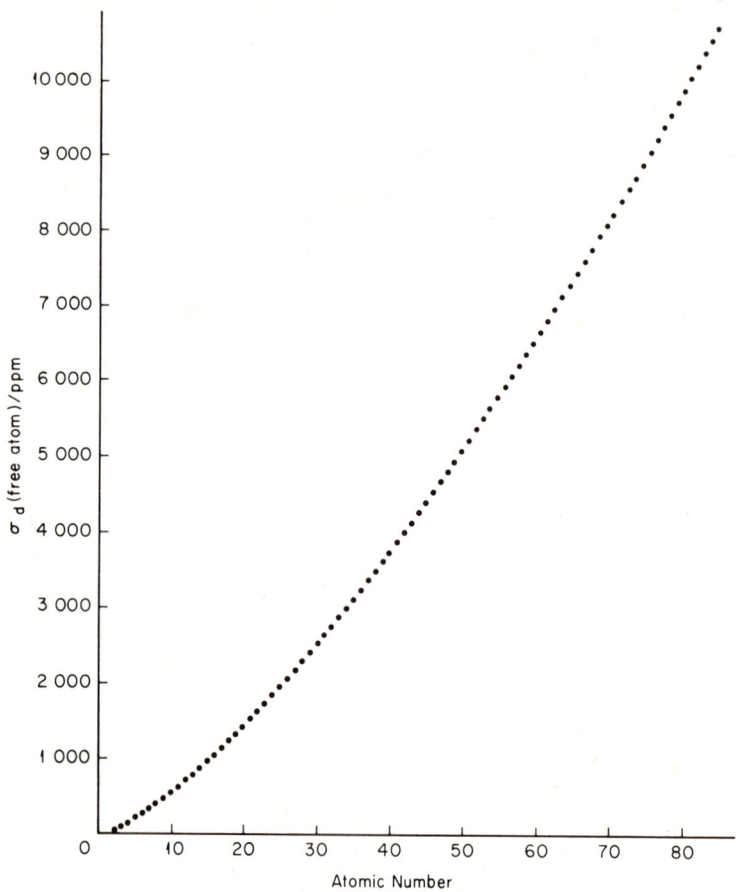

FIG. 2. Variation of σ_d (free atom) with atomic number Z (7, 55, 56).

A crude physical interpretation of Eq. (2) is that the factor ($\mu B/\Delta E$) $\langle \psi^0 | \mathbf{L} | \psi^k \rangle$ represents the applied field (B is the magnetic flux density and μ the Bohr magneton) driving the angular momentum and inducing a paramagnetic current, and the factor $\langle \psi^k | \mathbf{L} r^{-3} | \psi^0 \rangle$ represents the transmission of the magnetic field due to the current to the nucleus at a distance r (3).

The $\langle r^{-3}\rangle$ dependence imposes a periodicity on the variation of the paramagnetic term with atomic number Z of the nucleus if the molecular environments are similar. Figure 3 shows the variation with Z of

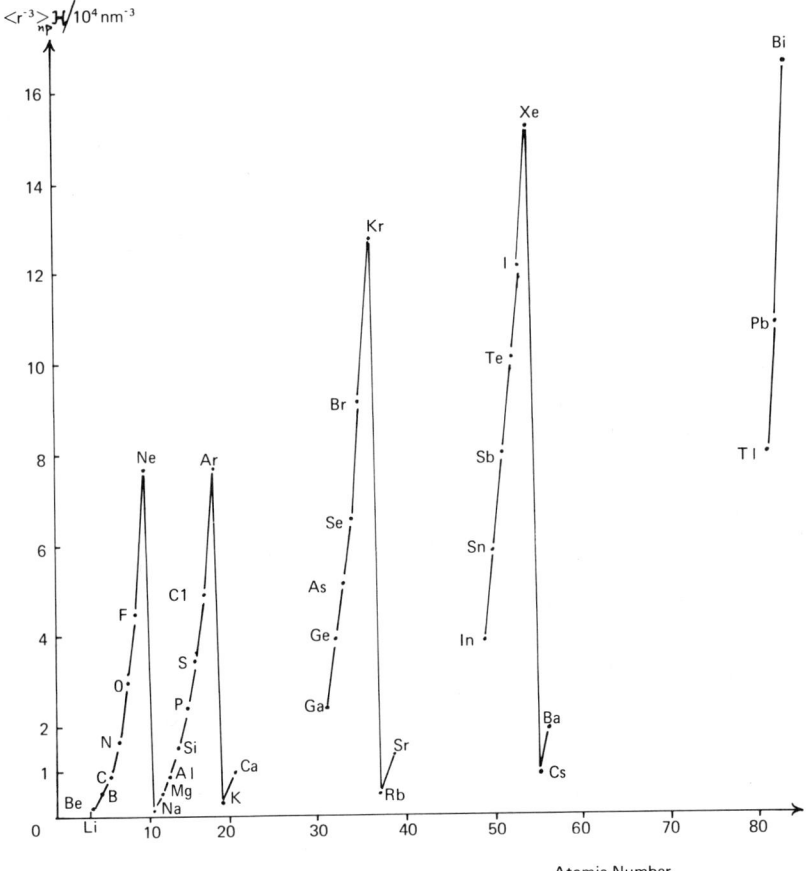

FIG. 3. Variation of $\langle r^{-3}\rangle_{np}\mathscr{H}$ for the valence p electrons with atomic number Z, as determined from spin-orbit splittings in atomic spectra (5). The $\mathscr{H}$ is a small relativistic correction, important mainly for heavy atoms ($Z > 50$). The values for the elements with the p^3 configuration (N, P, As, Sb) were obtained by interpolation. (Taken from Barnes and Smith (5), appeared also in Jameson and Gutowsky (35)].

$\langle r^{-3}\rangle$ for the valence p electrons, obtained from observed atomic spin-orbit splittings (5), to be compared in later sections with periodic variations of σ_p.

The calculation of σ_p is very difficult for molecules that are more complicated than hydrogen because of the summation over all excited states (including the continuum) and the sensitivity to the wave functions used. Since σ_d and σ_p are quite large for molecules of any size and are opposite in sign, errors in σ_p bulk large in the resultant shielding, and one is weighing the ship with and without the captain. For these reasons, direct measurement of σ_p, where this is possible, is particularly important; and local-term approximations (with or without an average-energy approximation) have been extensively used in discussions of nuclear magnetic shielding. Both of these topics are considered in detail below.

B. The Cornwell Effect

Although the diamagnetic term must be positive (shielding), the paramagnetic term is not always negative. The best-known example of a positive value for σ_p is that of the fluorine shielding in chlorine monofluoride ClF, for which the resonance is well upfield of the fluoride ion (*11*); σ_p has now been measured (through the spin-rotation interaction) as +68 ppm (*16*). This unusual observation has been explained by Cornwell (*11*), and also by Santry (*82*). In ClF, the $\pi^* \to \sigma^*$ excitation produces circulations in opposite senses on the two atoms. The circulation is normal on chlorine (for which σ_p is found to be negative) but reversed on fluorine because it arises from an antibonding orbital, in which the chlorine terms are dominant because fluorine is the more electronegative partner. The reversed circulation on fluorine results in a positive contribution to σ_p which is large because of the relatively low energy (36,400 cm^{-1}) of the $\pi^* \to \sigma^*$ excitation. In the bonding orbitals, the coefficients for the fluorine terms are the greater and the $\pi \to \sigma^*$ circulation is normal, but the resulting negative contribution to σ_p is now smaller because of the higher energy of excitation from the π than from the π^* orbital. It serves, however, to cancel part of the positive contribution.

This phenomenon of antiparamagnetic paramagnetic terms clearly needs a name and is called here the *Cornwell effect* (ideally the Cornwell-Santry effect). Positive contributions to σ_p (which may or may not be positive overall) are expected in heteronuclear diatomics if they have a π^* state; this excludes, e.g., HF, InF, and TlF. In homonuclear diatomics, the $\pi^* \to \sigma^*$ excitation is symmetry-forbidden. The possibility has been mentioned for XeF$_2$ (*34*), although, from the chemical shift and calculated values of σ_d, the resultant σ_p (^{19}F) is negative in XeF$_2$ and KrF$_2$ (cf. Fig. 7). Another candidate is FC≡CH, from the evidence of the fluorine chemical shift and spin-rotation interaction (*96*). According to this interpretation there should be a substantial upfield shift of the ^{13}C resonance for the carbon attached to fluorine, compared to carbon attached to hydrogen.

Low-energy circulations between π^* orbitals on fluorine and σ^* (e_g^*) d orbitals on the metal are possible in the (spin-paired) d^6 complex fluorides, NiF_6^{2-}, PdF_6^{2-}, and PtF_6^{2-}, and Cornwell's description of ClF has been invoked (62) in discussion of the fluorine resonance in these ions, which is at particularly high field compared with earlier transition metal fluorides (cf. Fig. 8). The ^{17}O resonance in $Co(acac)_3$, and the ^{14}N resonances in the hexammines of Co(III), Rh(III), and Ir(III) are at similarly high field.

C. The (Atomic) Local-Term Approximation

Because of the problems described above of the Ramsey shielding terms, a useful approximation is to take advantage of the rough cancellation of the long-range diamagnetic and paramagnetic contributions to the shielding of the nucleus A, and to represent this by local terms σ_d^A and σ_p^A, which are calculated by Ramsey's theory applied only to the electrons on A (81, 85). Contributions σ^{AB} from circulations on other atoms B are added as required. The σ_p^A term is obtained by valence bond (35, 81) or molecular orbital methods (35, 69).

For proton shielding, variations in all three terms σ_d^A, σ_p^A and σ^{AB} are important because of the small range of proton shifts, and because of the proximity of circulations on nearest neighbors, which are included in σ^{AB}; Fig. 6 shows this graphically (58). Well-known exceptions to generalizations about proton shifts can be traced to neglect of one or two of these terms. Contributions σ^{AB} may be obtained (51, 68) by replacing a "distant" atomic current by a point dipole at that distance; the strength of the dipole is proportional to the magnetic susceptibility, which may be estimated as the sum of atomic contributions. The resulting shielding contribution is then proportional to the anisotropy of the susceptibility. Here too high-resolution microwave spectroscopy can contribute. Zeeman effect measurement of the molecular susceptibility tensor in formamide has given an estimate of the neighboring group anisotropy effect of the α-proton in a polypeptide link, which was used to show that the chemical shift in the transition from the helix to the random coil in poly-(L-alanine) is a solvation effect (89).

Interatomic or ring currents also may contribute to σ^{AB} in hydrogen, and much work has been done on these (63).

For nuclei other than hydrogen, with a much larger range of chemical shifts, σ_d^A is not very different from the free-atom term, σ^{AB} terms are relatively small and often neglected, and chemical shift relationships are ascribed to variation in $\sigma_p{}^A$. (We discuss the limitations of this approach below.) A further more drastic approximation is to avoid the summation over excited states in $\sigma_p{}^A$ by the use of an average energy ΔE (38, 69):

$$\sigma_p{}^A = \frac{-e^2\hbar^2 \langle r^{-3}\rangle_{2p}}{2m^2c^2(\Delta E)}\left[Q_{AA} + \sum_{B\neq A} Q_{AB}\right] \quad (4)$$

Equation (4) applies to second-row elements, $\langle r^{-3}\rangle$ being the expectation value of the inverse cube of the radius of the $2p$ orbitals on atom A. The theory has been extended to lower rows of the periodic table by the inclusion of d orbitals (35, 48). The average energy ΔE is usually used as an adjustable parameter to rationalize observed shifts (note that this approximation presupposes that contributions from all excited states have the same sign). The Q terms express the imbalance (in the ground state) of the $2p$ electrons, which depends on the nature of the bonding and of the immediate ligands B. The Q_{AA} term depends on electron density, whereas Q_{AB} depends on π-bonding at A.

If we take ^{13}C magnetic resonance as an example, ΣQ_{AB} is 0 for C—C and C≡C, increasing to 0.44 for C=C and to 0.77 for C=C=C, and this helps to place the resonance at high field in alkanes and alkynes, at medium field in alkenes and arenes, and at low field for the middle carbon in allenes. In addition, σ_p tends to be increased for unsaturated relative to saturated carbon by the lower excitation energies in the former. Doubly bonded carbon is deshielded by $\sigma \to \pi^*$ and $\pi \to \sigma^*$ circulations (when the magnetic field is perpendicular to the double bond and in the molecular plane), the middle carbon in allene being doubly deshielded in this way. Corresponding arguments apply to other nuclei; e.g., the middle nitrogen in the azide ion is less shielded than the end nitrogens.

In ^{19}F NMR, Q_{AA} is usually the more important term, although the π electrons may be very important in shielding, e.g., in aromatic fluorine compounds, or as we have seen in ClF. Both Q_{AA} and $\langle r^{-3}\rangle_{2p}$ depend on the electron density on the atom A and increase with increasing electronegativity of the ligands. So, however, does ΔE, and this is dominant when there are low-lying excited states, as for the nitrogen shielding in nitroso, nitro or azo compounds (2, 60, 61).

This qualitative theory is helpful in the physical and chemical interpretation of chemical shifts in covalent molecules and ions and may be used semiquantitatively in some comparisons of closely related species. It is, however, a fairly gross approximation, predicting, for example, that all alkane carbons have the same chemical shift. For ^{14}N and ^{17}O shielding the average energy approximation may be less appropriate (than for ^{13}C shielding) because of the large variation in excitation energies with nonbonding electrons present. When Cornwell effects are large the average energy approximation breaks down, of course.

D. THE ATOM-PLUS-LIGAND LOCAL-TERM APPROXIMATION

The local-term approximation described in the previous section, without the hazards of the average energy approximation (52), can, in principle, give an accurate account of the shielding if the long-range diamagnetic and paramagnetic contributions cancel or if noncancelling contributions can be determined reliably. The first condition holds for contributions from sufficiently distant atoms as the paramagnetic contribution then becomes $\langle r^{-1}\rangle$-dependent (as is the diamagnetic contribution), since angular momentum about the origin increases with distance and there are two factors depending on angular momentum in Eq. (2); the $\langle r^{-2}\rangle$-dependent terms are 0 for distant dipoles (51). However, the inclusion with the distant atoms of the immediate ligands is a source of trouble since they may well be too close to be treated as dipoles. Near atom A the molecular diamagnetic term falls off more slowly than the paramagnetic term. As we see below, contributions by the immediate ligands to the molecular diamagnetic term in second-row elements may change by 100 ppm or so, even in compounds that are closely related chemically. But the local (atomic) term $\sigma_d{}^A$ varies by no more than about 20 ppm, often much less, for second-row elements (69). Occasional voices have been heard suggesting that variations in the local diamagnetic term should be taken more seriously, e.g., Davies says "The possibility of variations in σ_d for ^{19}F has been mentioned only to be dismissed" (13), and Chan says "for fluorine shielding, changes in the Lamb term are not as negligible as they are often assumed to be" (10). Lambert and Roberts (46) noted large changes in σ_d (^{14}N), e.g., 150 ppm from $NH_4{}^+$ to $NO_3{}^-$, when nearest neighbors were taken into account.

To avoid the confusion to which this topic is prone we should define a local-term approximation which gives the total shielding $\sigma(A)$ of the nucleus A in terms of the shielding σ^{AL} due to electronic circulations on A and on its directly bonded neighbors (ligands) L, i.e.,

$$\sigma(A) = \sigma_d{}^{AL} + \sigma_p{}^{AL} + \sigma^C$$

where (5)

$$\sigma_d{}^{AL} = \sigma_d{}^A + \sigma_d{}^L \quad \text{and} \quad \sigma_p{}^{AL} = \sigma_p{}^A + \sigma_p{}^L$$

where $\sigma_d{}^A$ and $\sigma_p{}^A$ are the atomic terms of the conventional local-term approximation, and $\sigma_d{}^A$ and $\sigma_p{}^L$ the diamagnetic and paramagnetic contributions to the screening of the nucleus A, from circulations on the ligand atoms. The σ^C term then represents the remaining long-range contributions, which we now with more confidence expect to be negligible or small. As before, the diamagnetic terms can be obtained with little difficulty, for example by Flygare and Goodisman's equation. It would be difficult to calculate $\sigma_p{}^L$, but less difficult to calculate $\sigma_p{}^{AL}$. Values of $\sigma_p{}^{AL}$ may be obtained, however, by subtracting $\sigma_d{}^{AL}$ from σ (A) obtained by referring the observed shift to an absolute scale (as discussed below). Alternatively, relative values can be used for the shielding parameters; the resonance of the simple hydride (CH_4, $NH_4{}^+$) may be a useful reference point since other resonances are normally downfield of this.

The atom-plus-ligand local-term approximation has been justified *a posteriori* for ^{14}N (30) and ^{13}C (57) shielding, since certain expected relationships of chemical shifts, such as additivity of substituent effects, were found to be absent for the observed shifts, but were fulfilled for $\sigma_p{}^{AL}$, i.e., after correction for the atom-plus-ligand diamagnetic term. Some examples are given in the following sections.

E. Additivity of Substituent Effects in Nuclear Magnetic Shielding

There are many examples of the linear additivity of substituent shifts. Shoolery's constants predict methylene proton shifts in disubstituted methanes for a wide range of substituents including the halogens, OR, SR, NO_2, and singly, doubly, and triply bonded carbon (67). Additivity is found extensively in ^{13}C shifts (88). Pairwise additivity, attributed to interaction between neighboring substituents, has been demonstrated in ^{11}B (87), ^{27}Al (53), ^{73}Ge (41), $^{47,49}Ti$ (40), and ^{93}Nb (42), as well as in ^{13}C resonance (54). The interactions have been expressed as perturbations of the McWeeny group wave functions of the substituents; the additivity rules fail when the perturbation is large, examples being drawn from ^1H, ^{11}B, and ^{19}F shifts (93).

Table II shows that the observed (29) ^{13}C shifts δ (taken as positive downfield) show a degree of linear additivity in the methyl-substituted methanes, with a shortfall for neopentane. Table II shows also the division of the shielding into atom-plus-ligand diamagnetic and paramagnetic parts, relative to methane for which σ_d is 295 ppm. The diamagnetic terms were calculated by Flygare's method (25). With each substitution of hydrogen by carbon, $\sigma_d{}^{AL}$ for the central carbon increases by 28 ppm but $|\sigma_p|$ increases by about 37 ppm, and the line moves 9 ppm downfield. Analogous relationships have been demonstrated for ^{14}N shielding in methyl-substituted NH_3 and $NH_4{}^+$ (30). Table II shows that the shortfall at neopentane is in the paramagnetic term.

Unexpected support for the atom-plus-ligand diamagnetic correction

TABLE II

^{13}C Shielding Terms for the Central Carbon in Methyl-Substituted Methanes[a]

	δ	σ_d^{AL} (ppm)	σ_p^{AL} (ppm)
CH$_4$	0	Say 0	Say 0
CH$_3$CH$_3$	8.0 (8)	28 (28)	−36 (−36)
CH$_2$(CH$_3$)$_2$	18.2 (10)	56 (28)	−74 (−38)
CH(CH$_3$)$_3$	27.3 (9)	85 (29)	−111 (−37)
C(CH$_3$)$_4$	29.7 (2.4)	112 (26)	−142 (−31)

[a] The numbers in parentheses are differences with successive substitution.

has come from arithmetic relationships of ^{13}C shifts in the hydrocarbons (57). Linear additivity in the observed shifts finds expression in the well-known substituent parameters (α—ε) which are used to assign ^{13}C resonances in alkanes and alkyl groups (29). The α parameter, about 9 ppm, gives the increase in chemical shift (positive downfield) for each substituent carbon α to the nucleus of interest, and similarly for the parameters (β—ε). However, eight additional parameters were found to be necessary in order to include carbons at or next to branching positions, making a total of thirteen independent parameters (of which two are small). After the atom-plus-ligand diamagnetic correction, the ^{13}C shifts relative to methane are given satisfactorily for linear and branched acyclic alkanes, and for simple cycloalkanes, by only four parameters (57) (Fig. 4).* Similarly, anomalies in ^{13}C resonances at bridgehead positions in polyphenyls and fused arenes, which are found to be well upfield of the positions predicted by molecular orbital theory, can be attributed to the diamagnetic part of the ligand contribution (57).

The empirical substituent parameters that give the values of the corrected shift for the unhindered alkanes are α = 35.4, β = 8.4, γ = −1.5, and δ = 1.4 (it is a general observation that the γ contribution is upfield). Support for the neglect of the long-range terms σ^C [Eq. (5)] for the γ- and δ-carbons is given by the small magnitude of these parameters, to which motional averaging may contribute as well. The rather large value of the β parameter demonstrates the noncancellation of the diamagnetic and paramagnetic contributions at that distance.

* Steric restrictions lead to irregularities; e.g., the methyl cyclohexane parameters vary with conformation, and the ^{13}C line in cyclopropane is significantly upfield of the predicted shift.

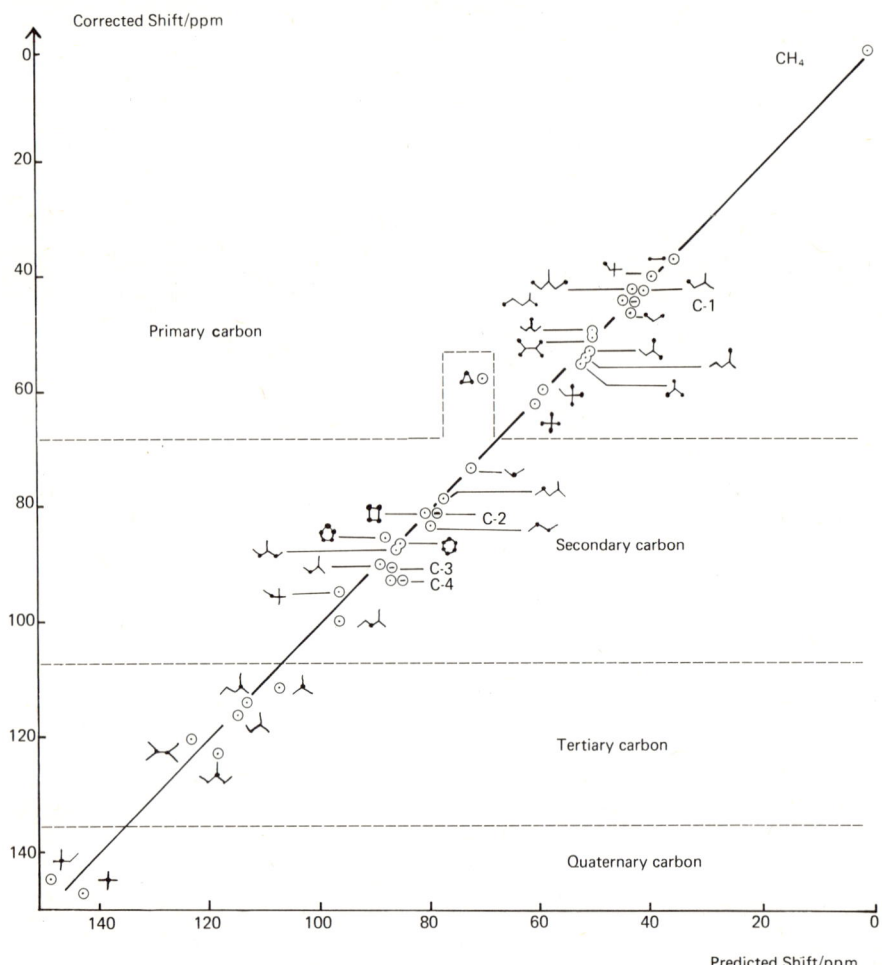

FIG. 4. ^{13}C Shifts in alkanes: plot of corrected shift against values predicted using four empirical parameters. [Appeared in Mason (57).]

Although the observed shifts are reasonably additive for methyl-substitution in methane, this is not true for other substituents, and the top half of Fig. 5 contains an up-dated version of the well-known plot (47) of ^{13}C shifts in substituted methanes against the number of substituents. The lower half of Fig. 5 shows that the large increase in σ_d^{AL} with successive substitution accounts for most of the upfield turn of the plots. The residual shortfall in σ_p^{AL} increases with degree of substitution and

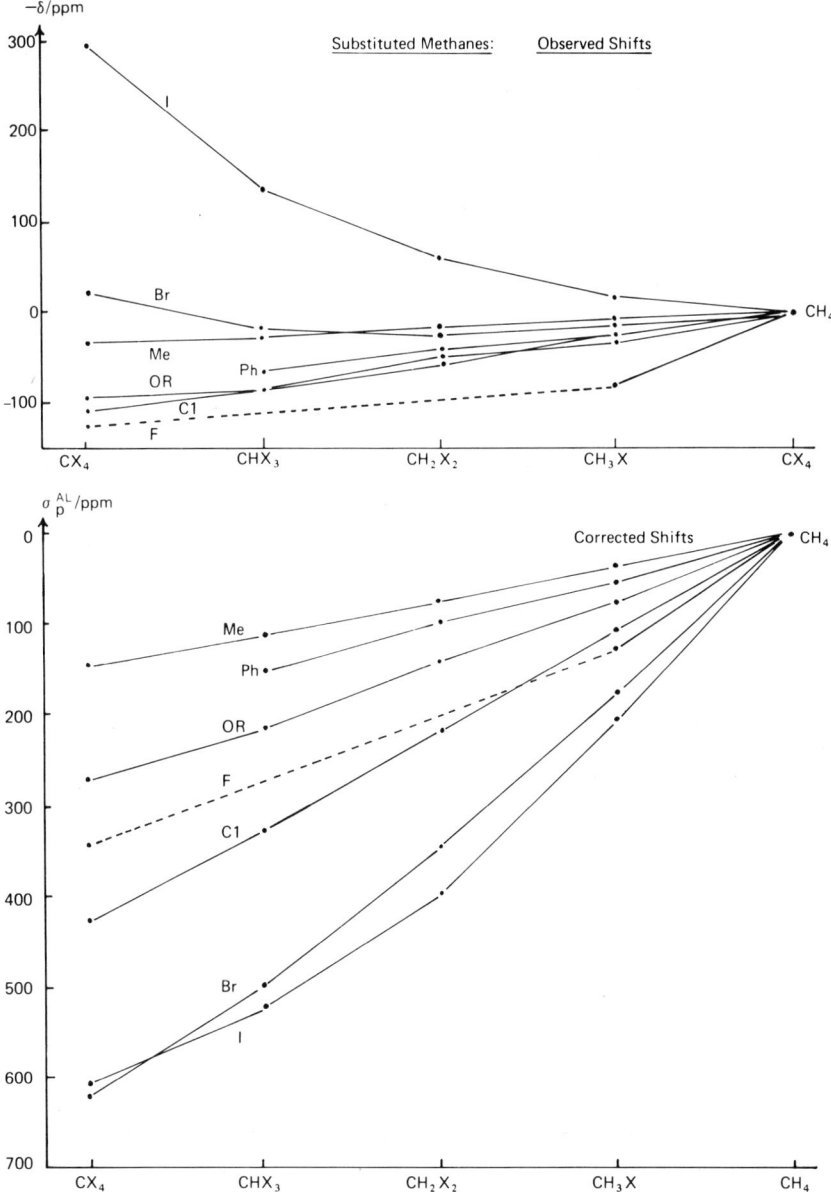

FIG. 5. Plot of ^{13}C shifts in substituted methanes against the number of substituents. The observed shifts are compared with these corrected for the atom-plus-ligand diamagnetic term. [Appeared in Mason (57).]

with size of the ligand and, presumably, involves electron delocalization; cf. the anomaly of cyclopropane in the alkane correlations.

Some other types of correlation are improved if observed shifts are corrected for σ_d contributions of the nearest ligands; e.g., the ^{14}N shielding in nitroso compounds XNO is roughly linear with the $n \to \pi^*$ wavelength [representing $(\Delta E)^{-1}$] when X is a second-row atom, as in RNO, R$_2$NNO, RONO, and NOF, but the ^{14}N line diverges to high field when X is a third- or fourth-row atom, as in RSNO, NOCl, and NOBr. The divergence is compensated by correction for $\sigma_d{}^{AL}$, which increases by 130 ppm from NOF to NOBr (2).

F. Ab Initio Calculations*

Ab initio molecular orbital calculations of ^{13}C, ^{14}N, ^{17}O, and ^{19}F shieldings in a wide range of organic molecules containing up to 3 second-row atoms, but including CF$_4$, have been reported by Ditchfield, Miller, and Pople (19). Because of the expense of large extended basis sets, even for these small molecules, the aim was for reasonable accuracy in shifts from molecule to molecule, since the absolute values depend on the choice of origin when the basis set is incomplete. A finite perturbation method was used to obtain the molecular diamagnetic and paramagnetic shielding parameters with the center of mass as origin. Good agreement with experiment was found so long as the nuclei being compared occupy similar positions relative to the origin, e.g., ^{13}C shifts in CH$_3$X compounds are fairly well described when X contains 1 second-row atom, but not the relative shifts in methane, ethane, and propane. There is reasonably good agreement, however, for methyl carbons in propane, dimethylamine, and dimethyl ether, or for the central carbons in propane, propene, propyne, allene, acetonitrile, acetaldehyde, and formamide.

The calculations allow some useful conclusions as to "chemical" influences on the shifts, e.g., a large part of the ^{13}C shifts in CH$_3$X compounds, when X is CH$_3$, NH$_2$, OH, or F, depends on the increase in the paramagnetic term with contraction of the carbon orbitals as charge is withdrawn by X (i.e., as $\langle r^{-3} \rangle$ increases).

G. Electronegativity Correlations

Correlations of chemical shifts with electronegativity of the substituent are of long standing (12, 72, 86). In general the ^{1}H, ^{13}C, ^{19}F (or etc.) line goes downfield with increasing electronegativity of the sub-

---

* More recent developments in this important field are described by R. Ditchfield and P. D. Ellis in "Topics in Carbon-13 NMR Spectroscopy" (G. C. Levy, ed.), Wiley, New York, 1974, and by W. T. Raynes in *Chem. Soc. Spec. Per. Rep. NMR* **3**, 8 (1974) and R. B. Mallion, *ibid.* **4**, 14 (1975).

stituent, but there are many anomalies. For ^{13}C shielding in Me$_n$X compounds the line goes downfield as X runs across the second or third row of the typical and post-transition elements, but this trend tends to be reversed in later rows, as shown by a plot of the observed ^{13}C shift against the atomic number of the ligand X (57). This variation was compared with the corresponding plot of $\sigma_p{}^{AL}$ (^{13}C), obtained by subtraction of the atom-plus-ligand diamagnetic terms from the observed shifts (all relative to CH$_4$); the latter plot was found to be more consistently periodic. The corrected shift ($\sigma_p{}^{AL}$) increases numerically as the ligand moves across the row of the periodic table (and its electronegativity increases), but it increases also as the ligand moves down the group (and its electronegativity decreases).

The corresponding periodic correlation of ^{19}F shifts (59) is more extensive and informative. This and the corresponding plot for proton shielding (58) have been constructed using absolute values of the molecular shielding terms, avoiding the hazards of local-term approximations. Discussion of these periodicities is postponed to Section V.

IV. Absolute Shielding

A. THE SPIN-ROTATION INTERACTION

Theory gives the nuclear magnetic shielding for the gaseous molecule relative to the bare nucleus. In conventional spectrometers the shielding is usually measured in fluid phases, so that intermolecular effects are motionally averaged, and relative to standard compounds because of the difficulty of accurate measurement of field strengths; frequencies are readily measured to 1 part in 10^8. Absolute measurements for comparison with theory can, however, be obtained under favorable circumstances from the spin-rotation interaction. The tumbling of a molecule generates magnetic fields at a nucleus because of the circular motions about it of the other nuclei (which can be calculated in classical terms) and of the electrons, just as the effect of a magnetic field on the electrons is equivalent to that of a molecular rotation (this is Larmor's theorem). In the absence of collisions the spin-rotation interaction can be observed, for simple molecules, as magnetic hyperfine structure either of the nuclear spin resonance or of the rotational spectrum.

This interaction was discovered in 1939 in the hydrogen molecule in studies deriving ultimately from the Stern–Gerlach experiment, which demonstrated the quantization of angular momentum, and was converted into a resonance method by Rabi and his collaborators in the 1930s for the measurement of magnetic moments (39). The various approaches to nuclear magnetic shielding were synthesized by Ramsey in 1950 (75). Spin-rotation coupling constants C_i are now measured by molecular beam magnetic resonance (*4, 9, 77*) or electric resonance (*99*), or by high-resolution microwave spectroscopy (*26, 31, 97*) or molecular beam maser, invented by Townes and his collaborators (*28, 90*); complex spectra are subjected to computer analysis. Perhaps because of the small interest taken in them so far by chemists (with some notable exceptions), spin-rotation constants, which may have been obtained in the pursuit of other molecular information, are often reported in the physical or chemicophysical literature without reference to the shielding information that they contain.

The importance of the spin-rotation interaction is that it monitors a paramagnetic electronic circulation divorced from the diamagnetic circulation that must accompany it when an external field is applied to the molecule. The physical basis for the interaction was examined by Wick (*98*) for the hydrogen molecule and then extended to polyatomic molecules (*21, 36*). In the rotating molecule, the inner-shell electrons rotate with their nuclei. Closed shells on the nucleus taken as origin produce no field there, nor do closed shells elsewhere (for which the field is cancelled by equivalent positive charges on their own nuclei); these shells have been compared to the chairs on a Ferris wheel that preserve their orientation as the frame rotates. If the valence electrons "slipped" in this way there would be no rotational paramagnetic effect; but since they are associated with more than one nucleus they rotate partially with the molecule. Their angular momentum relative to the molecular frame can be considered to arise, as before, from the mixing in of appropriate excited states, and the resultant field at the nucleus corresponds to the deshielding in an applied field. The important difference in the rotating molecule is that σ_p is opposed only by the effect of the rotating nuclei σ_{nuc}. The resultant is the σ_{sr} term due to the spin-rotation interaction:

$$\sigma_{sr} = \sigma_p - \sigma_{\text{nuc}}$$

compared with (6)

$$\sigma = \sigma_d + \sigma_p$$

in conventional NMR spectroscopy. The σ_{nuc} term due to the motion of the other nuclei about the origin is easily found (following the Biot–Savart law) from the molecular geometry. It is, in fact, the negative of the second term in Eq. (3), because of the balance of forces on a nucleus from the electrons and from the other nuclei. Thus the shielding in conventional NMR experiments can be estimated as the sum of σ_d (free atom) and σ_{sr}.

The value of σ_{sr} is obtained from the spin-rotation constants C_i, which are the components (in units of frequency) of the nuclear spin-rotational tensor, defined by the Hamiltonian

$$\mathscr{H} = -hC_i \mathbf{I} \cdot \mathbf{J}$$

where $\mathbf{I}$ and $\mathbf{J}$ are, respectively, the nuclear and rotational angular momenta (the spin-rotation interaction is sometimes called the "I dot J" interaction). This definition embodies Ramsey's sign convention, which is also followed by Flygare; many microwave spectroscopists use the opposite convention. Experiment gives the magnitude but not necessarily the sign of C_i, and this is found from chemical shifts, or by analogy, or from other evidence.

Equations relating σ_{sr} and the spin-rotation constants have been given by Ramsey (9, 77, 78) for linear molecules and by Flygare (24) for symmetric top, spherical top, and asymmetric molecules. A simple expression for the general case, and neglecting vibrational effects, is

$$\sigma_{sr} = \frac{e\pi \Sigma C_{\lambda\lambda} I_{\lambda\lambda}}{3m_e c \hbar \gamma} \tag{7}$$

where γ is the magnetogyric ratio of the nucleus of interest, the I's are the principal moments of inertia referred to axes λ, and the C's the corresponding components of the nuclear spin-rotational tensor. In linear molecules, two of these components are equal ($C^\perp$) and the third ($C^\parallel$) is zero ($\sigma_p{}^\parallel$ being zero) so the sum over $C_{\lambda\lambda} I_{\lambda\lambda}$ reduces to $\tfrac{2}{3} C^\perp I^\perp$.

Thus the spin-rotation constants contain the infinite sum over excited states that is so difficult to calculate. The relative contributions of σ_p and σ_{nuc} to σ_{sr} are of interest. For hydrogen, with its small nuclear charge and unusually spherical distribution of electrons, the slip effect is rather large in proportion, and the magnetic field due to the rotating nuclei exceeds that due to the electrons. For all other nuclei the electronic contribution outweighs the nuclear contribution, e.g. for ^{19}F all known values of C_i are negative (on Ramsey's sign convention) except for ClF.

B. Other Methods for the Measurement of Absolute Shielding

A method for obtaining the paramagnetic terms directly, which will become more important now that pulsed NMR provides accurate measurements of shielding anisotropies in solids (92, 94), is to take advantage of the zero parallel component of the paramagnetic term in linear molecules. Because of this the anisotropy $\Delta\sigma_p$ in the paramagnetic term gives the absolute (average) value:

$$\begin{aligned}\Delta\sigma_p &= \sigma_p{}^\parallel - \sigma_p{}^\perp = -\sigma_p{}^\perp \\ \sigma_p &= \tfrac{1}{3}(\sigma_p{}^\parallel + 2\sigma_p{}^\perp) = -\tfrac{2}{3}\Delta\sigma_p\end{aligned} \tag{8}$$

The anisotropy in the paramagnetic term is obtained by correcting the measured anisotropy for the anisotropy in σ_d, and Gierke and Flygare (27) have shown how this may be estimated. This correction is often neglected, but is rarely negligible; e.g., for F_2, $\Delta\sigma_d$ is -75 ppm. If the measured anisotropy for the fluorine molecule (65) is corrected in this way, then it gives a value of σ_p which agrees well with values obtained from the spin-rotation interaction and from the chemical shift (59).

Another way in which chemical shifts can be related to absolute shielding is by the comparison of the nuclear magnetic moment (measured by the NMR method for an atom in a molecule) with the moment for the free atom. This has been done with considerable accuracy for hydrogen (80) and also for lead (50).

C. Absolute Shielding Scales

In the past the difficulties of the absolute measurements have tended to limit their precision; but with modern techniques, and under favorable circumstances, the precision approaches that of the measurement of chemical shifts, and the information can be combined to set up absolute shielding scales. This has been done for hydrogen, as mentioned above, and for fluorine on the basis of spin-rotation measurements on HF (95), with vibrational correction of the spin-rotation and shielding constants (33). For ^{13}C, measurements on ^{13}CO (22, 64, 66) could be used, for ^{14}N,* measurements on NH_3 [by two-cavity molecular beam maser (43, 44)] or on HCN (17), for ^{31}P, measurements on PH_3 (15), and for the halogens, measurements on $H^{35}Cl$ (37) and on $H^{79}Br$ and $H^{127}I$ (18). This list is by no means exhaustive, and there is now a considerable body of information on absolute shielding that could yield interesting chemical comparisons.

V. Periodicity in Nuclear Magnetic Shielding

The study of periodicity in nuclear magnetic shielding has been neglected in the past, except for one or two isolated studies. Jameson and Gutowsky (35) showed that the range of chemical shifts for a particular nucleus (which resembles the diamagnetic term for the free atom) varies periodically with the atomic number of the nucleus, and traced this variation to the dependence of the paramagnetic term on $\langle r^{-3} \rangle$ for the bonding electrons. The periodic variation of $\langle r^{-3} \rangle_{np}$ is shown in Fig. 3.

* This has now been done (J. Mason and J. G. Vinter, *J. Chem. Soc., Dalton Trans.*, in press.

As mentioned above, the electronegativity correlation in nuclear magnetic shielding suggests a periodic dependence of nuclear shielding on the atomic number of the ligand. This was explored in a preliminary way (57) for ^{13}C shielding in methyl derivatives Me_nX, many of which have now been measured, and the periodicity of the ^{13}C shielding with the atomic number of the ligand X was found to be improved when the paramagnetic part of the shielding (relative to methane) was separated according to the atom-plus-ligand approximation.

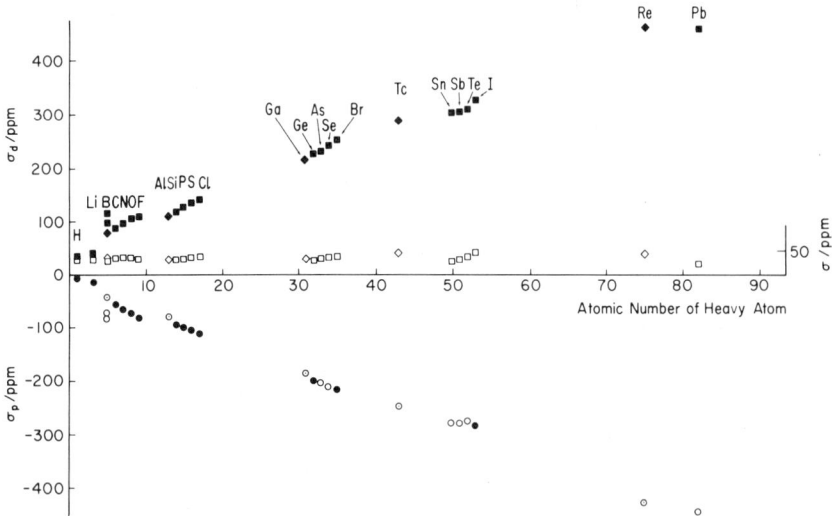

FIG. 6. Proton-shielding parameters for the binary hydrides of the elements plotted against the atomic number of the central atom. Explanation of symbols:

	Molecules	Ions
σ_d(calc.)	■	◆
$\sigma(\delta)$	□	◇
$\sigma_p(\delta)$	○	⊙
σ_p(obs.)	●	

For proton and ^{19}F shielding, a sufficiently large number of the paramagnetic shielding constants have been measured, mostly through the spin-rotation interaction, so that the periodicity can be examined (58, 59) for the Ramsey (all-atom) shielding terms σ_d and σ_p. Many calculated values of σ_d are available in the literature for the binary hydrides and some for the binary fluorides. The rest were calculated by Flygare and Goodisman's method (25).

Figure 6 shows the shielding terms σ_d, σ_p, and σ (the resultant shielding) for the proton in the binary hydrides plotted against the atomic

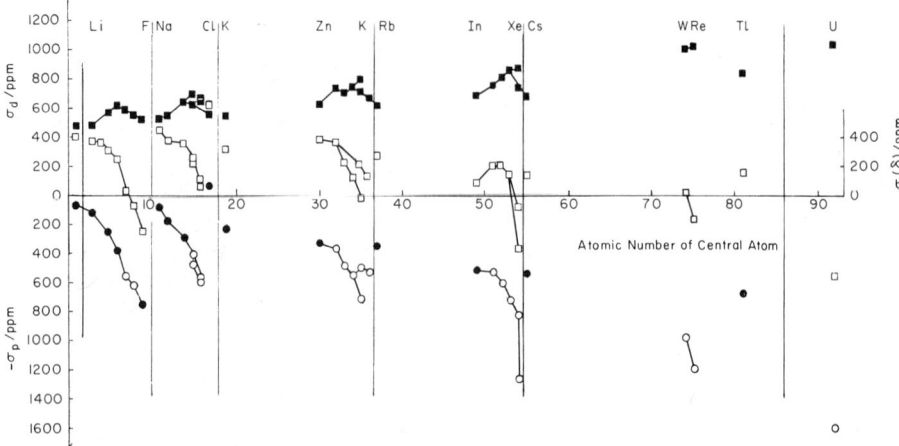

FIG. 7. ^{19}F Shielding parameters for the binary fluorides plotted against the atomic number of the central atom—molecular species. (■)σ_d(calc.); (□)$\sigma(\delta)$; (○)$\sigma_p(\delta)$; (●) σ_p(obs.).

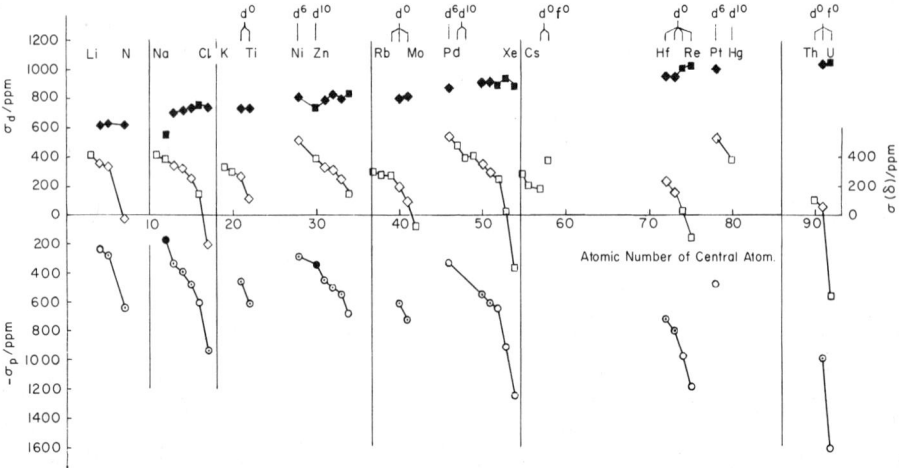

FIG. 8. ^{19}F Shielding parameters for the binary fluorides plotted against the atomic number of the central atom—fluoride anions, solids, and high-oxidation states of the central atom. Explanation of symbols:

	Molecules	Ions
σ_d(calc.)	■	◆
$\sigma(\delta)$	□	◇
$\sigma_p(\delta)$	☉	○
σ_p(obs.)	●	

number Z of the heavy atom (58), and Figs. 7 and 8 are analogous plots (59) for the ^{19}F shielding in the binary fluorides (in Fig. 7 are plotted molecular species, and in Fig. 8 mainly fluoride anions, solids, and high-oxidation states of the central atom). The values of σ_p(obs.) plotted as closed circles were obtained from the spin-rotation interaction, apart from some values for linear fluorides for which σ_p was obtained from the shielding anisotropy. The values of σ_d(calc.) are plotted as closed squares. The values of $\sigma(\delta)$, plotted as open squares, are obtained from conventional NMR by referring the chemical shift δ to an absolute scale, as described in the previous section.

It was found for the hydrides and for the fluorides that the values for the resultant shielding obtained by combining σ_p(obs.) and σ_d(calc.) agreed well with those obtained from the chemical shift. (Where agreement was not so good, e.g., for SiH_4 and GeH_4, the discrepancy corresponds to the uncertainty in the spin-rotation constants.) The paramagnetic term for the hydrides and fluorides for which no spin-rotation value was available was therefore estimated by subtracting σ_d(calc.) from $\sigma(\delta)$. The resulting values of $\sigma_p(\delta)$ are plotted as open circles in the Figs. 6 to 11.

The striking feature of the hydride correlation in Fig. 6 is the very similar periodicity in σ_d and σ_p. This is unique to hydrogen, for which all the electrons are valence electrons. The value of σ_d increases with the number of electrons on the heavy atom, and the periodic variation of σ_d with the number of ligands (which rises in the second row to 4 for carbon and then falls to 1 for fluorine) is not evident in the hydride plot, although it can be seen in the fluoride plot in Fig. 7. In Fig. 6, boron–hydrogen shieldings are shown for diborane and BH_4^-. The BH_4^- ion is seen to fit the periodic correlation, as do AlH_4^-, GaH_4^- (for which the chemical shift was estimated from that of related compounds), TcH_9^{2-} and ReH_9^{2-} (d^0). The shielding of hydrogen attached to boron correlates more with molecular topology than with chemical expectation since the electron-deficient bridge proton is not the least shielded, although the electron-affluent terminal proton is the most shielded. (The relative shielding of the bridge and terminal protons follows the increase in σ_d as the proton comes closer to the electrons on two borons, but the proton in BH_4^- is the most shielded because of the smallness of σ_p.)

As to the symmetry of the plot for σ_d and σ_p, we mentioned earlier that the effective r dependence of σ_p may be nearer $\langle r^{-1} \rangle$ than $\langle r^{-3} \rangle$ for "distant" contributions, because of the dependence of the moment of the electron on its distance r. The plot reflects (for σ_p) the increase in the r term across the row of the ligand and the decrease down the group, against the tendency of $(\Delta E)^{-1}$ to decrease across the row and increase down the group. Figures 6 and 9 illustrate the familiar alternation of properties down the group, due to irregularities in atomic structure and the periodic table. The relatively large increase in σ_d and σ_p from the first row (hydrogen) to the second reflects the relatively large increase in the number of electrons on the ligand, and similarly for the large increase from the third row to the post-transition elements of the fourth. However, σ_d and σ_p do not increase at the same rate, and given the steady increase in σ_d with the number of electrons in the molecule, we can ascribe the irregularities in the variation of the proton shielding down the

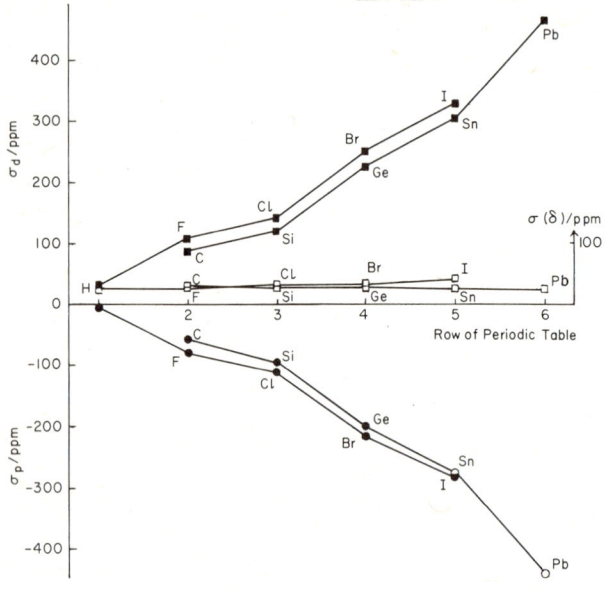

FIG. 9. Variation in the proton shielding parameters down the groups of the heavy atom for Groups IV and VII. (■) σ_d(calc.); (□) $\sigma(\delta)$; (○) $\sigma_p(\delta)$; (●) σ_p(obs.).

group of the ligand to the smaller increase in σ_p than in σ_d down the later groups.

Thus, although proton shielding is often described as dominated by the diamagnetic term, the periodic correlation shows that variations in the resultant shielding may be determined by changes in σ_p, as across the second row and down the group of the ligand. Across the third and subsequent rows, the diamagnetic term increases faster than the paramagnetic term, the more so the heavier the central atom, but the margin is relatively small.

Figure 7 shows clearly the periodic variation of σ_d for ^{19}F in fluorides

(59) with the number of electrons in the molecule. For the second-row ligands this reaches a maximum at CF_4, but in later rows the maximum is at the end of the row, at the highest fluoride of the halogen or xenon. Both σ_d and σ_p are larger for higher fluorides compared with lower fluorides of the same element. The resultant shielding, although it is higher in the higher-oxidation state for the fluorides of the elements of

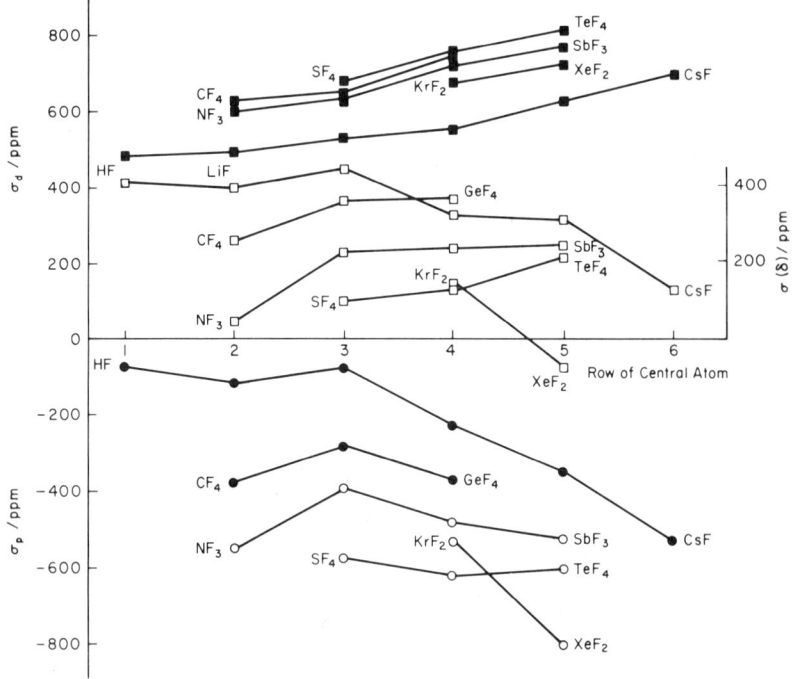

FIG. 10. Variation in ^{19}F shielding down the group of the central atom for Groups O, I, IV, V, and VI. (■) σ_d(calc.); (□) $\sigma(\delta)$; (○) $\sigma_p(\delta)$; (●) σ_p(obs.).

Groups V and VI, is higher for the *lower* fluorides of the halogens or xenon; a comparative study of Cornwell effects would be of interest here.

The most striking feature of Figs. 7 and 8, particularly as compared with the hydride plot, is the plunge downfield of the resultant shielding, following the paramagnetic term, across the row of the central atom for the typical elements, and also for the early transition metals (Fig. 8), with ready circulation of fluorine $2p$ electrons into empty t_{2g} orbitals in the d^0 complexes. Fluorine is highly shielded however in the d^6 molecules and ions, and this was discussed in Section III, B as a possible Cornwell effect (62).

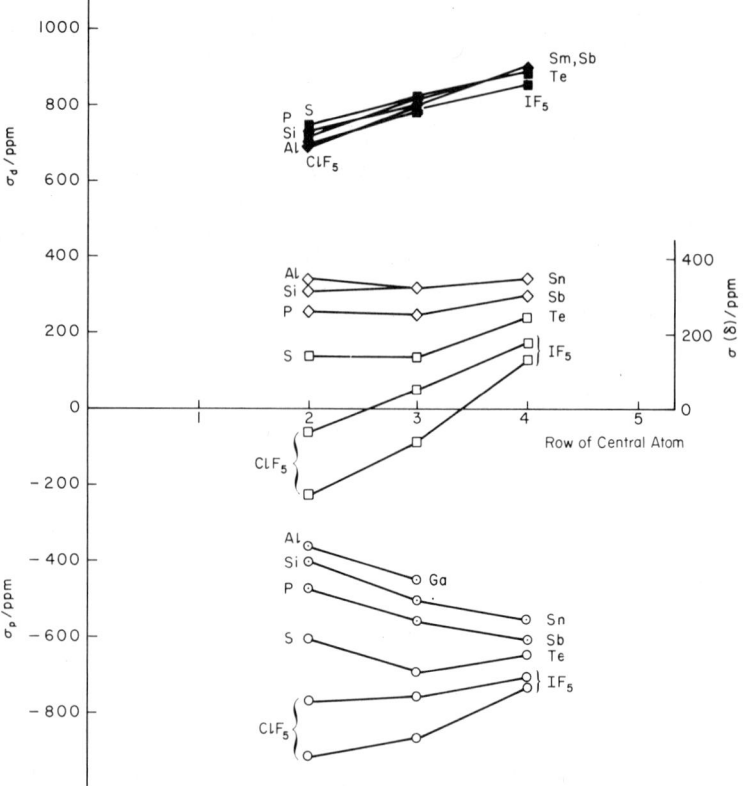

FIG. 11. Variation in ^{19}F shielding down the group of the central atom in hexafluoride ions and molecules and in halogen pentafluorides. Explanation of symbols:

	Molecules	Ions
σ_d(calc.)	■	○
$\sigma(\delta)$	□	◇
$\sigma_p(\delta)$	○	⊙
σ_p(obs.)	●	

The increase in σ_p across the row of the central atom follows the increasing imbalance of the fluorine $2p$ electrons with increasing covalency, the increase in $\langle r^{-3} \rangle_{2p}$ (cf. Fig. 3), and the tendency of $(\Delta E)^{-1}$ to increase [in contrast to the trend in the energy term for the proton (58) and ^{13}C shielding (57) across the row of the ligand]. Down the group of the central atom (Figs. 10 and 11), σ_p tends to increase with the energy term against the tendency of the radius term to decrease (for the typical elements), although there are important anomalies that, perhaps, show the influence of the radius term; e.g., the large values of σ_p when the central atom belongs to the second row and the radius term is large. Near the bottom of the later groups the increase in σ_p is slowed or reversed so that the resultant shielding increases, and this tendency is evident also in the proton and ^{13}C correlations (57, 58). Figure 11 shows that this is not an effect of change in the molecular shape across the row since it is very marked for the higher fluorides. Figure 10 shows that this tendency in Group VI and VII is not, however, shown by KrF_2 and XeF_2.

Thus, although the hydride and fluoride periodicities look very different at first sight, similar trends can be discerned overall as well as in many of the details, and these trends are evident also in the periodic correlations for the shielding of other nuclei, ^{13}C, for example.

The periodic correlations illustrate the basis in molecular electronic structure for the generalization, "more electrons, more shielding," and for the exceptions to this. Effects of ligand electronegativity in nuclear magnetic shielding are complex, since the excitation energies are influenced by both σ- and π-inductive effects, and so also are the orbital terms, including the radius terms. In the periodic correlations, as in the earlier examples, separation of the diamagnetic and paramagnetic parts often helps in understanding chemical shifts in groups of related compounds.

REFERENCES

1. Andersson, L.-O., and Mason, J., *Chem. Commun.* p. 99 (1968).
2. Andersson, L.-O., Mason, J., and W. van Bronswijk, *J. Chem. Soc., London* p. 296 (1970).
3. Atkins, P. W., "Molecular Quantum Mechanics," p. 441. Oxford Univ. Press (Clarendon), London and New York, 1970.
4. Baker, M. R., Anderson, C. H., and Ramsey, N. F., *Phys. Rev. A* **133**, 1533 (1964).
5. Barnes, R. G., and Smith, W. V., *Phys. Rev.* [2] **93**, 95 (1954).
6. Bloch, F., *Pure Appl. Chem.* **32**, 1 (1972).
7. Bonham, R. A., and Strand, T. G., *J. Chem. Phys.* **40**, 3447 (1964).
8. Bürges, C.-W., Koschmieder, R., Sahm, W., and Schwenk, A., *Z. Naturforsch. A* **28**, 1753 (1973).
9. Chan, S. I., Baker, M. R., and Ramsey, N. F., *Phys. Rev. A* **136**, 1224 (1964).
10. Chan, S. I., and Dubin, A. S., *J. Chem. Phys.* **46**, 1745 (1967).
11. Cornwell, C. D., *J. Chem. Phys.* **44**, 874 (1966); *Abstr. Columbus Symp. Molec. Spectrosc.*, p. 40 (1964).
12. Dailey, B. P., and Shoolery, J. N., *J. Amer. Chem. Soc.* **77**, 3977 (1955).
13. Davies, D. W., *Nature (London)* **207**, 75 (1965).
14. Davies, D. W., "The Theory of the Electric and Magnetic Properties of Molecules," p. 179. Wiley, London and New York, 1967.

15. Davies, P. B., Neumann, R. M., Wofsy, S. C., and Klemperer, W., *J. Chem. Phys.* **55**, 3564 (1971).
16. Davis, R. E., and Muenter, J. S., *J. Chem. Phys.* **57**, 2836 (1967).
17. DeLucia, F. C., and Gordy, W., *Phys. Rev.* **187**, 58 (1969).
18. DeLucia, F. C., Helminger, P., and Gordy, W., *Phys. Rev. A* **3**, 1849 (1971).
19. Ditchfield, R., Miller, D. P., and Pople, J. A., *J. Chem. Phys.* **54**, 4186 (1971).
20. Emsley, J. W., Feeney, J., and Sutcliffe, L. H., "High Resolution NMR Spectroscopy," Appendix A. Pergamon, Oxford, 1965.
21. Esbach, J. R., and Strandberg, M. W. P., *Phys. Rev.* [2] **85**, 24 (1952).
22. Ettinger, R., Blume, P., Patterson, A., and Lauterbur, P. C., *J. Chem. Phys.* **33**, 1547 (1960).
23. Figgis, B. M., Kidd, R. G., and Nyholm, R. S., *Proc. Roy. Soc., Ser. A* **269** 469 (1962).
24. Flygare, W. H., *J. Chem. Phys.* **41**, 793 (1964).
25. Flygare, W. H., and Goodisman, J., *J. Chem. Phys.*, **49**, 3122 (1968).
26. Flygare, W. H., and Weiss, V. W., *J. Chem. Phys.* **45**, 2785 (1966).
27. Gierke, T. D., and Flygare, W. H., *J. Amer. Chem. Soc.* **94**, 7277 (1972).
28. Gordy, W., and Cook, R. L., "Microwave Molecular Spectra," Chapters 9 and 11. Wiley (Interscience), New York, 1970.
29. Grant, D. M., and Paul, E. G., *J. Amer. Chem. Soc.* **86**, 2984 (1964).
30. Grinter, R., and Mason, J., *J. Chem. Soc.*, London p. 2196 (1970).
31. Gunther-Mohr, G. R., Townes, C. H., and Van Vleck, J. H., *Phys. Rev.* [2] **94**, 1191 (1954).
32. Herbison-Evans, D., and Richards, R. E., *Mol. Phys.* **6**, 191 (1964).
33. Hindermann, D. K., and Cornwell, C. D., *J. Chem. Phys.* **48**, 4148 (1968).
34. Hindermann, D. K., and Falconer, W. E., *J. Chem. Phys.* **50**, 1203 (1969); **52**, 6198 (ref. 31) (1970).
35. Jameson, C. J., and Gutowsky, H. S., *J. Chem. Phys.* **40**, 1714 (1964).
36. Jen, C. K., *Phys. Rev.* [2] **81**, 197 (1951); *Physica (Utrecht)* **17**, 379 (1951).
37. Kaiser, W., *J. Chem. Phys.* **53**, 1686 (1970).
38. Karplus, M., and Das, T. P., *J. Chem. Phys.* **34**, 1683 (1961); Karplus, M., and Pople, J. A., *ibid.* **38**, 2803 (1963).
39. Kellogg, J. M. B., Rabi, I. I., Ramsey, N. F., and Zacharias, J. R., *Phys. Rev.* [2] **56**, 728 (1939); **57**, 677 (1939).
40. Kidd, R. G., and Spinney, H. G., *J. Amer. Chem. Soc.* **94**, 6686 (1972).
41. Kidd, R. G., and Spinney, H. G., *J. Amer. Chem. Soc.* **95**, 88 (1973).
42. Kidd, R. G., and Spinney, H. G., *Inorg. Chem.* **12**, 1967 (1973).
43. Kukolich, S. G., *Phys. Rev.* [2] **156**, 83 (1967).
44. Kukolich, S. G., and Wofsy, S. C., *J. Chem. Phys.* **52**, 5477 (1970).
45. Lamb, W. E., *Phys. Rev.* [2] **60**, 817 (1941).
46. Lambert, J. B., and Roberts, J. D., *J. Amer. Chem. Soc.* **87**, 4087 (1965).
47. Lauterbur, P. C., *Ann. N.Y. Acad. Sci.* **70**, 840 (1958).
48. Letcher, J. H., and van Wazer, J. R., *J. Chem. Phys.* **44**, 815 (1966).
49. Logan, N., *in* "Nitrogen N.M.R." (M. Witanowski and G. A. Webb, eds.), p. 319. Plenum, New York, 1973.
50. Lutz, O., and Stricker, G., *Phys. Lett. A* **35**, 397 (1971).
51. McConnell, H. M., *J. Chem. Phys.* **27**, 226 (1957).
52. McLachlan, A. D., *J. Chem. Phys.* **32**, 1263 (1960).
53. Malinowski, E. R., *J. Amer. Chem. Soc.* **91**, 4701 (1969).

54. Malinowski, E. R., Vladimiroff, T., and Tavares, R. E., *J. Phys. Chem.* **70**, 2046 (1966).
55. Malli, G., and Fraga, S., *Theor. Chim. Acta* **5**, 275 (1966).
56. Malli, G., and Froese, C., *Int. J. Quantum Chem.* **1s**, 95 (1967).
57. Mason, J., *J. Chem. Soc., London* p. 1038 (1971).
58. Mason, J., *J. Chem. Soc., Dalton Trans.* p. 1422 (1975).
59. Mason, J., *J. Chem. Soc., Dalton Trans.* p. 1426 (1975).
60. Mason, J., and van Bronswijk, W., *J. Chem. Soc. A* p. 1763 (1970).
61. Mason, J., and van Bronswijk, W., *J. Chem. Soc., A* p. 791 (1971).
62. Matwiyoff, N. A., Asprey, L. B., Wageman, W. E., Reisfeld, M. J., and Fukushima, E., *Inorg. Chem.* **4**, 751 (1969).
63. Memory, J. D., "Quantum Theory of Magnetic Resonance Parameters," Chapter 6. McGraw-Hill, New York, 1968.
64. Neumann, D. B., and Moskowitz, J. W., *J. Chem. Phys.* **50**, 2216 (1969).
65. O'Reilly, D. E., Peterson, E. M., El Saffar, Z. M., and Scheie, C. E., *Chem. Phys. Lett.* **8**, 470 (1971).
66. Ozier, I., Crapo, L. M., and Ramsey, N. F., *J. Chem. Phys.* **49**, 2314 (1968).
67. Paudler, W. W., "Nuclear Magnetic Resonance," p. 27. Allyn & Bacon, Boston, 1971.
68. Pople, J. A., *Proc. Roy. Soc., Ser. A* **239**, 541 (1957).
69. Pople, J. A., *J. Chem. Phys.* **37**, 53 and 61 (1962); *Discuss. Faraday Soc.* **34**, 7 (1962); *Mol. Phys.* **7**, 301 (1963–1964).
70. Pople, J. A., Schneider, W. G., and Bernstein, H. J., "High Resolution Nuclear Magnetic Resonance," Appendix A. McGraw-Hill, New York, 1959.
71. Pople, J. A., Schneider, W. G., and Bernstein, H. J., "High Resolution Nuclear Magnetic Resonance," p. 170. McGraw-Hill, New York, 1959.
72. Pople, J. A., Schneider, W. G., and Bernstein, H. J., "High Resolution Nuclear Magnetic Resonance," Chapter 12. McGraw-Hill, New York, 1959.
73. Powles, J. G., *Rep. Progr. Phys.* **22**, 433 (1959).
74. Proctor, W. G., and Yu, F. C., *Phys. Rev.* [2] **77**, 717 (1950).
75. Ramsey, N. F., *Phys. Rev.* [2] **78**, 699 (1950); **90**, 232 (1953).
76. Ramsey, N. F., *Phys. Rev.* [2] **83**, 540 (1951); **86**, 243 (1952).
77. Ramsey, N. F., "Molecular Beams." Oxford Univ. Press (Clarendon), London and New York, 1956.
78. Ramsey, N. F., *Am. Sci.* **49**, 509 (1961).
79. Randall, E. W., and Gillies, D. G., *Progr. NMR Spectrosc.* **6**, 119 (1970).
80. Raynes, W. T., *Chem. Soc., Spec. Per. Rep. NMR* **2**, 4 (1973).
81. Saika, A., and Slichter, C. P., *J. Chem. Phys.* **22**, 26 (1954).
82. Santry, D. P., "Theoretical Chemical Research at the Carnegie Institute of Technology." 1965.
83. Schwenk, A., *Phys. Lett. A* **31**, 513 (1970); *J. Magn. Resonance* **5**, 376 (1971).
84. Silver, B. L., and Luz, Z., *Quart. Rev., Chem. Soc.* **21**, 458 (1967).
85. Slichter, C. P., *Ann. N. Y. Acad. Sci.* **70**, 769 (1958).
86. Spiesecke, H., and Schneider, W. G., *J. Chem. Phys.* **35**, 722 (1961).
87. Spielvogel, B. F., and Purser, J. M., *J. Amer. Chem. Soc.* **93**, 4418 (1971).
88. Stothers, J. B., "Carbon-13 NMR Spectroscopy," p. 390. Academic Press, New York, 1972.
89. Tigelaar, H. L., and Flygare, W. H., *J. Amer. Chem. Soc.* **94**, 343 (1972).
90. Townes, C., and Schawlow, A., "Microwave Spectroscopy," Chapter 8. McGraw-Hill, New York, 1955.

91. Van Vleck, J. H., "Theory of Electric and Magnetic Susceptibilities." Oxford Univ. Press, London and New York, 1932.
92. Vaughan, R. W., Elleman, D. D., Rhim, W.-K., and Stacey, L. M., *J. Chem. Phys.* **57**, 5383 (1972).
93. Vladimiroff, T., and Malinowski, E. R., *J. Chem. Phys.* **46**, 1830 (1967).
94. Waugh, J. S., Wang, C. H., Huber, L. M., and Vold, R. L., *J. Chem. Phys.* **48**, 662 (1968).
95. Weiss, R., *Phys. Rev.* [2] **131**, 659 (1963).
96. Weiss, V. M., Todd, H. D., Lo, M.-K., Gutowsky, H. S., and Flygare, W. H., *J. Chem. Phys.* **47**, 4021 (1967).
97. White, R. L., *Rev. Mod. Phys.* **27**, 276 (1955).
98. Wick, G. C., *Phys. Rev.* [2] **73**, 51 (1948).
99. Wofsy, S. C., Muenter, J. S., and Klemperer, W., *J. Chem. Phys.* **55**, 2014 (1971).

SOME APPLICATIONS OF MASS SPECTROSCOPY IN INORGANIC AND ORGANOMETALLIC CHEMISTRY

JACK M. MILLER

Department of Chemistry, Brock University, St. Catharines, Ontario, Canada

and

GARY L. WILSON

Department of Chemistry, John Abbott College, Ste. Anne de Bellevue, Quebec, Canada

I. Introduction	229
II. Instrumentation and Sample Handling	231
A. Mass Analyzers	231
B. Ion Sources	232
C. Basic Inlet Systems	235
D. Handling Air- and Moisture-Sensitive Compounds	237
III. Inorganic and Organometallic Mass Spectra: Practice and Pitfalls	239
A. Molecular Ions, Polymers, and Fragments	239
B. Factors Affecting Spectra	242
IV. Low- and Medium-Resolution Studies	247
A. Fragmentations	248
B. Rearrangements	257
C. Isotope Abundances	264
D. Negative Ions	267
E. Synthetic Models	268
V. High-Resolution Studies	268
VI. Metastable-Ion Techniques	270
VII. Coupled Gas Chromatography and Mass Spectrometry Applied to Inorganic and Organometallic Compounds	273
VIII. Conclusion	276
References	276

I. Introduction

Mass spectroscopic studies of organometallic compounds are almost as old as the field of mass spectrometry: nickel tetracarbonyl was studied by J. J. Thomson (1) and Aston (2, 3) in their work on the isotope ratios of nickel. Following this early flurry of specialized interest, however, inorganic and organometallic mass spectral studies were delayed until the late 1950s, at which time commercial "organic" mass spectrometers offered suitable inlet systems and extended mass range. Routine use of these instruments by inorganic chemists was postponed a further decade by the fear, unfortunately still prevalent today in many laboratories, that metal-containing compounds will decompose and thus contaminate the instrument. We hope to illustrate in this review, aimed at the inorganic or organometallic chemist considering the use of mass spectroscopy as a routine technique in his work, that this fear is groundless. Certainly these compounds may cause problems, but with care these are no worse than those encountered at times in the mass spectral studies of organic heterocycles, biological materials, or environmental samples.

The field of organometallic mass spectroscopy has been covered by a series of general reviews (4–6), whereas more specific reports include those on metallocenes (7), transition metal carbonyl derivatives (8), Group IV (8a), germanium compounds (9, 10), and boron compounds (11). King (12) reviewed his work with transition metal derivatives, whereas Müller (13) attempted to classify decomposition types for organometallic chemistry. More inorganic in nature are the proceedings of a 1966 American Chemical Society Symposium (14). A most comprehensive reference source for both inorganic and organometallic compounds is the recent book by Litzow and Spalding (15). Organometallic and coordination compounds are covered by Bruce in the Chemical Society's *Specialist Periodical Reports* (16, 17), and the monthly *Mass Spectrometry Bulletin* (18) indexes articles, books, and conference proceedings for mass spectrometry-related material, including organometallic and inorganic subsections. Mass spectrometry of metal-containing compounds has formed the topic of both ASMS (19) and international (20) symposia.

Several of the standard organic mass spectrometry reference works (21–24) also contain material useful to the inorganic/organometallic chemist, as do some of the more general mass spectrometry works, which also contain greater discussion of theory and instrumentation (25–31), Kiser's book (25) in particular referring to much of the early work in the field. The beginner in the field of inorganic/organometallic mass spectrometry should be familiar with the basics of organic mass spectra such as can be obtained from McLafferty's (31) and other books (32, 33).

In this review we point out briefly the various instrumentation factors important in obtaining mass spectra of inorganic and organometallic compounds and in affecting the quality of spectra observed. Selected examples are chosen, admittedly with a personal bias, to illustrate the various types of information available from mass spectra. We have deliberately not discussed elemental analysis by spark-source mass spectrometry (34, 35) nor detailed thermochemical studies possible with the Knudsen cell (36, 37), but we do point out some of the new ionization techniques which may be of potential use in the future.

II. Instrumentation and Sample Handling

A. Mass Analyzers

Most of the mass spectrometers used for the work that we are describing use sector magnetic mass analyzers; descriptions of the ion optics, etc., may be found in most of the standard works referenced in the Introduction. The minimum mass analyzer requirements for an instrument to be used routinely for inorganic and organometallic compounds are an upper mass range of m/e 1000, but preferably higher and medium resolution,* i.e., a resolution of 1000 to 5000, although some studies may require high resolution (commercial instruments now readily achieve resolutions of 10,000 to 150,000). Until recently only magnetic analyzers met the mass range and medium-resolution requirements.

The mass spectrometers most commonly used for the work described have been 60° and 90° single-focusing instruments (38), and double-focusing instruments of Nier-Johnson (39), reversed Nier-Johnson (40), and Mattauch-Herzog (41) geometry, the double-focusing instruments generally producing higher resolution. Older instruments all had nonlinear mass scans, but newer designs incorporate either Hall effect feedback loops in the magnet scan control or variable chart speed to give a pseudolinear spectrum presentation.

Quadrupole or radio-frequency mass analyzers have only recently become available with the mass range and resolution (42) to be of general use in inorganic and organometallic mass spectrometry, although they have been popular as small mass spectrometers built into specific systems as reaction monitors. They do have the advantage of essentially linear

* Resolution in mass spectrometry is defined as $R = M/\Delta M$, where M is the mass under consideration and ΔM the mass difference between adjacent peaks. Unlike most other forms of spectroscopy where width at half height would be used as ΔM, the mass spectrometry convention is the 10% valley definition, i.e., for two peaks of equal height ΔM would be the peak separation producing a valley between them, whose height above the base line is equal to 10% of the peak heights. To a first approximation, therefore, for an isolated peak, ΔM corresponds to the width at 5% maximum height. Some manufacturers still use width at half height, which gives an apparent resolution of about twice that of the 10% valley definition. Magnetic instruments have essentially constant resolution throughout the mass range, whereas quadrupole mass analyzers have variable resolution, e.g., $R = 250$ at m/e 250 and $R = 25$ at m/e 25.

mass scans. Time-of-flight instruments (*43*) have never really achieved the mass range or resolution to become popular in their field. However, much early work, especially that involving appearance potential studies (*25*) was facilitated by time-of-flight instruments where open ion source design offers particular adaptability to special problems or very dirty systems. The cycloidal double-focusing designs were not a commercial success (*44*) for routine work and are no longer manufactured. Of the various types, the magnetic analyzer is to be preferred for another reason, namely, the ready observation of metastable ions (see Section VI).

B. Ion Sources

The details of ion sources, recently reviewed by Chait (*45*), and adequately covered in the major mass spectrometry texts will just be mentioned briefly, and the potential mass spectrometer user is advised to investigate the facilities in his laboratories.

1. Electron Impact Sources

Most mass spectrometers available to the routine user are equipped with electron impact (EI) sources. In these sources, a heated filament of rhenium or tungsten produces electrons that are accelerated to a "trap" anode with energies typically variable from 5 to 100 eV, adjustable energies permitting appearance potential studies. Unless otherwise specified, most laboratories run spectra at 70 eV. Collimating slits and magnets pass the electron beam through the "cage" or ion source proper where ionization occurs by the following process:

$$M + e^- \longrightarrow M^{+\cdot} + 2e^- \tag{1}$$

The positive ions formed are extracted from the cage by a repeller plate with a small variable positive charge, collimated and focused by slits and electrostatic lenses, and finally accelerated to a high potential (1000–10,000 volts depending on the instrument and mass range). Decreasing the accelerating voltage increases the mass range of a particular magnetic analyzer but decreases both resolution and sensitivity.

Most common ion sources are categorized as "closed" or "tight," the inside of the cage typically being on the order of a 5-mm cube, the only passages to its interior being the electron entrance and exit slits, the positive ion exit slit, and one or more small holes for the sample inlet devices such as a solid probe, batch inlet, and gas chromatograph (GC) interface. These sources maximize sample pressure in the region of the electron beam and typically have sensitivities several orders of magnitude greater than "open" sources (46), where all components are open to the source housing and thus rapid pumping. Time-of-flight and quadrupole instruments are often designed with open sources, as are specialized units for studies in fast kinetics where pumping speed is important. In closed source instruments the user must be aware that the sample pressure in the ion source is not that read from the ion-source pressure gauge, which measures pressure in either the source housing or source pumping line, and it is necessary to calculate the actual pressure (47) in the cage, typically several orders of magnitude higher than the gauge reading.

2. Chemical Ionization

Chemical ionization (CI) sources (48, 49) use electron bombardment of a reagent gas at higher pressures than normally found in a mass spectrometer ion source, i.e., ~1 torr. Sample ionization follows via an ion–molecule reaction, often accompanied by a proton transfer to yield a "quasi-molecular ion":

$$CH_4 + e^- \xrightarrow[\text{1 torr}]{CH_4} CH_5^+ \text{ etc.} \quad (2)$$
reagent gas

$$CH_5^+ + M \longrightarrow MH^+ + CH_4 \quad (3)$$
sample $\qquad\qquad$ quasi-molecular ion

This method of ionization usually causes minimum fragmentation of the molecule under investigation, and thus can be useful in establishing molecular weights, but as little fragmentation occurs it may be of less use as an aid in structure elucidation.

Hunt's group (50, 51) have pioneered the application of the CI source to organometallics such as the iron tricarbonyl complex of heptafulvene, whose electron impact spectrum shows $(M-CO)^+$ as the heaviest ion, in contrast to the methane CI spectrum with the MH^+ ion as base peak. Boron hydrides (52) and borazine (53) have also been studied. The methane CI spectrum of arenechromium and -molybdenum (54) show protonation at the metal giving a protonated parent or molecular ion. Risby et al. have studied the isobutane CI mass spectra of lanthanide 2,2,6,6-tetramethylheptane-3,5-dionates[Ln(thd)$_3$] (55) and 1,1,1,2,2,3,3-heptafluoro-7,7-dimethyl-4,6-octanedione [H(fod)] lanthanide complexes (56). These latter complexes have been suggested as a means of analysis for the lanthanide elements.

3. Field Ionization/Field Desorption

Field ionization (FI) mass spectrometry (57, 58) (produced by a high electric field gradient between a sharp blade and the first source slit) is another technique useful for obtaining parent molecular ions without undue fragmentation. The sample, however, must still be volatilized into the source, so there may be no advantage for thermally sensitive materials. The recently developed field desorption (FD) technique of Beckey (59, 60) involves the use of a wire emitter, similar to that in an FI source, which has been activated by growing of carbon microneedles on its surface. The sample is deposited, usually from solution, onto the emitter, and then desorbed under the influence of a high field gradient. For example, biological species that cannot normally be thermally volatilized may be introduced into the ion source in this manner. The technique has been used as well in conjunction with liquid chromatography (61) and with heated emitters (62). To date, no inorganic or organometallic compounds have been investigated by this promising technique although inorganic salts have been alluded to (62a). The few commercial instruments (63) delivered to date have been used mainly for biochemical applications. If, as proposed, the double-beam MS-30 mass spectrometer can be fitted with an FD source on one side and a standard EI source on the other, the problem of mass marking and exact mass measurement, important at the high molecular weights to be expected for organometallic polymers, etc., will be greatly simplified (64, 65).

Hass et al. (65a) have observed no ion current from an emitter coated with hexaquocobalt(II)chloride, but on the admission to the ion source of gaseous acetylacetone (Hacac) the ion $[Co(acac)Cl]^+$ was observed. Trifluoroacetic acid produced an analogous ion and when $FeCl_2$ was coated on the emitter the corresponding iron chelate was observed. Sodium from sodium chloride has also been detected by volatilization with acetic acid.

Phthalocyanines and metal phthalocyanines which require high probe (and often high source) temperatures (>350°C) in conventional EI mass spectrometry, readily give molecular ions as the base peak (65b) in FD spectra and minimal fragmentation occurs, providing a rapid method for the qualitative examination of mixtures of phthalocyanines.

Organic complex salts of the type $[LM(CO)_3]^+BF_4^-$ [M = Fe, L = cyclo-C_6H_7, cyclo-C_7H_9, cyclo-$C_{10}H_{11}$, 2MeO-cyclo-C_6H_7; M = W, L = cyclo-C_7H_7], $[(Ar)Fe(C_5H_5)]^+PF_6^-$ [Ar = C_6H_6, $CH_3C_6H_5$, $(CH_3)_3C_6H_3$], and $[(cyclo-C_8H_{11})Co(C_5H_5)]^+BF_4^-$ have been studied by FD mass spectrometry (65c). They show molecular or quasimolecular ions, usually the base peak, and fragment ions corresponding to metal ligand cleavages. (The last 3 paragraphs were added in proof.)

4. Other Sources

Several other types of ion source under development should be watched closely for potential inorganic/organometallic use. Ionization by electrons from a ^{63}Ni source has been used in an external ion source that is at atmospheric pressure *(66, 67)* giving a reported sensitivity in the subpicogram range.

A second type of source worth watching is the electrohydrodynamic (EHD) ionization source *(68–70)*. This source, built into both AEI MS-7 and MS-902 mass spectrometers, was originally applied to the analysis of liquid metal systems *(68, 69)*, such as the Ga–In eutectic alloy, which showed not only Ga^+ and In^+ ions but also species such as Ga_2^+, $GaIn^+$, In_2^+, Ga_2In^+, Ga_4^+, and Ga_5^+ as well as five metallic impurities. Similar results were observed for GaInSn and Cerrolow 117 alloy containing Bi, In, Pb, Sn, Cd, and Hg. This work has now been extended to a nonmetallic liquid, glycerol *(70)*, which in the presence of dissolved metallic salts gives ions such as $[\text{glycerol} + \text{cation} - H]_n^+$. Complex inorganic salts may become amenable to study by EHD ionization.

Spark, photoionization, surface, thermionic, and laser ionization sources are not considered in this review.

C. Basic Inlet Systems

1. Direct Insertion or "Solids" Probe

As relatively high volatility (≈ 1 torr), thermal stability, and a fairly large sample size (1–10 μl liquid) are required of samples to be introduced via the batch inlet system, the development of the direct insertion probe makes possible much of modern mass spectroscopic work. The probe may be rapidly loaded and inserted through a sliding-seal vacuum lock such that the sample, usually contained in a glass or quartz cup, may be evaporated directly into the "tight" ion chamber. Older units were heated indirectly by the hot ion source, but newer probes are electrically heated with temperature control independent of source temperature, and in some cases coolable as well; the latter is useful in lowering the volatility of, for example, some metal carbonyls. Separate temperature control of the solids probe is a must for any new mass spectrometer. It should be noted that a solids probe can readily be used for relatively nonvolatile liquids, simply by transferring a small quantity to the sample cup with fine wire. For both solids or liquids, if the sample is visible, then there is enough in the probe.

2. *Gases/Liquids Probe*

Many manufacturers now offer other sample injection systems compatible with the vacuum lock used for the solids probe. These include small (e.g., 75-ml) heatable batch inlet systems, usually accessible via syringe (gas syringe or GC microliter syringe for liquids), which can be particularly useful as inlets for mass reference compounds. Other probes are designed as flexible, easily removed connections to a gas chromatograph via some form of interface.

3. *Batch Inlet System*

Batch inlet systems are in many ways the most convenient for gases and volatile, thermally stable liquids or solids, provided there is sufficient sample available. They consist of reservoirs, varying in volume from 20 ml to several liters, which are connected to the ion source via a "molecular leak," usually a porous ceramic material, or a pinhole in thin gold foil or glass. The leak serves to reduce the pressure from 10^{-2} torr in the inlet system to $\sim 10^{-5}$ torr in the ion source. Large ballast bulbs valved to the system allow a constant pressure feed to the source for long duration, such as appearance potential studies. Batch inlets for room temperature work were traditionally constructed of glass, whereas heated inlets were of metal, and later glass-coated metal. Today, however, most manufacturers offer all-glass heated inlet systems (AGHIS), valves being of the magnetically operated "optical flat" type with no lubricant or sealant. Gases are admitted via some form of pressure reduction manifold, gases and liquids may be admitted by syringe (*71, 72*), and liquids and solids by some form of removable sample cup.

4. *GC/MS Interfaces*

Combined gas chromatography/mass spectrometry (GC/MS), which is likely to be of increasing importance to inorganic and organometallic chemistry (Section VII), is discussed in detail in both a recent book (*30*) and review (*73*). Not only can complex reaction products be studied, but reactions may be monitored and trace by-products identified. The gas chromatograph has the advantage of concentrating a sample so that it has all eluted from the column and entered the mass spectrometer in a matter of seconds. This is no problem for modern fast-scanning mass spectrometers. The question of GC/MS interfacing is of prime importance, however, as the sample must be introduced into the source through a large pressure differential, from 1 atm to 10^{-6} torr, requiring the preferential removal of carrier gas. Early solutions involved capillary columns (most mass spectrometers will accept 1–3 ml/min of helium without undue pressure rise), stream splitters, and capillary "sniffers." Maximum detection limits (10^{-10}–10^{-11} gm) were obtained only when various molecular separators were employed as sample-enrichment devices.

The most common separators include the Ryhage or jet diffusion separator (74), the Watson–Biemann or pore diffusion separator (75), and the membrane solution diffusion separator originally developed by Llewellyn (76). The first two separators involve direct passage of the sample into the mass spectrometer; the low molecular weight helium diffuses more readily and is pumped away. The membrane separator involves diffusion of the sample through a silicone membrane while the carrier gas vents to the atmosphere; carrier gas is thus not confined to helium. There is no "best separator"; the choice depends on the nature of the compounds, the temperature range over which it will be operated, and most usually what is available in a particular laboratory. A convenient configuration for a double-beam mass spectrometer such as the AEI MS-30 is two different separators, one into each beam, which permits rapid evaluation of separator performance.

Of all the inlets, the GC is the most foolproof, since operation of high vacuum valving, the chief danger with probe and batch inlets, is not necessary. Thus, once the instrument is properly tuned and operating, even novices may run their own GC/MS with minimum supervision.

D. Handling Air- and Moisture-Sensitive Compounds

One problem arising with inorganic and organometallic samples is the air and/or moisture sensitivity of many of these compounds. With ingenuity such problems can usually be overcome. The greatest hurdle is to convince the person in charge of the mass spectrometer that the minor modifications will only enhance the capability of the instrument. To quote Cooks and Beynon (77): "commercial scientific instruments are anything but sacrosanct. They should be modified, rearranged, run backwards, prodded, and coaxed to perform tasks other than those envisaged by the designer. One can hardly be doing something new if one is using an instrument in a manner foreseen when the instrument was designed." Labels "organic" and "inorganic (spark-source)" mass spectrometer in no way justify limited use, to the exclusion of the organometallic sample or to the modification of the instrument to meet changing problems. Because various specialized inlets have been reviewed by Beynon (78) and others, we only indicate a few current techniques of particular use.

For many workers, the handling of samples under anhydrous or oxygen-free conditions is of primary importance. The speed and efficiency of solids probe sample introduction need not be lost, as two simple techniques are applicable to this problem. Since most probes have sample holders made of glass or quartz of similar cross-section to melting point tubing, samples loaded in a dry box into either actual sample cups or pieces of capillary may be transported to the mass spectrometer sealed under an inert atmosphere. The sample cup can then be transferred to the probe and the assembly inserted into the vacuum lock by the use of an inexpensive disposable plastic glove bag, which is readily erected over the vacuum lock and purged with nitrogen, aided, if necessary, by the vacuum pump of the lock. We have readily handled anhydrous

FIG. 1. Concentric sample cup for a direct probe. Typical dimensions: 1.7 mm O.D. × 9 mm long.

organotin nitrates (79) and spontaneously inflammable phosphines in this manner with no signs of hydrolysis or oxidation. Use of a concentric tube sample cup (Fig. 1), easily fabricated by a glassblower, can eliminate the need of the glove bag. If such tubes are loaded in a dry box and transported to the instrument under nitrogen, the few seconds required to transfer to the probe and pump down are insufficient for air to diffuse significantly through the narrow opening. Again, when tested with spontaneously inflammable phosphines, no phosphine oxide was observed in the mass spectra. The concentric tube is also useful in limiting sample quantity entering the ion source in the case of more volatile solids. Other workers have built modified probes (80) for handling specific compounds and describe (81) useful methods for concentrating microsamples for probe insertion.

For more volatile compounds, the simplest procedure is to build a small vacuum line to handle sensitive materials. Depending on the mass spectrometer this may either be permanently attached or portable, connecting through the batch inlet which also supplies the vacuum source. Modern greaseless stopcocks plus use of glass "break seals" allow the most sensitive compounds to be handled. One can also readily build small breakers for sealed capillaries for use with the AGHIS (*82*), so that the entire device fits into the sample-heating oven. Others have built simple heated inlet systems by replacing the column in a GC interfaced to the MS (*83, 84*). Sensitive liquids are readily handled via GC syringes (*72*) and even inflammable gases can be handled readily by syringe. An expensive gas syringe is useful, but we found that we could handle PH_3–organophosphine mixtures, sampled from an autoclave, very readily in 2-ml disposable plastic syringes (*85*), which for some purposes were more successful than the gas syringes. Syringes for injection of solids are also available and can be loaded in a dry box for transport to the mass spectrometer.

Spots from thin-layer chromatography (TLC) plates may often be used directly in the mass spectrometer without eluting thus for air-sensitive compounds, when TLC is done in a glove box, a "concentric" tube may be directly loaded with the unknown plus support; the concentric tube is recommended because it minimizes loss of silica gel into the ion source.

III. Inorganic and Organometallic Mass Spectra: Practice and Pitfalls

The initial stimulus to submit a sample for mass spectrometric analysis, is usually a desire for confirmation of the molecular weight and elemental composition of a compound. Although much additional structural information is contained in the fragmentation pattern, the novice may be discouraged by a few unsuccessful attempts at interpretation. Perseverance at this point often results in the discovery of interesting new phenomena and in an interest in mass spectrometry beyond confirmation of a molecular weight. Another potential area of interest, bond strength as determined by appearance potentials, is far more difficult and uncertain experimentally, but should not be neglected.

A. Molecular Ions, Polymers, and Fragments

In seeking to identify a newly prepared compound by mass spectrometry, the observation of the "parent" or "molecular" ion isotopic cluster is usually greeted with satisfaction, but observation of apparently lower or higher m/e values for the parent ion causes initial confusion. However, even in a series of closely related compounds, run in the same mass spectrometer under identical conditions, some compounds may show molecular ions and others only fragments as a result of cleavage. For example, loss of an R· group, from R_4M compounds, to give the R_3M^+ ion as base peak, is very common for hydrocarbon derivatives of Group IV elements (86, 87), but, in a series of pentafluorophenyl derivatives of tin (88), it is not possible to predict, *a priori*, which derivative will show a parent ion in its 70-eV spectrum (Table I) (88, 89). There is no

TABLE I

MOLECULAR IONS FOR A SERIES OF C_6F_5Sn DERIVATIVES

Ion	Abundance as percent of total positive ion current				
	$R=C_6H_5$[a]	$R=C_6F_5$[b]	$R=C_4H_9$[a]	$R=CH_3$[a]	$R=CH=CH_2$[a]
$(C_6F_5)_3SnR^{+\cdot}$	1.8	7.2	0	6.5	N.D.[c]
$(C_6F_5)_2SnR_2^{+\cdot}$	0.8	7.2	3.6	8.4	0
$C_6F_5SnR_3^{+\cdot}$	0.3	7.2	0	3.3	N.D.[c]

[a] Chivers et al. (88).
[b] Miller (89).
[c] Not determined.

immediately apparent reason why the parent ion should be lacking if there are one or three —C_4H_9 groups, but be present if there are two —C_4H_9 groups in the molecule.

There are many examples today of polynuclear metal complexes yielding a parent ion in the mass spectra, but initially their observation for high molecular weight compounds, such as di[bis(pentafluorophenyl)-phosphidoirontricarbonyl] (I), having a molecular ion of m/e 1010 (90) was considered unusual, as were the binuclear chromium complexes (II)

studied by Preston and Reed (91). Mass spectra proved invaluable in the determination of the molecular formula of carbide complexes of ruthenium (8, 92), since the differentiation between the actual compound,

$Ru_6C(CO)_{17}$ and $Ru_6(CO)_{18}$, would be difficult using classic analytical methods.

Examples of large polymetallic molecules that readily give molecular ions are $Co_3(CO)_{10}BH_2$—$N(CH_3)_3$ (93), $(CO)_5MnAs(CH_3)_2Cr(CO)_4$-$As(CH_3)_2Mn(CO)_5$ with m/e 764 (94), and $(C_5H_5)Fe(CO)_2SnX_2Ni(CO)$-$(C_5H_5)$ (where X = Cl or Br) (95). Interestingly, the germanium analog of the foregoing tin compound shows as its highest m/e peaks those assignable to the P—CO ion (P = parent ion; P—CO indicates the parent that has lost CO), another example of the unpredictability of the mass spectra of a homologous series. Even heavy atoms such as uranium give volatile complexes and parent ions, e.g., (salicylideneiminato)bis-(dipivaloylmethanato)uranium(IV), m/e 870 (96). The mass spectrum similarly proved invaluable in assigning one of the by-products in the polyrecombination of ferrocene to a 1,2-diferrocenylethene (97). The mass spectra do not show parent ions for phosphorus selenides such as $P_4Se_7 \cdot 2py$, the $P_2Se_5^+$ ion being the highest m/e ratio observed (98), although for the sulfur selenides, parent ions $S_xSe_{8-x}^+$ are observed (99) as they are for the simple phosphorus selenide P_4Se_3 and for the sulfides P_4S_3, P_4S_5, P_4S_7, and P_4S_{10} (100).

Many species also give dimeric or polymeric ions in the gas phase. For example, in the spectrum of BiF_5 and SbF_5 the highest m/e values observed correspond to $Bi_2F_9^+$ and $Sb_2F_9^+$, i.e., loss of a fluorine from the simple dimer (101). Copper (I) carboxylates, which show dimeric units in their crystal structures show dimers in their mass spectra, i.e., $Cu_2L_2^{+ \cdot}$ (102). Metal alkoxides form low polymers in solution and many alkoxy-bridged oligomers survive the impact of 70-eV electrons. Although not all compounds in the series $Ti_4(OMe)_{16}$, $Zr_4(OEt)_{16}$, $Nb_2(OEt)_{10}$, $Ta_2(OEt)_{10}$, and $MAl_3(OPr^i)_{12}$ (where M = Fe, La, Ce, Pr, Nd, Eu, Gd, Ho, and Yb) exhibited parent ions (103), $LaAl_3(OPr^i)_{12}^{+ \cdot}$ is observed in good abundance at m/e 928 (104).

Both monomers and polymers have been reported in the mass spectra of metal acetylacetonates (105, 106), depending on the nature of the metal, and its oxidation state; different laboratories have reported conflicting results. Undergraduates in our freshman laboratory (107) recently discovered $Cr_2(acac)_5^+$ ions in the spectrum of $Cr(acac)_3$ but found that this dimer was removed if the sample was carefully purified, thus casting doubt on the purity of some of the previously reported (acac) derivatives showing dimers.

Alkali metals may also form stable complexes for which parent ions are observed in the mass spectra; species such as $NaHo(C_8H_{10}O_2F_3)_4^{+ \cdot}$ ($m/e = 968$) (108) and metal β-diketones (109) have been observed. In the latter case, the 1:1 complexes have spectra complicated by polymeric ions such as $Li_2L_2^{+\cdot}$, $Li_3L_3^{+\cdot}$, $NaKL_2^{+\cdot}$, and $K_2L_2^{+\cdot}$. Thus, even the most unexpected species are sufficiently volatile for mass spectral study; in some cases no parent ion is observed, while in orders, one observes unexpected polymers.

B. Factors Affecting Spectra

Attention to the following factors is essential to success in the use of mass spectrometry.

1. Source and Inlet Temperatures

Perhaps the most common problem is that of thermal decomposition of the sample, either in the batch inlet, for which the cure is a lower inlet temperature or use of the direct probe, or in the source itself. A common misconception among mass spectroscopists, often promulgated by manufacturers, is that if the source is kept hot, the decomposition of contaminants is minimized. The ion source should routinely be run no hotter than 180°–200°C. A source at the common temperature of 250°C is much more likely to result in decomposition of sample and contamination of the source, and should be used only rarely. On our AEI MS-30 we run ~200 samples per month, many of which are organometallic or inorganic, and we are seldom forced to exceed 200°C more than once a month. If some sample condenses into the source it is far better to sublime it away slowly by carefully raising the temperature than to pyrolyze it.

An example of the effect of source temperature is seen for TiF_4oxH (*110*), for which, at 180°C, the highest m/e corresponds to TiF_3ox^+ (i.e., P—HF), whereas at 240°C the thermal decomposition product, $TiF_2ox_2^+$, is observed. Compound $Cu(NO_3)_2$ shows a parent ion (*111, 112*) [unlike $Sn(NO_3)_4$ (*79*)], but thermal decomposition occurs even at source temperatures of ~100°C resulting in much of the NO_2^+ and NO^+ observed. As samples are volatilized from the probe at temperatures of up to 350°C, serious thermal decomposition or polymerization may result (*8, 113–116*). Even with the source at a low temperature, there is still the very hot region in the vicinity of the filament that can cause pyrolysis.

2. Influence of Metal Parts

Glass or glass-lined heated batch inlet systems and transfer lines are to be preferred to the older metal systems since metals tend to accelerate decomposition, although metal deposited by decomposing compounds does have a particularly strong catalytic action (*8, 13, 114,*

116). Not only the complexes themselves, but potential ligands can be shown to rearrange in contact with hot metallic surfaces of an inlet system (117).

Unusual exchange reactions are observed with or on the metal surfaces of a source itself. Sherwood and Turner (118) in their studies of the spectra of iodine pentoxide observed peaks due to copper oxygen, copper iodine, and copper oxygen iodine species (Cu_2^+, Cu_2I^+, $Cu_3I_2^+$, $Cu_4I_3^+$, $Cu_4I_4^+$, ..., $Cu_6I_3^+$, CuO^+, Cu_2O^+, $Cu_5O_5^+$, $Cu_5IO_5^+$, etc.) as well as ions assignable to silver- and lead-containing species. These extraneous metal-containing species were observed in four different mass spectrometers with both open and closed sources. The authors report unsuccessful attempts to find a mass spectrometer with no trace of copper in its source system. Terlouw and de Ridder (119) observed metal–metal chelate exchange in their study of metal chelates of diethylthiocarbamate ligands. The pure iron chelate shows pronounced peaks due to the nickel complex. The amount of nickel complex was dependent on the source condition, being greatest for a clean source and diminished as the source was coated with a layer of decomposition products. The nickel spectrum was eliminated only by plating the stainless steel source components with layers of rhodium and gold. Similar effects were observed for β-diketonate complexes when a metal inlet system was used (120). Metal substitutions can be caused by source memory effects; for example, when $[Cr(CO)_4P(CF_3)_2]_2$ was run after $Zn(acac)_2$, a weak ion corresponding to the highly unusual $[Zn(CO)_4$-$P(CF_3)_2]_2$, was observed (121).

One of the most serious areas of exchange involves metal halides, for which there are definite source memory effects. After running many bromo compounds, followed by a chloro compound, the chloro compound will often show contamination by its bromo derivative. This can be minimized by source cleaning, or more simply by treating the source with a reactive halide, such as $SiCl_4$ for 15 to 30 min (122). One of the most severe cases of halogen exchange involves methyl mercuric halides, of particular importance in the identification of organomercury derivatives in environmental and biological materials. It is virtually impossible to eliminate CH_3HgI^+ from the spectrum of CH_3HgCl (122–124).

3. Sample Size and Instrument Background

Great care must be exercised when dealing with compounds that may decompose, so as to minimize the degree of source contamination. As indicated earlier, organometallics are no worse than any other type of compound, and we have direct evidence of this with our double-beam

MS-30 mass spectrometer (125). We normally operate with either perfluorokerosene (PFK) reference or the GC coupled to beam 2, whereas beam 1 is used for solid probe or batch inlet samples including organometallics. There is no evidence for source 1 being contaminated and requiring cleaning at any greater rate than beam 2. The key is to keep source temperatures as low as possible and to minimize the amount of sample. Most troubles arise from the use of too large a sample, resulting not only in greater contamination but in slow pumpout of residual materials between samples, and thus high background, whereas, with care, probe samples can be run at the rate of 8 per hour without background buildup. Background should be checked immediately prior to the running of a sample, and care must be taken that the probe tip is not contaminated. Large samples can also produce ion–molecule reactions, mentioned below.

4. Ion-Molecule Reactions

As the ion source pressure rises, ion–molecule reactions become possible, sample ions reacting with sample molecules. In the case of exact mass measurement, reaction can occur with the PFK mass reference (126). The observed reactions in the mass spectrum of ruthennium porphyrincarbonyl, yielding ions of the type $[M-CO + C_nF_{2n}]^+$ ($n = 1$–4), illustrate this problem. Similarly, in the spectrum of $Ni(PF_3)_4$ ion–molecule reactions result in species such as $Ni_2(PF_3)_n{}^+$ ($n = 2$–5) and $Ni_2(PF)_2(PF_3)_m{}^+$ ($m = 2$–4) and, in the $(CO)_5CrC(CH_3)OCH_3$ system, reactions of the following type are observed (127).

$$(CO)_5CrL^+ + (CO)_5CrL \longrightarrow (CO)_8Cr_2L_2{}^+ + 2CO \qquad (4)$$

In the substituted arenechromium tricarbonyl system (129), the following ion–molecule reactions are observed:

$$ArCr(CO)_3 + [ArCr(CO)_m]^+ \longrightarrow [Ar_2Cr_2(CO)_n]^+ + (m\text{-}n)CO \qquad (5)$$

In the above systems, it was suggested that the reactions occur at pressures of 5×10^{-6} to 2×10^{-5} torr (126, 128); however, at 10^{-5} torr the mean free path is too long to permit significant ion–molecule reactions. The pressures given are those read from the source pressure ionization gauge. One has to calculate back to determine the actual pressure in the source "cage" (47), several orders of magnitude higher.

5. Relative Volatilities

Gross volatility differences in the components of a submitted sample may give misleading results. Occluded solvent is often the worst offender, and since the operator of the mass spectrometer usually runs the spectrum when the "total-ion monitor" shows sufficient sample entering the source, a solvent spectrum is obtained. Sample submission forms should indicate solvents that were used so that an alert operator will permit the solvent to be pumped away and then obtain the desired spectrum. Traces of starting materials likewise cause problems if they are more volatile than the desired component. If there is uncertainty, it may be wise to run spectra as a function of time and temperature while the material is evaporated from the probe. The best example of this is Majer's detection and separation of geometrical isomers of metal chelates by fractional sublimation (129, 130). This is clearly shown in Fig. 2 for the

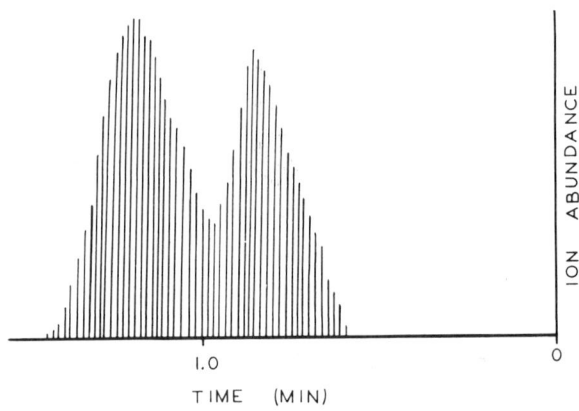

FIG. 2. Integrated ion current curve for iron 8-hydroxyquinolinate on the $FeOX_2^+$ peak. [Reproduced by permission from Majer and Reade (129)].

iron(II) 8-hydroxyquinoline complex. Ritter and Neuman (131) used similar techniques combined with mass chromatograms to identify the components of crude mixtures $(R_2Sn)_n$ from the reaction of R_2SnH_2 with Bu_2^+Hg.

6. Variation in Ionizing Energy of the Electron Beam

As earlier indicated, most spectra are routinely run at electron beam ionization energies of between 50 and 100 eV. In this range there is little change in the spectrum with changing electron voltage. However, at lower electron energies, pronounced changes occur, as is clearly shown in Fig. 3, which, using the "clastogram" presentation of Kiser (25), illustrates the variation of the fractional abundance of each species in the spectrum of $HCo(PF_3)_4$ with electron voltage (132). It is thus clearly seen that at 70 eV the parent ion is a small fraction of the total ion current, whereas below 20 eV it becomes a significant fraction of the total positive ionization. Thus, it is often advantageous, if no parent ion is observed,

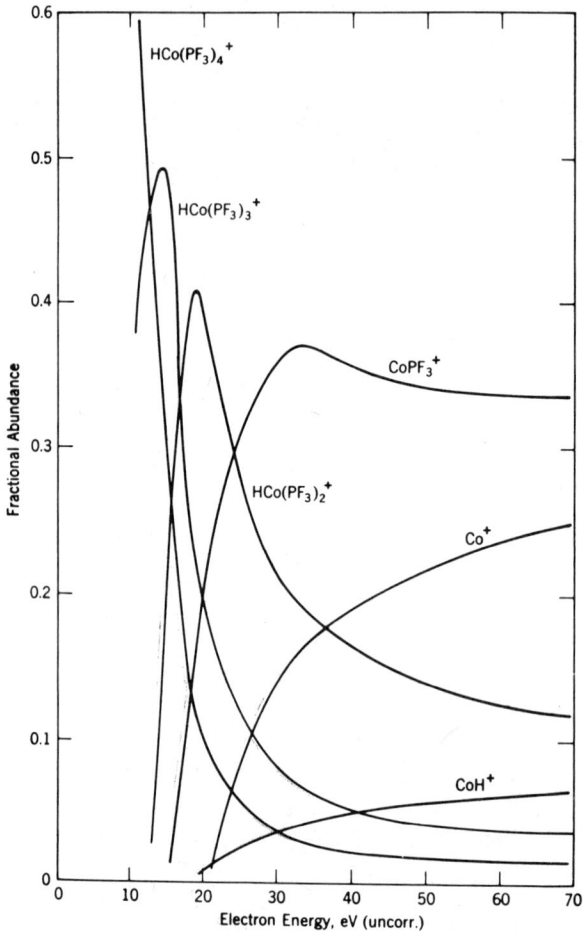

FIG. 3. Clastogram of $HCo(PF_3)_4$. [Reproduced by permission from Kiser (6).]

to attempt to run the spectrum at between 10 and 20 eV. It should be remembered that the absolute ion intensity drops drastically at lower electron voltages, so that, although the parent ion may be a large fraction of the total ionization, the latter is itself reduced many orders of magnitude, as shown in the ionization efficiency curves (Fig. 4) for $SnCl_4$ (133).

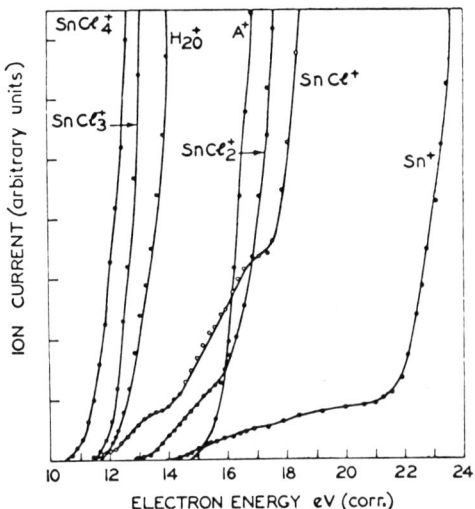

FIG. 4. Ionization efficiency curve's for SnCl₄. [Reprinted by permission from A. S. Buchanan, D. J. Knowles, and O. L. Swingler, *J. Phys. Chem.* 73, 4394 (1969); copyright by the American Chemical Society.]

IV. Low- and Medium-Resolution Studies

Most problems in organometallic and inorganic mass spectrometry are solved using resolutions of less than 3000. Compared to organic compounds, the polyisotopic nature of most metals and metalloids simplifies the task of identification of a particular species without resorting to high-resolution exact mass measurement. It is usually assumed that under electron impact (a) the decomposition of organometallic complexes is largely controlled by the central metal, (b) that free and complexed ligands decompose via different mechanisms, (c) that on decomposition of a complex ion, the positive charge usually remains on the metal-containing fragment, and (d) that ions with multiple positive charges occur relatively frequently (13). There are exceptions; for example, the fragmentation of R_4M species (M = Si, Ge, Sn; R = C_6F_5 or C_6H_5) shows greater differences on changing R than M (86, 87, 89) and more of the positive charge is carried by fluorocarbon species (73%) rather than metal-containing species for $(C_6F_5)_4Si$, whereas for the lead compound, 88% of the charge is carried by metal-containing species (89). It is such differences as these that make the subject interesting, but they undoubtedly complicate interpretations. Decomposition under electron bombardment falls into three categories as shown for the $(C_6F_5)_3Si^+$ ion:

$(C_6F_5)_4Si^{+\cdot} \longrightarrow (C_6F_5)_3Si^+ + C_6F_5^{\cdot}$

$\longrightarrow (C_6F_5)_2Si^{+\cdot} + C_6F_5^{\cdot}$ (6a)

$\longrightarrow C_{17}F_{13}Si^+ + CF_2$ (6b)

$\longrightarrow (C_6F_4)_3^{+\cdot} + SiF_3^{\cdot}$ (6c)

Equation (6a) illustrates cleavage of metal–ligand bond; Eq. (6b), rupture of a bond and/or rearrangements in the ligand; and Eq. (6c), a rearrangement reaction involving the metal atom in both bond breaking and bond making.

A common mass spectral fragmentation scheme observed for organic compounds involves the loss of a radical from an odd-electron parent ion, followed by loss of even-electron neutral fragments. Even-electron ions are, therefore, observed in high abundance. For organometallics, by contrast, successive loss of neutral ligands such as CO or PF_3 from the parent ion is a very common occurrence. Inorganic compounds also have another common source of odd-electron ions, the metal ion itself, for all even-electron metals. Normally, as one goes down a group in the periodic table, the bare metal ion becomes more common in the spectrum. These and other trends are illustrated for selected examples in the following sections.

A. Fragmentations

1. Halides and Related Compounds

Among the simplest systems studied by mass spectrometry are metal halides and related compounds. For the simple binary halides, metal–ligand cleavages are the important breakdown mechanism, although for silicon tetrahalides, metastable ion studies show simultaneous as well as consecutive loss of halogen (*134*). Under ideal conditions, Group IV tetrahalides show all ions in the series MX_n^+ (where $n = 0$–4), but parent ions may be weak or nonexistent as shown in Table II for the spectrum of $(GeCl_3)_2CCl_2$ and $GeCl_3CCl_3$ (*135*). Except for $GeCl_4^+$ rearrangement ions, these compounds have undergone simple fragmentation, and mass spectroscopy proved invaluable in their identification among the reaction products of molecular carbon with $GeCl_4$. Similarly the preparations of B_8Cl_8 (*136*) and B_9Cl_9 (*137*) were confirmed by their mass spectra. In both cases simple fragmentation occurs, although the spectra are complicated by the isotopes of boron and chlorine. Both PF_3 and PF_2CN undergo the

TABLE II

MASS SPECTRUM OF $(GeCl_3)_2CCl_2$ AND $GeCl_3CCl_3$

Ion+	$(GeCl_3)_2CCl_2$ (Relative abundance)	$GeCl_3CCl_3$ (Relative abundance)
CGe_2Cl_8	0.02	—
CGe_2Cl_7	0.2	—
CGe_2Cl_6	0.04	—
CGe_2Cl_5	0.05	—
CGe_2Cl_4	0.02	—
$CGeCl_6$		Parent not observed
$CGeCl_5$	1.5	4.7
$CGeCl_4$	41.1	0.3
$GeCl_4$	0.9	Very weak
$CGeCl_3$	6.7	0.4
$GeCl_3$	16.0	5.0
$CGeCl_2$	2.7	0.4
$GeCl_2$	3.5	1.8
CCl_3	54.5	100.0 (plus GCeCl)
$GeCl$	100.0	12.2
CCl_2	32.0	94.3 (plus CGe)
Ge	6.4	0.3
CCl	20.5	12.0

simple central atom–ligand cleavages expected (138), but PF_2CN shows cleavage of the cyano ligand as well, with species such as PFC^+, PC^+, C^+, and N^+ appearing in the spectrum.

Molecular beam mass spectrometry has been used to study pentafluorides of Nb, Ta, Mo, Re, Os, Ir, Ru, Rh, Pt, Sb, and Bi (139), indicating dimeric, trimeric, and even tetrameric ions (139). Simple fragmentations of the type

$$Rh_4F_{20}^+ \longrightarrow Rh_4F_{19}^+ \longrightarrow Rh_4F_{18}^+ \qquad (7)$$

and

$$Rh_3F_{15}^+ \longrightarrow Rh_3F_{14}^+ \longrightarrow Rh_3F_{13}^+ \longrightarrow Rh_3F_{12}^+ \longrightarrow Rh_3F_{11}^+ \quad (8)$$

are observed, and the polymeric ions are shown to arise from the neutral vapors rather than from ion–molecule reactions. In further studies it has been shown that only the lighter pentafluorides and interhalogen pentafluorides are not associated, with the simple spectrum expected for a binary pentafluoride (140). Dioxygenyl salts of the type $O_2^+MF_6^-$ (M = As, Sb, Bi, Ru, Rh, Au, and Pt) have been studied by mass spectrometry (*141*), and although no parent ion is detected, O_2^+ and MF_n^+ ($n = 0$–6) are observed. For gold, the spectrum more closely resembles that of a pentafluoride with $Au_2F_9^+$ and $Au_3F_{12}^+$ species observed, suggesting the independent existence of AuF_5. Similar studies have been carried out with transition metal oxo- and thiohalides and their fragmentation patterns indicate essentially monomeric species (*142*). The spectra of the lanthanide triiodides have been measured using Knudsen cell techniques and show simple fragmentation (*143*).

The mass spectra of the boron trihalides and mixed trihalides have been studied (*144*), the latter being formed on mixing the simple halides. All species undergo simple fragmentation. The corresponding trimethylamine adducts show both metal–ligand and ligand–ligand cleavages (*145*):

$$X_2B \cdot NMe_3^+ \longrightarrow NMe_3^{+\cdot} + X_2B\cdot \qquad (9)$$

$$X_2B \cdot NMe_3^+ \longrightarrow X_2B \cdot NMe_2^{+\cdot} + Me\cdot \qquad (10)$$

$$I_2B \cdot NMe_3^+ \longrightarrow BNMe_3^+ + I_2 \qquad (11)$$

Ion abundances are consistent with increasing B–N and decreasing B–X bond strength as X varies from F to I.

2. Main Group Organometallic Compounds

The mass spectra of main group organometallic compounds have been discussed in many of the reviews listed in the Introduction, and here we only summarize some recent examples.

Cragg and Weston (*11*) in their review of the mass spectra of boron compounds have shown the great amount of work being done with cyclic boron compounds. This trend has continued. Straight-chain trialkylborates fragment by C—O cleavage reactions accompanied by hydrogen transfers,

$$(C_7H_{15}O)_3B^{+\cdot} \xrightarrow[-C_7H_{13}\cdot]{} (C_7H_{15}O)_2BO^+H_2 \qquad (12)$$

(13)

whereas tri-(sec-alkyl)borates tend toward α-cleavage reactions (cleavage of C—C bond between carbons α and β to the oxygen) (146).

A B—O bond is broken in the case of phenylboron derivatives of o-hydroxybenzyl alcohol (147) where the loss of PhBO is observed,

$$\text{[benzodioxaborine-Ph]}^{+\cdot} \xrightarrow{-\text{PhBO}} \text{[cyclohexadienone-CH}_2\text{]}^{+} \xrightarrow{-\text{CO}} C_6H_6^{+\cdot} \qquad (14)$$

rather than the expected process leading to formation of a tropylium ion (148–150). In the case of chelates involving aromatic amines and N-oxides with Ph_2B or F_2B (151), the base peak is usually a result of loss of either Ph or F, and from the base peak fragment, neutral OBPh or OBF may be eliminated. Boronsulfur heterocycles also exhibit stability under electron bombardment (152): triphenylborthiin $(PhBS)_3$ gives an intense parent ion (28.5%)* and doubly charged parent (4%), but no PhBS or $(PhBS)_2$ type ions were observed.

The mass spectra of Group IV derivatives have received much attention, the most recent review being that of Orlov (8a). Current areas of interest vary from organopolysilanes (153) to detailed studies of the simple tetramethyl derivatives (154). For the latter compounds, the high intensity of the R_3M^+ ion is explained on the basis that it is "even-electron," or isoelectronic with the R_3N compounds (N = Group IIIA metal); the increasing intensity of RM^+ ions at the expense of R_3M^+ ions as the central atom gets heavier is explained by the increasing stability of the lower oxidation state.

Mass spectra of disilanes (155–157) and, more recently, polysilanes (153) up to dodecamethylpentasilane have been investigated; the important fragmentation is shown in Scheme 1. Lageot et al. (158) have investigated the expulsion of neutral fragments in the mass spectra of germacyclopentenes (III); this process occurs in addition to the simple cleavages of these compounds. Fragmentations of organogermanium hydrides (159) are quite straightforward, but their interpretation is

* For comparison purposes, mass spectral relative abundances are best described as a percent of the total positive ionization, rather than the traditional percent of base peak height, where the base or largest peak is normalized to 100%. If, in comparing spectra in a series of compounds, the base peak differs very confusing results may be obtained. Percent of the total positive ionization gives a much clearer picture of the nature of the charge-carrying species; for example, knowing that the base peak is 45% of the total positive ionization is more informative than knowing it simply as the largest peak.

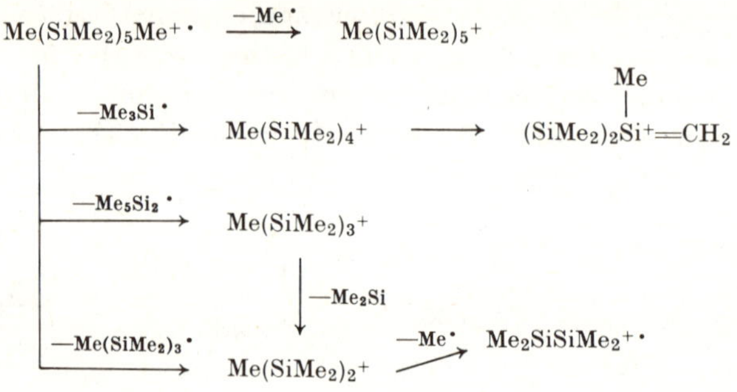

$Me(SiMe_2)_5Me^{+\cdot} \xrightarrow{-Me^{\cdot}} Me(SiMe_2)_5^+$

```
           ┌─ −Me₃Si·  → Me(SiMe₂)₄⁺  →  (SiMe₂)₂Si⁺=CH₂
           │                              |
           │                              Me
           │
           ├─ −Me₅Si₂· → Me(SiMe₂)₃⁺
           │                │ −Me₂Si
           │                ↓
           └─ −Me(SiMe₂)₃· → Me(SiMe₂)₂⁺ ──−Me·──→ Me₂SiSiMe₂⁺·
```

SCHEME 1

complicated both by the germanium isotope pattern, especially for the digermanes, and by many successive losses of hydrogens. The base peaks

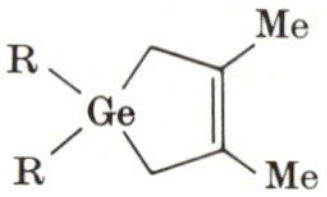

(III)

of diethylmonogermanes and diethyldigermanes occur in the GeC_2H_n ($n = 0$–7) region, rather than the parent ion region. The monogermane also has well-defined envelopes in the regions of GeH_n^+ ($n = 0$–3) and $GeC_4H_n^+$ ($n = 0$–12). In addition the digermane has peaks in the region of $Ge_2H_n^+$ ($n = 0$–4), $Ge_2C_2H_n^+$ ($n = 0$–9), and $GeC_4H_n^+$ ($n = 0$–11).

A recent area of interest has been the mass spectra of organotin pesticides (*160*, *161*), for example, triphenyltin acetate that fragments as follows:

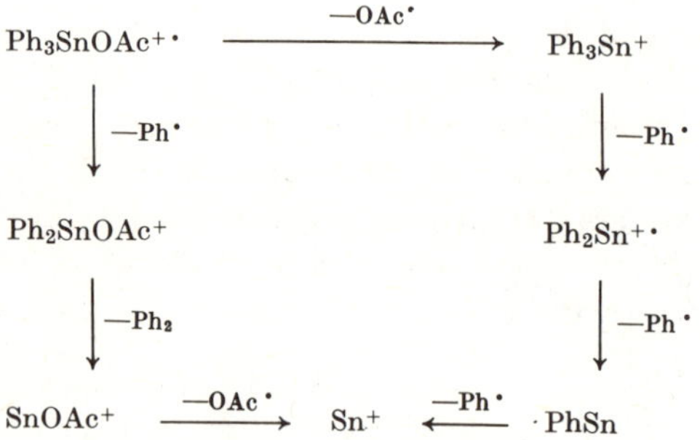

$$Ph_3SnOAc^{+\cdot} \xrightarrow{-OAc^{\cdot}} Ph_3Sn^+$$

$$\downarrow -Ph^{\cdot} \qquad\qquad \downarrow -Ph^{\cdot}$$

$$Ph_2SnOAc^+ \qquad\qquad Ph_2Sn^{+\cdot} \qquad\qquad (15)$$

$$\downarrow -Ph_2 \qquad\qquad \downarrow -Ph^{\cdot}$$

$$SnOAc^+ \xrightarrow{-OAc^{\cdot}} Sn^+ \xleftarrow{-Ph^{\cdot}} \cdot PhSn$$

Relatively few rearrangements are noted in the spectra of these tin pesticide compounds which fragment in a simple manner.

Group V has also been an area of interest, the mass spectrometry of phosphorus compounds having recently been reviewed (*162*). The mass spectra of phosphorus-containing pesticides have attracted interest

(163–164), the studies including phosphates (165), phosphorothiolates (163), phosphorothionates (165, 166), phosphorodithionates (166), and metabolites of organophosphorus pesticides (167–170).

In a recent study (171) the Group V acetylides $M(C{\equiv}C{-}CH_3)_3$ (M = P, As, Sb) are compared with their Group IV analogs (Table III).

TABLE III

COMPARISON OF THE PARTIAL MASS SPECTRA OF PROPYNYL DERIVATIVES OF GROUPS IV AND V

Metal M	Abundances as percent of total positive ionization			
	Parent ion	(P—15)$^+$	$M(C{\equiv}CCH_3)_2{^+}^{\cdot}$	$M(C{\equiv}CCH_3)^+$
Si	12.9	15.1	3.2	15.4^a
Ge	3.7	1.1	0.3	8.2^a
Pb	—	—	—	7.3
P	9.5	10.6^a	0.6	5.9
As	2.9	4.2	2.0	10.5^a
Sb	1.1	0.2	1.3	21.2^a

^{a} Base peak.

The $M(C{\equiv}C{-}CH_3)^+$ ion is the base peak except for the lead compound (base peak = $Pb^{+\cdot}$) and the phosphine [base peak = $(P{-}CH_3)^+$]. For Group V divalent metal species are of lower abundance than either mono- or trivalent ions. As expected, metal-containing ions carry the bulk of the ion current. Rearrangements were also observed corresponding simply to loss of a neutral metalloid atom.

Organoarsenic and -arsenate compounds have been studied by Froyer and Moller (172–173). In the initial stages, $(CH_3CH_2O)_3AsO$ fragments by simple cleavage (6) (Scheme 2), after which the $(RO)_2AsO^+$

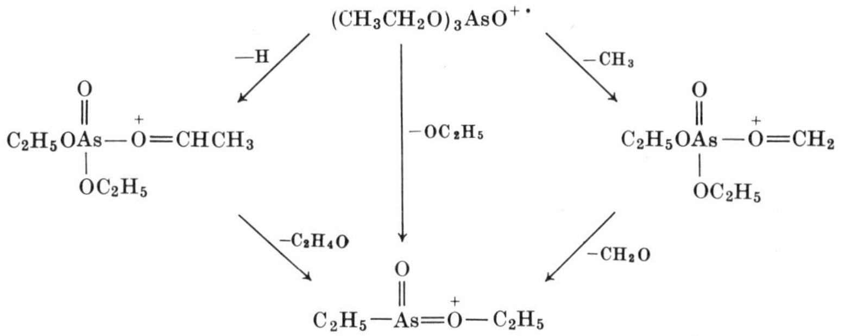

SCHEME 2

ion fragments with extensive rearrangement. A study of a series of
O-phenylenediarsines (*174*) shows stepwise loss of alkyl substituents
from the arsenic, but, for halogen- or phenyl-substituted arsenic
compounds, rearrangements and migrations analogous to those for
phenyl derivatives of Group V are observed (*90, 175–180*).

Perfluoroaromatic derivatives of Group VI (*180–182*) primarily
undergo simple fragmentation rather than the rearrangements usually
observed for other fluoroaromatic derivatives (*182*). Aliphatic and aromatic
selenides have been extensively studied by Rebane (*183–184*).
Compounds R_2SeCl_2 usually show no molecular ions, the highest m/e
corresponding to $(P-X)^+$. The aryl derivatives eliminate the second
halogen to give $R_2Se^{+\cdot}$ ions, whereas the dialkyl or alkyl/aryl derivatives
show alkyl elimination producing $(RSeX)^{+\cdot}$ ions. Further simple
fragmentation then occurs, although some rearrangement species are
also observed.

The mass spectra of organomercurials were neglected for some years,
the first comprehensive study of these compounds being by Byrant and
Kinstle (*185*). They observed that for R_2Hg and $RHgCl$ ($R = $ alkyl)
there was a surprising tendency toward charge retention on carbon rather
than mercury, although this could be varied with appropriate substituents.
Extrusion of a mercury atom is also favored in diaryl derivatives
(*186*), of which the fragmentation routes are

$$\begin{array}{ccc} Ar-Hg-Ar^{+\cdot} & \xrightarrow{-Ar\cdot} & HgAr^+ \\ \Big\downarrow {\scriptstyle -Hg} & \xrightarrow{-ArHg} & \Big\downarrow {\scriptstyle -Hg} \\ Ar-Ar^{+\cdot} & & Ar^+ \end{array} \qquad (16)$$

Exchange of organic groups was also detected, since if pairs of compounds,
R_2Hg and R'_2Hg, were introduced simultaneously, then
$RHgR'^{+\cdot}$ and $R'R^{+\cdot}$ were observed. Recent concern over environmental
matters has led to the mass spectral study of methylmercury compounds
from fish (*123*), and also to an interest in fungicidal organomercury
compounds (*161, 187*) and the use of the mass spectrometer in their
detection and identification. The important ions observed in the mass
spectrum of CH_3HgI include CH_3HgI^+, HgI^+, CH_3Hg^+, Hg^+, I^+, and
the rearrangement ions $I\text{-}CH_3^{+\cdot}$ and $HI^{+\cdot}$. Organomercury carboxylates
also show a rearrangement corresponding to loss of CO_2 from the parent
ion.

3. *Transition Metal Complexes*

The mass spectra of metal β-diketonates have been extensively
studied (*105–106, 188–192*). The fragmentation of these complexes has

been rationalized by Shannon (*193–194*) in terms of the "valence change concept," the modes of dissociation being dependent on the oxidation states normally assumed by the metal concerned. They postulate electron transfers between metal atom and ligand, such that the odd- or even-electron character of the ion is interchangeable, accounting for the ready consecutive loss of two radicals, normally improbable for organic compounds. Thus two "acac" ligands are lost from the $Fe^{III}(acac)_3$ molecular ion, but not from the analogous Al(III) compound, since iron but not aluminum may normally assume a lower oxidation state (*193*). They consider $[M^{III}(acac)_2]^+$ and $[M^{II}(acac)_2]^{+\cdot}$ as canonical contributions to a resonance hybrid, and the latter odd-electron species can eliminate a second ligand radical to become even-electron.

Recent studies have involved some fluorinated monothio-β-diketones with zinc and nickel (*195*) and palladium and platinum (*196*). Both nickel and zinc show valence change of the metal,

$$[M^{II}L_2]^+ \longrightarrow L^+ + M^IL \qquad (17)$$

The number of metal-containing fragments were found to be small for Zn and Pd, but large for Ni and Pt chelates, whereas the $(ML)^+$ ion is very strong for zinc, moderate for platinum, and weak for nickel and palladium chelates. Free metal is observed for the nickel complex only.

Volatile metal chelates are also useful in determining isotope ratios of geological interest, e.g., the Zr/Hf ratio (*197*). This method proved invaluable as a microtechnique for chromium isotope analysis of lunar samples from the Apollo program (*198*).

The mass spectra of metal phthalocyanines and porphyrins have also been studied: Eley et al. (*199*) compared the spectra of free ligand and transition metal phthalocyanines. Both the free ligand and the complexes are among the most stable compounds toward dissociation by electron impact, the parent ion being the base peak in all cases, and the doubly charged parent the second most abundant species. Significant amounts of triply charged parent ions are observed for some complexes. The various parent ion species carry from 54 to 90% of the total positive ionization. The perfluorophthalocyanine complex of zinc has an even more stable molecular ion (*200*). The limited ligand fragmentation that does occur involves either halves or quarters of the ligand.

Studies of copper(I) (*102, 210, 202*) and copper(II) carboxylates (*203*) show that the copper(II) derivatives have the general formula $[Cu_2R_4] \cdot 2H_2O$, where $R = CH_3CO_2$, $C_2H_5CO_2$, $n\text{-}C_3H_7CO_2$, and $n\text{-}C_4H_9CO_2$. The samples dehydrate and the fragmentations are essentially based on Cu_2R_4. Very intense ions are observed for $Cu_2R_2^+$ and Cu_2R^+ ions with higher molecular weight species such as $Cu_3R_2^+$, $Cu_3R_3^+$, $Cu_3R_5^+$, and $Cu_4R_3^+$. Ogura and Fernando reported that the copper(I) derivatives do not show peaks above the parent ion $Cu_2R_2^+$ (*102*), but Edwards and Richards (*201*) described weak peaks for trimers and tetramers. Extensive fragmentation of the bridging ligands is observed. Alkyl carboxylic acid salts of copper(I) fragment via loss of $RCO_2\cdot$ from the molecular ion, whereas aryl carboxylates have a parallel pathway involving loss of CO_2 from the molecular ion and migration of the aryl group to the metal (*202*).

The importance of metal coordination compounds in biological systems has led to the study of polydentate Schiff base complexes of cobalt(II), nickel(II), and copper(II) (*204, 205*). Dimers have been observed in the spectra of complexes of both tri- and tetradentate ligands [e.g., salicylaldehyde and N,N-bis(salicylidene)ethylenediamine]. The parent ions form the base peaks, and the spectra are characterized

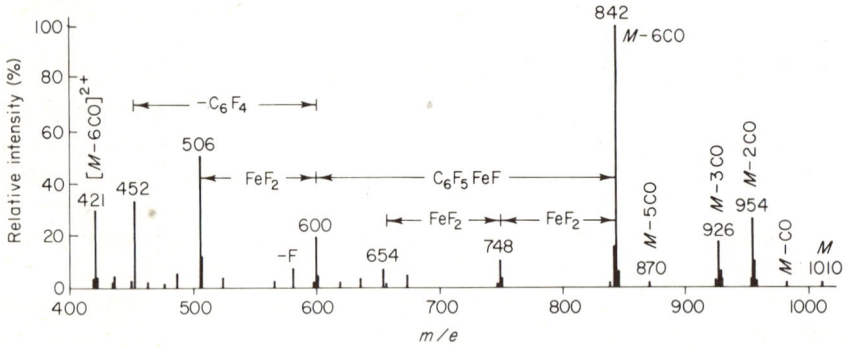

FIG. 5. Mass spectrum of $[(C_6F_5)_2PFe(CO)_3]_2$. [Reproduced by permission from Miller (*90*).]

by complex ligand fragmentations. It is for complexes such as these, of potential biological interest, that field desorption mass spectrometry may prove valuable.

The fragmentation of metal carbonyl derivatives is characterized by successive loss of carbonyl groups, but in some cases all possible losses of CO are not observed. For $[(C_6F_5)_2PFe(CO)_3]_2$, in addition to the parent ion, there are peaks corresponding to the loss of 1, 2, 3, 5, and 6-carbonyl groups, but for reasons which are not clear, the loss of 4 is not observed (*90*) (Fig. 5).

Innorta *et al.* (*206*), in a study of carbonyl complexes containing organic rings, have rationalized the loss of small stable molecules, rather than radicals, in terms of the activation energies for fragmentation pathways possible from the molecular ion. Thus loss of H_2 after CO has been removed can become common if it results in aromatization of a ring.

In the spectrum of π-tetrahydronaphthalenenonacarbonyltetracobalt, (P-nCO)$^+$ ions become less intense with increasing n, whereas [P-(nCO + H$_2$)]$^+$ and [P-(nCO + 2H$_2$)]$^+$ predominate for $n \geqslant 4$. The free ligand itself shows loss of atomic hydrogen, and complexed ligand predominantly has ions due to the loss of H$_2$ or 2H$_2$. Unsaturation of the cobalt or loss of CO is not applicable as the driving force, as was suggested for loss of H$_2$ from ions in the spectrum of π-C$_6$H$_8$Fe(CO)$_3$ *(207)*. It is, therefore, the influence of π-bond formation on the loss of H$_2$ that lowers the activation energy sufficiently for this process to compete with CO loss *(208)*.

In the spectra of some iron carbonyl derivatives of CF$_3$-substituted polyphosphines *(209)*, cleavage of CO groups and fragmentation of the phosphine are competitive. The main fragments from (CF$_3$)$_2$P$_2$C$_2$(CF$_3$)$_2$-Fe(CO)$_4$ include successive loss of all 4 carbonyls, or loss of a fluorine from the ligand. For (CF$_3$)$_2$P$_2$C$_2$(CF$_3$)$_2$Fe$_3$(CO)$_{10}$, loss of carbonyl is observed, as well as the loss of 4 carbonyls and P$_2$F. Similarly, P$_4$F and 4 carbonyls have been split from (CF$_3$P)$_4$Fe$_2$(CO)$_6$.

B. REARRANGEMENTS

Bombardment with 70-eV electrons is not a gentle process: the molecular ions formed have sufficient activation energy to undergo not just simple cleavage but rearrangements involving both cleavage and bond formation. Organic ligands attached to a metal atom may undergo rearrangement reactions similar to those observed for simple organic molecules. Of far more interest are those reactions that involve the central metal atom itself, either accompanied by the transfer of an atom or group to the central metal in a fragment ion, or the expulsion of the neutral metal atom or a neutral metal-containing species from an ion. Perhaps one of the most interesting of these reactions in organometallic chemistry, as pointed out by Kiser *(6)*, involves transfer of fluorine to the central metal in the mass spectra of fluorocarbon-containing organometallics and complexes *(90)*. Aided by metastable ion studies one can execute a detailed study of unimolecular gas phase reactions having potential synthetic analogies.

1. M—F Bond Formation

Miller *(90)*, King *(209, 210)*, and Bruce *(211)* first observed the formation of neutral metal fluoride species and metal fluoride-containing ions in the mass spectra of pentafluorophenyl derivatives of phosphorus, germanium, silicon, and phosphido-bridged iron carbonyls *(90)* and aliphatic and aromatic fluorocarbon derivatives of iron, cobalt *(209–211)*, and manganese (*212–213*). Trifluoromethyl derivatives exhibit similar behavior as shown by Cavell and Dobbie (*214–217*). The nature of the exotic ions and transitions observed are clearly illustrated in Fig. 5, where after the loss of 6 carbonyls from $[(C_6F_5)_2PFe(CO)_3]_2$, the following transitions are observed from the base peak (*90*):

$$C_{24}F_{20}P_2Fe_2^{+\cdot} \xrightarrow{-FeF_2} C_{24}F_{18}P_2Fe^{+\cdot}$$

$$\downarrow -C_6F_5FeF \qquad\qquad \downarrow -FeF_2$$

$$C_{18}F_{14}P_2Fe^{+\cdot} \qquad\qquad C_{24}F_{16}P_2^{+\cdot} \qquad (18)$$

$$\downarrow -FeF_2 \quad \searrow -C_6F_4$$

$$C_{18}F_{12}P_2^{+\cdot} \qquad C_{12}F_{10}P_2Fe^{+\cdot}$$

All of these ions are of high abundance and all are odd-electron species. Further work has been carried out in the authors' laboratory on fluoroaromatic derivatives of Groups III (*217–218*), IV (*88, 89, 157, 181, 182, 219*), V (*176–178*), and VI (*180–182*). McGlinchey has recently studied chromium derivatives (*220*) whereas Clark and Rake (*211*) have studied in detail the system $Ph_{3-n}(C_6F_5)_nMMn(CO)_5$ ($n = 0–3$; M = Si, Ge, Sn); species corresponding to the loss of F_3SiMnF and MnF_3 from $(C_6F_5)_3$-$SiMn^{+\cdot}$ were observed. Thus in all of these systems, extensive fluorine transfers occur. Scheme 3 illustrates the complexities of the spectra for $(C_6F_5)_3B$, whereas the analogous $(C_6H_5)_3B$ shows little proton migration (Scheme 4) corresponding to the B—F bond formation (*217*). Postulated structures for some of the ions concerned are shown in Scheme 5, in particular the polyphenylenes and fused-ring systems.

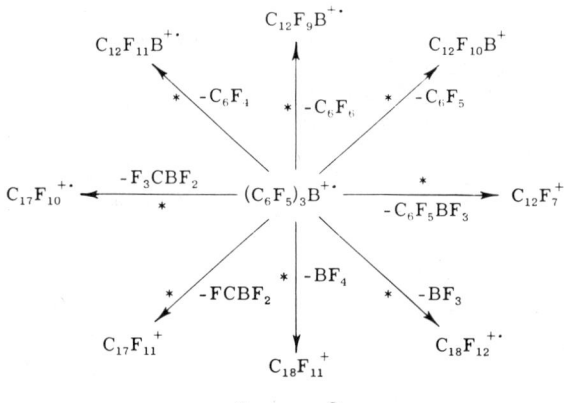

SCHEME 3

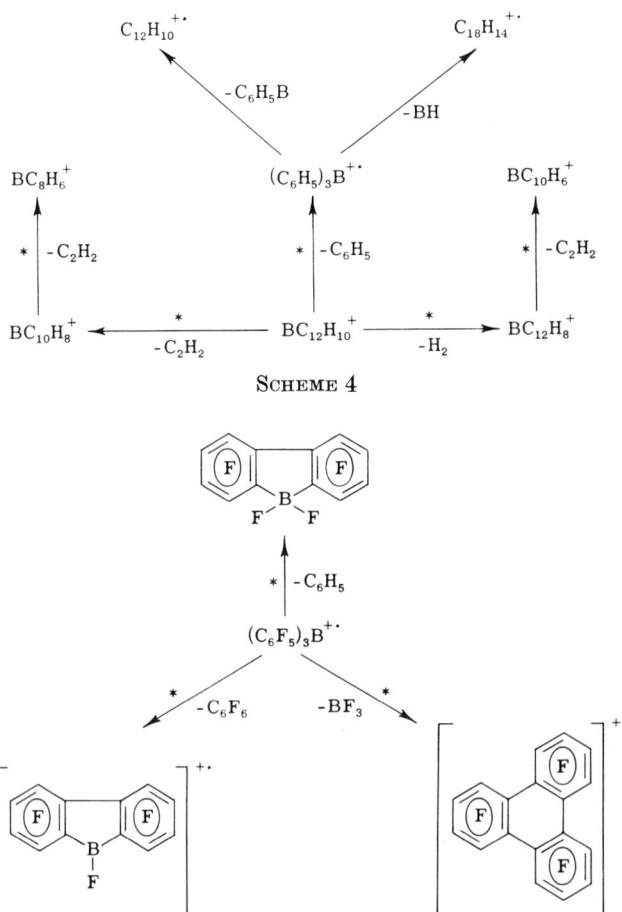

SCHEME 4

SCHEME 5

In studies of the mechanism of fluorine transfer from a fluoroaromatic ring to the central atom, no such migrations are observed if the central atom is carbon (89), nitrogen (222), oxygen (180), or sulfur (180), whereas selenium fluorides are formed to only a slight extent (180). Although electronegativity arguments cannot be ruled out, it would appear that the transfer requires vacant orbitals on the central atom, such orbitals being absent for the C, N, and O derivatives, whereas for sulfur, sulfur–fluorine bond formation may not be particularly favorable energetically. Of the first-row elements, pentafluorophenyl derivatives of trigonal boron, which is both electropositive and electron deficient, exhibit extensive loss of BF_3, BF_2, and BF species in their mass spectra (217, 223).

Cavell and Dobbie (*214–216*) have suggested that halogen transfer rearrangements in trifluoromethylphosphines arise from interactions of nonbonding fluorine p orbitals with vacant d orbitals on phosphorus. Such an explanation is consistent with observations for the Groups IV and V pentafluorophenyl derivatives, exclusive of carbon and nitrogen, and similarly fits the behavior of boron with its vacant p orbital.

The involvement of d orbitals can also explain the existence of fluorine transfer to transition metals (*90, 209–213, 220–221, 224–227*). Hawthorne et al. (*212*) have suggested that when there are unsaturated

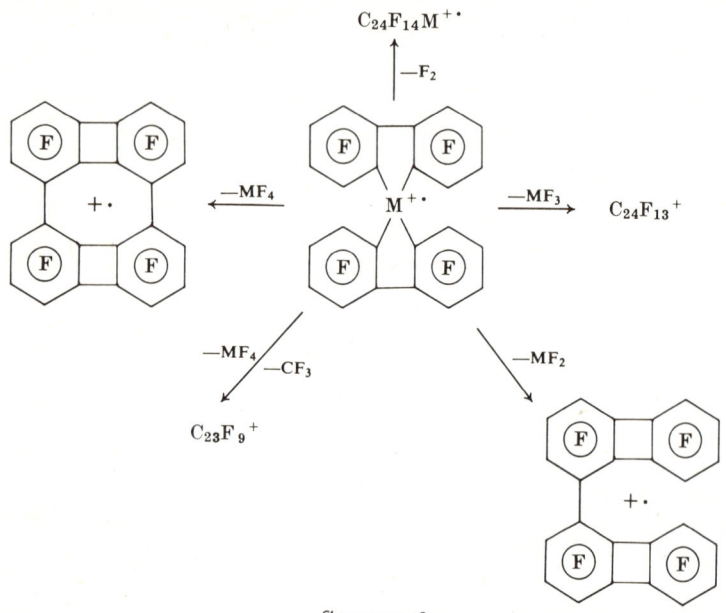

SCHEME 6

σ-bonded organic ligands the rearrangement ion can be rationalized by assuming that the ligand becomes "π"-bonded to the metal during the fragmentation. They based this suggestion on fluorine migrations to the metal atom occurring from all the o-, m-, and p-substituted monofluorophenyl manganese pentacarbonyls, although *ortho* migrations are most favored. The well-established scrambling of aromatic substituents under electron bombardment could also account for their observations, however.

It is difficult to conceive of these σ–π type transfers in compounds such as $(C_{12}F_8)_2Si$, as shown in Scheme 6. These fused-ring compounds are analogous to the intermediates proposed for the boron compounds (Scheme 5) and for the mass spectra of diarylphosphinic acids in which

Haake et al. (228) postulate that the structures of both the $[(C_6H_5)_2PO]^+$ and $[(C_6H_5)_2P]^+$ ions involve cyclized systems with four-coordinate phosphorus. Rake and Miller (176–177) have suggested that the loss of MF_2 from $[(C_6F_5)_2M)^+$ (M = P, As, Sb) occurred via similar structures, as shown for ions (IV) and (V), also consistent with the structures proposed by Williams et al. (175), Bowie and Nussey (179), and Chow and McAuliffe (229), all involving expansion of the coordination shell of the central metal.

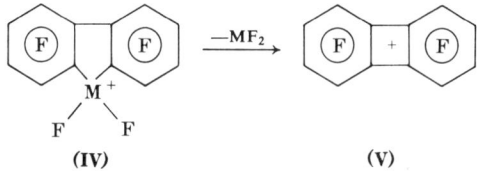

(IV)   (V)

Therefore by starting with heterocyclic ring systems analogous to structure IV, e.g., $(C_6H_5)_2Ge(C_{12}F_8)$, it was hoped analogous transitions might be observed (182), but this compound shows loss of HF or GeF, but not $(C_6H_5)_2Ge$. It was thus postulated that expansion of the coordination shell of the central atom from 4 to 6 in systems such as Scheme 6 was the only mechanism that could explain all observations, a species such as structure VI being a possible intermediate. Such an ion could then lose MF_2 or undergo further bond breaking and reforming rearrangements to allow the elimination of MF_3 or MF_4 as shown in Scheme 6.

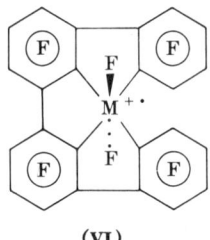

(VI)

When looking at the $(C_{12}F_8S)_2M$ species, it can again be postulated that in the parent ion (also stable and highly abundant as for Structure VI) the metal expands its coordination number to six, resulting in elimination of MF_2, MF_3, and MF_4.

Although by no means conclusive, the work on cyclic fluoroaromatic organometallic systems suggests that various fused-ring ions, postulated as intermediates in [MF] elimination from pentafluorophenylorganometallics, are entirely plausible. Similarly, if valence shell expansion were not involved, systems such as $(C_{12}F_8)_2M$ would also split out the metal atom alone to yield the very stable, previously isolated perfluorotetraphenylene ion $[(C_6F_4)_4]^+$. This is not observed, the metal being eliminated only in company with fluorines. Thus, although $(C_6F_5)_4Si$ eliminates SiF_4 to give $(C_6F_4)_4{}^{+\cdot}$, compound $(C_{12}F_8)_2Si$ eliminates SiF_4 to give $(C_6F_3)_4{}^{+\cdot}$.

As in all mass spectrometric systems, the postulated ion structures cannot be directly verified, but recent advances in ion kinetic energy spectroscopy (see Section VI) allow one to prove that two ions of the same mass from different sources have either identical or differing structures. Ionic structures tend to be drawn based on common sense and chemical intuition.

2. Rearrangements in B—N and B—O Heterocycles

Borazines have been shown to undergo some remarkable rearrangements. Lanthier (*217*) and Wilson (*230*) observed ions corresponding to the loss of ammonia from various fragment ions. This was more strikingly demonstrated for B-trispentafluorophenyl borazine (*217, 218*):

$$(C_6F_5BNH)_3{}^{+\cdot} \xrightarrow{-C_6F_5} B_3N_3H_3C_{12}F_{10}{}^+ \xrightarrow{-NH_3} B_3N_2C_{12}F_{10}{}^+$$

$$\downarrow -C_6F_4$$

$$FB_3N_3H_3C_6F_5{}^+ \xrightarrow{-NH_3} FB_3N_2C_6F_5{}^+ \quad (19)$$

$$\downarrow -C_6F_4$$

$$F_2B_3N_3H_3{}^+ \xrightarrow{-NH_3} F_2B_3N_2{}^+$$

After the first loss of a $C_6F_5{}^\cdot$ group, neutral NH_3 may be lost, or, after loss of one or two C_6F_4 groups (presumably resulting in the formation of a B—F bond), ammonia may also be ejected. A drastic rearrangement is necessary for a single nitrogen to remove all three hydrogen atoms.

The rearrangements involved in fragmentation of $(C_6H_5)_2C_6F_5B_3O_3$ are less drastic, since there is only one example each of H—B and F—B bond formation:

$$(C_6H_5)_2C_6F_5B_3O_3{}^{+\cdot} \xrightarrow{-C_6F_4} (C_6H_5)_2FB_3O_3{}^{+\cdot}$$

$$\downarrow -C_6F_5BO \quad \searrow -C_6H_5BO$$

$$(C_6H_5)(C_6F_5)B_2O_2{}^{+\cdot} \quad (C_6H_5)_2B_2O_2{}^{+\cdot} \quad (20)$$

$$\downarrow -BO_2 \qquad\qquad \downarrow -HBO_2$$

$$C_6H_5BC_6F_5{}^+ \qquad\qquad C_{12}H_9B^{+\cdot}$$

This is also an example of a ligand-transfer rearrangement—two phenyls ending up on one boron. Such ligand transfers are very common, for example, the spectrum of $Ph_2PCH=CHPPh_2$ shows an abundant ion corresponding to $Ph_3P^{+\cdot}$ (229).

3. Other Rearrangements

The mass spectra of π-cyclopentadienyl metal compounds usually give abundant parent ions and show varying degrees of ring fragmentation. Despite this, however, processes are observed corresponding to the extrusion of the metal ion (231) or neutral metal atom (232), e.g.,

$$(C_5H_5)_2Fe^{+\cdot} \longrightarrow Fe^{+\cdot} + C_{10}H_{10} \tag{21}$$

$$C_5H_5FeC_5H_4^+ \longrightarrow C_{10}H_9^+ + Fe \tag{22}$$

Compound II mentioned in Section III, A, fragments with the loss of CrS_2 from the parent ion, leaving $(C_5H_5)_2Cr^{+\cdot}$ (91). In some substituted ferrocenes, transfer of a substituent to the iron has been noted:

$$(C_5H_5FeC_5H_4C(O)R)^{+\cdot} \longrightarrow C_5H_5FeR^{+\cdot} \quad (R = C_6H_5, OH, OCH_3, NHCN_3) \tag{23}$$

Similar arrangements occur with alkylhydroxy side chains, the OH group migrating to the metal (233). Müller (13) has shown that migration of electronegative groups is favored if stable ions such as the tropylium ion can be formed:

$$CrC_7H_7CN^+ \xrightarrow{-CrCN} C_7H_7^{+\cdot} \tag{24}$$

$$CrC_6H_5CH_2OH^+ \xrightarrow{-CrOH} C_7H_7^{+\cdot} \tag{25}$$

Migrations of oxygen to the metal from nitroso derivatives have been observed (234), e.g.,

$$C_5H_5V(NO)_2CO^+ \xrightarrow{-NO,\ -CO} C_5H_5VNO^+ \longrightarrow VO^+ + (C_5H_5N) \tag{26}$$

Rearrangements involving hydrogen transfer, either within the ligand or to the metal, have been discussed extensively by Müller (13) and Chambers et al. (5). These include main group elements for which alkene elimination from even-electron alkyl-containing ions,

$$Et_3M^+ \xrightarrow{-C_2H_4} Et_2MH^+ \xrightarrow{-C_2H_4} EtMH_2^+ \xrightarrow{-C_2H_4} MH_3^+ \tag{27}$$

proceeds via a β-elimination,*

$$M \overset{CH_2}{\underset{H}{\diagup\!\!\diagdown}} CH_2 \longrightarrow MH^+ + C_2H_4 \qquad (28)$$

Several recent studies have shown the ease of oxygen migration to tin, OH groups from hydroxylic ligands (235) [Eq. (29)] as well as oxygen from nitro groups becoming bonded to tin (79):

$$(CH_3)_3Sn^{+\cdot}\cdot(CH_2)_nOH \xrightarrow{-CH_3} (CH_3)_2Sn^+\underset{(CH_2)_n}{\diagup\!\!\diagdown}OH \longrightarrow (CH_3)_2Sn^+OH \qquad (29)$$

$$Sn^+O_2^{\cdot} + NO^{\cdot} \longleftarrow Sn^+NO_3 \longrightarrow Sn^+O^{\cdot} + NO_2 \qquad (30)$$

Glockling et al. (236) have recently investigated platinum alkyl and aryl complexes in which organic groups are transferred to and from phosphorus and platinum. The $(C_6H_5)_3P^{+\cdot}$ ion is detected in the spectrum of $Pt(C_6H_5)_2[((C_6H_5)_2P)_2CH_2]$, and, by labeling the phenyl group on platinum with fluorine (m-C_6H_4F), it is shown by formation of both $(C_6H_5)_3P^{+\cdot}$ and $(C_6H_5)_2PC_6H_4F^{+\cdot}$ that phenyl migration occurs from Pt to P. The apparent insertion of C_6H_4, arising from a $P(C_6H_5)$ group, into Pt—$C_6H_4F^+$ is also observed.

C. Isotope Abundances

In inorganic and organometallic mass spectroscopy, one encounters the presence of polyisotopic atoms in most molecules. Although, as a result, the spectra appear to be somewhat complicated, the presence of isotope clusters is usually of great advantage in providing positive identification of high molecular weight species without exact mass verification. A good example is compound I (Section III, A): the envelope of the molecular ion cluster, centered about m/e 1010, demands the presence of two iron atoms. If, for example, the species contained only one iron but two more carbonyls (a chemically implausible species), although the nominal molecular weight would be unchanged $[-58 + (2 \times 28)]$, the pattern would be considerably different (Table IV). For systems such as B_8Cl_8 (136) (Fig. 6) and B_9Cl_9 (137), the confirmation by isotopic species is as definitive as an exact mass measurement, perhaps more so, because exact mass measurement can be made

---

* The mass spectrometric convention is that a double-barbed arrow (→) implies transfer of two electrons, whereas a single barb (⇀) indicates a one-electron transfer.

TABLE IV

Observed and Calculated Isotopic Pattern for the Parent Ion Cluster of $[(C_6F_5)_2PFe(CO)_3]_2$

m/e	Observed intensity[a]	Calculated intensities[b]	
		$C_{24}F_{20}P_2Fe_2(CO)_6$	$C_{24}F_{20}P_2Fe(CO)_8$
1008	14.0	12.6	6.3
1009	6.2	4.6	2.3
1010	100.0	100.0	100.0
1011	40.0	38.5	38.5
1012	10.0	9.1	9.1
1013	3.8	1.6	1.6
1914	1.0	0.2	0.2

[a] Miller (237).
[b] Calculated using computer program BMASROS (182).

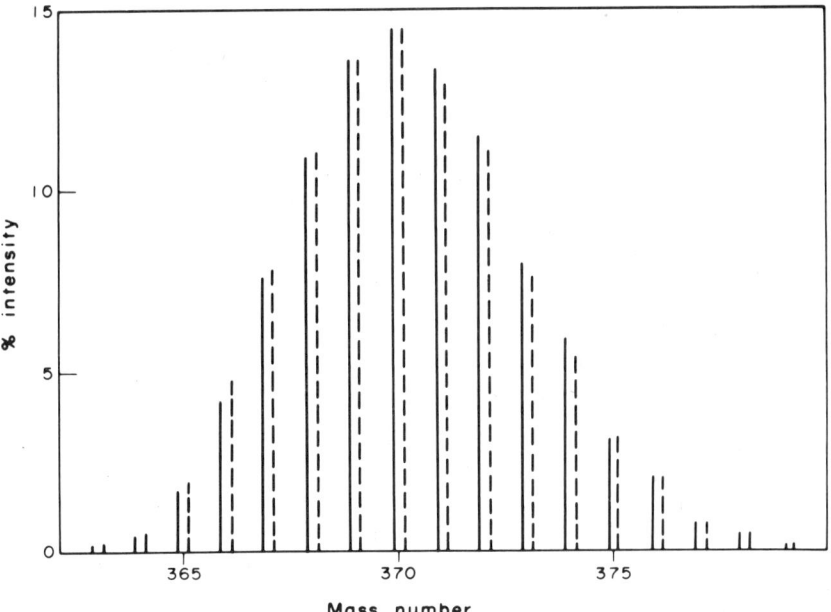

Fig. 6. Comparison of observed (solid line) and theoretical (broken line) isotopic intensities for the molecular ion of B_8Cl_8. [Reproduced by permission from Lanthier (136).]

difficult by many species of the same nominal mass and by the absence of good mass references at high m/e values.

A complication is introduced for systems that readily lose protons, or have two different coincident species overlapping in the same m/e region. In this case, deconvolution of the overlapping spectra becomes necessary. Once begun, however, deconvolution must be carried to completion. As shown in Table V, for $B_3N_3H_6$ serious errors can occur if

TABLE V

CONTRIBUTIONS TO THE PARENT ION CLUSTER IN
THE MASS SPECTRUM OF BORAZINE

	Relative intensity		
m/e	P + (P—H)[a]	P + (P—nH) $n = 1$–6[b]	Observed[c]
75	0.0	4.1	4.1
76	0.0	9.3	9.3
77	1.1	13.3	13.3
78	14.0	23.2	23.2
79	61.5	64.8	64.8
80	100.0	100.0	100.0
81	32.9	35.5	35.5
82	0.4	0.4	0.0

[a] Melcher et al. (238): P = 29.4%; P—H = 70.6%.

[b] Computer fit of the observed data shows the following relative contributions: P = 26.5%, P—H = 57.3%, P—2H = 3.2%, P—3H = 4.8%, P—4H = 5.3%, P—5H = 2.4%, P—6H = 0.4% (239).

[c] Melcher et al. (238).

only P$^+$ and (P—H)$^+$ contributions are considered (238) rather than all contributing species (239). Note that the (P—H)$^+$ contribution drops from 70.6 to 57.3% on consideration of all possible overlapping species.

Although isotopic patterns and deconvolutions may be calculated manually (6), to achieve full potential a computer is virtually a necessity. Manual calculations often omit the 1.1% contribution for ^{13}C; for molecules with large ligands, thirty carbons are not unusual, and these would give a 33% contribution to the m/e value, greater than that from the nominal mass by one mass unit.

Various computer programs are mentioned in the literature for calculating isotopic patterns (182, 240–242) and the least-squares deconvolution of overlapping species, i.e., the calculation of monoisotopic mass spectra. McLaughlin et al. (241, 243) have used their programs to generate monoisotopic mass spectra of complex boranes and to establish the existence of heptaboranes.

D. NEGATIVE IONS

Although negative-ion mass spectra are easily obtained in most mass spectrometers by relatively simple reversal of magnetic and electric field polarities, frequent absence of molecular ions and low negative-ion intensities [about 10^{-3} that of positive ionization (244)] have meant, until recently, few negative-ion studies.

A series of papers has appeared on the negative-ion spectra of binary fluorides of boron (245, 246), Group IVA (247, 248), phosphorus (245), and Group VIA (249). The Group III and IV fluorides, as well as showing fragment negative ions, showed species such as BF_4^-, SiF_5^- and GeF_5^-, presumably the result of ion–molecule reactions. Pignataro's (244, 250–254) and Kiser's (6, 255, 256) groups have been active in the applications of negative-ion mass spectroscopy to organometallic metal carbonyl systems. They find that in general, if a positive ion is abundant, the corresponding negative ion is not, and vice versa. In contrast to the general rule that relatively few ions are observed in the negative-ion spectra, they found the negative-ion spectrum of $W(CO)_5PPh_3$ to show more peaks than its positive-ion spectrum. Thus positive- and negative-ion spectra are highly complementary.

Further negative-ion mass spectral studies include Cohen's work (257, 258) on organomercurials, Pant and Sacher's (258) on tetrakis-(3,3,3-trifluoropropynyl) silane, Onak's et al. (259) on some closocarboranes as shown in (15), whereas Fraser et al. (260) have looked at the negative-ion spectra of some fluorinated β-diketonates of various metals. Molecular ions and ligand ions predominate in these latter spectra. Thus, it would appear that negative-ion mass spectrometry is an area offering great potential growth, provided more mass spectrometry laboratories would offer negative-ion spectra as a routine service. In the future the atmospheric pressure ion source discussed earlier (66, 67) may prove to be an excellent source of negative ions:

$$C_2B_4H_6 \xrightarrow{e^-} C_2B_4H_6^{-\cdot} \xrightarrow{-[BH_3 + H^\cdot]} C_2B_3H_2^-$$
$$\downarrow -H^\cdot \qquad \swarrow -BH_3$$
$$C_2B_4H_5^- \qquad \qquad \qquad (31)$$

E. Synthetic Models

Mass spectrometry can serve as a model for chemical synthesis since pyrolysis or thermolysis of compounds often leads to products analogous to the rearrangement ions observed. One of us (*90*) had suggested on the basis of mass spectral rearrangements that pyrolysis of the $(C_6F_5)_4M$ compounds of Group IV could serve as a synthetic route to the unknown perfluoro di-, tri-, and tetraphenylenes. These compounds were in fact synthesized (*261, 262*) by the decomposition of unstable C_6F_5 derivatives of titanium. Clinton and Kochi (*263*) used selectivities in the mass spectral cracking patterns of tetraalkyllead compounds as models for the selective cleavages of cation radicals in solution. Bryant and Kinstle (*185*) have observed loss of methylene ($:CH_2$) from the molecular ion of 2-hydroxycyclooctyl mercuric chloride and suggest that this compound might be a useful carbene source under thermal conditions.

Orlov (*8a*) has compared the dissociative ionization and thermal decomposition of a series of Group IV compounds and discusses the use of the mass spectrometer to study the unstable or "nonexistent" compounds that are presumably thermolysis intermediates. These include bivalent compounds of Group IV such as silicon(II) derivatives and compounds with multiple bonds. Not only are ions observed important in such correlations, but the mass of the neutral fragment, as confirmed by the decomposition of metastable ions, is of equal importance.

V. High-Resolution Studies

Whereas the exact measurement of mass is a necessity for the unambiguous confirmation of the empirical formulas of organic ions, isotopic patterns can often serve the same purpose in inorganic and organometallic chemistry (Section IV, C). Most exact-mass measurements have been made manually using a peak-matching accessory, which permits the accelerating voltage to be changed, alternately focusing an unknown and reference peak at the exit slit. The two peaks are displayed on an oscilloscope and adjusted for superposition. This technique, although accurate to about 1 ppm is slow and tedious and requires considerable quantities of sample. More recently, on-line computers have been used to determine exact mass "on the fly" with scan rates of 10 sec per mass decade, or faster, using a continuous interpolation beween internal reference peaks. Unfortunately, most of the on-line data acquisition systems are based on minicomputers with restricted core capacity. Although they may be advertised as capable of handling up to ten elements simultaneously when calculating "element maps," each isotope counts as one element. Thus heteronuclear metal clusters or elements such as tin or mercury can cause confusion. A second problem for the computer is a suitable mass marker with high molecular weight; PFK, most commonly used, is good only to about m/e 800. Various reference compounds have been prepared for special purposes, the highest having masses up to 3628 (264); others of lower mass are available commercially (265). Hexakis(pentafluorophenyl)benzene, itself a product of an organometallic reaction (226), is a useful reference. Most reference compounds have been fluorocarbon-containing compounds since (a) monoisotopic fluorine results in simple spectra, (b) the molecular weights increase rapidly while volatility remains high, and (c) the fluorine mass defect permits resolution from hydrocarbon compounds. This latter point may well be a disadvantage for organometallics that also display a mass defect.*

An alternative to observing sample and reference together at high resolution involves the use of a double-beam mass spectrometer such as the AEI MS-30. If closely spaced mass multiplets need not be resolved, one can determine the exact mass of each peak in the spectrum at resolutions of ~1500, the computer acquiring data from the independent sample and reference beams simultaneously (226). Thus, source conditions can be optimized for compound and reference at the higher sensitivities available at lower resolutions, the mass measuring accuracy still being in the order of 10 ppm. This is possible, since if mass doublets are not involved, the centroid of a peak is as well defined at resolution 1500 as at 10,000.

In Table VI we illustrate the mass-measuring accuracy attainable even for polyisotopic species such as decaboranes (267) showing resolution of mass doublets, and uranium(IV) N,N'-ethylenebis(salicylideneimiminato) (Salen) complexes (96) at higher molecular weights. In this latter case the parent ion at m/e 702 was not determined due to lack of a suitable reference. The values recorded for the iron(II) dimethylglyoxime clathrochelates with boron compounds are examples of less accurate mass measurement, and for which the same information might have been obtained by isotope abundance measurements (268). The corresponding cobalt compounds (268a) were determined to even lesser precision.

* The mass defect is the deviation of exact monoisotopic masses from integral whole numbers. Hydrocarbons have a positive mass defect ($^{12}C = 12.0000$ and $^{1}H = 1.0078$) resulting in exact masses slightly greater than the integral value, whereas the elements in the center of the periodic table (from oxygen to bismuth) usually have negative mass defects resulting in compounds with exact masses somewhat less than the nearest integral value.

TABLE VI

Some Examples of Exact-Mass Measurement

Ion+	Observed m/e	Calculated m/e	Error (amu)	Error (ppm)
$^{10}B_2{}^{11}B_8{}^1H_9{}^{35}Cl$	152.1396	152.1396	0.0000	0.0
$^{11}B_{10}{}^1H_7{}^{35}Cl$	152.1184	152.1167	−0.0017	−11.2
$^{10}B_3{}^{11}B_7{}^1H_9{}^{127}I$	243.0784	243.0788	0.0004	1.6
$^{10}B^{11}B_9{}^1H_7{}^{127}I$	243.0551	243.0559	0.0008	3.3
U(Salen)acac (i.e., M-acac)	603.2000	603.1996	0.0007	0.7
$[Fe(DMG)_3(BF)_2]$	458.079	458.080	0.001	2.2
$[Fe(DMG)_3(BOH)_2]$	454.087	454.081	−0.006	−13.2
$[Fe(DMG)_3(BOC_4H_9)_2]$	566.212	566.155	−0.057	−100.7
$^{59}Co^{12}C_8{}^1H_{12}{}^{14}N_4{}^{16}O_4{}^{11}B_2{}^{19}F_2$	347.04	347.03	−0.01	−28.8
$^{59}Co^{12}C_{12}{}^1H_{18}{}^{14}N_6{}^{16}O_6{}^{11}B_2{}^{19}F_2$	461.08	461.08	0.00	0.0

Haegele et al. (269) have used exact isotope masses and isotope abundances together in determining the detailed fragmentation patterns of square planar rhodium(I) β-diketonate complexes. They found that some species postulated by other workers were in error. High resolution is needed to distinguish the 28 mass units for loss of CO (27.9949) from C_2H_4 (28.0313) (269) or the 69 mass units for PF_2 (68.9906) from CF_3 (68.9952) (90).

VI. Metastable-Ion Techniques

To the novice, the rearrangements discussed in Section IV, B may appear to be alchemical figments (or fragments) of the imagination. However, the modern magnetic-sector mass spectrometer provides a method for the verification of these transitions in the form of "metastable ions," and all of the exotic transitions discussed in Section IV, B were so verified as being due to unimolecular dissociation and not source pyrolysis products. "Metastable-ion peaks" is perhaps a misnomer, since the low-intensity broad diffuse peaks actually observed are normal ions arising from the decomposition of metastable ions in field-free regions of the mass spectrometer, but the term metastable or metastable ion is often attributed to them. Metastable ions have been the subject of a comprehensive monograph (270), but applications to organometallics have not been included.

"Normal" metastable ions are formed between the source and magnet of single-focusing sector instruments. In double-focusing instruments of Nier-Johnson geometry, these arise in the second field-free region between the electric and magnetic sectors, where a precursor ion m_1 decomposes to give a daughter ion m_2, now carrying only a fraction m_2/m_1 of its original energy. These are brought to a focus by the magnet at an apparent mass $m^* = m_2{}^2/m_1$ (see Ref. 77 for sample calculation). Normally then, decomposition pathways are verified by measuring the position of m^* and by trial-and-error fit of m_1 and m_2 (almost invariably abundant fragments). Since m^* appears at a lower apparent mass then either m_1 or m_2, transitions between high m/e fragments are much easier to assign, as the choices for m_1 and m_2 are limited (only those peaks appearing above m^*); metastable ions appearing at lower m/e values may be found to have more than one set of precursor-daughter couples satisfying the measured m^*. The importance of normal metastable ions has long been recognized (77, 270, 271), and most of the rearrangements discussed have been studied with the aid of normal metastables.

More recently, several metastable "defocusing" techniques have been introduced, which gives enhanced metastable-ion sensitivity by removing the normal ions, as well as removing much, if not all, of the ambiguity involving trial-and-error fits of m_1 and m_2. The first of these, the Barber and Elliott technique and its refinements (272–274) permits metastable ions formed in the first field-free region between the source and electrostatic analyzer (ESA) to be focused on the collector. After tuning the magnet to a particular ion, the accelerating voltage is increased while the ESA voltage is held constant, and the collector output is recorded as a function of accelerating voltage. A discussion of the ion optics of these techniques may be found in Cooks et al. (270). By this means the precursors to the ion tuned in by the magnet are observed, i.e., $m_2/m_1 = V_0/V_1$, where m_1 and m_2 are the masses of the precursor and daughter ions and V_0 and V_1 are the initial accelerating voltage and that required to tune the precursor ion to the energy of the ESA. A later development of this technique, referred to as "ion kinetic energy (IKE) spectroscopy" by Beynon (77, 270, 275, 276), involves an ESA voltage scan with the accelerating voltage held constant. Now all the first drift-region metastables are recorded at the total-ion monitor between the ESA and magnet. Interpretation of these results requires trial fits to all possible transitions.

The most recent metastable defocusing technique, which is referred to as "mass-analyzed ion kinetic spectroscopy" (MIKES) by Beynon (77, 270, 277) and as "direct analysis of daughter ions" (DADI) by Maurer (278) requires the interchange of the source and collector positions in a double-focusing mass spectrometer of Nier–Johnson geometry. With this configuration, if the magnet is tuned to a particular ion and the ESA voltage scanned while keeping the accelerating voltage constant, the collector output as a function of ESA voltage gives all the daughter ions formed in the region between the magnet and ESA. To obtain MIKES involves either extensive mechanical modification of conventional commercial instruments (277, 279), i.e., physically interchanging the source and collector ends of the instrument or the purchase of a machine constructed with reverse geometry (278). Such reversed geometry instruments have the disadvantage, for a normal, mass spectrometry service laboratory of suppressing "normal" metastables. If the nonexpert does not see normal metastables, he has nothing to entice him to study further metastable ions.

We have observed that the AEI MS-30 mass spectrometer has the potential for all of the above methods included in its design and have recorded MIKE spectra (280) by scanning the beam-2 deflector plates, which are located between the magnet and the beam-2 collector, in an instrument of otherwise conventional Nier–Johnson geometry. Normal metastables are observed in good intensity in beam 1. An extension of the MIKES or DADI technique permits direct observations of "metastable ions arising from metastable ions in ion energy spectra" (MAMIES) (281) for which it was possible to observe in the spectrum of N-trimethyl borazole, the following consecutive transitions from m/e 122: 122 → 81 → 65; 122 → 81 → 54; and 122 → 81 → 40.

Results of a low-resolution MIKES study of $C_6F_5Si(CH_3)_3$ (280) are shown in Fig. 7, for transitions from the parent ion, the base peak, $C_6F_5Si(CH_3)_2^+$, and from the rearrangement ion $(CH_3)_2SiF^+$. Only the more intense peaks are also observed as normal metastable ions ($m^* =$ 69.4 for the transition $C_6F_5Si(CH_3)_2^+ \rightarrow C_6F_2CH_3^+ + CH_3SiF_3$).

Various other techniques are available for the observation of normal metastable ions in the absence of the much more intense normal ions, and these can be of use for photographic instruments of Mattauch–Herzog geometry (282, 283), for instruments with a Daly detector (284), and for instruments able to observe MIKE spectra (285). Metastable ions have also been used to establish the existence of unobserved parent ions (286). The development of high-resolution MIKE spectrometers, studies of kinetic energy release, and ion–molecule reactions are all areas of development of metastable-ion studies with great practical application to inorganic and organometallic chemistry. Cooks et al. (287) have shown that one can detect charge-stripping transitions both from the ground and excited states for $NO^+ \rightarrow NO^{2+}$, but to date high-resolution MIKES have not been applied to organometallic compounds. Now that negative-ion studies on organometallics are becoming common, another area ripe

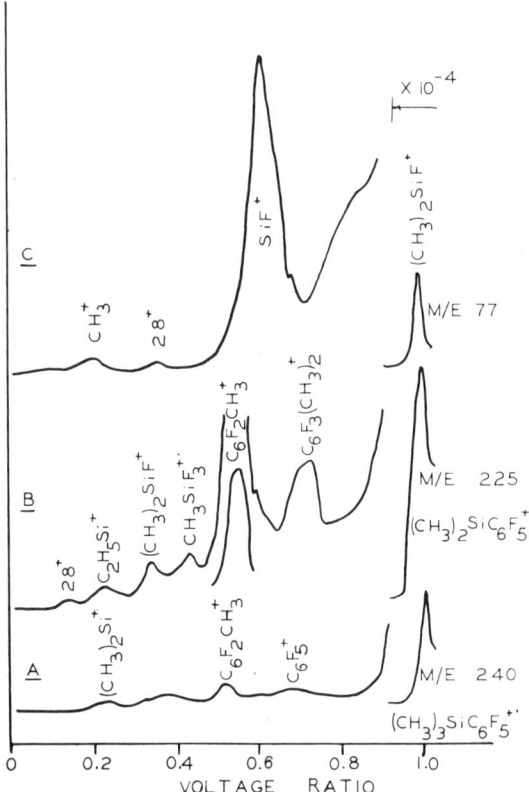

FIG. 7. Mass-analyzed ion kinetic energy (MIKE) spectra of $C_6F_5Si(CH_3)_3$. [Reproduced with permission from J. M. Miller, J. Ross, J. Rustenberg, and G. L. Wilson, *Anal. Chem.* **45**, 627 (1973); copyright by the American Chemical Society.]

for development will be the applications of both normal metastables and MIKES techniques to these negative-ion spectra.

VII. Coupled Gas Chromatography and Mass Spectrometry Applied to Inorganic and Organometallic Compounds

Application of the integrated GC/MS system to problems in organometallic chemistry currently falls far short of potential. Interestingly, one of the major applications of GC/MS in organometallic systems has been made with compounds of biochemical interest. The widespread use of trimethylsilyl derivatives for GC (*288, 289*) was extended to MS (*290*) and GC/MS spectrometry (*291, 292*). Similarly, the use of GC for the detection and analysis of organomercury compounds in environmental samples is widespread (293), and concern over possible decomposition has led to GC/MS studies of such systems (123, 124, 161, 293, 294).

The following examples (217, 295) illustrate the routine use of GC/MS in the solution of day-to-day laboratory problems, solving in an hour or 2, problems that may either be ignored and bypassed or that would require more difficult and time-consuming methods to solve. It is possible to check the purity of starting materials and products and to identify unwanted components, which often is a great aid in their elimination during synthesis or in deciding on purification techniques. The monitoring of exploratory syntheses by GC/MS permits a rapid identification of GC peaks so that GC alone may be used as an aid, as the synthetic procedure is refined for optimal yield, purity, etc. Similarly, the products of unexpected reactions, as may occur between solute and solvent in an NMR tube, may be monitored rapidly with minimal sample consumption (1 μl); air- and moisture-sensitive compounds are readily handled using microsyringes.

Recently, using as a starting material a commercial sample of borazine, $B_3N_3H_6$, purported to be "pure," inexplicable products were detected by GC/MS, suggesting impure starting materials. The borazine was thus subjected to GC/MS analysis. Although the first and largest GC peak corresponds to $B_3N_3H_6$, the impurities predominate. A peak of about equal intensity was readily identified as chlorobenzene (m/e 112:114 in a 3:1 ratio), a solvent used in one preparation of $B_3N_3H_6$ (although the manufacturers initially suggested that chlorobenzene had not been used, later correspondence admitted its probable presence), and trimethylamine, another possible precursor in the synthesis was also found. Other peaks corresponded to the formula $B_5N_5H_8$ (cluster at m/e 133, 132, 131, etc.), a previously identified material (296), i.e., the BN naphthalene analog, the biphenyl analog (296) $(B_3N_3H_5)_2$ (cluster at m/e 160, 159, 158, 157, etc.), and two higher polymers, although no mass spectra have been previously reported for either species, $B_6N_7H_{11}$ (m/e 175, 174, 173, etc.) and $B_7N_7H_{10}$ (m/e 185, 184, 183, etc.) the latter being the anthracene analog. In a matter of minutes, the nature of the impurities were determined and the presence of unexpected, but interesting species revealed. Analysis via the GC/MS system, therefore, may offer a saving in time where the purity of starting materials is suspect, as well as potentially revealing new and interesting compounds previously discarded as unwanted impurities.

Another example involves a sample of $(C_6F_5)_2Si(CH_3)_2$ which was 99%+ pure by GC and "pure" by the criterion of ^{1}H NMR spectra, but which showed peaks in the mass spectrum to high mass of the expected parent at m/e 392. It was therefore subjected to a GC/MS analysis, and the compounds detected are shown in Table VII. Peaks 6 and 7 correspond to two different isomers.

During a proton NMR examination of a CCl_4 solution of $Me_3N_3B_3F_3$, an extraneous peak was observed in the spectrum, although the borazine was known to be analytically and mass spectrometrically pure. One microliter of the solution was subjected to GC/MS analysis. Peaks identified as trimethylsilane (TMS) and CCl_4 were obtained followed by the expected $Me_3N_3B_3F_3$ (m/e 177, 176, etc.). An additional component

TABLE VII

COMBINED GAS CHROMATOGRAPHY/MASS SPECTROMETRY RESULTS FOR $(C_6F_5)_2Si(CH_3)_2$[a]

Peak No.	Retention time (min)	Quantity	Parent ion (m/e)	Identification
1	4.0	Trace	240	$C_6F_5Si(CH_3)_3$
2	8.0	Trace	281	?
3	17.5	99%+	392	$(C_6H_5)_2Si(CH_3)_2$
4	21.0	0.2%	440/438/436	$C_{12}F_7HBr_2$
5	22.0	0.2%	450	$(C_6F_5)_2Si_2(CH_3)_4$
6	23.0	0.3%	454/452	$(C_6F_5)(C_6F_4Br)Si(CH_3)_2$
7	23.7	0.2%	454/452	$(C_6F_5)(C_6F_4Br)Si(CH_3)_2$
8	25.0	Trace	416/414/412/410/408	C_6HFBr_4
9	27.0	Trace	482	$(C_6F_5)_2Si_2O_2(CH_3)_4$

[a] Conditions: 175°C; 25 ml/min He; 5 ft × 0.25 in. 3% SE-30.

eluted after 15 min was readily identified by its mass spectrum as $Me_3N_3B_3F_2Cl$, undoubtedly the extraneous NMR peak. It appears therefore that B-trifluoro-N-trimethyl borazine undergoes halogen exchange with CCl_4, another example of an "inert" NMR solvent acting as a good chlorinating agent for metal–halogen or metal–hydrogen bonds. The rapid identification of the impurity ($\sim$10%) permitted the NMR investigation to proceed, as well as suggesting possible synthetic routes to mixed fluorochloro-substituted borazines.

Sowinski and Suffet (297) have used GC/MS to detect boron hydrides at trace levels, whereas Blum and Richter (298) have used capillary columns in the combined GC/MS of a series of phenylboronate derivatives. There have also been recent applications of GC/MS to the TMS derivatives of inorganic anions (299). The TMS derivatives of ammonium arsenates, phosphates, vanadates, borates, carbonates, oxalates, and sulfates gave satisfactory GC separation and satisfactory mass spectra, but the corresponding sodium or potassium salts gave unsatisfactory results. Coupled GC/MS has also been satisfactorily applied to the separation of β-diketonates of Ni(II), Pt(II), and Pd(II) (*300*) and to the determination of chromium and beryllium at the picogram level (*301*).

The ready availability of GC/MS instrumentation as well as the speed, specificity, and small sample requirements are likely to result in a great increase in the applications of this technique to inorganic and organometallic chemistry.

VIII. Conclusion

We hope to have given some insight into the present usefulness of mass spectrometry to the inorganic and organometallic chemist, but more important, we would hope to have stimulated thought toward future applications. The possibilities for the chemist implicit in the development of new ion sources, in metastable-ion and ion kinetic energy studies, and in the interfacing of such powerful separation techniques as GC and HPLC, would suggest that mass spectrometry will continue to be of growing interest in the future.

Acknowledgments

We thank our colleagues for many useful discussions and in particular Dr. J. S. Hartman and Dr. P. H. Bickart for their critical reading of the manuscript.

References

1. Aston, F. W., "Mass Spectra and Isotopes," 2nd ed., p. 150ff. Arnold, London 1942.
2. Aston, F. W. *Phil. Mag.* [6] **45**, 936 (1923).
3. Aston, F. W., *Proc. Roy. Soc., Ser. A* **149**, 396 (1935).
4. Bruce, M. I., *Advan. Organomental. Chem* **6**, 273 (1968).
5. Chambers, D. B., Glockling, F., and Light, J. R. C., *Quart. Rev., Chem. Soc.* **22**, 317 (1968).
6. Kiser, R. W., in "Characterization of Organometallic Compounds Part I" (M. Tsutsui, ed.), p. 137 ff. Wiley (Interscience), New York, 1969.
7. Cais, M., and Lupin, M. S., *Advan. Organometal Chem.* **8**, 211 (1970).
8. Lewis, J., and Johnson, B. F. G., *Accounts Chem. Res.* **1**, 245 (1968).
8a. Orlov, V. Yu., *Russ. Chem. Rev.* **42**, 529 (1973).
9. Glockling, F., "The Chemistry of Germanium." Academic Press, New York, 1969.
10. Lesbre, M., Mazerolles, P., and Satge, J., "The Organometallic Compounds of Germanium." Wiley (Interscience), New York, 1971.
11. Cragg, R. H., and Weston, A. F., *J. Organometal. Chem.* **67**, 161 (1974).

12. King, R. B., *Fortschr. Chem. Forsch.* **14**, 92 (1970).
13. Müller, J., *Angew. Chem., Int. Ed. Engl.* **11**, 653 (1972).
14. Margrave, J. L., ed., "Mass Spectrometry in Inorganic Chemistry," Advan. Chem. Ser. No. 72, Amer. Chem. Soc., Washington, D.C., 1968.
15. Litzow, M. R., and Spalding, T. R., "Mass Spectrometry of Inorganic and Organometallic Compounds." Elsevier, Amsterdam, 1973.
16. Bruce, M. I., *Chem. Soc. Rev.* **1**, 182 (1972).
17. Bruce, M. I., *Chem. Soc. Rev.* **2**, 193 (1973).
18. "Mass Spectrometry Bulletin," Spectrometry Data Centre, AWRE, Aldermaston, Reading, RG7 4PR, U.K.
19. American Society for Mass Spectrometry, *19th Annu. Conf. Mass Spectrom. Allied Top.* (1971).
20. "International Symposium on the Mass Spectrometry of Metal-Containing Compounds." The Polytechnic of North London, London, 1973.
21. Beynon, J. H., "Mass Spectrometry and its Applications to Organic Chemistry." Elsevier, Amsterdam, 1960.
22. Beynon, J. H., Saunders, R. A., and Williams, A. E., "The Mass Spectra of Organic Molecules." Elsevier, Amsterdam, 1968.
23. Budzikiewicz, H., Djerassi, C., and Williams, D. H., "Mass Spectrometry of Organic Compounds." Holden-Day, San Francisco, California 1967.
24. Hamming, M. C., and Foster, N. G., "Interpretation of Mass Spectra of Organic Compounds." Academic Press, New York, 1972.
25. Kiser, R. W., "Introduction to Mass Spectrometry and its Applications." Prentice-Hall, Englewood Cliffs, New Jersey, 1965.
26. Reed, R. I., "Modern Aspects of Mass Spectrometry." Plenum, New York, 1968.
27. Roboz, J., "Introduction to Mass Spectrometry, Instrumentation and Techniques." Wiley (Interscience), New York, 1968.
28. Reed, R. I., ed., "Recent Topics in Mass Spectrometry." Gordon & Breach, New York, 1971.
29. McDowell, C. A., ed., "Mass Spectrometry." McGraw-Hill, New York, 1963.
30. McFadden, W., ed., "Techniques of Combined Gas Chromatography/Mass Spectrometry." Wiley (Interscience), New York, 1973.
31. McLafferty, F. W., "Interpretation of Mass Spectra," 2nd ed. Benjamin, New York, 1973.
32. Shrader, S. R., "Introductory Mass Spectrometry." Allyn & Bacon, Boston, Massachusetts, 1971.
33. Silverstein, R. M., Bassler, G. C., and Morrill, T. C., "Spectrometric Identification of Organic Compounds." Wiley, New York, 1974.
34. Ahearn, A. J., ed., "Mass Spectrometric Analysis of Solids." Elsevier, Amsterdam, 1966.
35. Ahearn, A. J., ed., "Trace Analysis by Mass Spectrometry." Academic Press, New York, 1972.
36. Drowart, J., and Goldfinger, P., *Advan. Mass Spectrom.* **3**, 923 (1966).
37. Redman, J. D., "A Literature Review of Mass Spectrometric—Thermochemical Technique Applicable to the Analysis of Vapor Species Over Solid Inorganic Materials," ORNL, TM 989. 1966.
38. Commonly used instruments of this type include the AEI MS-12; Atlas CH-4; Varian MAT CH-5 and CH-7; Hitachi RMU-6; Dupont 21-490 and LKB 9000.

39. The standard instrument of this type has been the AEI MS 9/MS 902 now superseded by the MS-50, MS-30 and the new MS-1073, others include the Hitachi RMU-7, Dupont 21-491 and 492, and VG instruments.
40. Varian MAT CH-5 DF and 311 and the new 112, and a Nuclide instrument.
41. The long time standard, now discontinued, Mattach-Herzog instrument was the CEC (Dupont) 21-110, while JEOL, Varian MAT and AEI still build instruments of this geometry.
42. Finnigan, Hewlett-Packard, EAI, Extranuclear and others.
43. Bendix now superseded by CVC.
44. Varian M-66.
45. Chait, E. M., *Anal. Chem.* **44**, 77A (1972).
46. McFadden, W., ed., "Techniques of Combined Gas Chromatography/Mass Spectrometry," Chapter 2. Wiley (Interscience), New York, 1973.
47. McFadden, W., ed., "Techniques of Combined Gas Chromatography/Mass Spectrometry," Chapter 4. Wiley (Interscience), New York, 1973.
48. Munson, M. S. B., and Field, F. H., *J. Amer. Chem. Soc.* **88**, 2621 (1966).
49. Munson, M. S. B., *Anal. Chem.* **43**, 28A (1971).
50. Rodeheaver, G. T., Farrant, G. C., and Hunt, D. F., *J. Organometal. Chem.* **30**, C22 (1971).
51. Hunt, D. F., Russell, J. W., and Torian, R. Z., *J. Organometal. Chem.* **43**, 175 (1972).
52. Solomon, J. J., and Porter, R. F., *J. Amer. Chem. Soc.* **94**, 1443 (1972).
53. Porter, R. F., and Solomon, J. J., *J. Amer. Chem. Soc.* **93**, 56 (1971).
54. Anderson, W. P., Hsu, N., Stranger, C. W., and Munson, B., *J. Organometal. Chem.* **69**, 249 (1974).
55. Risby, T. H., Jurs, P. C., Lampe, F. W., and Yergey, A. L., *Anal. Chem.* **46**, 161 (1974).
56. Risby, T. H., Jurs, P. C., Lampe, F. W., and Yergey, A. L., *Anal. Chem.* **46**, 726 (1974).
57. Beckey, H. D., "Field Ionization Mass Spectrometry." Pergamon, Oxford, 1971.
58. Anbar, M., and Aberth, W. H., *Anal. Chem.* **46**, 59A (1974); Robertson, A. J. B., *J. Phys. E.* **7**, 321 (1974).
59. Schulten, H.-R., and Beckey, H. D., *Org. Mass Spectrom.* **6**, 885 (1972).
60. Beckey, H. D., *Int. J. Mass Spectrom. Ion Phys.* **2**, 500 (1969).
61. Schulten, H.-R., and Beckey, H. D., *J. Chromatogr.*, **83**, 315 (1973).
62. Schulten, H.-R., Beckey, H. D., Boerboom, A. J. H., and Meuzelaar, H. L. C., *Anal. Chem.* **45**, 2385 (1973).
62a. Schulten, H.-R., and Beckey, H. D., *Amer. Soc. Mass Spectrom. 22nd Annu. Conf.* Paper 0-7 (1974); Games, D. E., Games, M. P., Jackson, A. H., Olavesen, A. H., Ressiter, M., and Winterburn, P. J., *Tetrahedron Lett.* p. 2377 (1974).
63. FD sources are available commercially from Varian MAT and have been tentatively announced by AEI.
64. Burlingame, A. L., Cox, R. E., and Derrick, P. J., *Anal. Chem.* **46**, 248R (1974).
65. Elliott, R. M., at the AEI Users Meeting, Livingston, New Jersey (1974).
65a. Hass, J. R., Sammons, M. C., Bursey, M. M., Kukuch, B. J., and Buck, R. P., *Org. Mass Spectrom.* **9**, 952 (1974).
65b. Games, D. E., Jackson, A. H., and Taylor, K. T., *Org. Mass Spectrom.* **9**, 1245 (1974).

65c. Games, D. E., Jackson, A. H., Kane-Maguire, L. A. P., and Taylor, K. J., *J. Organomet. Chem.* **88**, 345 (1975).
66. Horning, E. C., Horning, M. G., Carroll, D. I., Dzidic, I., and Stillwell, R. N., *Anal. Chem.* **45**, 936 (1973).
67. Carroll, D. I., Dzidic, I., Stillwell, R. N., Horning, M. G., and Horning, E. C., *Anal. Chem.* **46**, 706 (1974).
68. Evans, L. A., and Hendricks, C. D., *Rev. Sci. Instrum.* **43**, 1527 (1972).
69. Colby, B. N., and Evans, C. A., *Anal. Chem.* **45**, 1884 (1973).
70. Simons, D. S., Colby, B. N., and Evans, C. A., *Amer. Soc. Mass Spectrom., 22nd Annu. Conf.* Paper S-12 (1974).
71. Duna, W. G., and Hooper, J. B., *Anal. Chem.* **45**, 216 (1973).
72. Miller, J. M., Vandenhoff, J., and Wilson, G., *Can. J. Spectrosc.* **18**, 134 (1973).
73. Junk, G., *Int. J. Mass Spectrom. Ions Phys.* **8**, 1 (1972).
74. Ryhage, R., and von Sydow, E., *Acta Chem. Scand.* **17**, 2025 (1963); Ryhage, R., *Ark. Kemi* **26**, 305 (1966).
75. Watson, J. T., and Biemann, K., *Anal. Chem.* **36**, 1135 (1964).
76. Llewellyn, P. M., and Littlejohn, D. P., *Pittsburgh Conf. Anal. Chem. Appl. Spectrom.* (1966).
77. Cooks, R. G., and Beynon, J. H., *J. Chem. Educ.* **51**, 437 (1974).
78. Beynon, J. H., "Mass Spectrometry and its Applications to Organic Chemistry," Chapter 5. Elsevier, Amsterdam, 1960.
79. Potts. D., and Miller, J. M., *J. Chem. Soc., Dalton Trans.* p. 1975 (1975); *Annu. Meet. Chem. Inst. Can.* (1974).
80. Copperthwaite, R. G., and Cook, P., *Chem. Ind. (London)* No. 18 (1973).
81. Wharton, D. R. A., and Bazinet, M. L., *Anal. Chem.* **43**, 623 (1971).
82. Tse, R. S., and Wong, S. C., *Anal. Chem.* **46**, 967 (1973).
83. Thomas, C. B., and Davis, B., *Chem. Ind. (London)* p. 413 (1969).
84. Padrta, F. G., and Donohue, J. J., *Anal. Chem.* **42**, 950 (1970).
85. Jones, T. R. B., and Miller, J. M., unpublished results.
86. Glockling, F., and Light, J. R. C., *J. Chem. Soc., A* p. 717 (1968).
87. Chambers, D. B., Glockling, F., and Weston, M., *Chem. Commun.* p. 281 (1966); *J. Chem. Soc., A* p. 1759, (1967).
88. Chivers, T., Lanthier, G. F., and Miller, J. M., *J. Chem. Soc., A* p. 2556 (1971).
89. Miller, J. M., *Can. J. Chem.* **47**, 1613 (1969).
90. Miller, J. M., *J. Chem. Soc., A* p. 828 (1967).
91. Preston, F. J., and Reed, R. I., *Chem. Commun.* p. 51 (1966).
92. Johnson, B. F. G., Johnston, R. D., and Lewis, J., *Chem. Commun.* p. 1057 (1967).
93. Klanberg, F., Askew, W. B., and Guggenberger, L. J., *Inorg. Chem.* **7**, 2265 (1968).
94. Ehrl, W., Rinck, R., and Vahrenkamp, H., *J. Organometal. Chem.* **56**, 285 (1973).
95. Thompson, L. K., Eisner, E., and Newlands, M. J., *J. Organometal. Chem.* **56**, 327 (1973).
96. Calderazzo, F., Pasquali, M., and Salvatori, T., *J. Chem. Soc., Dalton Trans.* p. 1102 (1974).
97. Neuse, E. W., *J. Organometal. Chem.* **56**, 323 (1973).
98. Mickey, C. D., and Zingaro, R. A., *Inorg. Chem.* **12**, 2115 (1973).
99. Schrobilgen, G. J., private communication to J. M. Miller.
100. Penney, G. J., and Sheldrick, G. M., *J. Chem. Soc., A* p. 243 (1971).
101. Lawless, E. W., *Inorg. Chem.* **10**, 2084 (1971).

102. Ogara, T., and Fernando, Q., *Inorg. Chem.* **72**, 2611 (1973).
103. Bradley, D. C., and Vasishta, S., *Pap. Int. Symp. Mass Spectrom. Metal-Containing Compounds, 1973* Paper No. C10 (1973).
104. Oliver, J. G., and Worrall, I. J., *J. Chem. Soc., A* p. 845 (1970).
105. MacDonald, C. G., and Shannon, J. S., *Aust. J. Chem.* **17**, 1545 (1966).
106. Macklin, J., and Dudek, G., *Inorg. Nucl. Chem. Lettr.* **2**, 403 (1966).
107. Miller, J. M., Potts, A., Richardson, M. F., and Rothstein, S. M., *Can. Chem. Educ.* **10**, 5 (1974).
108. Belcher, R., Majer, J. R., Perry, R., and Stephen, W. I., *J. Inorg. Nucl. Chem.* **31**, 471 (1968).
109. Belcher, R., Majer, J. R., Perry, R., and Stephen, W. I., *Anal. Chim. Acta* **45**, 305 (1969).
110. Frazer, M. J., Newton, W. F., and Rimmer, B., *Chem. Commun.* p. 1336 (1968).
111. Porter, R. F., Schoonmaker, R. C., and Addison, C. C., *Proc. Chem. Soc., London* p. 61 (1959).
112. Potts, D., and Miller, J. M., unpublished observations.
113. Edgar, K., Johnson, B. F. G., Lewis, J., and Wilson, J. M., *J. Chem. Soc., A* p. 379 (1967).
114. Pignataro, S., and Lossing, F. P., *J. Organometal. Chem.* **11**, 571 (1968).
115. King, R. B., *Inorg. Chem.* **5**, 2227 (1966).
116. Svec, H. J., and Junk, G. A., *Inorg. Chem.* **7**, 1688 (1968).
117. Alpin, R. T., and Frearson, M. J., *Chem. Ind. (London)* p. 1663 (1969).
118. Sherwood, P. M. A., and Turner, J. J., *J. Chem. Soc., A* p. 2349 (1970).
119. Terlouw, J. K., and de Ridder, J. J., *Org. Mass Spectrom.* **5**, 1127 (1971).
120. Dobias, M., *Pap. Int. Conf. Mass Spectrom. Metal-Containing Compounds, 1973* Paper No. C17 (1973).
121. Dobbie, R. C., and Lockhart, J. C., private communication to J. M. Miller.
122. Wilson, G., and Miller, J. M., unpublished observations.
123. Johansson, B., Ryhage, R., and Westoo, G., *Acta Chem. Scand.* **24**, 2349 (1970).
124. Glockling, G., *Discuss., Int. Conf. Mass Spectrom. Metal-Containing Compounds, 1973* (1973).
125. Miller, J. M., unpublished observations based on $3\frac{1}{2}$ years operating experience with the Brock University AEI MS-30.
126. Rosenthal, D., Hopf, F. R., Whitten, D. G., and Bursey, M. M., *Org. Mass Spectrom.* **7**, 497 (1973).
127. Müller, J., and Goll, W., *J. Organometal. Chem.* **69**, 123 (1974).
128. Gilbert, J. R., Leach, W. P., and Miller, J. R. *J. Organometal. Chem.* **56**, 295 (1973).
129. Majer, J. R., and Reade, M. J. A., *Chem. Commun.* p. 58 (1970).
130. Majer, J. R., and Perry, R., *J. Chem. Soc., A* p. 822 (1970).
131. Ritter, H.-P., and Neuman, W. P., *J. Organometal. Chem.* **56**, 199 (1973).
132. Kiser, R. W., *in* "Characterization of Organometallic Compounds, Part I" (M. Tsutsui, ed.), Wiley (Interscience), New York, 1969.
133. Buchanan, A. S., Knowles, D. J., and Swingler, D. L., *J. Phys. Chem.* **73**, 4394 (1969).
134. Svec, H. J., and Sparrow, G. R., *J. Chem. Soc., A* p. 1163 (1970).
135. McGlinchey, M. J., Odom, J. D., Reynoldson, T., and Stone, F. G. A., *J. Chem. Soc., A* p. 31 (1970).

136. Lanthier, G. F., Kane, J., and Massey, A. G., *J. Inorg. Nucl. Chem.* **33**, 1569 (1971).
137. Lanthier, G. F., and Massey, A. G., *J. Inorg. Nucl. Chem.* **32**, 1807 (1970).
138. Harland, P. W., Rankine, D. W. H., and Thynne, J. C. J., *Int. J. Mass Spectrom. Ion Phys.* **13**, 395 (1974).
139. Vasile, M. J., Jones, G. R., and Falconer, W. E., *Int. J. Mass Spectrom. Ion Phys.* **10**, 457 (1972/1973).
140. Falconer, W. E., Jones, G. R., Sunder, W. A., Vasile, M. J., Muenter, A. A., Dyke, T. R., and Klemperer, W., *J. Fluorine Chem.* **4**, 213 (1974).
141. Edwards, A. J., Falconer, W. E., Griffiths, J. E., Sunder, W. A., and Vasile M. J., *J. Chem. Soc., Dalton Trans.* p. 1129 (1971).
142. Singleton, D. L., and Stafford, F. E., *Inorg. Chem.* **11**, 1208 (1973).
143. Hiragama, C., and Castle, P. M., *J. Phys. Chem.* **77**, 3110 (1973).
144. Lappert, M. F., Litzow, M. R., Pedley, J. B., Riley, P. N. K., and Tweedale, A. *J. Chem. Soc., A* p. 3105 (1968).
145. Lanthier, G. F., and Miller, J. M., *J. Chem. Soc., A* p. 346 (1971).
146. Hammerum, S., and Djerassi, C., *Org. Mass Spectrom.* **8**, 217 (1974).
147. Cragg, R. H., and Nazerty, M., *J. Chem. Soc. Dalton Trans.* p. 162 (1974).
148. Cragg, R. H., and Todd, J. F. J., *Chem. Commun.* p. 386 (1970).
149. Cragg, R. H., Gallahger, D. A., Husband, J. P. N., Lawson, G., and Todd, J. F. J., *Chem. Commun.* p. 1562 (1970).
150. Cragg, R. H., Lawson, G., and Todd, J. F. J., *J. Chem. Soc., Dalton Trans.* p. 878 (1972).
151. Hohaus, E., and Riepe, W., *Z. Naturforsch. B* **28**, 440 (1973).
152. Cragg, R. H., and Weston, A. F., *Chem. Commun.* p. 22 (1974).
153. Nakadaira, Y., Kobayashi, Y., and Sakurai, H., *J. Organometal. Chem.* **63**, 79 (1973).
154. Heuman, K. G., Bachmann, K., Kubassek, E., and Lieser, K. H. *Z. Naturforsch. B* **28**, 107 (1973).
155. Chambers, D. B., and Glockling, F., *J. Chem. Soc., A* p. 735 (1968).
156. Sakurai, H., Kira, M., and Santo, T., *J. Organomental. Chem.* **42**, C24 (1972).
157. Jones, T. R. B., and Miller, J. M., *Pap. Organosilicon Symp. Amer. Chem. Soc. Reg. Meet. Detroit* (1974).
158. Lageot, C., Maire, J. C., Rivere, P., Massol, M., and Barrau, J., *J. Organometal. Chem.* **66**, 49 (1974).
159. Pinson, J. W., and Khandelual, J. K., *Spectrosc. Lett.* **6**, 745 (1973).
160. Jamieson, W. D., Safe, S., and Hutzinger, O., *Pap. Chem. Inst. Can. Conf. 1971*, Paper No. Anal. VII-3 (1971).
161. Safe, S., and Hutzinger, O., "Mass Spectrometry of Pesticides and Pollutants," Chapter 20, CRC Press, Cleveland, Ohio, 1973.
162. Gillis, R. G., and Occoclowitz, J. L., *in* "Analytical Chemistry of Phosphorus Compounds" (M. Halmaun, ed.), Chapter 6, p. 295. Wiley (Interscience), New York, 1972.
163. Safe, S., and Hutzinger, O., "Mass Spectrometry of Pesticides and Pollutants," Chapter 21. CRC Press, Cleveland, Ohio, 1973.
164. Subramaniam, K. S., MSc. Thesis, Brock University, St. Catharines, Ontario (1971).
165. Damico, J. N., *J. Ass. Offic. Anal. Chem.* **49**, 1027 (1966).
166. Jorg, J., Houriet, R., and Spiteller, G., *Monatsh. Chem.* **97**, 1064 (1966).

167. Bonelli, E. J., and Cornelius, J., "Applications Tips," Nos. 15 and 19, Finnigan Corp., Sunnyvale, California 1970.
168. Mucke, W., Alt, K. O., and Esser, H. O., *J. Agr. Food. Chem.* **18**, 208 (1970).
169. Gatterdam, P. E., Wozniak, L. A., Bullock, M. W., Parks, G. L., and Boyd, J. E., *J. Agr. Food Chem.* **15**, 845 (1967).
170. Gardner, A. M., Damico, J. N., Hansen, E. A., Lustig, E., and Storherr, R. W., *J. Agr. Food Chem.* **18**, 1181 (1969).
171. Bazinet, M. L., Merritt, C., and Sacher, R. E., *22nd Conf. Mass Spectrom. Applied Top. 1974* Paper Q7 (1974).
172. Froyen, P., and Moller, J., *Org. Mass. Spectrom.* **7**, 73 and 691 (1973).
173. Froyen, P., and Moller, J., *Org. Mass Spectrom.* **8**, 132 (1974).
174. Henrick, K., Kepert, D. L., Shewchuk, E., Trigwell, K. R., and Wild, S. B., *Aust. J. Chem.* **27**, 727 (1974).
175. Williams, D. H., Ward, R. S., and Cooks, R. G., *J. Amer. Chem. Soc.* **90**, 966 (1968).
176. Rake, A. T., and Miller, J. M., *J. Chem. Soc., A* p. 1881 (1970).
177. Rake, A. T., and Miller, J. M., *Org. Mass. Spectrom.* **3**, 237 (1970).
178. Jones, T. R. B., and Miller, J. M., unpublished results.
179. Bowie, J. H., and Nussey, B., *Org. Mass. Spectrom.* **3**, 933 (1970).
180. Cohen, S. C., Massey, A. G., Lanthier, G. F., and Miller, J. M., *Org. Mass Spectrom.* **6**, 373 (1972).
181. Lanthier, G. F., Miller, J. M., and Oliver, A. J., *Can. J. Chem.* **51**, 1945 (1973).
182. Lanthier, G. F., Miller, J. M., Cohen, S. C., and Massey, A. G., *Org. Mass Spectrom.* **8**, 235 (1974).
183. Rebane, E., *Acta Chem. Scand.* **27**, 2861 and 2870 (1973).
184. Rebane, E., *Chem. Scripta* **5**, 5 (1974).
185. Bryant, W. F., and Kinstle, T. H., *J. Organometal. Chem.* **24**, 573 (1970).
186. Breuer, S. W., Fear, T. E., Lindsay, P. H., and Thorpe, F. G., *J. Chem. Soc., C* p. 3519 (1971).
187. Hutzinger, O., Jamieson, W. D., and Safe, S., *Int. J. Environ. Anal. Chem.* **1**, 85 (1971).
188. Sasaki, S., Itagaki, Y., Kurokawa, T., and Nakanishi, K., *Bull. Chem. Soc. Jap.* **40**, 76 (1967).
189. Reichert, C., Westmore, J. B., and Gesser, H. D., *Chem. Commun.* p. 782 (1967).
190. Bancroft, G. M., Reichert, C., and Westmore, J. B., *Inorg. Chem.* **7**, 870 (1968).
191. Bancroft, G. M., Reichert, C., Westmore, J. B., and Gesser, H. D., *Inorg. Chem.* **8**, 474 (1969).
192. Reichert, C., and Westmore, J. B., *Inorg. Chem.* **8**, 1012 (1969).
193. Shannon, J. S., and Swan, J. M., *Chem. Commun.* p. 33 (1965).
194. Lacey, M. J., and Shannon, J. S., *Org. Mass Spectrom.* **6**, 931 (1972).
195. Das, M., and Livingston, S. E., *Aust. J. Chem.* **27**, 53 (1974).
196. Das, M., and Livingston, S. E., *Aust. J. Chem.* **27**, 749 (1974).
197. Leary, J. J., Tsuge, S., and Isenhour, T. L., *Anal. Chem.* **45**, 1269 (1973).
198. Frew, N. M., Leary, J. J., and Isenhour, T. L., *Anal. Chem.* **44**, 665 (1972).
199. Eley, D. D., Hazeldine, D. J., and Palmer, T. F., *J. Chem. Soc., Faraday. Trans. 2* **69**, 1808 (1973).
200. Wilson, G. L., and Miller, J. M., unpublished observations.
201. Edwards, D. A., and Richards, R., *J. Chem. Soc., Dalton Trans.* p. 2463 (1973).
202. Lin, D. C. K., and Westmore, J. B., *Can. J. Chem.* **51**, 2999 (1973).

203. Khariton, K. S., Ablov, A. V., and Popovich, G. A., *Dok. Akad. Nauk SSSR* **204**, 1374 (1972).
204. Gilbert, W. C., Taylor, L. T., and Dillard, J. D., *J. Amer. Chem. Soc.* **95**, 2477 (1973).
205. Malek, A., and Fresco, J. M., *Can. J. Spectrosc.* **18**, 43 (1973).
206. Innorta, G., Pignataro, S., and Natile, G., *J. Organometal. Chem.* **65**, 391 (1974).
207. Haas, M. A., and Wilson, J. M., *J. Chem. Soc.*, B p. 104 (1968).
208. Whitesides, T. H., and Arhart, R. W., *Tetrahedron Lett.* p. 297 (1972).
209. King, R. B., *J. Amer. Chem. Soc.* **89**, 6368 (1967).
210. King, R. B., and Korenowski, T. F., *Chem. Commun.* p. 771 (1966).
211. Bruce, M. I., *J. Organometal. Chem.* **10**, 495 (1967).
212. Hawthorne, J. D., Mays, M. J., and Simpson, R. N. F., *J. Organometal. Chem.* **12**, 407 (1968).
213. Mays, M. J., and Simpson, R. N. F., *J. Chem. Soc.*, A p. 1936 (1967).
214. Dobbie, R. C., and Cavell, R. G., *Inorg. Chem.* **6**, 1450 (1967).
215. Cavell, R. G., and Dobbie, R. C., *Inorg. Chem.* **7**, 101 and 690 (1968).
216. Cavell, R. G., and Dobbie, R. C., *J. Chem. Soc.*, A p. 1406 (1968).
217. Lanthier, G. F., and Miller, J. M., *Pap., 5th Int. Conf. Organometal. Chem. 1971* (1971).
218. Wilson, G. L., and Miller, J. M., in preparation.
219. Miller, J. M., Ross, J., Rustenburg, J., and Wilson, G. L., *Anal. Chem.* **45**, 627 (1973).
220. McGlinchey, M. J., and Tan, T. S., *Can. J. Chem.* **52**, 2349 (1974).
221. Clark, H. C., and Rake, A. T., *J. Organometal. Chem.* **82**, 159 (1974).
222. Lanthier, G. F., and Miller, J. M., *Org. Mass Spectrom.* **6**, 89 (1972).
223. Cohen, S. C., and Massey, A. G., *Advanc. Fluorine Chem.* **6**, 185 (1970).
224. Bruce, M. I., *J. Organometal. Chem.* **21**, 415 (1970).
225. Bruce, M. I., *Org. Mass Spectrom.* **2**, 997 (1969).
226. Rausch, M. D., Andrews, P. S., and Gardner, S. A., *Organometal. Chem. Syn.* **1**, 289 (1971).
227. Roe, D. M., and Massey, A. G., *J. Organometal. Chem.* **12**, 407 (1968).
228. Haake, P., Frearson, M. J., and Diebert, C. E., *J. Org. Chem.* **34**, 788 (1969).
229. Chow, K. K., and McAuliffe, C. A., *J. Organometal. Chem.* **59**, 247 (1973).
230. Wilson, R. G., *J. Appl. Phys.* **44**, 5056 (1973).
231. Schumacher, E., and Taubenest, R., *Helv. Chim. Acta* **47**, 1525 (1964).
232. Mendelbaum, A., and Cais, M., *Tetrahedron Lett.* p. 3847 (1964).
233. Egger, H., *Monatsh. Chem.* **97**, 602 (1966).
234. Müller, J., *J. Organometal. Chem.* **23**, C38 (1970).
235. Kingston, D. G. I., Tannenbaum, H. P., and Kuivila, H. G., *Org. Mass Spectrom.* **9**, 31 (1974).
236. Glockling, F., McBride, T., and Pollock, R. J. I., *Inorg. Chim. Acta* **8**, 81 (1974).
237. Miller, J. M., Thesis, p. 132, Cambridge University (1966).
238. Melcher, L. A., Adcock, J. L., Anderson, G. H., and Lagowski, J. J., *Inorg. Chem.* **12**, 601 (1973).
239. Miller, J. M., and Wilson, G. L., *Inorg. Chem.* **13**, 498 (1974).
240. McLaughlin, E., and Rozett, R. W., *J. Organometal. Chem.* **52**, 261 (1973).
241. McLaughlin, E., Hall, L. H., and Rozett, R. W., *J. Phys. Chem.* **77**, 2984 (1973).
242. Crawford, L. R., *Int. J. Mass Spectrom. Ion Phys.* **10**, 279 (1972).

243. McLaughlin, E., and Rozett, R. W., *Inorg. Chem.* **11**, 2567 (1972).
244. Pignataro, S., Torroni, S., Innorta, G., and Foffani, A., *Gazz. Chim. Ital.* **104**, 97 (1974).
245. MacNeil, K. A. G., and Thynne, J. C. J., *J. Phys. Chem.* **74**, 2257 (1970).
246. Stockdale, J. A., Nelson, D. R., Davis, F. J., and Compton, R. N., *J. Chem. Phys.* **56**, 3336 (1972).
247. MacNeil, K. A. G., and Thynne, J. C. J., *Int. J. Mass. Spectrom. Ion Phys.* **3**, 455 (1970).
248. Craddock, S., Harland, P. W., and Thynne, J. C. J., *Inorg. Nucl. Chem. Lett.* **6**, 425 (1970).
249. Brion, C. E., *Int. J. Mass Spectrom. Ion. Phys.* **3**, 197 (1969).
250. Pignataro, S., Foffani, A., Grasso, F., and Cantone, B., *Z. Phys. Chem. (Frankfurt am Main)* [N.S.] **47**, 106 (1965).
251. Foffani, A., Pignataro, S., Cantone, B., and Grasso, F., *Z. Phys. Chem. (Frankfurt am Main)* [N.S.] **45**, 79 (1965).
252. Foffani, A., Pignataro, S., Distefano, G., and Innorta, G., *J. Organometal. Chem.* **7**, 473 (1967).
253. Pignataro, S., Distefano, G., Nencini, G., and Foffani, A., *Int. J. Mass Spectrom. Ion Phys.* **3**, 479 (1970).
254. Bonati, F., Distefano, G., Innorta, G., Minghetti, G., and Pignataro, S., *Z. Anorg. Chem.* **386**, 107 (1971).
255. Kiser, R. W., in "Recent Developments in Mass Spectroscopy" (K. Ogata and I. Hayakawa, eds.), p. 844. Univ. of Tokyo Press, Tokyo, 1970.
256. Sullivan, R. E., and Kiser, R. W., *J. Chem. Phys.* **49**, 1978 (1968).
257. Cohen, S. C., *Inorg. Nucl. Chem. Lett.* **6**, 757 (1970).
258. Pant, B. C., and Sacher, R. E., *Inorg. Nucl. Chem. Lett.* **5**, 549 (1969).
259. Onak, T., Howard, J., and Brown, C., *J. Chem. Soc., Dalton Trans.* p. 76 (1973).
260. Fraser, I. W., Garnett, J. L., and Gregor, I. K., *J. Chem. Soc., Chem. Commun.* p. 365 (1974).
261. Cohen, S. C., and Massey, A. G., *J. Organometal. Chem.* **10**, 471 (1967).
262. Smith, V. B., and Massey, A. G., *Tetrahedron* **25**, 5495 (1969).
263. Clinton, N. A., and Kochi, J. K., *J. Organometal. Chem.* **56**, 243 (1973).
264. Fales, H. M., *Anal. Chem.* **38**, 1058 (1966).
265. PCR Inc., Gainesville, Florida.
266. Compson, K. R., Aspinal, M. L., Elliott, R. M., and Wolstenholme, W. A., *Amer. Soc. Mass Spectrom., 22nd Annu. Conf.* Paper A-1 (1974).
267. Sprecher, R. F., Aufderheide, B. E., Luther, G. W., III, and Carter, J. C., *J. Amer. Chem. Soc.* **96**, 4404 (1974).
268. Jackels, S. C., and Rose, N. J., *Inorg. Chem.* **12**, 1232 (1973).
268a. Boston, D. R., and Rose, N. J., *J. Amer. Chem. Soc.* **95**, 4163 (1973).
269. Haegele, K. D., Schurig, V., and Desiderio, D. M., *Inorg. Chem.* **13**, 1960 (1974).
270. Cooks, R. G., Beynon, J. A., Caprioli, R. M., and Lester, G. R., "Metastable Ions." Elsevier, Amsterdam, 1973.
271. Beynon, J. H., *Anal. Chem.* **421**, 97A (1970).
272. Barber, M., and Elliott, R. M., *Amer. Soc. Test. Mater. Conf. Mass Spectrom., 1964* E-14 (1964).
273. Barber, M., Wolstenholme, W. A., and Jennings, K. R., *Nature (London)* **214**, 664 (1967).

274. Shannon, T. W., Mead, T. E., Warner, C. G., and McLafferty, F. W., *Anal. Chem.* **39**, 1748 (1967).
275. Beynon, J. H., Amy, J. W., and Baitinger, W. E., *Chem. Commun.* p. 723 (1969).
276. Beynon, J. H., Caprioli, R. M., Baitinger, W. E., and Amy, J. W., *Int. J. Mass Spectrom. Ion Phys.* **3**, 313 (1969).
277. Beynon, J. H., and Cooks, R. G., *Res/Develop.* p. 26 (1971).
278. Maurer, K. H., Brunee, C., Kappus, G., Habfast, K., Schroder, U., and Schulze, P., *19th Conf. Mass Spectrom., 1971* Paper K-9 (1971).
279. Wachs, T., Bente, P. F., and McLafferty, F. W., *Int. J. Mass Spectrom. Ion Phys.* **9**, 333 (1972).
280. Miller, J. M., Ross, J., Rustenburg, J., and Wilson, G. L. *Anal. Chem.* **45**, 627 (1973).
281. Miller, J. M., and Wilson, G. L. *Int. J. Mass Spectrom. Ion Phys.* **12**, 225 (1973).
282. Watanabe, E., Itagaki, Y., Aogama, T., and Yamauchi, E., *Anal. Chem.* **40**, 1000 (1968).
283. Shannon, T. W., Mead, T. E., Warner, C. G., and McLafferty, F. W., *Anal. Chem.* **39**, 1748 (1967).
284. Daly, N. R., McCormick, A., Powell, R. E., and Hayes, R., *Int. J. Mass Spectrom. Ion Phys.* **11**, 255 (1973).
285. Miller, J. M., and Wilson, G. L., *Pittsburgh Conf. Anal. Chem. Appl. Spectrosc. 1973* Paper No. 241 (1973); *Anal. Chem.* **47**, 191 (1975).
286. Shadoff, L. A., *Anal. Chem.* **39**, 1902 (1967).
287. Cooks, R. G., Beynon, J. H., and Ast, T., *J. Amer. Chem. Soc.* **94**, 1004 (1972).
288. Sweeley, C. C., Bentley, R., Makita, M., and Wells, W. E., *J. Amer. Chem. Soc.* **85**, 2497 (1963).
289. "Handbook of Silylation," GPA-3a. Pierce Chemical Co., Rockford, Illinois, 1972.
290. Sharkey, A. G., Friedel, R. A., and Langer, S. H., *Anal. Chem.* **29**, 770 (1957).
291. De Jongh, D. C., Radford, T., Hribar, J. D., Hannessian, S., Bieber, M., Dawson, G., and Sweeley, C. C., *J. Amer. Chem. Soc.* **91**, 1728 (1969).
292. Watson, J. T., in "Ancillary Techniques of Gas Chromatography" (L. S. Ettre and W. H. McFadden, eds.), Chapter 5, p. 157. Wiley (Interscience), New York, 1969.
293. Baughman, G. L., Carter, M. H., Wolf, N. L., and Zepp, R. G., *J. Chromatogr.* **76**, 471 (1973), and references contained therein.
294. Ohkoshi, S., Takahashi, T., and Sato, T., *Bunseki Kagaku* **22**, 593 (1973).
295. Jones, T. R. B., Wilson, G. L., Lin, S., and Miller, J. M., unpublished observations.
296. Laubengayer, A. W., Moeus, P. C., and Porter, R. F., *J. Amer. Chem. Soc.* **83**, 1337 (1971).
297. Sowinski, E. J., and Suffet, I. H., *Anal. Chem.* **46**, 1218 (1974).
298. Blum, W., and Richter, W. J., *Finnigan Spectra* **4**, No. 1 (1974).
299. Butts, W. C., and Rainey, W. T., Jr., *Anal. Chem.* **43**, 538 (1971).
300. Bayers, E., Müller, H. P., and Sievers, R. E., *Anal. Chem.* **43**, 2012 (1971).
301. Wolf, W. R., Taylor, M. L., Hughes, B. M., Tiernan, T. O., and Sievers, R. E., *Anal. Chem.* **44**, 616 (1972).

THE STRUCTURES OF ELEMENTAL SULFUR

BEAT MEYER

Chemistry Department, University of Washington, Seattle, Washington, and
Inorganic Materials Research Division, Lawrence Berkeley Laboratory,
University of California, Berkeley, California

I. Introduction 287
  Structural Considerations 288
II. Well-Established Allotropes of Sulfur 291
  A. Molecules with Less Than 6 Atoms 291
  B. Cyclohexasulfur, Engel-Aten's Rhombohedral S_6 . . . 293
  C. Cycloheptasulfur, Schmidt's S_7 294
  D. Cyclooctasulfur 296
  E. Cycloenneasulfur, Schmidt's S_9 300
  F. Cyclodecasulfur, Schmidt's S_{10} 301
  G. Cycloundecasulfur, Schmidt's S_{11} 301
  H. Cyclododecasulfur, Schmidt-Wilhelm's S_{12} . . . 301
  I. Cyclooctadecasulfur, Schmidt's S_{18} 302
  J. Cycloicosasulfur, Schmidt's S_{20} 303
  K. Polycatenasulfur, Fibrous Sulfur, S_{∞} 304
  L. Cyclocatenasulfur: Charge-Transfer Complexes . . . 305
  M. Molecular Ions 305
  N. Rings Containing Other Elements 306
III. Incompletely Characterized Allotropes and Mixtures . . 308
  A. From Solutions of Cyclooctasulfur 311
  B. From Solutions Containing Other Sulfur Compounds . . 311
  C. From Solid Sulfur 312
  D. From Liquid Sulfur 312
  E. From Sulfur Vapor 313
  References 314

I. Introduction

Elemental sulfur is commonly known and widely used. Its properties are already described in the Old Testament and in Homer's Odyssey. It exists in several different forms.

Thermodynamically stable sulfur forms deep yellow, nonodorous orthorhombic crystals with a space group $Fddd$-D_{2h}^{24}, containing 16 molecules, i.e., 128 atoms in the unit cell. It has a density of 2.069 gm/cm³ and is well soluble in CS_2. Its molecular unit is S_8, cyclooctasulfur, a crown-shaped molecule with a symmetry of D_{4d}. The pale "flowers of sulfur," prepared by alchemists by distillation, are insoluble in CS_2, and the structure is not yet fully understood. Another form, plastic sulfur, is obtained by melting sulfur to about 180°C, where it forms as highly viscous syrup. To date, some forty-five different forms of sulfur have been described (25, 61, 62), and for twelve of these detailed structural data are available. These forms differ in intramolecular and intermolecular structure.

Sulfur can form rings and chains of various shapes and sizes. In addition to the crown-shaped cyclo-S_8, nine other types of rings can be prepared: the first of these, cyclohexasulfur, also called Engel's or Aten's sulfur, or rhombohedral sulfur, S_ρ, was first described by Engel in 1891 (28). The other eight types of rings were all discovered during the last 10 years by Schmidt and his co-workers (87, 88) notably Wilhelm (125).

Sulfur chains of uniform length cannot be made. All catenallotropes are mixtures, containing molecules of varying lengths. There is still much question whether pure polymeric chains can exist at room temperature and whether the terminal atoms of long chains are bound to impurities, to sulfur rings, forming charge-transfer complexes of catenasulfur with cyclosulfur or whether they curl up and form long chains (50).

Three polymorphs of S_8 are well established and recipes for preparing some twenty others have been published and much tried. The knowledge of sulfur allotropes up to 1965 has been reviewed in detail (61, 62). Recently, a detailed summary of much of the crystallographic work on solids, showing the history of structure determination, has been published in an excellent book by Donohue (25). Many data on the preparation, stability and structure of rings can be found in recent review papers by Schmidt (87, 88). As will be seen below, our knowledge of sulfur allotropes is still very incomplete, because the chemistry of elemental sulfur is far more complex than anyone suspected. During the last 100 years, as our knowledge and techniques of chemistry became more sophisticated, more and more riddles appeared, but it seems now that we might have reached a turning point, because we know now that many different types of sulfur rings can exist, but that they all show the basic characteristics of the S—S bond. It is noteworthy here that progress in this field was not caused by the availability of advanced technology, but based on ingenious use of chemical knowledge, thought, and intuition. In this review heavy emphasis is given to recent work.

STRUCTURAL CONSIDERATIONS

The S—S bond distances observed in various compounds range from 1.89 Å in S_2 (5) and S_2F_2 (56) to 2.20 Å in S_8O (105). The bond length depends on molecular geometry and substitution. Normally, the bond distance is close to 2.05 Å (Table I). The S—S—S bond angle lies between

TABLE I

STRUCTURAL PARAMETERS OF MOLECULES WITH S—S BOND

Molecule	S—S bond length (Å)	Ref.
S_8	2.060	*17*
S_{12}	2.053	*57*
S_∞	2.066	*105*
H_2S_2	2.0	*101*
S_2F_2	1.89	*56*
Me_2S_2	2.038	*101*
$S_3(CF_3)_2$	2.04	*101*
$S_3(Me)_2$	2.04	*101*
S_8O	2.04–2.20	*105*
Diphenyl disulfide	2.03	*82*
2,2′-Biphenyl disulfide	2.03	*82*
α-Cystine hydrochloride	2.05	*82*
α-Cystine	2.03	*82*

101° in S_8O (*105*) and 108° in monoclinic γ-sulfur (*120*). In cyclooctasulfur, it is 108°; in polymeric sulfur, it is 106°. The S—S—S—S dihedral angle can be as small as 66°, in S_{20} (*98*), or as large as 98° and S_8. Normal bond values are shown in Fig. 1a.

Experimental and theoretical ramifications of S—S bond properties observed in S_8, S_6, and polymeric sulfur have been reviewed by Pauling

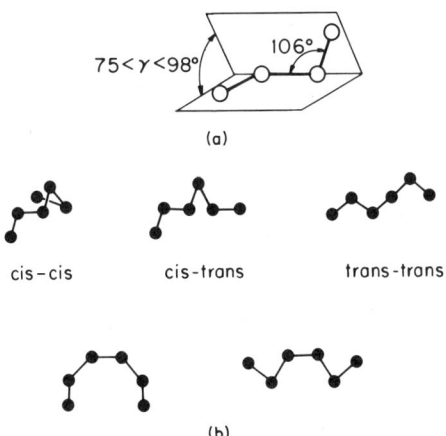

FIG. 1. (a) S—S bond parameters; (b) S—S bond configurations.

in 1949 in a paper (79) in which he elucidated much of what had been previously observed. He proposed that the S—S bond had natural, "free" values close to those found in cyclo-S_8, the thermodynamically stable form. He concluded that the stability of this molecule precluded rings of other than S_8, or a very similar size, such as S_6. Pauling's paper discounted, in advance, future claims of Skjerven (102) and others, that various experimental observations indicated the existence of large rings (61). The distrust of large rings was proven unfounded prejudice when Schmidt and Wilhelm, in 1966, demonstrated (95) simple and reliable ways for making S_{12}. The existence of S_{12}, which is amazingly stable, showed that molecules other than S_8 can fulfill the S—S bond requirements, but its structure confirmed that Pauling's assumption about bond characteristics is correct. The S_{12} structure (Fig. 2) is surprisingly simple. Since the time that S_{12} was discovered, Schmidt has prepared seven other types of rings. Bond properties and characteristics

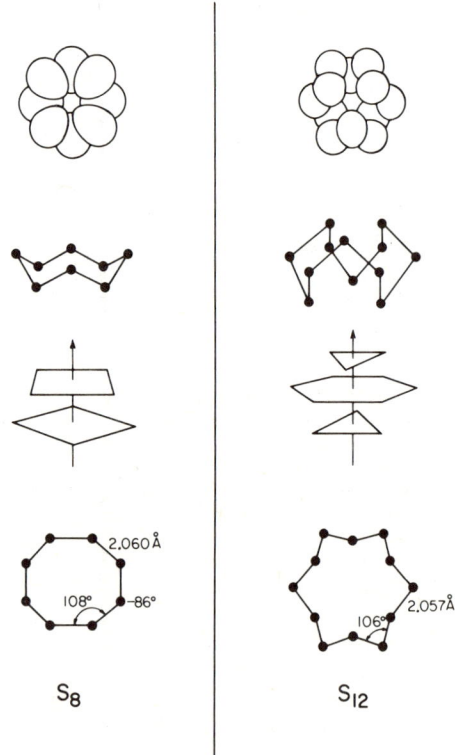

FIG. 2. Different views of cyclooctasulfur and cyclododecasulfur.

of some of these have been reviewed (88). Properties and characteristics of the S—S bond in other sulfur compounds have been discussed by Foss (34–36), and others. Reviews of structural considerations were written by Tuinstra (115) and Rahman (82).

In sulfur chains, subsequent sulfur atoms can take either cis or trans positions. Figure 1b shows three enantiomers: cis–cis, cis–trans, and trans–trans configurations. Cis–cis occurs in the S_{12} rings, cis–trans in cyclo-S_8, and trans–trans yields the helices of fibrous sulfur. Pauling discussed the structure of sulfur helices. Semlyen (101) updated and enlarged the models for the unperturbed dimensions of catenapoly sulfur. Potassium barium hexathionate crystallizes as cis–cis (35), whereas the hexathionate of trans-dichlorobis(ethylenediamine)cobalt(III) dihydrate (36) and cesium hexasulfide crystallize in trans–trans configurations (82). The electron configuration and electron density at various positions in rings and chains have been computed using various semi-empirical and ab initio models. Work by Cusachs and his group (20, 68–70) and others (65) has given convincing evidence that ground-state properties of sulfur compounds are not influenced by d-orbital participation. However, properties connected with excited states of sulfur depend on d-orbital considerations. Miller (70), Wiewiorowski (124), and Cusachs (20) have used their model to develop an acid–base model for charge-transfer complexes between rings and chains. This model provides a very useful approach to explaining the ESR spectrum (50) and other properties of viscous liquid sulfur. These calculations will be mentioned in later sections. They helped explain the color of liquid sulfur (65). The nature of sulfur–sulfur bonds in compounds and minerals has been reviewed in recent papers based on the K_α or K_β X-ray spectrum of the sulfur atom (73, 123). The thermodynamics of S—S bonds have been reviewed by Tobolski (48) and Jensen (46).

II. Well-Established Allotropes of Sulfur

A. Molecules with Less Than 6 Atoms

Owing to the S—S bond geometry, rings with less than 6 atoms are highly strained and unstable. Pure S_4 and S_5 have not yet been produced, or at least not yet identified. Chains, neutral or charged, are structurally allowed, but they have unstable electron configurations at their terminal atoms. Thus, they have a tendency to rearrange and quickly to form equilibrium mixtures of molecules, all of which are unstable at room temperature and which convert to polymeric chains and then, eventually, to cyclo-S_8. Of the small species, only S_2 is well-known: S_2 exists at a temperature above 800°C and pressures below 1 torr in high purity.

With a bond distance of 1.89 Å in the ground state, S_2 is paramagnetic, like O_2, and has a ground state $^3\Sigma_g^-$. The electron configuration of over eighteen excited states has been established by Barrow and his group (*4, 127*), who studied a large number of electronic spectra using isotopes.

Thiozone (S_3) has a ground state of $^1\Sigma$, like ozone. It is found together with S_4 in sulfur vapor at a temperature around 450°C and pressures around 1 torr. Thiozone can be prepared in low-temperature glasses and rare gas matrices, by photolysis of S_3Cl_2 (*64, 75*). The electronic structure was investigated by Spitzer (*65*).

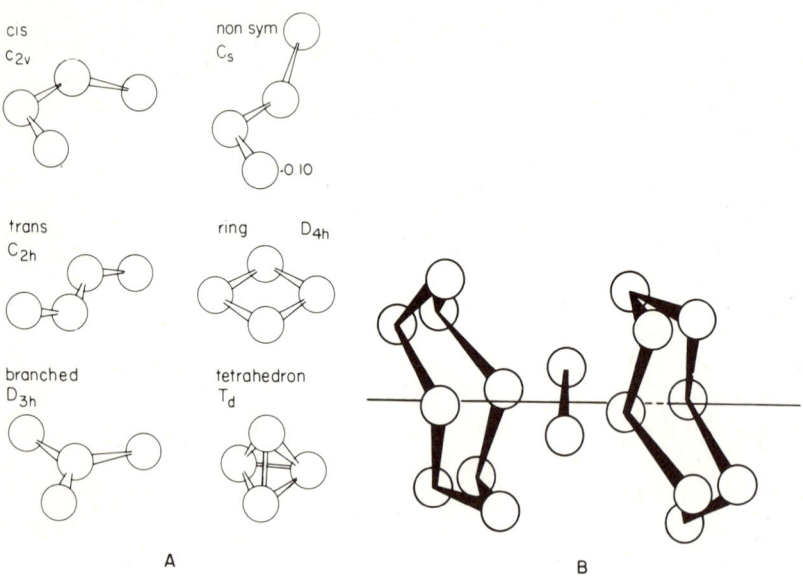

FIG. 3. (A) Isomers and conformers of S_4 (*65*); (B) Charge-transfer complex S_8–S_2–S_8 (*65*).

Species S_4 occurs in a mixture with S_3 in the gas phase, as indicated above. It can be synthesized in matrices by photolysis of S_4Cl_2 or S_2Cl_2 (*67*), or simply by recombination diffusion of S_2 (*66*). This molecule has been identified by its UV and IR spectra. A variety of structures are feasible (Fig. 3a). The planar ring is also the structure proposed for S_4^{2+} (*41*) which is considered to be aromatic in character. Calculations indicate that the branched structure, analogous to the SO_3 structure, should be most stable (*65*). The most likely structure for this very unstable molecule is the trans chain. Species S_4, as well as S_3 and S_2, have been identified by mass spectroscopy (*7*).

The S_5 also occurs in sulfur vapor (7, 24). It is not clear whether the vapor contains cyclopentasulfur or catenapentasulfur, or both. Schmidt (87) proposes that S_5 can be prepared from bis-π-cyclopentadienyl-molybdenum tetrasulfide by reaction with monosulfurdichloride:

$$(C_5H_5)_2MoS_4 + SCl_2 = S_5 + (C_5H_5)_2MoCl_2$$

The reaction product is a liquid at room temperature.

B. Cyclohexasulfur, Engel-Aten's Rhombohedral S_6

Engel (28) prepared cyclohexasulfur in 1891, and Aten (2) identified, in 1914, the rhombohedral crystals, but ingrained notions that S_8 should

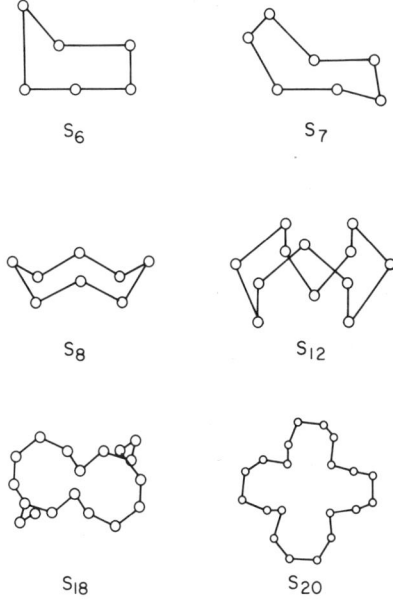

FIG. 4. Six well-established sulfur allotropes.

be the only possible molecule stifled acceptance of their work. The existence of S_6 was disbelieved or ignored by most chemists, until Frondel and Whitfield (37a), Donnay (24a), and, especially, Donohue (14), presented a series of papers and all details of the structure were determined. The molecule has a chair form (Fig. 4), and the following parameters:

$$\text{S—S bond length} = 2.057 \pm 0.018 \text{ Å}$$
$$\text{S—S—S bond angle} = 102.2 \pm 1.6°$$
$$\text{S—S—S—S torsion angle} = 74.5 \pm 2.5°$$

The substance can be crystallized from toluene or CS_2. It forms orange-red rhombohedral crystals with a density of 2.209 gm/cm^3, the highest density of any known sulfur allotrope. Eighteen S_6 molecules occupy a unit cell with the space group $R3\text{-}C_{3i}{}^2$. The lattice constants are

$$a = 10.818 \text{ Å}$$
$$c = 4.280 \text{ Å}$$
$$c/a = 0.3956$$

The crystals decompose above 50°C. Engel and Aten prepared the substance by the reaction

$$NaHS_2O_3 + HCl(conc) = S_6 + S_8 + NaCl + H_2O$$

In their quest for new methods for preparing new sulfur rings (94), Wilhelm and Schmidt (125) studied the simultaneous addition of sulfane and chlorosulfane to cold dry ether,

$$S_2Cl_2 + H_2S_4 = S_6 + 2HCl$$

and succeeded in obtaining S_6 in 87% yield. This synthesis is the basis for the preparation of S_{12}, and other hitherto inaccessible, molecules. Cyclohexasulfur is sensitive to light (it decomposes, leaving S_8 and small amounts of S_{12}), and reacts about 10^4 times faster than S_8 with nucleophilic agents. This molecule exists also in sulfur vapor. Its mass spectrum and IR spectrum have been investigated by Berkowitz (6) and others (74). The packing of S_6, which is extremely efficient, as demonstrated by the density referred to above, is beautifully visualized in the reference by Donohue (25).

C. Cycloheptasulfur, Schmidt's S_7

Schmidt (90) succeeded in 1968 in preparing S_7, the first sulfur ring with an odd number of atoms. The synthetic path employed for preparing S_6 and S_{12} does not provide high yields, thus a new method had to be found. The reaction

$$(C_5H_5)_2TiS_5 + S_2Cl_2 = S_7 + (C_5H_5)_2TiCl_2$$

was made possible after synthesis of the cyclopentadienyl titanium pentasulfide became available (52, 53), as the result of work based on experiences gained with $(NH_4)_2PtS_{15}$, a compound which was first described by Hoffmann *(44)* in 1903. In these compounds, and in the corresponding MoS_4 derivative *(51)*, the sulfur atoms form rings in which 1 sulfur atom is replaced by the central metal ion (Fig. 5). The S_7 molecules have the structure *(47)* shown in Fig. 4: 4 sulfur atoms lie on one plane; the IR spectrum indicates that the S—S bond is unequal in different positions *(38)*. The existence of the compound was confirmed by mass spectroscopy *(128)*. Cycloheptasulfur forms light yellow needles

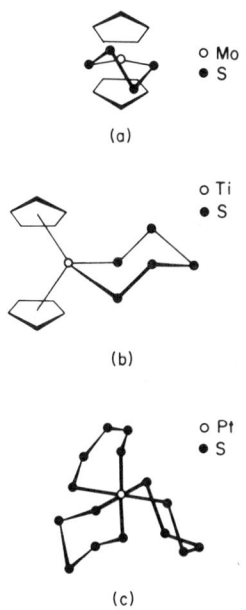

FIG. 5. Molecular structures of (a) $(C_5H_5)_2MoS_4$; (b) $(C_5H_5)_2TiS_5$; and (c) $(NH_4)_2PtI_5$.

with a density of $d = 2.090$ gm/cm³. The lattice constants of this unstable compound had to be determined at $-80°C$ *(47)*:

$$a = 21.77 \text{ Å}$$
$$b = 20.97 \text{ Å}$$
$$c = 6.09 \text{ Å}$$
$$\alpha = \beta = \gamma = 90°$$

The space group is not yet known. Sixteen molecules, i.e., 112 atoms, occupy the unit cell. The molecule decomposes at 39°C. It must be stored in the dark and at a low temperature.

D. CYCLOOCTASULFUR

The crown-shaped cyclooctasulfur molecule has been thoroughly studied. The distances and angles within the molecule are (25)

$$S—S \text{ bond length} = 2.060 \pm 0.003 \text{ Å}$$
$$S—S—S \text{ bond angle} = 108.0° \pm 0.7°$$
$$S—S—S—S \text{ torsion angle} = 98.3° \pm 2.1°$$

The molecule was first studied in orthorhombic solid sulfur by Bragg (10) in 1914. Its properties have been well-reviewed (25). It is remarkable

TABLE II

STRUCTURAL PARAMETERS OF ELEMENTAL SULFUR MOLECULES

Molecule	S—S bond length (Å)	S—S—S bond angle (deg)	S—S—S—S torsion angle (deg)	Ref.
S_2	1.890			4
$S_{8\alpha}$	2.060 ± 0.003	108.0 ± 0.7	98.3 ± 2.1	17
S_{12}	2.053 ± 0.007	106.5 ± 1.4	86.1 ± 5.5	57
S_{18}	2.059	106.3	84.4	22, 98
S_{20}	2.047	106.5	83.0	98
S_∞	2.066	106.0	85.3	39, 59
$S_{8-\text{ion}}$	2.04	—	—	41
S_8O	(1.483 : S—O)	—	—	105
	2.20 (bond 1,8 and 1,2)	101.8	101.4	—
	2.04	108	—	—

that the light sensitivity of this molecule, which remains one of the riddles of elemental sulfur, has not been investigated. There is not one report on experimental work on the determination of the lowest triplet state, which very likely accounts for the dissociation of the ring in room light, a process which probably accounts for the formation of many of the poorly defined allotropes of sulfur. The symmetry of an isolated molecule is D_{4h} as shown in Fig. 2. In orthorhombic sulfur, the symmetry is $\bar{8}2m–D_{4d}$ (109). The above-mentioned bond data are compared with that of other sulfur species in Table II. The S_8 data are carefully reviewed by Donohue (25).

Three of the polymorphs of cyclooctasulfur are firmly established: orthorhombic α-sulfur, monoclinic β-sulfur, and monoclinic γ-sulfur. Their structure, along different axes, is shown in Fig. 6. Over a dozen of

1. Orthorhombic α-Sulfur

The first accurate structure was presented by Abrahams (1). It was refined by Caron and Donohue (15), Pawley and Rinaldi (81) restudied it and confirmed the earlier work (72). The most precise lattice constants are those of Cooper (17), computed for 24.8°C:

$$a = 10.4646 \text{ Å}$$
$$b = 12.8660 \text{ Å}$$
$$c = 24.4860 \text{ Å}$$
$$\alpha = \beta = \gamma = 90°$$

The space group is $Fddd\text{-}D_{2h}^{24}$; the unit cell contains 16 molecules, i.e. 128 atoms, and the density is 2.069 gm/cm³. The stacking of molecules has been explained by Donohue (15). The molecules are not stacked along an axis, but follow a "crankshaft" arrangement. Thus, projections along the axes are quite involved (26) (see Fig. 6a).

Since α-sulfur converts into monoclinic β-sulfur upon heating, one cannot grow single crystals from the melt. Very recently a method was described for growing very perfect single crystals from CS_2 solution. It is claimed that they are "almost free" of solvent (43). This claim is in conflict with the observation that CS_2 forms a stable, well-defined inclusion in α-sulfur (62). However, for many purposes the method might provide ideal single crystals. In studying reports of the IR and Raman spectra of α-sulfur (76, 77, 109, 118), it should be remembered that a strong band, believed to be a fundamental of S_8, was identified as belonging to CS_2 (103). Thermal analysis shows that single crystals of sulfur do not convert to S_β, even after an hour, but melt at 112°C (19).

2. Monoclinic β-Sulfur

The first structure determination of β-monoclinic sulfur, which forms from β-orthorhombic by phase transition at 94.2°C, was performed by Trillat and Forestier (113). Burwell (13) and Sands (85) refined the structure:

$$a = 10.778 \text{ Å}$$
$$b = 10.844 \text{ Å}$$
$$c = 10.924 \text{ Å}$$
$$\beta = 95.80°$$

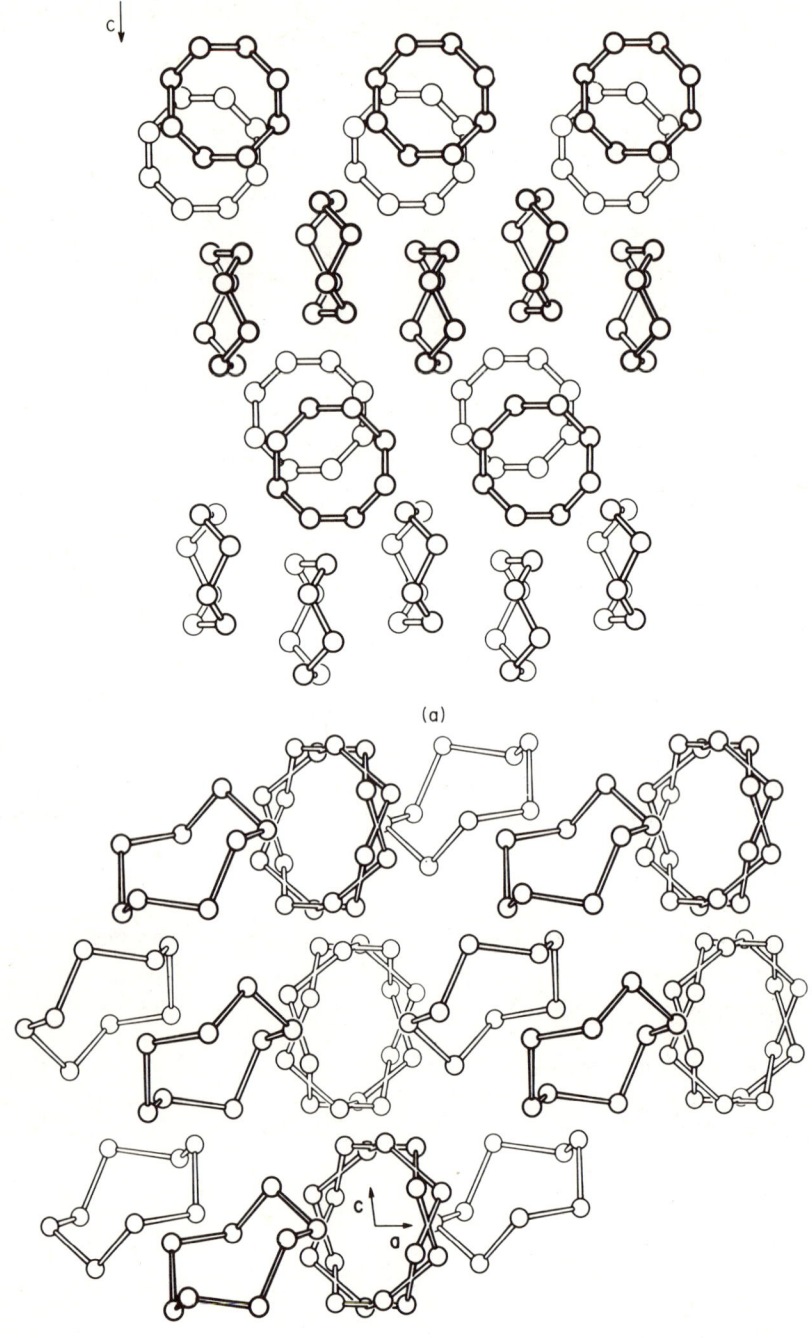

(a)

(b)

(c)

Fig. 6. Structures of (a) S_α; (b) S_β; and (c) S_γ.

The density of this form is 1.94 gm/cm^3. Six S_8 molecules, i.e., 48 atoms, occupy the unit cell; the space group is $P2_1/a-C_{2h}{}^5$. The ideal melting point has been recently computed to be 133°C (92), but the crystals melt around 128° because of decomposition of the S_8 ring, and subsequent melting point depression by solution of the resulting rings and chains.

There has been much controversy over the structure of β-sulfur, and the question of whether it is a true allotrope. It has been suggested that it constitutes merely a thermally distorted lattice expansion of orthorhombic sulfur. Furthermore, phase transition, at 101°C, has been described by various authors (32), but it has been shown that this effect was due to traces of water in the lattice (65). However, recently a true anomaly in the heat capacity has been found (71) at −75°C.

3. Monoclinic γ-Sulfur; Muthmann's Sulfur (III)

Until very recently, the existence of γ-sulfur as a pure allotrope of cyclooctasulfur remained in doubt. Despite the fact that it was first described by Muthmann (72) in 1890, it was very difficult to prepare until 1974, when Watanabe (120) reported a new method. It forms best not from cyclo-S_8 solutions, but from cuprous ethylxanthate which decomposes upon dissolving in pyridine, leaving a brownish solution from which the light yellow γ-sulfur needles crystallize. They decompose when dried. Watanabe (120) determined the molecular constants of S_8 in monoclinic γ-sulfur and found them to be very similar to those of orthorhombic α-sulfur. The lattice constants are:

$$a = 8.442 \text{ Å}$$
$$b = 13.025 \text{ Å}$$
$$c = 9.356 \text{ Å}$$
$$\beta = 124°98'$$

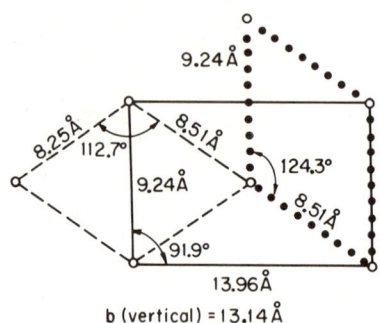

FIG. 7. Correlation between different unit cells of S_γ.

Four S_8 molecules, i.e., 32 atoms occupy one unit cell; the space group is $P2/c$, the density is 2.19 gm/cm^3, i.e., larger than that of either α- or β-sulfur. Thus, γ-sulfur contains more efficiently packed cyclooctasulfur than any other crystal. The crystals decompose upon standing.

The discussion of the structure of this formerly elusive sulfur form had been long hampered by three different choices of axes for labeling. Figure 7 shows the correlation between the three different unit cells. The molecular packing proved to be that proposed by deHaan (23). He called it a "sheared penny roll" structure, it is shown in Fig. 6c.

E. CYCLOENNEASULFUR, SCHMIDT'S S_9

Schmidt and Wilhelm in 1970 prepared cycloenneasulfur (96) by the reaction:

$$(C_5H_5)_2TiS_5 + S_4Cl_4 = (C_5H_5)_2TiCl_2 + S_9$$

The molecules are apparently of comparable stability with S_6, but the structure has not yet been published. The substance forms deep yellow needles.

F. Cyclodecasulfur, Schmidt's S_{10}

Wilhelm and Schmidt prepared lightly yellow–green solids, containing S_{10} rings by reactions of various matching combinations of sulfanes and chlorosulfanes (94, 125), e.g.,

$$H_2S_6 + S_4Cl_2 = S_{10} + 2HCl$$

The yellow S_{10} has since been prepared in far better yield (90) according to:

$$(C_5H_5)_2TiS_5 + 2SO_2Cl_2 = S_{10} + 2SO_2 + (C_5H_5)_2TiCl_2$$

At $-78°C$, 35% yield can be obtained. The compound must be stored at $-40°C$. The structure has not yet been published; the mass spectrum has been difficult to obtain (128).

G. Cycloundecasulfur, Schmidt's S_{11}

Schmidt and Wilhelm prepared this compound by the reaction

$$(C_5H_5)_2TiS_5 + S_6Cl_2 = (C_5H_5)_2TiCl_2 + S_{11}$$

Details of the properties and structure have not yet been released (97). The structure of the titanium–sulfur compound is shown in Fig. 5b.

H. Cyclododecasulfur, Schmidt-Wilhelm's S_{12}

Publications of Schmidt and Wilhelm's careful work (125) on new synthetic methods for preparing sulfur rings opened a new chapter in our knowledge of elemental sulfur in 1966. Cyclododecasulfur (S_{12}) was discovered 75 years after S_6. It is the third elemental sulfur ring discovered. Its preparation and identification was followed in the 7 subsequent years by the synthesis of another six new types of rings by the same group. The existence of S_{12} shows that large rings can exist and are far more stable than anyone would have thought.

The S_{12} is prepared by reaction of sulfanes with chlorosulfanes (95):

$$H_2S_4 + S_2Cl_2 = S_6(50\%) + S_{12}(3\%) + 2HCl$$

or (93)

$$H_2S_8 + S_4Cl_2 = S_{12}(18\%) + 2HCl$$

It is amazingly stable. Apparently, S_{12} exists in liquid sulfur, because it can be found in quenched melts (91), and it is formed in solutions of S_6 in toluene during decomposition under the influence of light (92). The vapor is, however, unstable; thus, mass spectroscopic identification proves to be difficult (12). The molecules have $2/m$–C_{2h} symmetry in the crystal, but are close to $3m$–D_{3d}. Table II shows that the bond properties lie between those of S_6 and S_8:

$$S\text{—}S \text{ bond length} = 2.053 \pm 0.007 \text{ Å}$$
$$S\text{—}S\text{—}S \text{ bond angle} = 106.5 \pm 1.4°$$
$$S\text{—}S\text{—}S\text{—}S \text{ torsion angle} = 86.1 \pm 5.5°$$

The atoms are stacked in three parallel planes (see Fig. 2) yielding a highly symmetric molecule which forms such a stable solid that the melting point is 148°C, i.e., almost 20° higher than that of orthorhombic cyclooctasulfur. Separation of S_{12} from S_8 is aided by the low solubility of the first: S_{12} is about 150 times less soluble (93). The lattice constants are (57):

$$a = 4.730 \text{ Å}$$
$$b = 9.104 \text{ Å}$$
$$c = 14.7574 \text{ Å}$$

The unit cell contains two molecules, i.e., 24 atoms. The density is 2.036 gm/cm³. The space group is $Pnmm$–D_{2h}^{12}.

A comparison of the stability of different sulfur allotropes will be made possible with S_{12}. The order of reactivity toward diphenyl-o-tolylphosphine is

$$S_6 > S_7 > S_{12} > S_8$$

A recent report deals with the IR spectrum (107) of S_{12} and confirms that very pure crystals free of S_6 and S_8 can be prepared.

I. Cyclooctadecasulfur, Schmidt's S_{18}

Very recently, details of the preparation, properties, and structure of S_{18} have become available (22, 98). It is best prepared by the reaction:

$$H_2S_8 + S_{10}Cl_2 = S_{18} + 2HCl$$

As S_8H_2 and $S_{10}Cl_2$ are both not easily prepared in the pure state, the starting materials are specially designed mixtures containing impurities of the correct chain length. The lemon-yellow crystals are separated from S_{20} by recrystallization from CS_2. The solubility of S_{18} is 240 mg/

100 ml CS_2 at 20°C. This molecule is unexpectedly stable. It melts at 128°, and, if stored in the dark, displays an unchanged X-ray diffraction pattern after 5 years. The bond parameters are (22, 98):

$$S-S \text{ bond length} = 2.059 \text{ Å}$$
$$S-S-S \text{ bond angle} = 106.3°$$
$$S-S-S-S \text{ torsion angle} = 84.4°$$

The bond data are very similar to those of fibrous sulfur, and the values are intermediate between those of S_6 and S_8, and similar to those of S_{12}. The lattice constants are

$$a = 21.152 \text{ Å}$$
$$b = 11.441 \text{ Å}$$
$$c = 7.581 \text{ Å}$$

Four molecules, i.e., 72 atoms, are contained in a unit cell. The density is 2.090 gm/cm³, the space group is $P2_12_12_1$.

The discovery of this large ring will undoubtedly lead to reexamination of the properties of melts around the melting point: Krebs and many others proposed long ago that liquid sulfur contained a large ring, S_π, which was to be responsible for the melting point depression during melting.

J. CYCLOICOSASULFUR, SCHMIDT'S S_{20}

Schmidt synthesized S_{20} from sulfanes and chlorosulfanes (98) by using carefully designed mixtures of compounds and catalyzing the reaction with HCl.

$$H_2S_{10} + S_{10}Cl_2 = S_{20} + 2HCl$$

The S_{20} was described in a very recent paper. It melts at 124°C, but in solution, where it is more soluble than S_{18}, the molecule decomposes already at 35°C. In this molecule, 4 atoms are each in a plane (see Fig. 4). The density is the lowest of any known allotrope, $d = 2.016$ gm/cm³. The pale yellow crystals have the following lattice parameters (98):

$$a = 18.580 \text{ Å}$$
$$b = 13.181 \text{ Å}$$
$$c = 8.600 \text{ Å}$$

Four molecules, i.e., 80 atoms fill the unit cell. The space group is *Pbcn*. The bond characteristics are similar to those of S_{18}, S_{12}, and $S_{fibrous}$, and lie between the values for S_6 and S_8:

$$\text{S—S bond length} = 2.047 \text{ Å}$$
$$\text{S—S—S bond angle} = 106.5°$$
$$\text{S—S—S—S torsion angle} = 83.0°$$

K. Polycatenasulfur, Fibrous Sulfur, S_∞

Polycatenasulfur is also called fibrous sulfur, polymeric sulfur, plastic sulfur, or S_∞. It is formed by quenching a viscous sulfur melt. Liquid polymeric sulfur is a kinetic equilibrium mixture containing a very large number of molecules with an average chain length of up to 10^5. The chain length depends on temperature, and on the concentration of rings. Schmidt (*87*) has shown that 2% S_6 suffice to induce polymerization at 150°C, i.e., 10° below the normal temperature. He also demonstrated that stable sulfur-containing rings, added at 200°C, can reduce the average chain length drastically.

Crystalline samples are produced by stretching polymeric sulfur during or after chilling (*25*). Solid samples contain always other sulfur allotropes, among them rings. It is now believed that the long sulfur helices contain 10 atoms for every three turns. The best presently available bond data (*25*) are

$$\text{S—S bond length} = 2.066 \text{ Å}$$
$$\text{S—S—S bond angle} = 106°$$
$$\text{S—S—S—S torsion angle} = 85.3°$$

These values are often taken as the "free" sulfur–sulfur bond characteristics (*101*). The recent preparation of rings with 12, 18, and 20 atoms has shown that the bond angle and torsion angle given above are closely observed in all these compounds. The first structure determinations were authored by Trillat and Forestier in 1931 (*113*), and Meyer and Go (*67a*) in 1934. The present data were provided by Donohue (*27*), Tuinstra (*114, 115*), and Geller (*39, 59*). The structure is not as certain as that for the above-mentioned rings. The lattice constants are

$$a = 13.8 \text{ Å}$$
$$b = 4 \times 8.10 \text{ Å}$$
$$c = 9.25 \text{ Å}$$
$$\gamma = 85.3°$$

One hundred sixty atoms fill a unit cell that has the space group $Ccm2_1$-$C_{2v}{}^{12}$. The density is $d = 2.01$ gm/cm^3.

As mentioned, freshly drawn fibrous sulfur is a mixture, containing S_8 and possibly other rings. Geller prepared a substance with an identical X-ray pattern by applying a pressure of 27 kbar (*57*).

The arrangement of left- and right-turn helices and the overall structure still remain to be elucidated.

Fibrous sulfur is probably rarely pure. Impurities can influence the structure in many ways. As long as only mixtures are available, further, apparently contradictory reports on similar structures are to be expected. We discuss some of these forms in Section III.

L. CYCLOCATENASULFUR: CHARGE-TRANSFER COMPLEXES

Charge-transfer complexes between cyclooctasulfur and other compounds have been proposed for some time. Iodine complexes were described by Jander (45) and Meyer (65) and complexes with CHI_3 by Bjorvatten (8). Wiewiorowski proposed a complex of sulfur with itself, cyclo-S_8–catena-S_8–cyclo-S_8 (124), in order to explain the low ESR intensity of liquid sulfur (50). Figure 3b shows a charge-transfer complex of cyclosulfur with catenasulfur. Semiempirical Hückel calculations indicate that such complexes are stable (65). The recent work of Koningsberger (50) has lent support to this postulate. Schmidt (87) discovered that 2% S_6, added to sulfur at 150°C causes polymerization lasting 20 min, and that sulfur rings, e.g., 2% trithiane or trimeric thioacetaldehyde (which are not known to react, as they can be recovered unchanged even at 200°C), reduce the viscosity. This also supports the belief that rings and chains interact to form a charge-transfer complex, probably containing two rings and a chain. It is also possible for a chain to complex both sides of a ring. The stability of such complexes is not yet known. It may be that photosulfur, prepared by illuminating solutions of cyclooctasulfur in benzene or toluene, consists of such a substance. Photosulfur is insoluble and, in the dark, slowly converts to cyclooctasulfur (61, 62).

M. MOLECULAR IONS

Negative ions of sulfur exist in aqueous polysulfide solutions (112a) and nonaqueous solutions (16). The structure of these ions is presumed to be similar to that of Feher's sulfanes (87) and their corresponding salts. Figure 1 shows that in these compounds cis–cis or trans–trans configurations are possible. Some bond data are given in Table II. Typical values are (101)

$$\text{S—S bond length} = 2.048 \text{ Å}$$
$$\text{S—S—S bond angle} = 107°55'$$
$$\text{S—S—S—S torsion angle} = 90°$$

Positive ions of sulfur have been identified by Gillespie (*41, 42*) and others in concentrated acids, in minerals (*99, 100*), and in doped salts (*40*). Three ions, S_4^{2+}, S_8^{2+}, and S_{16}^{2+} (*41*) seem to exist. The S_4^{2+} is pale yellow, diamagnetic, and, as X-ray studies indicate, very likely planar, with aromatic character. Solutions containing S_8^{2+} are deep blue. The structure of the S_8^{2+} ring is shown in Fig. 8c. It is similar to cyclooctasulfur, but the crown-structure, exo-exo is changed to exo-endo. The nature of various singly charged ions S_n^+, with $4 < n < 8$ is not yet understood. An excellent review of recent work is given by Gillespie (*41*).

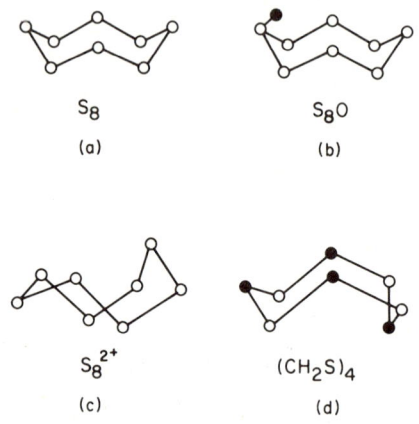

FIG. 8. Eight-membered sulfur rings: (a) S_8; (b) S_8O; (c) S_8^{2+}; and (d) $(CH_2S)_4$.

N. RINGS CONTAINING OTHER ELEMENTS

This is not the place for a review of sulfur-containing rings. We mention merely some selected species which have been described in the recent literature.

1. S_8O

By using the synthetic method of Feher modified by Schmidt and Wilhelm, Steudel and Rentsch (*106*) prepared a sulfur ring S_8O (Fig. 8b) by the reaction:

$$H_2S_n + SOCl_2 = S_8O + 2HCl$$

The S_8O is stable if stored at low temperature. This molecule contains an oxygen attached to the normal cyclooctasulfur molecule. The branching of the sulfur ring is quite surprising. The presence of the oxygen, as expected, influences the bonding: the S—S bond distance in neighboring S atoms increases to 2.20 Å, the highest yet known value, indicating a greatly weakened bond. The average S—S bond distance is 2.04. The bond characteristics (105) are

S—S bond length = 2.04 Å (2.20 Å for bonds 1, 8 and 1, 2)
S—S—S bond angle = 108° (101° for bonds 1, 8 and 1, 2)
S—S—S—S torsion angle = 101°

The lattice constants (105) of the yellow crystals, with space group $Pca2$ and density $d = 2.13$ gm/cm^3 are

$$a = 13.197 \text{ Å}$$
$$b = 7.973 \text{ Å}$$
$$c = 8.096 \text{ Å}$$

The substance decomposes above room temperature, but the IR and Raman spectra have been reported (108).

2. Selenium–Sulfur Rings

Cooper (18) prepared mixed S_nSe_{8-n} rings, as did Schmidt and others (65); forty isomers are possible. Compound S_7Se has been described (121) together with a lemon-yellow S_6Se_2 and an orange-red S_5Se_2 seven-ring (122). Synthesis proceeds as follows:

$$H_2S_n + SeCl_4 = Se_nS_m + 4HCl$$

In this way, mixed crystals containing both S_{12} and Se_2S_{10} were prepared (121). Their lattice parameters are

$$a = 4.774 \text{ Å}$$
$$b = 9.193 \text{ Å}$$
$$c = 14.68 \text{ Å}$$

These values are similar to those of S_{12}. The space group is $Pnmmm$-D_{2h}^{12}. Bond data are as follows:

S—S bond distance = 2.10 Å
S—S—S bond angle = 106°
S—S—S—S torsion angle = ~100°

3. Tellurium–Sulfur Rings

Weiss (122) prepared S_7TeCl_2, using the following reaction:

$$H_2S_n + TeCl_4 = S_7TeCl_2 + 2HCl$$

The orange crystals decompose at 95°C and melt at 110°C. The tellurium atom is coordinated as a distorted bipyramid, with Cl occupying the apices of the pyramid.

4. Organic Compounds

A variety of rings of formula $C_nH_mS_x$, containing S—S bonds, can be prepared. Several seven-, eight-, nine-, ten-, eleven-, and twelve-membered rings containing 3–8 sulfur atoms, as polysulfide, have been prepared (63) by the reaction:

$$(CH_2)_n - (SH)_2 + S_mCl_2 = (CH_2)_n - S_{m+2} \text{ ring}$$

Fehér (33) used a similar technique to make pentathiepin, benzopentathiepin, and similar compounds (125).

Of further interest are eight-membered rings containing alternating S—C bonds. Frank and Degen (37) published the structure of 1,3,5,7-tetrahiocane, $(CH_2S)_4$, which is shown in Fig. 8d. Its lattice constants are

$$a = 20.340 \text{ Å}$$
$$b = 8.747 \text{ Å}$$
$$c = 13.466 \text{ Å}$$

The unit cell contains 12 molecules in three crystallographically independent sites. The space group is $P2_1/c$. The bond characteristics are

$$\text{S—C bond length} = 1.817 \text{ Å}$$
$$\text{C—S—C bond angle} = 103°$$
$$\text{S—C—S bond angle} = 119°$$

Torsion angles are about 80°, except for C(1)–S(2) and S(8)–C(1) = 109° and for C(3)–S(4) and S(6)–C(7) = 49°. Schmidt prepared thiaformaldehyde (89) and other cyclic compounds during his studies of sulfur polymers (86). The structures of all allotropes are compared in Table III.

III. Incompletely Characterized Allotropes and Mixtures

In Section II, the structures of fourteen well-defined allotropes have been described. About forty other allotropes have been reported, but for these structural data are incomplete or contradictory. The reason for this varies. Some allotropes, such as τ-sulfur that forms only in 2 out of 5000 experiments (29), are hard to prepare or are unstable. Others, such as ω-sulfur are easy to prepare and reasonably stable, but X-ray data are contradictory and have led to much confusion and controversy.

THE STRUCTURES OF ELEMENTAL SULFUR

TABLE III
Structural Data for Solid Sulfur Allotropes

Molecule	Color	Unit cell (Å) a	Unit cell (Å) b	Unit cell (Å) c	Angle β (deg)	Space cell[a]	Space group	Density (gm/cm³)	Decomp. or melting pt. (°C)	Ref.
S_6	Orange-red	10.818	(c/a = 0.3956)	4.280 ± 0.001	—	3; 18	$R\bar{3}-C_{3i}^2$	2.209	50–60	14
S_7	Yellow	21.77	20.97	6.09	—	16; 112	?	2.090	39	47
$S_{8-\alpha}$	Yellow	10.4646	12.8660	24.4860	—	16; 128	$Fddd-D_{2h}^{24}$	2.069	94(112)	17
$S_{8-\beta}$	Yellow	10.778	10.844	10.924	95.80	6; 48	$P2_1/a-C_{2h}^5$	1.94	133	85
$S_{8-\gamma}$	Light yellow	8.442	13.025	9.356	124.98	4; 32	$P2_1/c-C_{2h}^4$	2.19	~20	120
S_{12}	Pale yellow	4.730	9.104	14.574	—	2; 24	$Pnnm-D_{2h}^{12}$	2.036	148	57
S_{18}	Lemon-yellow	21.152	11.441	7.581	—	4; 72	$P2_12_12_1-D_2^4$	2.090	128	22, 98
S_{20}	Pale yellow	18.580	13.181	8.600	—	4; 80	$Pbcn$	2.016	124–125	98
S_∞	Yellow	13.8	4 × 8.10	9.25	85.3	160[b]	$Ccm2_1-C_{2v}^{12}$	2.01	104	59
S_8O	Yellow	13.197	7.973	8.096	—	4; 32	$Pca2-C_{2v}^5$	2.13	20–78	105
S_7TeO_2	Orange	8.82	9.01	13.28	—	4; 28	$Pmnb-D_{2h}^{16}$	2.65	—	122

[a] First number = number of molecules; second number = number of atoms in unit cell.
[b] Ten atoms for three turns.

The reason for the lack of progress on this allotrope, which was first reported in 1939, and on which over a dozen publications have appeared, is often misunderstood. The difficulty is intrinsic. The existence of a great variety of intramolecular and intermolecular allotropes proves that many species must have comparable stability, i.e., that they can coexist. Thus, many contradictory allotropes are mixtures. We know now that the challenge of preparing and identifying a new allotrope does not end with its synthesis, but includes isolation of the metastable species from a mixture of equally metastable compounds, a process that

TABLE IV

Summary of Allotropes

Reagent	Product		
	Molecular species	Well-established	Inconclusive or mixture
Cyclo-S_8 solution	Cyclo-S_8	$\alpha, \beta, \gamma,$	$\gamma, \delta, \xi, \psi, \mu, \psi, \eta,$ $o, \chi, \kappa, \zeta, \theta, \tau, \psi$
Sulfur compounds in solution	Cyclo-(S_5), S_6, S_7, S_8, S_9, S_{10}, S_{11}, S_{12}, S_{18}, S_{20}, S_∞	S_6, S_7, S_8, S_9, S_{10}, S_{11}, S_{12}, S_{18}, S_{20}, S_∞	$\epsilon, \nu, \omega,$ red, orange
Solid sulfur	Cyclo-S_8 catena-S_∞	S_ψ	Laminar, ω, orange, metallic, cubic
Liquid sulfur	Cyclo-S_6, S_{12} Catena C $< n < 10^5$	α, β, S_{12} $\psi(=\mu)$	π, ι, ν, ψ
Sulfur vapor	$2 < n < 12$	α, S_∞	Crystex, ω, red, green, blue

can be very difficult. Separation is impossible if it leads to chemical reactions, as is the case with ω-sulfur, and others, which even decompose during low-temperature chromatography (30). Thus, not all new allotropes are pure; many allotropes are complex mixtures, and many materials that were believed to be new allotropes are in reality merely new mixtures of well-established allotropes.

In the following sections a short summary of the better known of the incompletely characterized allotropes is given. The data are summarized in Table IV; for details, we refer to reviews (25, 65) or the original literature. The study of the latter is, however, often hard, as different authors use different conventions and nomenclature, and many researchers are not familiar with all earlier literature. We use the preparation method as a basis for this review, as it constitutes the most reliable identification.

A. FROM SOLUTIONS OF CYCLOOCTASULFUR

Below 90°C, solutions of sulfur in organic solvents contain normally cyclooctasulfur. Muthmann (72) in 1890, Korinth (54) in 1928, and Erämetsä (29–31) in 1953 are among the many who tried to produce new types of sulfur allotropes from solution. Korinth (54) used various solvents and additives, such as selenium, rubber, and nitrobenzene to induce formation of unusual crystal forms. He succeeded in finding recipes for four new allotropes which were reinvestigated by Erämetsä (29–31). The latter found that several of these could be separated into various fractions. In the process of doing so, Erämetsä identified twelve new allotropes. None is stable for more than 15 min. As structural data are lacking, we will not further dwell on them. The case of γ-sulfur, first prepared by Muthmann in 1890, shows that some of these allotropes may be confirmed in the future but probably only after better synthetic methods for preparing them have been found.

Some attempts have been made to prepare new allotropes from hot solutions. The phase equilibria of solutions of various organic solvents with liquid sulfur have been reported (58), but the upmost caution is in order in dealing with these systems, as sulfur reacts quite smoothly with with all solvents above 120°C.

B. FROM SOLUTIONS CONTAINING OTHER SULFUR COMPOUNDS

In Section II it is shown that many new allotropes can be very elegantly prepared if matched reagents are used to build up the molecular unit in dilute solution, before the metastable molecules come in contact with each other. Difficulty arises if the reagents are not selective. This is the case if sulfane mixtures, called "raw oil," are used, or if chains are built up by continued addition of a sulfur group, as is the case in the classic synthesis of Engel (28). A further problem arises if large species are produced in aqueous solution, because in that environment polythionates are also stable, and the resulting product might contain 99% sulfur, but the terminal group of chains might be another element. The work by Watanabe (120) and Schmidt (88, 125) proves that this synthetic approach to the preparation for new sulfur allotropes has not yet been exhausted.

Das (21) and others have used reagents without solvents for synthesizing new materials. ω-Sulfur is such an example, showing that the resultant mixtures are very hard to analyze. Details on the controversy about ω-sulfur are given in Donohue (*25*).

C. From Solid Sulfur

Irradiation of sulfur by various means yields discoloration of elemental sulfur, but it is doubtful that these forms will ever lead to pure allotropes. Above 92°C, α-sulfur converts into monoclinic sulfur. The kinetics of the transformation depend on various factors. Thermal analysis of single crystals of α-sulfur shows that the conversion is so slow that α-sulfur can be heated to 112°C, where it melts, before β-sulfur is formed (*19*).

Exposure of solid sulfur to pressure has proven more fruitful (*60*) than was assumed when Bridgman reported that sulfur did not display any new properties at high pressure. The work of Bååk (*3*), who reported the preparation of cubic sulfur by application of pressure to α-sulfur, stimulated much interest and led Geller (*39, 59, 60*) and others to pay attention to this system. Geller made several new allotropes. One has the same X-ray pattern as fibrous sulfur (*39*). Another is similar to laminar sulfur, and new forms were reported by Das (*21*) and Tuinstra (*114*). Much structural elucidation (*60, 84*) is still going on, but after careful study of the literature, it seems that all these allotropes probably are mixtures, from which a component cannot be separated unharmed. Nevertheless, these substances demand attention, because the structural patterns might yield new information on the nature of the sulfur–sulfur bond. Tuinstra's suggestion (*115*) of a "cross-grained plywood structure" for his ω-2-sulfur shows that not all steric considerations have been properly explored: If the pitch of sulfur helices is, indeed, equal to the distance between adjacent helix axes (*25*), chains and rings might convert into each other with ease. This might explain why polymeric sulfur, under dynamic stress, is far more stable than if it is not worked.

High-pressure investigations led to reports of over twelve new phases by Vezzoli (*116, 117*), Ward (*119*), Susse (*111*), and Pankov (*78*). In evaluating this work it should be known that equilibria are established very slowly, especially at high temperatures where viscous species are produced. Block (*9*) reported that above 235°C equilibrium is not nearly reached after 3 days. Thus the properties of samples reflect their history.

D. From Liquid Sulfur

Fibrous sulfur, described in Section II, is prepared by quenching liquid sulfur. Aten and Erämetsä's π-sulfur is obtained from the melt at

120°C. Schmidt (91) showed that the melt contains rings other than S_8, for example S_{12}. Above 156°C, the well-known polymerization sets in, and species with up to 10^5 sulfur atoms are in equilibrium with various rings and chains. Allotropes prepared from such a complex mixture cannot be expected to be pure. Much more needs to be known about the liquid before systematic attempts can be made to exploit the species in the liquid for synthetic purposes. However, many new data are becoming available quickly. Bröllos studied the thermal analysis of sulfur at various pressures (11), the optical absorption at various pressures (55), and its influence on the polymerization. Kuballa (55) and Klement (49) also studied the thermal analysis, and determined the kinetics of intramolecular conversion, which was found to be slow. Baur (5) reports that molar polarization indicates the presence of a new species in liquid sulfur, below the polymerization temperature. He assigned it tentatively to the chair form of S_8, which would be similar to that observed for the S_8^{2+} ion (Fig. 8c). Hot liquid sulfur loses its viscosity and turns very dark, due to the formation of small species such as S_3 and S_4 (75). Freezing of such species leads to a red glass from which, with ingenuity, various species might conceivably be isolated. Wiewiorowski has proposed that charge-transfer complexes (124) are present in liquid sulfur and account for much of its behavior, as has been explained in Section II. Obviously, much remains to be learned about the complex chemical system called liquid sulfur.

E. From Sulfur Vapor

Sulfur vapor contains small molecules, of which only S_8, at low temperature, and S_2, at high temperature (4, 6, 7, 83), can be separated in the pure state. Both have been described in Section II. In an intermediate pressure and temperature range, mixtures of molecules with the formula S_n ($2 < n < 12$) have been identified with mass spectroscopy. The components of sulfur vapor at low temperature and very high pressure are not yet known. It is not even known whether gas-phase molecules are present as rings or as chains.

Quenched sulfur vapor has been studied for a long time. Slowly quenched to room temperature it yields flowers of sulfur, which were prepared by alchemists. The flowers can be separated into several phases by elution with CS_2. Thermal analysis (19) shows that α-sulfur, β-sulfur, and ω-sulfur are present. The latter melts at 104°C. The X-ray structure of ω-sulfur, prepared in this way, and that of Crystex (104), a widely used insoluble form of sulfur, also called supersublimation sulfur, are not fully established. As explained above, these and similar substances are probably mixtures, and their structural data depends on the method of preparation, history of the reagents and products, and the age of the sample. Several of these allotropes are stable only if traces of additives are present. In conclusion, we want to point out that many of the doubtful allotropes are not pure elemental sulfur. It is difficult to purify the element, although there has been much progress in this work (*80, 112*).

Acknowledgments

The work reported here was done under the auspices of the U.S. Atomic Energy Commission through the Inorganic Materials Research Division of the Lawrence Berkeley Laboratory.

The author wishes to thank Dr. L. Brewer for valuable suggestions and support.

References

1. Abrahams, S. C., *Acta Crystallogr.* **8**, 661 (1955).
2. Aten, A. H. W., *Z. Phys. Chem.* **88**, 321 (1914).
3. Bååk, T., *Science* **148**, 1220 (1965).
4. Barrow, R. F., and duPark, R., in "Elemental sulfur" (B. Meyer, Ed.), p. 251. Wiley (Interscience), New York, 1965.
5. Baur, E. M., and Horsma, D. A., *J. Phys. Chem.* **78**, 1670 (1974).
6. Berkowitz, J., Chupka, W. A., Bromels, E., and Belford, R. L., *J. Chem. Phys.* **47**, 4320 (1967).
7. Berkowitz, J., and Marquart, J. R., *J. Chem. Phys.* **39**, 275 (1963).
8. Bjorvatten, T., *Acta Chem. Scand.* **16**, 749 (1962).
9. Block, S., and Piermarini, G. J., *High Temp.—High Pressures* **5**, 567 (1973).
10. Bragg, W. H., *Proc. Roy. Soc., Ser. A* **89**, 575 (1914).
11. Bröllos, K., and Schneider, G. M., *Ber. Bunsenges. Phys. Chem.* **78**, 296 (1974).
12. Buchler, J., *Angew. Chem.* **78**, 1021 (1966); *Angew. Chem., Int. Ed. Engl.* **5**, 965 (1966).
13. Burwell, J. T., II, *Z. Kristallogr., Kristallgeometric, Kristallphys.* **97**, 123 (1937).
14. Caron, A., and Donohue, J., *J. Phys. Chem.* **64**, 1767 (1960).
15. Caron, A., and Donohue, J., *Acta Crystallogr.* **18**, 562 (1965).
16. Chivers, T., and Drummond, I., *Inorg. Chem.* **11**, 2525 (1972).
17. Cooper, A. S., *Acta Crystallogr.* **15**, 578 (1962).
18. Cooper, R., and Culka, J. V., *J. Inorg. Nucl. Chem.* **29**, 1217 (1967); **32**, 1857 (1970).
19. Currell, B. R., and Williams, A. J., *Thermochim. Acta* **9**, 255 (1974).
20. Cusachs, L. C., and Miller, D. J., *Advan. Chem. Ser.* **110**, 154 (1972).
21. Das, S. R., *Indian J. Phys.* **12**, 163 (1938).
22. Debaerdemaeker, T., and Kutoglu, A., *Naturwissenschaften* **60**, 49 (1973).
23. deHaan, J. M., *Physica (Utrecht)* **24**, 855 (1958).
24. Detry, D., Drowart, J., Goldfinger, P., Keller, H., and Rickert, H., *Z. Phys. Chem. (Frankfurt am Mein)* [N.S.] **55**, 314 (1967).
24a. Donnay, J. D. H., *Acta Cryst.* **8**, 245 (1955).

25. Donohue, J., "The Structures of the Elements," p. 324. Wiley, New York, 1974.
26. Donohue, J., Caron, A., and Goldish, E., *J. Amer. Chem. Soc.* **83**, 3748 (1961).
27. Donohue, J., Goodman, S. H., and Crisp, M., *Acta Crystallogr., Sect. B* **25**, 2168 (1969).
28. Engel, M. R., *C. R. Acad. Sci.* **112**, 866 (1891).
29. Erämetsa, O., *Suom Kemistilehti B* **32**, 15, 47, 97, and 233 (1959).
30. Erämetsa, O., *Suom. Kemistilehti B* **35**, 154 (1962).
31. Erämetsa, O., *Suom. Kemistilehti B* **36**, 213 (1963).
32. Erämetsa, O., and Niinisto, L., *Suom. Kemistilehti B* **42**, 471 (1969).
33. Fehér, F., and Lauger, M., *Tetrahedron Lett.* **24**, 2125 (1971).
34. Foss, O., Furberg, S., and Zachariasen, H., *Acta Chem. Scand.* **12**, 1700 (1958).
35. Foss, O., and Johnsen, K., *Acta Chem. Scand.* **19**, 2207 (1965).
36. Foss, O., and Marøy, K., *Acta Chem. Scand.* **19**, 2219 (1965).
37. Frank, G. W., and Degen, P. J., *Acta Crystallogr., Sect. B* **29**, 1815 (1973).
37a. Frondel, C., and Whitfield, R. E., *Acta Cryst.* **3**, 242 (1950).
38. Gardner, M., and Rogstad, A., *J. Chem. Soc., Dalton Trans.* p. 599 (1973).
39. Geller, S., and Lind, M. D., *Acta Crystallogr., Sect. B* **25**, 2166 (1969).
40. Giggenbach, W., *Inorg. Chem.* **10**, 1308 (1971).
41. Gillespie, R. J., and Passmore, J., *Accounts Chem. Res.* **4**, 413 (1971).
42. Gillespie, R. J., and Ummat, P. K., *Inorg. Chem.* **11**, 1674 (1972).
43. Hampton, E. M., Shaw, B. S., and Sherwood, J. N., *J. Cryst. Growth* **22**, 22 (1974).
44. Hoffmann, K. A., and Hochtlen, F., *Ber. Deut. Chem. Ges.* **36**, 3090 (1903).
45. Jander, J., and Turk, G., *Chem. Ber.* **97**, 25 (1964).
46. Jensen, D., in "Selected Values of Thermodynamic Properties of the Elements" (R. Hultgren et al., eds.), p. 333. Amer. Soc. Metals, 1973.
47. Kawada, I., and Hellner, E., *Angew. Chem.* **82**, 390 (1970); *Angew. Chem., Int. Ed. Engl.* **9**, 379 (1970).
48. Kende, J., Pickering, T. L., and Tobolski, A. V., *J. Amer. Chem. Soc.* **87**, 5582 (1965).
49. Klement, W., *J. Polym. Sci.* **12**, 815 (1974).
50. Koningsberger, D. C., Ph.D. Thesis, Tech. University, Eindhoven (1971).
51. Köpf, H., *Angew. Chem.* **81**, 332 (1969), *Angew. Chem., Int. Ed.* **8**, 375 (1969).
52. Köpf, H., *Angew Chem.* **81**, 875 (1969); *Angew. Chem., Int. Ed. Engl.* **8**, 962 (1969).
53. Köpf, H., Block, B., and Schmidt, M., *Chem. Ber.* **101**, 272 (1968).
54. Korinth, E., Dissertation, Jena (1928).
55. Kuballa, M., and Schneider, G. M., *Ber. Bunsenges. Phys. Chem.* **75**, 6 (1971).
56. Kuczkolski, R. L., *J. Amer. Chem. Soc.* **86**, 3617 (1964).
57. Kutoglu, A., and Hellner, E., *Angew. Chem.* **78**, 1021 (1966); *Angew. Chem., Int. Ed. Engl.* **5**, 965 (1966).
58. Larkin, J. A., Katz, J., and Scott, R. L., *J. Phys. Chem.* **71**, 352 (1967).
59. Lind, M. D., and Geller, S., *Science* **152**, 644 (1966).
60. Lind, M. D., and Geller, S., *J. Chem. Phys.* **51**, 348 (1969).
61. Meyer, B., *Chem. Rev.* **64**, 429 (1964).
62. Meyer, B., "Elemental Sulfur", Wiley (Interscience), New York, 1965.
63. Meyer, B., and Gotthard, B., unpublished results.
64. Meyer, B., Oomen, T. V., and Jensen, D., *J. Phys. Chem.* **75**, 912 (1971).

65. Meyer, B., and Spitzer, K., *J. Phys. Chem.* **76**, 2274 (1972).
66. Meyer, B., and Stroyer-Hansen, T., *J. Phys. Chem.* **76**, 3968 (1972).
67. Meyer, B., Stroyer-Hansen, T., and Oomen, T. V., *J. Mol. Spectrosc.* **42**, 335 (1972).
67a. Meyer, K. H., and Go, Y., *Helv. Chim. Acta* **17**, 1081 (1934).
68. Miller, D. J., Ph.D. Thesis, Tulane University (1970).
69. Miller, D. J., and Cusachs, L. C., *Chem. Phys. Lett.* **3**, 501 (1969).
70. Miller, D. J., and Cusachs, L. C., "Quantum Aspects of Heterocyclic Compounds in Chemistry and Biochemistry", Jerusalem Symposia on Quantum Chemistry and Biochemistry, II. Acad. Sci. Humanities, Jerusalem, 1970.
71. Montgomery, R. L., *Science* **184**, 562 (1974).
72. Muthmann, W., *Z. Kristallogr., Kristallgeometrie, Kristallphys.* **17**, 336 (1890).
73. Narkuts, K. I., *Izv. Akad. Nauk SSSR, Ser. Fiz.* **38**, 548 (1974).
74. Nimon, L. A., Neff, V. D., Cantley, R. E., and Buttlar, R. O., *J. Mol. Spectrosc.* **22**, 105 (1967).
75. Oomen, T. V., *Diss. Abstr. Int. B* **31**, 3904 (1971).
76. Ozin, G. A., *Chem. Commun.* p. 1325 (1969).
77. Ozin, G. A., *J. Chem. Soc.*, *A* p. 116 (1969).
78. Pahkov, I. E. P., Tonkov, E. Y., and Mirinski, D. S., *Dokl. Akad. Nauk SSSR* **164**, 588 (1965).
79. Pauling, L., *Proc. Nat. Acad. Sci. U.S.* **35**, 495 (1949).
80. Pavlov, V. I., and Kirillov, L. N., *Tr. Tol'yattinsk. Politekh. Inst.* No. 1, p. 78 (1969).
81. Pawley, G. S., and Rinaldi, R. P., *Acta Crystallogr., Sect. B* **28**, 3605 (1972).
82. Rahman, R., Safe, S., and Taylor, A., *Quart. Rev. Chem. Soc.*, **24**, 208 (1970).
83. Rau, H., Kutty, T. R. N., and Guedes de Carvalho, J. R. F., *J. Chem. Thermodyn.* **5**, 833 (1973).
84. Roof, R. B., *Aust. J. Phys.* **25**, 335 (1972).
85. Sands, D. E., *J. Amer. Chem. Soc.* **87**, 1395 (1965).
86. Schmidt, M., *Inorg. Macromol. Rev.* **1**, 101 (1970).
87. Schmidt, M., *Chem. unserer Zeit* **7**, 11 (1973).
88. Schmidt, M., *Angew. Chem.* **85**, 474 (1973); *Angew. Chem., Int. Ed. Engl.* **12**, 334 (1973).
89. Schmidt, M., Blaettner, K., and Kochendörfer, *Z. Naturforsch. B* **21**, 622 (1966).
90. Schmidt, M., Block, B., Block, H. D., Köpf, H., and Wilhelm, E., *Angew. Chem.* **80**, 660 (1968); *Angew. Chem., Int. Ed. Engl.* **7**, 632 (1968).
91. Schmidt, M., and Block, H. D., *Angew. Chem.* **79**, 944 (1967).
92. Schmidt, M., and Block, H. D., *Z. Anorg. Allg. Chem.* **385**, 119 (1971).
93. Schmidt, M., Knippschild, G., and Wilhelm, E., *Chem. Ber.* **101**, 381 (1968).
94. Schmidt, M., and Wilhelm, E., *Inorg. Nucl. Chem. Lett.* **1**, 39 (1965).
95. Schmidt, M., and Wilhelm, E., *Angew. Chem.* **78**, 1020 (1966); *Angew. Chem., Int. Ed. Engl.* **5**, 964 (1966).
96. Schmidt, M., and Wilhelm, E., *Chem. Commun.* p. 1111 (1970); *J. Chem. Soc., D* p. 17 (1970).
97. Schmidt, M., and Wilhelm, E., unpublished.
98. Schmidt, M., Wilhelm, E., Debaerdemaeker, T., Hellner, E., and Kutoglu, A., *Z. Anorg. Allg. Chem.* **405**, 153 (1974).
99. Seel, F., and Güttler, H. J., *Angew. Chem. Int., Ed. Engl.* **12**, 420 (1973).
100. Seel, F., and Simon, G., *Z. Naturforsch. B* **27**, 1110 (1972).

101. Semlyen, J. A., *Trans. Faraday Soc.* **63**, 743 (1967).
102. Skjerven, O., *Z. Anorg. Allg. Chem.* **314**, 206 (1962).
103. Srb, I., and Vasco, A., *J. Chem. Phys.* **37**, 1892 (1962).
104. Stauffer Chemical Corp., U.S. Patent 2,460,365 (1954).
105. Steudel, R., Luger, P., Bradaczek, H., and Rentsch, M., *Angew. Chem.* **85**, 452 (1973).
106. Steudel, R., and Rentsch, M., *Angew. Chem.* **84**, 344 (1972).
107. Steudel, R., and Rentsch, M., *J. Mol. Spectrosc.* **51**, 189 (1974).
108. Steudel, R., and Rentsch, M., *J. Mol. Spectrosc.* **51**, 334 (1974).
109. Strauss, H. L., and Greenhouse, J. A., *in* "Elemental Sulfur" (B. Meyer, ed.), Wiley (Interscience), p. 241. New York, 1965.
110. Susse, C., and Epain, R., *C. R. Acad. Sci., Ser. C* **263**, 613 (1966).
111. Susse, C., Epain, R., and Vodar, B., *C. R. Acad. Sci.* **258**, 4513 (1964).
112. Suzuki, H., Higashi, K., and Miyake, Y., *Bull. Chem. Soc. Jap.* **47**, 759 (1974).
112a. Teder, A., *Acta Chem. Scand.* **25**, 1722 (1971).
113. Trillat, J. J., and Forestier, J. J., *C. R. Acad. Sci.* **192**, 559 (1931).
114. Tuinstra, F., *Physica (Utrecht)* **34**, 113 (1967).
115. Tuinstra, F., "Structural Aspects of the Allotropy of Sulphur and other Divalent Elements," U. Waltman, Delft, 1967.
116. Vezzoli, G. C., and Dachille, F., *Inorg. Chem.* **9**, 1973 (1970).
117. Vezzoli, G. C., Dachille, F., and Roy, R., *Science* **116**, 218 (1969).
118. Ward, A. T., *J. Phys. Chem.* **72**, 744 (1968).
119. Ward, K. B., Jr., and Deaton, B. C., *Phys. Rev.* **153**, 947 (1967).
120. Watanabe, Y., *Acta Crystallogr., Sect. B* **30**, 1396 (1974).
121. Weiss, J., and Bachtler, W., *Z. Naturforsch. B* **28**, 523 (1973).
122. Weiss, J., and Pupp, M., *Angew. Chem.* **82**, 447 (1970)
123. Whitehead, H. C., and Andermann, G., *J. Phys. Chem.* **77**, 721 (1973).
124. Wiewiorowski, T. K., and Touro, F. J., *J. Phys. Chem.* **70**, 3528 (1966).
125. Wilhelm, E., Ph.D. Thesis, Philips University, Marburg (1966).
126. Winnewasser, M., and Haase, J., *Z. Naturforsch. A* **23**, 56 (1968).
127. Yee, K. K., Barrow, R. F., and Rogstad, A., *J. Chem. Soc., Faraday Trans 2.* **68**, 1808 (1972).
128. Zahorszky, U.-I., *Angew. Chem.* **80**, 661 (1968).

CHLORINE OXYFLUORIDES

K. O. CHRISTE and C. J. SCHACK
Rocketdyne, Division of Rockwell International, Canoga Park, California

I. Introduction	319
II. General Aspects	321
A. Geometry	322
B. Ligand Distribution	323
C. Relative Bond Strengths	323
D. Amphoteric Nature, Tendency to Form Adducts, and Reactivity	327
III. Specific Compounds	328
A. Chlorine Monofluoride Oxide	328
B. Chlorine Trifluoride Oxide	331
C. Difluorooxychloronium(V) Cation	340
D. Tetrafluorooxychlorate(V) Anion	343
E. Chlorine Pentafluoride Oxide	345
F. Chloryl Fluoride	347
G. Chloryl Cation	356
H. Difluorochlorate(V) Anion	359
I. Chlorine Trifluoride Dioxide	361
J. Difluoroperchloryl Cation	367
K. Perchloryl Fluoride	371
L. Chlorine Fluoride Oxide Radicals	385
M. Miscellaneous	386
IV. Appendix: Tables of Thermodynamic Properties for Some Chlorine Oxyfluorides	386
References	390

I. Introduction

This review is limited to compounds containing both oxygen and fluorine atoms bonded directly to a common chlorine central atom. Therefore, compounds, such as fluorine perchlorate (O_3ClOF), have not been included. Data on O_3ClOF were summarized in a recent review on inorganic hypofluorites (*180*).

The subject of chlorine oxyfluorides* has been reviewed in 1963 by Schmeisser and Brändle (*253*) and in 1969 in "Gmelin's Handbuch der

---
* For the sake of clarity, we have not followed a rigid system of nomenclature, such as starting formulas with the central atom followed by the ligands. If, for example, FClO is written as ClOF, an uninformed reader might be induced to think of the compound as a hypofluorite.

Anorganischen Chemie" (122). However, both reviews deal only with chloryl fluoride ($FClO_2$) and perchloryl fluoride ($FClO_3$). Since the writing of these reviews, three of the four possible remaining chlorine oxyfluorides i.e., FClO, F_3ClO, and F_3ClO_2, have been characterized, and claims have been made for the synthesis of the fourth one, ClF_5O. Additional information on chlorine oxyfluorides can be found in various monographs and textbooks (32, 75, 77, 95, 156, 169, 244) and in particular in "Comprehensive Inorganic Chemistry" in the chapter on the halogens written by Downs and Adams (84). A second area of significant recent progress comprises ions derived from chlorine oxyfluorides. Therefore, these are also discussed in detail.

For the present review, the literature cited in *Chemical Abstracts* (Jan. 1965–Dec. 1973) was used in addition to more recent work pub-

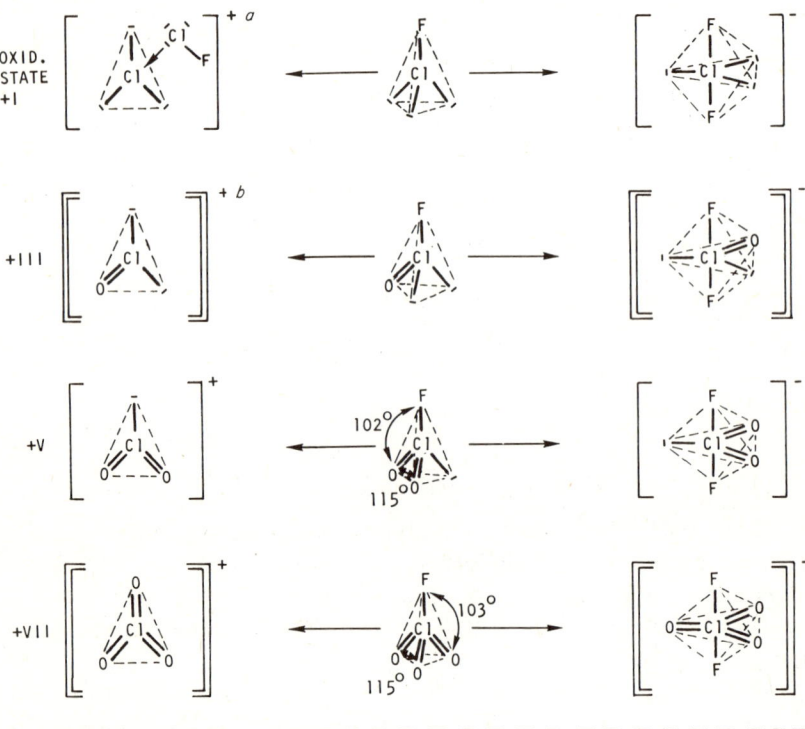

FIG. 1. Geometries of the chlorine oxyfluoride molecules and their ions compared to those of the corresponding chlorine fluorides. [a] Since the Cl^+ cation would possess only an electron sextet, it is stabilized by a ClF molecule to form the Cl_2F^+

CHLORINE OXYFLUORIDES 321

lished during 1973 and 1974. For literature before 1965, we have relied mainly on Schmeisser's review (253) and Gmelin's handbook (122).

In addition to a discussion of the individual compounds, a section was added correlating the physical and chemical properties of the chlorine oxyfluorides with their structure. In the Appendix, full tables of thermodynamic properties are given for each compound, where known.

II. General Aspects

Since most of the physical and chemical properties of the chlorine oxyfluorides can be readily correlated with their molecular structure, we shall discuss briefly some of the more general aspects.

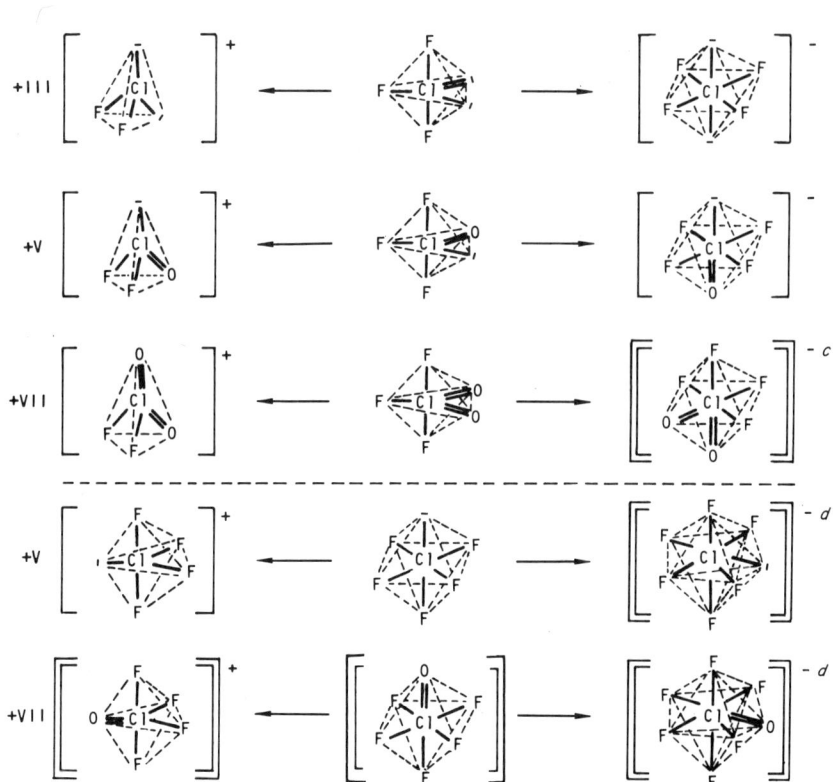

cation. [b] Double brackets indicate yet unknown ions. [c] See text for preference of cis model. [d] These compounds with a coordination number of 7 are unlikely to exist.

A. GEOMETRY

As can be seen from Fig. 1, the structures of all the chlorine oxyfluoride molecules and ions can be derived from those of the corresponding binary chlorine fluorides (53) by replacing a free chlorine valence electron pair by a doubly bonded oxygen atom without significant rearrangement of the rest of the molecule.

The only possible exception to this rule could be the yet unknown (68) $ClF_4O_2^-$ anion. By comparison with the known structures of the pseudoisoelectronic $IF_4O_2^-$ (45, 93) and $TeF_4O_2^{2-}$ (260) anions, the 2 oxygens in $ClF_4O_2^-$ should also be in cis and not in trans position. In these and similar oxyfluoride anions, such as SF_5O^- (65) or CF_3O^- (59), the negative charge is located mainly on the most electronegative ligands, i.e., fluorine. Furthermore, in pseudo-octahedral species not containing a free valence electron pair on the central atom, such as XF_5O^-, the fluorine trans to the less electronegative ligand appears to be more weakly bonded than the remaining fluorines. This is plausible from molecular orbital arguments. Therefore, for $XF_4O_2^-$ the structure with 2 oxygen atoms trans to 2 fluorines and cis with respect to each other, should favor the resonance structures having the negative charge located on the fluorine ligands.

Since the degree of mutual repulsion decreases in the order, free valence electron pair > double-bonded oxygen > fluorine, the observed bond angles deviate somewhat from those expected for the ideal geometries. Typical examples are $FClO_2$ and $FClO_3$ (Fig. 1).

The structure of radicals and radical ions can also readily be predicted by treating an unpaired electron in the same manner as a free valence

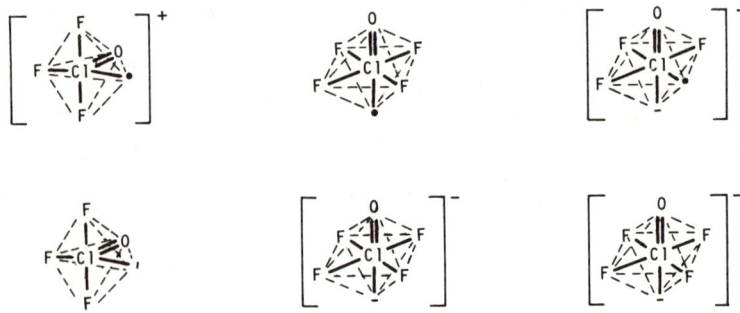

FIG. 2. Structures of the yet unknown ClF_3O^+, ClF_4O, and ClF_3O^- radicals (upper row) predicted by comparison with the known structures (bottom row) of ClF_3O and ClF_4O^-.

electron pair. Therefore, a chlorine oxyfluoride free radical should have the same geometry as the corresponding anion with identical fluorine and oxygen ligands (Fig. 2). Similarly, the structure of a radical cation should be analogous to that of the corresponding molecule having the same ligands. For a radical anion, however, the additional sterically active valence electron will increase the coordination number around the central atom by 1. This should result in a geometry resembling that of the anion containing one F ligand more, but in which one F ligand is replaced by the sterically active free electron. The exact spin distribution would have to be determined experimentally and is not necessarily the same as shown in Fig. 2.

B. Ligand Distribution

As can be seen from Fig. 1, the structures are simple and can be logically predicted if one keeps in mind that free valence electron pairs on the central atom are sterically active and behave as a ligand. For 3, 4, 5, and 6 ligands, always the sterically most favorable arrangements are observed, namely, the triangular plane, tetrahedron, trigonal bipyramid, and the octahedron, respectively (*119*). Based on the information available for halogen oxyfluorides and related compounds such as xenon or chalcogen oxyfluorides the following conclusions concerning the ligand distribution can be reached. In a triangular plane and a tetrahedron all positions are equivalent. In a trigonal bipyramid the two axial positions are occupied by the most electronegative ligands, i.e., F atoms. In octahedrons of the type XF_5A only one arrangement is possible. For XF_4AB, however, the A or B ligands are trans if A and B are either two free electron pairs or one free electron pair and one oxygen ligand. When A and B are 2 O atoms, the cis arrangement appears more favorable (see above). The case of the pentagonal bipyramid is not of practical interest since it appears that the coordination number around a high oxidation state, chlorine central atom is limited to a maximum of 6.

C. Relative Bond Strengths

Unfortunately, exact bond lengths are known only for $FClO_2$ (*220*) and $FClO_3$ (*72*). However, complete vibrational spectra have been published for essentially all of the chlorine oxyfluorides. These can be used for the evaluation of the corresponding force constants. Since the latter are a good measure for the relative strengths of these bonds, their comparison is interesting. As can be seen from Table I, the ClO bonds all

TABLE I

Stretching Force Constants of Some Chlorine Oxyfluorides

Oxidation state	Compound	f_{ClO} (mdyn/Å)	f_{ClF} (mdyn/Å)			Ref.
			I[a]	II[b]	III[c]	
+VII	$ClF_2O_2^+$	12.1	4.46	—	—	(69)
+V	ClF_2O^+	11.20	3.44	—	—	(58)
+VII	$FClO_3$	9.4	3.9	—	—	(174)
+V	ClF_3O	9.37	3.16	2.34	—	(55)
+VII	ClF_3O_2	9.23	3.35	2.70	—	(57)
+V	ClF_4O^-	9.13	—	1.79	—	(56)
+V	$FClO_2$	9.07	—	—	2.5	(270)
+V	ClO_2^+	8.96	—	—	—	(66)
+V	$ClF_2O_2^-$	8.3	—	1.6	—	(54)
+III	$FClO$	6.85	—	—	2.59	(5)

[a] Mainly covalent bonds.
[b] Mainly semi-ionic 3c–4e bonds
[c] Special case of highly polar $(p-\pi^*)\sigma$ bonds.

possess more or less double-bond character. The variation in the values of the ClO-stretching force constants is mainly due to the combination of several effects. For example, a formal positive charge (i.e., in cations), a high oxidation state of the central atom, and a high number of fluorine ligands tend to increase the ClO-stretching force constant (57). In contrast to the ClO bonds, the ClF bond strengths are subject to much

Fig. 3. Schematic bonding in ClF_2^- as explained by a semi-ionic 3c–4e bond model.

larger changes. These strong variations cannot be explained by effects such as listed above for the ClO bonds or by the Gillespie–Nyholm valence shell, electron pair repulsion (VSEPR) theory (*119*) alone. By analogy with the halogen fluorides (*53*), it is necessary to assume contributions from two different kinds of bonding. In addition to the normal covalent bonds possessing a bond order of about 1, the occurrence of semi-ionic 3 center–4 electron bonds (*130, 232, 243*) must be invoked. The principle of a semi-ionic 3c–4e bond is demonstrated in Fig. 3. For simplicity, ClF_2^- (*63*) was chosen as an example. Ideally, the two F ligands form two semi-ionic 3c–4e [p—p] σ-bonds with one p electron pair of the chlorine central atom, whereas the free Cl valence electron pairs form an sp^2 hybrid.

Instead of using this semiempirical molecular orbital model, the bonding in ClF_2^- can also adequately be described in the valence-bond representation (*76*) as a resonance hybrid of the following canonical forms: (F—Cl) F$^-$ and F$^-$(Cl—F). This results in the same average charge distribution as in the molecular orbital model, i.e., $^{-1/2}$F-Cl-F$^{-1/2}$. Another and the simplest bond model, proposed by Bilham and Linnett (*29*) for XeF_2 which is pseudoisoelectronic with ClF_2^-, assumes single electron bonds for each X—F bond. It is relatively immaterial, which of these three descriptions is preferred since all of them result in the same charge distribution and a Cl—F bond order of about 0.5.

As can be seen from Table I, these weak ClF bonds occur only when the central atom has a coordination number in excess of 4 and possesses at least one free Cl valence electron pair. In addition to Gillespie's simple VSEPR theory, the following general rule has been proposed by Christe (*53*), which permits the prediction of whether, and how many, semi-ionic bonds are to be formed:

The free valence electron pairs on the central atom seek high *s*-character; i.e., sp^n hybridization. If the number of ligands is larger than 4 and one or more of them are free valence electron pairs, then as many F ligands form linear semi-ionic 3 center–4 electron bonds as are required to allow the free electron pairs to form an sp^n hybrid with the remaining F ligands. These semi-ionic 3c–4e bonds are considerably weaker and longer than the mainly "covalent" sp^n hybrid bonds.

This rule also holds for the chlorine oxyfluorides as well as for the chlorine fluorides for which it was originally formulated.

An additional effect, however, must be invoked to be able to rationalize fully the experimental data. Inspection of Table I reveals that the ClF-stretching force constants of $FClO_2$ and FClO are significantly lower than expected from the above discussion. In particular, if the known ClF-stretching force constants and bond distances within the pseudo-

TABLE II

Comparison of ClF-Stretching Force Constants and Bond Lengths Within the Pseudotetrahedral Series FCl, FClO, FClO$_2$, FClO$_3$

Molecule	f_{ClF} (mdyn/Å)	r_{ClF} (Å)	Ref.	
FCl	4.56	1.628	(122)	(122)
FClO	2.59	—	(5)	—
FClO$_2$	2.5	1.697	(270)	(220)
FClO$_3$	3.9	1.610	(174)	(72)

tetrahedral series FCl, FClO, FClO$_2$, FClO$_3$ are compared (see Fig. 1 and Table II), it becomes obvious that the ClF bonds in FClO and FClO$_2$ are abnormally long and weak. Application of the rules discussed above is of no help in explaining the observed trends. However, if a simple molecular orbital description, similar to that proposed by Spratley and Pimentel (274) for FNO and F$_2$O$_2$, is used, the data can be rationalized. Molecules FCl, FClO, FClO$_2$, and FClO$_3$ can be thought of as being derived from the combination of an F atom with the Cl, ClO, ClO$_2$, and ClO$_3$ radicals, respectively. This hypothetical bond formation involves a 2p electron of the fluorine atom and the unpaired electron of the Cl-containing radical. If according to the example of (NO)$_2$ and (CN)$_2$, given by Spratley and Pimentel (274), the unpaired electron occupies an antibonding (π^*) orbital, the resulting bond is very weak. On the other hand, if the unpaired electron occupies a bonding orbital the resulting bond is strong. Since the unpaired electron in Cl and ClO$_3$ occupies a bonding orbital, the resulting Cl—F bond in FCl and FClO$_3$, respectively, should be strong, whereas those in FClO and FClO$_2$, derived from ClO and ClO$_2$, respectively, with an antibonding (π^*) electron (193), should be weak. These predictions are in excellent agreement with the data of Table II. As a consequence of the high electronegativity of fluorine, most of the electron density in the antibonding (π^*) orbital of ClO or ClO$_2$ is transferred to the F atom. For FClO and FClO$_2$, this results in a long and highly polar ClF bond with a significant negative charge located on F. Since at the same time electron density is removed from an antibonding orbital of the ClO$_n$ part of the molecule, the bond strength of these ClO bonds is increased. As pointed out by Chi and Andrews (47) for ClClO, there is a marked difference in behavior between radicals with a first-row element central atom and those with a second-row element central atom. Owing to their larger size and polarizability, the second-row elements facilitate a charge transfer and the XY_n stretching frequencies usually increase upon combination of XY_n with a halogen radical. For first-row element central atoms, the corresponding frequencies usually show a slight decrease. It should be pointed out, however, that in both cases a highly polar and weak bond of the type $\overset{\delta-}{F}-\overset{\delta+}{XY_n}$ results, provided the unpaired electron in the XY_n parent radical occupies an antibonding orbital. Supporting evidence for the above postulated charge transfer from XY_n to F was recently given by Parent and Gerry (220) for $FClO_2$.

In summary, three types of bonding are invoked to rationalize the remarkable differences in Cl—F bonds encountered for chlorine fluoride oxides. These are (a) conventional, mainly covalent bonds, (b) weak semi-ionic 3 center–4 electron bonds, and (c) weak highly polar (p—π*)σ bonds. It must be kept in mind, however, that all these bond descriptions are idealized extremes, used mainly for didactic reasons. The actual bonds may contain significant contributions from more than one kind of bonding and, as a consequence, there is little black and white, but many shades of gray. Obviously, other bond models can also be used, so long as they adequately account for the experimental data. The steady increase in our knowledge about these compounds is bound to result in significant improvements of these rather empirical and intuitive bond models.

D. Amphoteric Nature, Tendency to Form Adducts, and Reactivity

In many respects the chlorine oxyfluorides resemble the chlorine fluorides. For example, they exhibit little or no self-ionization, but are amphoteric. With strong Lewis acids or bases they can form stable adducts. The tendency to form adducts was found (64) not to be so much a function of the relative acidity of the parent chlorine oxyfluoride but rather to depend on the structure of the amphoteric molecule and of that of the anion or the cation formed. The preferred structures are the energetically favored tetrahedron and octahedron. Consequently, a trigonal bipyramidal molecule, such as ClF_3O (64), exhibits a pronounced tendency to form either a stable pseudotetrahedral cation or a pseudo-octahedral anion:

On the other hand, tetrahedral $FClO_3$ does not form an adduct with either Lewis acids or bases *(167, 209, 224)*:

$$\left[O=\overset{O}{\underset{}{\overset{\parallel}{Cl}}}=O \right]^+ \xleftarrow{-F^-} \quad O=\overset{F}{\underset{O}{\overset{|}{Cl}}}=O \quad \xrightarrow{+F^-} \left[O=\overset{F}{\underset{F}{\overset{|}{Cl}}}=O \right]^-$$

Similarly, the chemical reactivity of these two chlorine oxyfluorides differs vastly: whereas ClF_3O is extremely reactive and cannot be handled even in a well-dried glass vacuum system, $FClO_3$ reacts only slowly with water.

III. Specific Compounds

A. CHLORINE MONOFLUORIDE OXIDE

According to Ruff and Krug *(242)*, FClO is formed during hydrolysis of ClF_3 as a solid melting at $-70°C$ to a red liquid which is unstable in the gas phase. However, no conclusive proof for the existence of FClO was given, and it appears that the red color observed may have been due to the presence of chlorine oxides. Heras and co-workers *(137)* have proposed the formation of FClO as an intermediate in the thermal decomposition of $FClO_2$. More recent studies by Bougon and co-workers on the hydrolysis of ClF_3 *(9, 36)*, by Christe on the reaction of ClF_3 with $HONO_3$ *(51)* and on the reaction of ClF_3O with SF_4 *(60)*, by Pilipovich *et al.* on the photochemical synthesis of ClF_3O *(228)*, and by Schack *et al.* on the reaction chemistry of ClF_3O *(246)* all point to the formation of FClO as an intermediate that is unstable with respect to disproportionation:

$$2FClO \longrightarrow FClO_2 + ClF$$

Attempts to stabilize the FClO formed as an intermediate by complexing with a strong Lewis acid, such as AsF_5 to give $ClO^+AsF_6^-$, were also unsuccessful. Thus the controlled hydrolysis of $ClF_2^+AsF_6^-$ with stoichiometric amounts of H_2O in HF solution resulted only in the formation of $ClO_2^+AsF_6^-$ *(51)*. This is not surprising since Lewis acids are known to catalyze such disproportionation reactions.

Recently, Cooper and co-workers *(74)* succeeded in obtaining direct evidence for the existence of free FClO in the gas phase. During a study of the hydrolysis of excess ClF_3 in a flow reactor, a novel species was observed in the infrared spectrum showing a PQR band centered at 1032 cm^{-1}. The species causing this band was found to decompose at ambient

TABLE III

Observed and Calculated Frequencies for the FClO Species[a]

Isotope	Assignment	Obsd. (cm^{-1})	Calcd. (cm^{-1})
F^{35}Cl^{16}O	ν_1	1038.0	1038.3
	ν_2	593.5	593.9
	ν_3	315.2	316.0
F^{37}Cl^{16}O	ν_1	1029.0	1028.9
	ν_2	587.5	588.4
	ν_3	315.2	313.8
F^{35}Cl^{18}O	ν_1	999.2	999.5
	ν_2	593.5	592.6
	ν_3	307.0	308.3
F^{37}Cl^{28}O	ν_1	990.1	989.6
	ν_2	587.5	587.1
	ν_3	307.0	306.2

[a] Data from Andrews et al. (5).

temperature with a half-life of about 25 sec into FClO$_2$ and ClF. If an excess of H$_2$O was used in the hydrolysis, no FClO but the expected (9, 36) ClO$_2$ was observed as the main product.

The results of Cooper et al. were confirmed by a matrix isolation study by Andrews and associates (5). The latter authors observed the same species during the photolysis (2200–3600 Å) of argon matrix-isolated ClF and O$_3$ in the temperature range 4°–15°K. All three fundamentals expected for a bent FClO molecule were observed, and their assignment to FClO was confirmed by the measurement of the ^{18}O and ^{37}Cl isotopic

TABLE IV

Force Field of FClO Assuming a Bond Angle of 120° and All Interaction Constants to be Zero[a]

$f_{ClO} = 6.85$ mdyn/Å
$f_{ClF} = 2.59$ mdyn/Å
$f_\alpha = 0.92$ mdyn Å/rad^2

[a] Data from Andrews et al. (5).

shifts (Table III) and by force field computations (Table IV). For the force field computation, an FClO bond angle of 120° was assumed. However, on the basis of the increased repulsion from the chlorine free valence electron pairs (see Section II, A), we would expect this angle to be less than the tetrahedral angle of 109°, but larger than that found for ClF_2^+ [103.17° in $ClF_2^+AsF_6^-$ (*181*) and 95.9° in $ClF_2^+ SbF_6^-$ (*88*)]. The small size of the molecule, its high dipole moment, the naturally occurring ^{37}Cl isotope, and its half-life at ambient temperature make it ideally suited for a structure determination by microwave spectroscopy in a flow system.

The force field reported (*5*) for FClO allows some conclusions concerning the strength of the bonds in this molecule. Comparison of the ClO-stretching force constant of FClO with those of the higher oxidation state species listed in Table I makes the FClO value appear surprisingly low. However, when compared to species of similar oxidation state and

TABLE V

Comparison of the ClO-Stretching Force Constants and Bond Orders of FClO with Those of Related Pseudotetrahedral Species Having a Comparable Oxidation State

Species	Oxidation state	f_{ClO} (mdyn/Å)	Bond order	Ref.
ClO_2^-	+III	4.26	1.5	(*266*)
FClO	+III	6.85	2	(*5*)
ClO_2	+IV	7.02	2	(*161*)

geometry (Table V), FClO exhibits a value very much in line with our expectations for a ClO double bond. The ClF bond is relatively weak,

indicating that contributions from resonance structures, such as II, are significant as is also the case in the related $FClO_2$ molecule. The high ionicity of the Cl—F bond in these two chlorine fluoride oxides has been discussed above (Section II, C) in terms of a (p—π^*)σ bond.

B. CHLORINE TRIFLUORIDE OXIDE

Chlorine trifluoride oxide,

$$\text{F}-\underset{\underset{\text{F}}{|}}{\overset{\overset{\text{F}}{|}}{\text{Cl}}}\diagdown_{\text{O}}$$

(III)

was discovered in 1965 at Rocketdyne by Pilipovich et al. (226, 231). However, these results were not published until 1972 owing to classification. The same compound was independently discovered in 1970 by Bougon and co-workers (37, 39). A minor modification of Bougon's synthesis by Züchner and Glemser also produced (300) ClF_3O. All the data on ClF_3O, except for the short note by Züchner et al., were obtained either at Rocketdyne or at the Centre d'Etudes Nucléaires de Saclay.

Owing to its pseudotrigonal bipyramidal structure with two highly polar Cl—F bonds, ClF_3O possesses only low kinetic stability. This renders it a powerful fluorinating and oxygenating agent requiring the use of metal or Teflon or Kel-F equipment for its handling.

1. Synthesis

Several synthetic routes to ClF_3O were developed at Rocketdyne (226, 228–231, 240). One of these involves the fluorination of Cl_2O at −78°C:

$$2F_2 + Cl_2O \longrightarrow ClF_3O + ClF$$

$$3F_2 + Cl_2O \longrightarrow ClF_3O + ClF_3$$

When no catalyst is used or if KF and NaF are present as catalysts, ClF is the main by-product. When the more basic alkali metal fluorides, RbF and CsF, are used, ClF_3 is the favored coproduct. The formation of ClF_3 rather than ClF is presumably associated with the more ready formation of ClF_2^- intermediates with RbF and CsF. Yields of ClF_3O from Cl_2O are rather variable and may be affected by the particular alkali fluoride present. Yields of over 40% have been consistently obtained and have reached over 80% using either NaF or CsF. Since NaF does not form an adduct with ClF_3O (64), stabilization of the product by complex formation does not seem to influence the ClF_3O yields strongly.

Owing to unpredictable explosions experienced with liquid Cl_2O, attempts were made to circumvent the Cl_2O isolation step. For this purpose, the crude Cl_2O, still absorbed on the mercuric salts, was directly fluorinated. Again, ClF_3O was formed, but its yield was too low to make this synthetic route attractive.

The fluorination of solid Cl_2O to ClF_3O proceeded at temperatures as low as $-196°C$ provided the fluorine was suitably activated by methods such as glow discharge. Unactivated fluorine did not interact with Cl_2O at $-196°C$. The relatively low yield of ClF_3O (1–2%) makes this modification impractical.

The low-temperature fluorination of $NaClO_2$ produced ClF_3O in low yield (*175*, *226*). However, the low yields and poor reproducibility make this route unattractive.

The method (*176*, *226*, *229*) most suitable for the preparation of ClF_3O on a larger scale involves the fluorination of chlorine nitrate at $-35°C$ according to:

$$2F_2 + ClONO_2 \longrightarrow ClF_3O + FNO_2$$

The main advantages of this process are (*a*) less fluorine is required than in the fluorination reactions of Cl_2O yielding ClF_3 as a coproduct, (*b*) the great difference in the volatilities of products FNO_2 and ClF_3O ($\Delta T_{bp} \sim 100°C$) permits an easy separation by fractional condensation, and (*c*) chlorine nitrate can be made more conveniently and, most importantly, does not appear to be hazardous in its handling. Yields of ClF_3O using $ClONO_2$ as a starting material are somewhat higher than those from Cl_2O.

In the fluorination of both Cl_2O and $ClONO_2$, side reactions compete with the actual fluorination step. These are caused by thermal decomposition of the starting materials due to inefficient removal of the heat of reaction. Hence, the rate of the competing reactions is markedly affected by the reaction temperature. At reaction temperatures near or above ambient, the decomposition of the hypochlorite appears to be favored and little or no ClF_3O is formed, resulting in rapid, rather uncontrolled reactions. Apparently, thermal decomposition preceding the fluorination step yields only intermediates incapable of producing ClF_3O. Thus, in order to maximize the desired fluorination reaction, long reaction times at low temperature ($T < 0°C$) are indicated.

A convenient laboratory method for the synthesis of ClF_3O involves UV photolysis of systems containing Cl, F, and oxygen starting materials. At Rocketdyne (*228*, *230*, *240*), ClF_3O was prepared from seven different systems, including a direct synthesis from the elements Cl_2, F_2, and O_2. Bougon *et al.* (*37*, *39*) obtained ClF_3O in high yield from $ClF_3 + OF_2$. The latter synthesis was modified by Züchner *et al.* (*300*) by replacing ClF_3 with ClF_5.

In small-scale operations, ClF_3O can conveniently be purified by complexing it with KF at room temperature. Impurities, such as $FClO_2$, that do not form an adduct under these conditions can be pumped off.

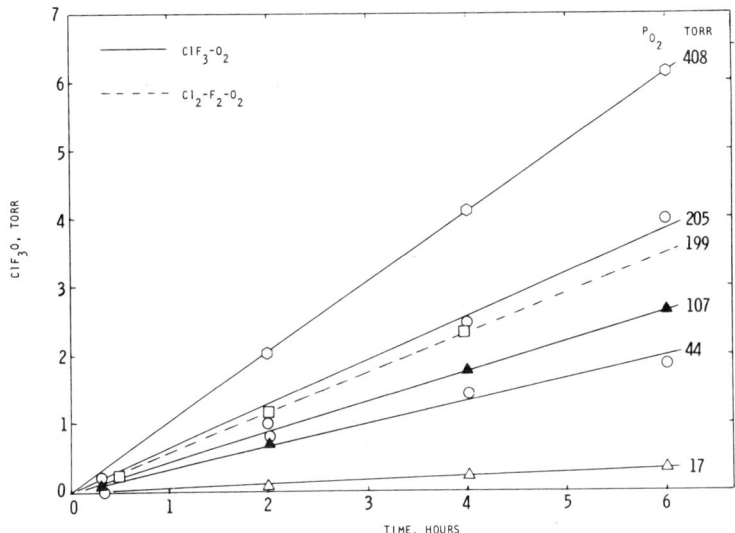

FIG. 4. Chlorine trifluoride oxide formed as a function of time and oxygen partial pressure ($P_{ClF_3} = 10$ torr).

Pure ClF_3O can be obtained by vacuum pyrolysis at 50° to 70°C, whereas compounds, such as ClF_3, which form a more stable KF adduct remain complexed (226).

A detailed kinetic study of the photolyses of the ClF_3—O_2 and of the Cl_2—F_2—O_2 systems was carried out by Axworthy et al. (10). Contrary to the original report (228), the rate of ClF_3O formation was demonstrated to be the same for both systems, to increase with O_2 concentration, and to be independent of irradiation time (Fig. 4). Furthermore, the rate of ClF_3O formation was shown to be proportional to the intensity of the 1847 Å band of the Hg spectrum indicating that the dissociation of O_2 to two ground-state, ^3p, oxygen atoms is the primary photochemical process. The following mechanism was proposed which requires the photochemical dissociation of ClF_3 as well:

$$O_2 \xrightarrow{h\nu\ (1847\ \text{Å})} O + O$$
$$ClF_3 \xrightarrow{h\nu\ (2000-3500\ \text{Å})} ClF_2 + F$$
$$O + ClF_2 \longrightarrow ClF_2O$$
$$ClF_2O + F_2 \longrightarrow ClF_3O + F$$

The photolysis of ClF_3 was investigated under similar conditions. A photochemical steady state was quickly achieved, where $[F_2] = [ClF] = \alpha[ClF_3]$, and α has a value of about 1 at low and of about 3 at high pressures. These results together with the known photochemical decomposition of OF_2 (113) explain why ClF_3O can be readily generated by the photolysis of so many different starting materials.

2. Molecular Structure

Although the exact geometry of ClF_3O has not yet been established, its approximate structure is known from vibrational and ^{19}F NMR spectroscopy. Its UV spectrum has also been reported (228).

The ^{19}F NMR spectrum of ClF_3O was studied by several groups. A single signal at $\phi = -262$ (226) or -253 (300) ppm was reported for liquid ClF_3O. For the gas, a singlet at $\phi = -327$ ppm was observed (226).

TABLE VI

VIBRATIONAL SPECTRA OF ClF_3O GAS AND LIQUID AND THEIR ASSIGNMENT IN POINT GROUPS $C_s{}^a$

Observed frequencies (cm^{-1}) and relative intensities					Approximate description of mode
Gas		Solid matrix IR	Liquid Raman	Assignment	
IR	Raman				
1228 ⎫ 1224 ⎬ s 1218 ⎪ 1213 ⎭	1222 (1.5) p 1211 (0.5) p	1223 s 1212 m	1224 (1.0) p	$\nu_1(A')$ $\nu_1(A')$	$\nu(^{35}Cl{=}O)$ $\nu(^{37}Cl{=}O)$
701 ⎫ 684 ⎪ 676 ⎬ vs 666 ⎭	694 (2.6) p 686 sh, p	686 s 678 m 652 vs 641 s 499 m	689 (2.7) p	$\nu_2(A')$ $\nu_2(A')$ $\nu_7(A'')$ $\nu_7(A'')$ $\nu_8(A'')$	$\nu(^{35}Cl{-}F')$ $\nu(^{37}Cl{-}F')$ $\nu_{as}(F^{35}ClF)$ $\nu_{as}(F^{37}ClF)$ $\delta_{rock}(O^{35}ClF')$
501 ⎫ 491 ⎬ ms 481 ⎭	500 (1) 489 (1) 482 (10) p	498 sh 486 mw 484 w 478 mw	497 sh 466 (10)	$\nu_8(A'')$ $\nu_3(A')$ $\nu_3(A')$ $\nu_4(A')$	$\delta_{rock}(O^{37}ClF')$ $\delta_{sciss}(O^{35}ClF')$ $\delta_{sciss}(O^{37}ClF')$ $\nu_s FClF$
412 w	414 (0.2) dp	414 w	405 (0.5) sh	$\nu_9(A'')$	τ
323 ⎫ m 313 ⎭	319 (0.1)	323 mw	316 (0.3) p	$\nu_5(A')$	$\delta_s FClF$ out of FClF plane $\equiv$ $\delta_{wag} OClF'$
230 mw	224 (0.4) p		227 (1.2) p?	$\nu_6(A')$	$\delta_s FClF$ in FClF plane

^{a} Data from Christe and Curtis (55).

From nuclear relaxation time measurements, Alexandre and Rigny (*3*) were able to determine the chemical shift difference between the equatorial and the 2 axial fluorine atoms as 50 ± 2 ppm. They also obtained a value of 195 Hz for the mean Cl—F coupling constant and values for the exchange time between the fluorine atoms.

Vibrational spectroscopy (*37, 55, 300*) provided the best evidence for ClF_3O possessing a pseudotrigonal bipyramidal structure of symmetry C_s, in which 2 fluorines occupy the axial and 1 fluorine, 1 oxygen, and a sterically active free valence electron pair occupy the equatorial positions (see structure III). At Rocketdyne (*55*), a thorough spectroscopic study was carried out including the infrared spectra of gaseous, solid, and matrix-isolated ClF_3O and the Raman spectra of the gas and the liquid.

TABLE VII

Internal Force Constants of ClF_3O[a,b]

f_D	9.37	f_{rr}	0.26
f_R	3.16	$f_{\beta\beta}$	0.11
f_r	2.34	$f_{\gamma\gamma}$	0.13
f_α	1.84	$f_{r\beta} = -f_{r\beta'}$	0.25
f_β	1.69	$f_{\beta\gamma} = f_{\beta\gamma'}$	0.22
f_γ	1.87		

[a] Data from Christe and Curtis (*55*).
[b] Stretching constants in mdyn/Å, deformation constants in mdyn Å/radian², and stretch–bend interaction constants in mdyn/radian.

The observed spectra agree well with those reported by the other groups (*37, 300*), although the latter was incorrectly assigned. The best assignment (*55*) is given in Table VI. A normal coordinate analysis was also carried out for ClF_3O and a modified valence force field was computed (*55*) using the observed ^{35}Cl—^{37}Cl isotopic shifts. Table VII summarizes the internal force constants thus obtained. The geometry of ClF_3O assumed for this computation was $D(ClO) = 1.42$, $R(ClF_{eq}) = 1.62$, and $r(ClF_{ax}) = 1.72$ Å based on the known geometry of ClF_3 and Robinson's correlation between bond length and stretching frequency (*236, 237*). In the absence of exact structural data, the following ideal bond angle values were assumed: α (OClF′) = 120° and β (OClF) = γ (FClF′) = 90°. However, increased repulsion from the free valence electron pair on chlorine and the double-bonded oxygen should cause some deviations from this ideal structure (see Section II, A).

The force constants of greatest interest are the stretching force constants. The value of 9.37 mdyn/Å obtained for $f_{Cl=O}$ is similar to those computed for $FClO_2$ and ClO_2^+ (see Table I) indicating double-bond character. The value of 2.34 mdyn/Å computed for the axial Cl-F stretching force constant f_r is almost identical with that of 2.34 mdyn/Å, previously calculated (63) for ClF_2^-. The corresponding interaction constant, f_{rr}, is also very similar for both species. The relatively low value of f_r in ClF_2^- has previously been interpreted (63) in terms of semi-ionic 3 center–4 electron bonds. The same reasoning holds for the axial ClF bonds of ClF_3O. It should be pointed out, however, that in ClF_3O, enhancement of the ionic character of the axial ClF bonds is due to oxygen substitution, whereas in ClF_2^- it is due to the formal negative charge. The value of 3.16 mdyn/Å computed for the equatorial ClF bond of ClF_3O is considerably larger than that of the axial bonds, indicating predominantly covalent bonding. These results are in excellent agreement with a generalized bonding scheme discussed in Section II, C and suggest that the overall bonding in ClF_3O might be described by the following approximation. The bonding of the three equatorial ligands (including the free electron pair on Cl as a ligand and ignoring the second bond of the Cl=O double bond) is mainly due to a sp^2 hybrid, whereas the bonding of the two axial ClF bonds involves mainly one delocalized p-electron pair of the chlorine atom for the formation of a semi-ionic 3 center–4 electron pσ bond.

3. Physical Properties

Chlorine trifluoride oxide is colorless as a gas or liquid and white in the solid state. Some of its properties are summarized in Table VIII. The vapor pressure of the liquid can be described according to the Rocketdyne study (226) by the equation

$$\log P(\text{mm}) = 8.433 - \frac{1680}{T(°K)}$$

or, according to Bougon et al. (31), by

$$\log P(\text{mm}) = 8.394 - \frac{1655}{T(°K)}$$

Vapor density measurements (37, 226) and mass spectroscopy (226, 300) were used to show that ClF_3O is monomeric in the gas phase. The relatively high boiling point and Trouton constant of ClF_3O imply its association in the liquid phase. More specific evidence about the nature of this association was obtained from the vibrational spectra

TABLE VIII

Some Properties of ClF_3O

Property	Value	Ref.
Melting point	$-42°$ to $-44.2°C$	(16, 37, 226)
Boiling point	$29°$ or $27°C$	(37, 226)
ΔH_{fusion}	1.975 kcal mole^{-1}	(16)
ΔS_{fusion}	8.63 e.u.	(16)
ΔH_{vap}	7.7 or 7.57 kcal mole^{-1}	(37, 226)
Trouton constant	25.4 or 25.2 e.u.	(37, 226)
Density(l; 20°C)	1.865 gm ml^{-1}	(226)
$\Delta H°_{f\,298}(g)$	-36.5^a or -35.3^b kcal mole^{-1}	(15, 16, 269)
$\Delta H°_{f\,298}(l)$	$-44.1^{a,\,c}$, $-42.9^{b,c}$, or -38.7^b kcal mole^{-1}	(16, 152, 269)

^{a} Corrected for $\Delta H°_{f\,HF(g)} = -65.14$ kcal mole^{-1} (83).

^{b} Corrected for $\Delta H°_{f\,HF(soln)(75H_2O)} = -77.04$ kcal mole^{-1} (151).

^{c} Using the $\Delta H°_{f\,298(g)}$ values of Barberi (16) and Sinke (269) for the gas and the above listed $\Delta H_{vap} = 7.6$ kcal mole^{-1}.

recorded for the liquid and the solid and from a controlled diffusion experiment carried out for matrix-isolated ClF_3O. It was concluded (55) that association appears to involve exclusively the axial fluorine atoms. This finding agrees with the association proposed by Frey et al. (102) for the structurally related, trigonal bipyramidal molecules SF_4 and ClF_3.

The thermodynamic properties were computed with the molecular geometry and vibrational frequencies given above assuming an ideal gas at 1 atm pressure and using the harmonic-oscillator rigid-rotor approximation. These properties are given for the range 0–2000°K in the Appendix (Table AI).

4. Chemical Properties

Chlorine trifluoride oxide is stable at ambient temperature and can be stored and handled in well-passivated metal, Teflon, or Kel-F containers without decomposition. Its thermal stability is intermediate between that of ClF_3 and ClF_5. When heated to 280–300°C in a Monel cylinder (37, 226), or to 200°C in a stainless steel cylinder, or to 350°C in a flow system (226), ClF_3O decomposes:

$$ClF_3O \longrightarrow ClF_3 + \tfrac{1}{2}O_2$$

It reacts rapidly with glass or quartz and, therefore, cannot be handled in standard glass vacuum systems (226). It reacts with numerous materials causing oxidation through both fluorination and oxygenation. With hydrogen-containing species, these reactions may occur at quite low temperature and with hydrocarbon type compounds are generally explosive. However, many chlorine-, fluorine-, or oxygen-substituted compounds, even with lower valent central atoms, react only slowly at ambient temperature, or not at all. Thus, no reaction was observed at room temperature between ClF_3O and chlorine, chlorine fluorides, chlorine oxyfluorides, and the nitrogen fluorides, FNO, FNO_2, NF_3, and N_2F_4 (246). However, elevated temperatures or UV photolysis have resulted in appreciable reaction of all compounds examined. With Cl_2 no interaction was detected at 25°C, but at 200°C the following reaction occurred:

$$ClF_3O + Cl_2 \longrightarrow 3ClF + 0.5 O_2$$

Chlorine monoxide and ClF_3O reacted slowly at room temperature (246):

$$ClF_3O + Cl_2O \longrightarrow 2ClF + FClO_2$$

Similarly, $ClOSO_2F$ interacts with ClF_3O (246):

$$ClF_3O + 2ClOSO_2F \longrightarrow S_2O_5F_2 + FClO_2 + 2ClF$$

and

$$ClF_3O + ClOSO_2F \longrightarrow SO_2F_2 + FClO_2 + ClF$$

All these reactions can be rationalized in terms of a reduction of ClF_3O to the unstable FClO (see Section III, A) which readily decomposes to $FClO_2$ and ClF. At elevated temperature, $FClO_2$ may decompose further to $ClF + O_2$ (24, 137, 183).

Several reaction systems were discovered in which, in addition to fluorination, oxygenation also occurred. These include SF_4 (60); N_2F_4, HNF_2, and F_2NCFO (246, 248); and MoF_5 (35). In the following equations, the end products observed for the SF_4–ClF_3O reaction are underlined:

$$ClF_3O + SF_4 \xrightarrow[25°]{CsF} \underline{SF_6} + FClO$$

$$2FClO \longrightarrow FClO_2 + ClF$$

$$SF_4 + ClF \xrightarrow{CsF} \underline{SF_5Cl}$$

$$SF_4 + ClF_3O \longrightarrow \underline{SF_4O} + \underline{ClF_3}$$

$$SF_4 + ClF_3 \longrightarrow \underline{SF_6} + \underline{ClF}$$

In the MoF_5–ClF_3O system, both MoF_6 and MoF_4O were formed, followed by adduct formation. With N_2F_4, an appreciable reaction rate was observed only above 100°C:

$$\text{ClF}_3\text{O} + 2\text{N}_2\text{F}_4 \longrightarrow 3\text{NF}_3 + \text{FNO} + \text{ClF}$$

In addition to these products, small amounts of NF_3O were obtained. The yield of NF_3O from this reaction system could be increased to about 5% when UV irradiation was used. Higher yields of NF_3O (~70%) could be obtained at low temperature from HNF_2 and ClF_3O:

$$\text{ClF}_3\text{O} + 2\text{HNF}_2 \longrightarrow [\text{FClO}] + 2\text{HF} + 2\text{NF}_2\cdot$$

$$2\text{NF}_2\cdot + [\text{FClO}] \longrightarrow \text{NF}_2\text{Cl} + \text{NF}_3\text{O}$$

with the side reactions,
$$2\text{NF}_2\cdot \longrightarrow \text{N}_2\text{F}_4$$

$$2[\text{FClO}] \longrightarrow \text{FClO}_2 + \text{ClF}$$

$$\text{ClF} + \text{HNF}_2 \longrightarrow \text{HF} + \text{NF}_2\text{Cl}$$

The reaction between difluoraminocarbonyl fluoride, F_2NCFO, and ClF_3O yielded again NF_3O and ClNF_2 in nearly equimolar amounts. However, the yields were much lower (20% based on ClF_3O consumed) with N_2F_4 being the main N—F containing product.

One reaction was discovered (246) in which ClF_3O did not act as an oxidizing but rather as a reducing agent. With the powerful oxidizer PtF_6, it reacted according to

$$\text{ClF}_3\text{O} + \text{PtF}_6 \longrightarrow \text{ClF}_2\text{O}^+\text{PtF}_6^- + 0.5\text{F}_2$$

The interaction of ClF_3O with HF, resulting in a fluoride ion abstraction to give the ClF_2O^+ cation (38), will be discussed below. With H_2O, an excess of chlorine trifluoride oxide hydrolyzes (226) according to

$$\text{ClF}_3\text{O} + \text{H}_2\text{O} \longrightarrow \text{FClO}_2 + 2\text{HF}$$

Mixtures of ClF_3O and ClF_5 (225) hold promise as an oxidizer in rocket propulsion.

As discussed in Section II, D, the compound ClF_3O has an energetically unfavorable pseudotrigonal bipyramidal structure. Consequently, it exhibits a pronounced tendency to form adducts with both strong Lewis acids and bases. Adducts containing the ClF_2O^+ cation (see Section III, C) were obtained (33–35, 38, 58, 64, 246, 300) with the following Lewis acids: BiF_5, SbF_5, AsF_5, PF_5, TaF_5, NbF_5, VF_5, PtF_5, UF_5, MoF_4O, SiF_4, BF_3, and HF. With WF_4O and UF_4O, no stable ionic products were formed (35) in spite of the fact that WF_4O is a stronger Lewis acid than MoF_4O. This is caused by the increased tendency of WF_4O to enter the following oxygen–fluorine exchange reaction:

$$\text{ClF}_3\text{O} + \text{MF}_4\text{O} \longrightarrow \text{FClO}_2 + \text{MF}_6 \quad (\text{M} = \text{W or U})$$

Adducts containing the ClF_4O^- anion (see Section III, D) were prepared (*56, 64, 300*) by reaction of ClF_3O with the Lewis bases CsF, RbF, and KF. With the weaker bases FNO and FNO_2, it does not interact even at $-95°C$ (*64*).

C. DIFLUOROOXYCHLORONIUM(V) CATION

Compounds containing the ClF_2O^+ cation with the following counterions are known: BiF_6^-, SbF_6^-, $Sb_2F_{11}^-$, AsF_6^-, PF_6^-, TaF_6^-, NbF_6^-, VF_6^-, PtF_6^-, UF_6^-, SiF_6^{2-}, BF_4^-, HF_2^-, MoF_5O^-, and $Mo_2F_9O_2^-$ (*33–35, 38, 58, 64, 246, 300*).

1. Synthesis

With the exception of the PtF_6^- salt which was prepared from ClF_3O and PtF_6 [(*246*), Section III, B, 4], all the other salts were prepared by direct combination of ClF_3O with the corresponding Lewis acid. When the Lewis acid is a solid at the reaction temperature, or nonvolatile, it is advisable to use either a large excess of ClF_3O or anhydrous HF as a solvent to avoid polyanion formation (*33–35, 64*).

2. Molecular Structure

The ionic nature of $ClF_3O \cdot$ Lewis acid adducts was established by vibrational (*33–35, 38, 58, 300*) and ^{19}F NMR (*61*) spectroscopy.

The NMR spectrum of $ClF_2O^+AsF_6^-$ in anhydrous HF showed (*61*) the characteristic quadruplet of AsF_6^- at $\phi = 67.5$ ppm in addition to a single signal due to rapidly exchanging HF and ClF_2O^+. Upon acidification of the HF solvent with AsF_5, a separate signal at $\phi = -272$ ppm was observed for ClF_2O^+ in addition to a single signal due to HF, AsF_6^-, and AsF_5. For $ClF_2O^+PtF_6^-$ in HF the ClF_2O^+ signal was also found at $\phi = -272$ ppm. The observation of a singlet for ClF_2O^+ shows the magnetic equivalence of the 2 fluorine atoms.

The vibrational spectra were reported (*33–35, 38, 58, 300*) for all of the above-listed ClF_2O^+ salts. In addition to the bands characteristic of the anions, all spectra exhibited bands with frequencies and relative intensities similar to those shown in Table IX. These are characteristic for the ClF_2O^+ cation. The vibrational spectrum of ClF_2O^+ closely resembles that of isoelectronic SF_2O and, therefore, could be readily assigned. The only ambiguity in the assignment existed (*34, 58*) for the two deformation modes occurring in the 380–400 cm^{-1} region. Recent Raman polarization measurements (*34*) have shown that the 400-cm^{-1} band belongs most likely to ν_4 (A'), and the 380-cm^{-1} band to ν_6 (A").

TABLE IX

VIBRATIONAL SPECTRUM OF THE ClF_2O^+ CATION

Raman (HF solution) (cm^{-1})	IR (solid) (cm^{-1})	Assignment in point group C_s	Approx. description of mode
1333 (4) ⎫ p 1322 sh ⎭	1334 s ⎫ 1323m ⎭	ν_1 (A′)	νClO
741 (10) p	734 m	ν_2 (A′)	ν_sClF$_2$
715 (1)	694 s	ν_5 (A″)	ν_{as} ClF$_2$
512 (2) p	512 s	ν_3 (A′)	δ_s OClF$_2$
404 (2) p	405 m	ν_4 (A′)	δ_{sciss} ClF$_2$
383 (1)	383 m	ν_6 (A″)	δ_{as} OClF$_2$

The spectroscopic evidence is consistent with the following structure of symmetry C_s for ClF_2O^+:

$$\left[\begin{array}{c} \overset{|}{F-Cl\!\!\!\diagdown_O} \\ F \end{array} \right]^+$$

(IV)

A normal coordinate analysis was carried out (58) for ClF_2O^+ assuming the following geometry: $R_{ClO} = 1.41$ Å; $r_{ClF} = 1.62$ Å, $\beta(OClF) = 108°$; and $\alpha(FClF) = 93°$. A modified valence force field was computed, and the results are given in Table X. As can be seen from Table I, the ClO-

TABLE X

VIBRATIONAL FORCE CONSTANTS OF $ClF_2O^{+\,a,b}$

f_R	11.20
f_r	3.44
f_β	1.65
f_α	1.78
$f_{\beta\beta}$	0.21
f_{rr}	0'39

[a] Data from Christe et al. (58).
[b] Stretching constants in mdyn/Å and deformation constants in mdyn Å/radian^2.

stretching force constant of ClF_2O^+ exhibits a high value, implying that the positive charge in ClF_2O^+ is partially located on the oxygen atom and that contributions from resonance structures, such as VI,

$$\begin{array}{cc} F-\overset{\overset{|}{Cl^+}}{\underset{F}{}}\!\!\!\diagdown\!\!O/ & \longleftrightarrow & F-\overset{\overset{|}{Cl}}{\underset{F}{}}\!\!\!\Longequal\!O/^+ \\ (V) & & (VI) \end{array}$$

are significant. The ClF-stretching force constant of ClF_2O^+ is within the range expected for a predominantly covalent ClF bond (see Table I and discussion in Section II, C).

3. *Properties*

Except for the following salts, the above-listed ClF_2O^+ salts are stable, white, crystalline solids. The UF_6^- salt is blue-green and of marginal stability at ambient temperature. In HF solution or during exposure of the solid to a laser beam, the UF_6^- anion is slowly oxidized by ClF_2O^+ to UF_6 (*33*). For MoF_5 this instability of the pentavalent metal toward oxidation to the hexavalent state is even more pronounced. When ClF_3O and MoF_5 are combined, no stable MoF_6^- salt is formed, but MoF_6 and MoF_4O are the products with the latter being capable of forming stable adducts (*35*). The $ClF_2O^+PtF_6^-$ salt is a canary yellow solid (*246*). The VF_5 and PF_5 adducts exhibit dissociation pressures of 2.5 and 3.5 mm, respectively, at room temperature (*33, 34*). The $(ClF_2O^+)_2\, SiF_6^{2-}$ salt is unstable at room temperature. It reaches a dissociation pressure of 760 mm at 31°C and its dissociation pressure can be represented (*64*) by the equation

$$\log P(\text{mm}) = 11.8018 - \frac{2712.3}{T(°K)}$$

From these data, the heat of dissociation, $\Delta H_d^0 = 37.24\,\text{kcal mole}^{-1}$, and the heat of formation of the solid adduct, $\Delta H^0_{f298} = -495.7$ kcal mole^{-1} were obtained. For the latter the literature value was corrected by using the more precise value of -35.9 kcal mole^{-1} for the heat of formation of gaseous ClF_3O (see Table VIII). The adduct melts under its own vapor pressure at 50.5°C (*300*).

The Raman spectrum of a solution of ClF_3O in anhydrous HF shows no bands due to ClF_3O but only those of ClF_2O^+ in agreement with the following ionization scheme (*38*):

$$\text{ClF}_3\text{O} + \text{HF} \longrightarrow \text{ClF}_2\text{O}^+ + \text{HF}_2^-$$

However, no attempts were reported to isolate the neat solid at low temperature and to examine its thermal stability.

The thermal stability of the adducts depends on the strength of the Lewis acids and decreases for the ClF_2O^+ salts in the following order: $\text{SbF}_5 > \text{AsF}_5 > \text{BF}_3 > \text{VF}_5 > \text{PF}_5 > \text{SiF}_4 > \text{HF}$ (33–35, 38, 64). The $\text{ClF}_2\text{O}^+\text{MoF}_5\text{O}^-$ salt, when heated in vacuum to 75–80°C or when dissolved in anhydrous HF, is converted to ClF_2O^+, $\text{Mo}_2\text{F}_9\text{O}_2^-$, and ClF_3O. It was shown by Raman spectroscopy that this reaction is reversible. Heating of these compounds to higher temperatures results in decomposition to MoF_4O and ClO_2^+ salts of MoF_4O (35).

The X-ray powder patterns were reported for the XF_6^- type (34, 64, 246) and the BF_4^- (64) adduct and were tentatively indexed in the orthorhombic system.

D. TETRAFLUOROOXYCHLORATE(V) ANION

The existence of adducts between ClF_3O and CsF (56, 64, 300), RbF (64), and KF (64) has been reported. It was shown (56, 300) by vibrational spectroscopy that these adducts are ionic and contain the ClF_4O^- anion.

1. Synthesis and Properties

Chlorine trifluoride oxide was found (64) to combine readily with the alkali metal fluorides, CsF, RbF, or KF, at room temperature to form white stable adducts. High conversion to the 1:1 adduct appears to be easiest for CsF. The use of a large excess of ClF_3O, agitation, and extended contact times are conducive to nearly complete conversions. These alkali metal ClF_4O^- salts have found use in the purification of ClF_3O (226). The thermal stability of the adducts decreases in the order CsF > RbF > KF. For example, the KClF_4O salt can be decomposed by vacuum pyrolysis at 50–70°C (226), whereas a much higher temperature is required for the pyrolysis of CsClF_4O.

2. Molecular Structure

The ionic nature of these adducts and the structure of the ClF_4O^- anion were established by vibrational spectroscopy (56, 300). It was shown (56) that the observed vibrational spectrum (Table XI) is consistent with the following structure of symmetry C_{4v}:

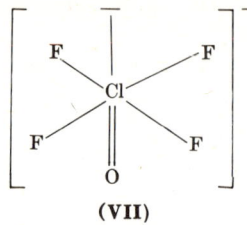

(VII)

TABLE XI

VIBRATIONAL SPECTRA OF $Rb^+ClF_4O^-$ AND $Cs^+ClF_4O^-$ AND THEIR ASSIGNMENT[a]

Observed frequencies (cm^{-1}) and relative intensities				Assignment for XZF_4 in point group C_{4v}		Type of vibration
$Rb^+ClF_4O^-$		$Cs^+ClF_4O^-$				
IR	Raman	IR	Raman			
1216 s	1211 (0.6)	1201 s	1203 (0.6)	A_1	ν_1	ν XZ
462 w	461 (10)	457 w	456 (10)		ν_2	ν_{sym} in-phase XF_4
339 s	[350][b]	339 s	[345][b]		ν_3	δ_{sym} out-of-plane XF_4
	350 (4.3)		345 (4)	B_1	ν_4	ν_{sym} out-of-phase XF_4
					ν_5	δ_{asym} out-of-plane XF_4
283 vw	285 (0.4)	280 vw	283 (0.4)	B_2	ν_6	δ_{sym} in-plane XF_4
600 ⎫ vs	599 (0.1)	600 ⎫ vs	594 (0.2)	E	ν_7	ν_{asym} XF_4
500 ⎭	557 (0.4)	560 ⎭	564 (0.3)			
415 ⎫ s	416 (1.4)	415 ⎫ s	416 (1.4)			
394 ⎭	395 (0.1)	396 ⎭	397 (0.1)		ν_8	δ ZXF
	213 (0.6)		204 (0.7)		ν_9	δ_{asym} in-plane XF_4

[a] Data from Christe and Curtis (56).
[b] Calculated frequency.

A normal coordinate analysis was carried out (56) for ClF_4O^- assuming the following geometry: $D(ClO) = 1.42$ Å; $r(ClF) = 1.75$ Å; and all bond angles are 90°. The internal force constants obtained are listed in Table XII. Comparison with the stretching force constants of other chlorine fluorides oxides (see Table I) shows that the ClO bond in ClF_4O^- has full double-bond character, but that the ClF bond is a rather weak semi-ionic 3 center–4 electron bond. This implies that the formal negative charge in ClF_4O^- is distributed almost exclusively over the four fluorine

TABLE XII

Force Constants of ClF_4O^- [a,b]

f_R	9.13		
f_r	1.79	$f'_{\beta\beta}$	0.29
f_β	1.33	$f'_{\alpha\alpha}$	0.08
f_α	0.61	$f_{r\beta}$	0.15
f_{rr}	0.25	$f''_{r\beta}$	−0.15
f'_{rr}	0.04		

[a] Data from Christe and Curtis (56).
[b] Stretching force constants in mdyn/Å and deformation force constants in mdyn Å/radian².

ligands. Resonance structures of the following type can be used to describe this effect:

$$\underset{(VIII)}{\overline{|F|}\overset{\overline{F}\diagdown\underset{\parallel}{Cl}\diagup F}{\underset{O}{}}F} \longleftrightarrow \underset{(IX)}{F\overset{\overline{F}\diagdown\underset{\parallel}{Cl}\diagup F}{\underset{O}{}}\overline{|F|}^{(-)}} \text{ etc.}$$

E. Chlorine Pentafluoride Oxide

The synthesis of ClF_5O was claimed in 1972 by Züchner and Glemser (300) by UV photolysis of a mixture of ClF_5 and OF_2 in a nickel vessel fitted with a sapphire window. Although the authors failed to isolate a pure product, they "identified" ClF_5O in the product mixture by negative-ion mass spectroscopy and ^{19}F NMR spectroscopy. However, the following properties attributed to ClF_5O do not agree with the general trends observed for the remaining chlorine fluorides and oxyfluorides: (a) low volatility at −78°C, (b) a ^{19}F NMR resonance between −146 and −103 ppm relative to $CFCl_3$, and (c) exchange broadening in the NMR spectrum even at −76°C. For ClF_5O, we would expect (a) a volatility comparable to that of ClF_5 (227) or SF_6 (279), (b) an averaged ^{19}F NMR chemical shift of about −390 ppm (61), and (c) the absence of intramolecular exchange owing to chlorine having its maximum coordination number and no free valence electron pair, and owing to the lack of a plausible exchange mechanism.

Attempts to duplicate Züchner and Glemser's experiment (300) at Rocketdyne and the Centre d'Etudes Nucleaires de Saclay did not result in any evidence for ClF_5O. In the Rocketdyne study, the progress of the ClF_5–OF_2 photolysis in the temperature range $-78°$ to $30°C$, using both unfiltered and Pyrex-filtered UV radiation, was continuously monitored by gas chromatography. At the end of an experiment, the products were also separated by fractional condensation in a Teflon–stainless steel vacuum system and were characterized by vibrational spectroscopy. It was shown that in the ClF_5–OF_2 system, when exposed to unfiltered UV radiation, ClF_5 rapidly decomposes to ClF_3 and F_2 and, therefore, yields only the same products obtainable from the photolysis of ClF_3–OF_2 mixtures, i.e., mainly ClF_3O.

The ^{19}F NMR spectrum observed by Züchner and Glemser (300) might be rationalized in terms of a rapidly exchanging mixture of ClF_3 and ClF_3O as was pointed out to us by Dr. Bougon. To verify this, we have recorded the NMR spectra of ClF_3–ClF_3O mixtures over the temperature range $40°$ to $-102°C$. It was found that mixtures of pure ClF_3 and ClF_3O yield separate signals for ClF_3 (at about $\phi = -118$ and -10 ppm) and ClF_3O (at about -269 ppm) over the whole temperature range studied. However, upon addition of about 5 mole% of HF, one single signal is observed for all three species with a chemical shift corresponding to the averaged chemical shifts of the three components. The temperature dependence of these spectra is similar to that reported by Züchner and Glemser (300). At $40°C$ the signal was rather broad, became narrower with decreasing temperature, but broadened below $-60°C$ and shifted to higher field. At $-102°C$ a new and relatively narrow signal appeared significantly shifted upfield. The observed shifts and temperature dependence of these spectra are strongly dependent on the exact composition of the mixture.

The only remaining piece of evidence presented (300) for ClF_5O was the presence of a low-intensity fragment due to ClF_4O^- in the negative ion mass spectrum of the crude reaction product. However, this fragment might be attributed to a recombination process in the mass spectrometer since a 52% peak was also reported for F_2^- which can form only by recombination. Furthermore, negative-ion spectra frequently show species of higher mass than that of the parent molecule due to attachment of other atoms or groups (28), as was recently also demonstrated for BrF_5, of which the negative-ion spectrum shows a rather intense BrF_6^- fragment (194). In agreement with the preceding NMR interpretation, the observed (300) negative-ion mass spectrum is best ascribed to a mixture of ClF_3O, ClF_3, $FClO_3$, and some ClF_5, with several higher mass peaks and F_2^- being due to recombination in the spectrometer.

Based on the cited evidence it appears unlikely that Züchner and Glemser (300) had indeed observed ClF_5O.

F. Chloryl Fluoride

Chloryl fluoride,

$$\underset{O}{\overset{\displaystyle |}{\underset{\|}{O=Cl}}}\!\!-\!F$$

(X)

was first obtained in 1942 by Schmitz and Schumacher (256) by the low-temperature fluorination of ClO_2 with F_2. The compound itself has not been studied very intensively although it is the most frequently encountered reaction product in systems involving reactions of chlorine mono-, tri-, or pentafluorides with oxides or hydroxides. Its structure can be derived from a tetrahedron with a free valence electron pair of chlorine occupying one of the four corners. Compared to $FClO_3$, this structure is less symmetric, kinetically less stable, and contains a highly polar, long $(p\!\!-\!\!\pi^*)\sigma$ (see Section II, C) bond. Therefore at moderate temperatures, $FClO_2$ is far more reactive than $FClO_3$ in spite of its lower oxidation state.

1. Synthesis

In our experience (70), $FClO_2$ is most conveniently prepared by combining $NaClO_3$ with an about equimolar amount of ClF_3 at $-196°C$ in a stainless steel cylinder and holding the mixture at room temperature for a day. Chloryl fluoride (bp $= -6°C$) is thus obtained in high yield and can be separated from the by-products O_2, Cl_2 (bp $= -33.8°C$), and unreacted ClF_3 (bp $= 11.75°C$) either by fractional distillation or by repeated fractional condensation through a series of traps maintained at $-95°$, $-112°$, and $-126°C$. This procedure is safe and does not involve the handling of any shock-sensitive materials. It is based on the previous reports by Engelbrecht and Atzwanger (92) and Smith and co-workers (270) that gaseous ClF_3 reacts with $KClO_3$ to give $FClO_2$ in high yield. The substitution of $KClO_3$ by $NaClO_3$ is significant since the product NaF does not form an adduct with ClF_3, whereas KF does. This decreases by 60% the amount of ClF_3 required for the reaction. By analogy with the known $KClO_3 + BrF_3$ reaction (296), the idealized stoichiometry of the above reaction is

$$6\,NaClO_3 + 4\,ClF_3 \longrightarrow 6\,NaF + 2\,Cl_2 + 3\,O_2 + 6\,FClO_2$$

The use of larger than stoichiometric amounts of ClF_3 is advisable to avoid the possible formation of shock-sensitive chlorine oxides.

Woolf's original method (296) involved the use of $KClO_3$ and BrF_3, according to

$$6KClO_3 + 10BrF_3 \longrightarrow 6KBrF_4 + 2Br_2 + 3O_2 + 6FClO_2$$

Although the yield of $FClO_2$ is high, it is very difficult to obtain pure colorless $FClO_2$ by this method. When $KClO_3$ is replaced by $KClO_4$ (285), $FClO_2$ is obtained in 97% yield:

$$3KClO_4 + 5BrF_3 \longrightarrow 3KBrF_4 + Br_2 + 3O_2 + 3FClO_2$$

The product purification problem for this system is analogous to that encountered for the $KClO_3$–BrF_3 system. Direct fluorination of $KClO_3$ with F_2 (31, 89, 92, 265) is not synthetically useful for preparing $FClO_2$, since the main product is always $FClO_3$. The interaction of $HOSO_2F$ with $KClO_3$ was reported (99) to produce $FClO_2$ in 30% yield. However, a study of this system carried out at Monsanto (198) failed to produce $FClO_2$, probably owing to formation of chloryl fluorosulfate.

An alternative route to $FClO_2$ involves the fluorination of chlorine oxides. The resulting $FClO_2$ is usually very pure but the handling of the shock-sensitive chlorine oxides renders these methods unattractive, particularly for the production of larger amounts of material. The original synthesis of $FClO_2$ by Schmitz and Schumacher in 1942 (256) involved direct fluorination of ClO_2. When F_2 was added at $-80°C$ to a quartz vessel containing ClO_2, followed by slow warm-up to $20°C$, $FClO_2$ was formed in a moderate reaction. The most favorable conditions were a reaction time of 2 min, a reaction temperature of $0°C$, and the use of a mixture consisting of 25.6 mm ClO_2, 54.0 mm F_2, and 540.7 mm air. The reaction was found to be homogeneous and bimolecular (12). Modifications of this reaction involve passing gaseous F_2 through liquid ClO_2 at $-50°$ to $-55°C$ (264) or, preferably, using $CFCl_3$ as a solvent at $-78°C$ (162, 254). Chlorine dioxide can also be fluorinated to $FClO_2$ by passing ClO_2 diluted with N_2 at room temperature over AgF_2 or CoF_3 or by passing ClO_2 through liquid BrF_3 at $30°C$ (255).

The fluorination of chlorine oxides other than ClO_2 also produces $FClO_2$. Thus, $FClO_2$ was obtained in yields of up to 75% by fluorination of Cl_2O_6 with F_2 between $22°$ and $48°C$ (7, 8). The high yield of $FClO_2$ coupled with the absence of $FClO_3$ indicates that the primary step is the decomposition of Cl_2O_6 to $2ClO_2 + O_2$ followed by the fluorination of ClO_2 to $FClO_2$. Similarly, the reaction between Cl_2O_6 and FNO_2, when carried out in $CFCl_3$ solution at $0°C$, produces $FClO_2$ in addition to $NO_2^+ClO_4^-$ (255). Chloryl fluoride is also formed during the fluorination of Cl_2O_6 at $-40°C$ with BrF_3 or BrF_5 (294) or with HF (252). During thermal decomposition of Cl_2O_7 in the presence of F_2 at $100°-120°C$ in quartz or Pyrex, $FClO_2$ is formed in addition to $FClO_3$ and ClF (98).

The risk of explosions is somewhat reduced in the ClO_2–AgF_2 reaction when the ClO_2 is replaced by the less dangerous Cl_2O. The yield of $FClO_2$ was 35% (117, 182). Similarly, Cl_2O can be fluorinated at −78°C with either ClF (51),

$$2Cl_2O + ClF \longrightarrow FClO_2 + 2Cl_2$$

or ClF_3O (246),

$$Cl_2O + ClF_3O \longrightarrow FClO_2 + 2ClF$$

Oxygenation of a chlorine fluoride, if possible, would be more attractive than fluorination of the shock-sensitive chlorine oxides. A process for $FClO_2$ has been claimed by Faust et al. (97) furnishing $FClO_2$ in about 50% yield by simply heating a mixture of ClF and O_2 to 80°–90°C. However, attempts in our laboratory (70) to verify this synthesis failed. It appears, that the $FClO_2$ observed by Faust et al. (97) in their experiments was due to hydrolysis of ClF (9, 36, 70).

Numerous reactions have been reported in which $FClO_2$ is formed as a product. Most of these involve the interaction between a chlorine fluoride or oxyfluoride with an oxide or hydroxide. The oxidation state of the chlorine fluoride is not important since +I, +III, and +V compounds all yield $FClO_2$ owing to the tendency of the lower oxyfluorides, such as FClO, to disproportionate. The presence of excess chlorine fluoride is important to avoid formation of chlorine oxides. The following equations are typical examples for these types of reactions:

$$5ClF + 2H_2O \longrightarrow 4HF + FClO_2 + 2Cl_2 \quad (9, 36)$$
$$2ClF_3 + 2H_2O \longrightarrow 4HF + FClO_2 + ClF \quad (9, 36)$$
$$ClF_5 + 2H_2O \longrightarrow 4HF + FClO_2 \quad (227)$$
$$ClF_3O + H_2O \longrightarrow 2HF + FClO_2 \quad (226)$$
$$2ClF_3 + 2HONO_2 \longrightarrow 2HF + FClO_2 + ClF + 2FNO_2 \quad (51)$$
$$ClF_5 + 2HONO_2 \longrightarrow 2HF + FClO_2 + 2FNO_2 \quad (51)$$
$$ClF_3O + 2ClOSO_2F \longrightarrow S_2O_5F_2 + FClO_2 + 2ClF \;\Big\}$$
$$ClF_3O + ClOSO_2F \longrightarrow SO_2F_2 + FClO_2 + ClF \;\Big\} \quad (246)$$
$$2ClF_3 + 3COF_2 \longrightarrow 2CF_4 + FClO_2 + CF_3OCl \quad (288)$$
$$2ClF_3 + UO_2F_2 \longrightarrow UF_6 + FClO_2 + ClF \quad (178, 179, 263)$$
$$6ClF + TeO_2 \longrightarrow TeF_5Cl + FClO_2 + 2Cl_2 \quad (168)$$

2. Molecular Structure

The exact structure of $FClO_2$ was determined by Parent and Gerry (219, 220) using microwave spectroscopy. The molecule was shown to have C_s symmetry with the following internuclear parameters:

r(Cl-O)
1.418 ± 0.002 Å

r(Cl-F)
1.697 ± 0.003 Å

∠ (OClO)
$115.23 \pm 0.05°$

∠ (FClO)
$101.72 \pm 0.03°$

(XI)

Values were also reported for the rotational constants, centrifugal distortion constants, and the chlorine nuclear quadrupole coupling constants of the three isotopic species $^{19}F^{35}Cl^{16}O_2$, $^{19}F^{37}Cl^{16}O_2$, and $^{19}F^{35}Cl^{16}O^{18}O$. The molecular dipole moment was found to be 1.722 ± 0.03 D.

The pyramidal structure of symmetry C_s for $FClO_2$ was also confirmed by vibrational spectroscopy. E. A. Smith et al. (271) and Arvia and Aymonino (6) reported the infrared spectrum of the gas. D. F. Smith et al. (270) studied the infrared spectrum of the gas, measured the ^{35}Cl–^{37}Cl and ^{16}O–^{18}O isotopic shifts, recorded the Raman spectrum of the liquid, and carried out a normal coordinate analysis. The observed frequencies and their assignment are summarized in Table XIII.

Andrews and co-workers have recently reported (5) ^{35}Cl–^{37}Cl and ^{16}O–^{18}O isotopic shifts for the infrared spectrum of argon matrix-isolated $FClO_2$. Tantot (282) has studied in his thesis work the infrared and Raman spectra of the gas, the Raman spectrum of the neat liquid and of HF solutions, and the infrared and Raman spectra of the solid.

Force fields for $FClO_2$ were computed by D. F. Smith et al. (270), Robinson et al. (238), So and Chau (272), and Tantot (282). The force fields computed by Smith et al. (270), So and Chau (272), and Tantot (282) agree relatively well for the two stretching force constants, suggesting values of about 9.0 and 2.5 mdyn/Å for f_{ClO} and f_{ClF}, respectively. Except for Tantot's computation (282) which did not give plausible values for the deformation constants ($f_\alpha > f_\beta$), estimates that significantly deviate from the actual (220) geometry of $FClO_2$ were used for these computations. Since the deformation constants are more likely to be angle-dependent, a recomputation using the exact geometry and the observed (270) isotopic shifts is desirable.

Mean square amplitudes of vibration were calculated by Baran (14) based on the frequencies and estimated geometry reported by Smith et al. (270). The UV absorption spectrum of $FClO_2$ was studied by Sicre and Schumacher (264) and Pilipovich et al. (228). From a mass spectro-

TABLE XIII
Fundamental Vibrational Frequencies of $FClO_2$[a]

	Assignment	$F^{35}ClO_2$ (cm^{-1})	$F^{37}ClO_2$ (cm^{-1})	$F^{35}Cl^{18}O_2$ (cm^{-1})	$F^{37}Cl^{18}O_2$ (cm^{-1})	$F^{35}Cl^{16}O^{18}O$ (cm^{-1})	$F^{37}Cl^{16}O^{18}O$ (cm^{-1})
A'	ν_1 sym ClO_2 stretch	1105.8	1098.4	1060.4	1052.6	1080.7	1072.4
	ν_2 ClF stretch	630.2	621.6	624.7	616.0	628.6	618.6
	ν_3 ClO_2 scissor	546.5	543.0	529.0	—	537.8	534.0
	ν_4 FClO bend	401.6	—	—	—	—	—
A"	ν_5 asym ClO_2 stretch	1271.4	1258.6	1229.6	1215.0	1253.6	—
	ν_6 FClO bend	367.0	—	—	—	—	—

[a] Data from Smith et al. (270).

scopic study of $FClO_3$ (82) and using a value of 57 kcal mole^{-1} for the Cl—O bond energy, the electron affinity of $FClO_2$ was estimated to be ⩾2.7 eV.

The ^{19}F NMR spectrum of liquid $FClO_2$ at −80°C was recorded by Carter et al. (43) and Christe et al. (61) and consisted of a single peak at $\phi = -328$ or −315 ppm, respectively. A signal at $\phi = -332$ ppm was tentatively assigned by Alexakos and Cornwell (2) to gaseous $FClO_2$.

The weak and highly polar Cl—F bond in FClO can be rationalized in terms of either a (p—π*)σ bond (see Section II, C) or a simple valence bond model (66) resulting in a resonance hybrid of the following canonical forms: $FClO_2 \leftrightarrow F^- + ClO_2^+$. It has been discussed in detail by Parent and Gerry (220), by Carter et al. (43), and in Section II, C of this review.

3. Physical Properties

Chloryl fluoride is colorless as a gas and liquid, and white as a solid. It is stable under normal conditions and some of its physical properties are summarized in Table XIV. Although precise measurements of some

TABLE XIV

Some Properties of $FClO_2$

Property	Value	Ref.
Melting point	−115° or −123.0° ± 0.4°C	(15, 16, 256)
Boiling point	∼ −6°C	(256)
ΔH_{fusion}	1.440 kcal mole^{-1}	(15, 16)
ΔS_{fusion}	9.60 e.u.	(15, 16)
ΔH_{vap}	6.2 kcal mole^{-1}	(256)
Trouton constant	23.2 e.u.	(256)
$\Delta H^0_{f\ 298\ (g)}$	−8.1 ± 2.5 kcal mole^{-1} ^{a}	(15, 16)
Dipole moment (g)	1.722 ± 0.03 D	(220)

^{a} Corrected for $\Delta H^0_{fHF(g)} = -65.14$ kcal mole^{-1} (83).

of its spectroscopic properties have recently been undertaken (220, 282), most of its physical properties are either still unknown or were determined (256) at a time when corrosion-resistant metal–Teflon vacuum systems were not yet available. It was shown by vibrational spectroscopy (282) that solid $FClO_2$ between −263°C and its melting point exists only in one phase. Neutron diffraction data obtained for this phase at −196°C (282) were tentatively indexed based on a monoclinic unit cell with $a = 8.7$, $b = 6.2$, $c = 4.7$ Å, $\beta = 96°$, and $Z = 4$, similar to that of ClF_3.

Tantot et al. (*190*, *282*, *283*) also studied association effects in the liquid phase using vibrational spectroscopy, pulse ^{19}F NMR spectroscopy, and conductometric measurements. They suggest a dipolar dynamic interaction resulting in short-lived associated forms and, possibly, a short-range local order observable on a vibrational but not on an NMR time scale. The specific conductivity of $FClO_2$ in the temperature range $-120°$ to $23°C$ varies according to Martin and Tantot (*190*) from 1.2 to 3.12 μS cm^{-1} (=10^{-6} ohm^{-1}cm^{-1}). The observed conductivity was taken as evidence for self-ionization:

$$2FClO_2 \rightleftharpoons ClO_2^+ + ClO_2F_2^-$$

However, more conclusive evidence is required in view of the reluctance of $FClO_2$ to form $ClO_2F_2^-$ anions (see Section III, F, 4) and of its known reactivity which renders the preparation and handling of very pure $FClO_2$ quite difficult. The vapor pressure of $FClO_2$ as a function of the temperature was measured by Schumacher et al. (*8*, *256*), and is listed in Table XV. It can be described by the equation log P(mm) = 8.23 −

TABLE XV

Vapor Pressures of Chloryl Fluoride

°C	−78	−65.5	−55	−45.5	−38	−30.2	−23.8	−17.2	−9.7	−6.3
mm of Hg	8.8	25.2	55.9	103.8	161.4	244	338	459	645	740

[1412/T(°K)]. Several thermodynamic properties of $FClO_2$ have been estimated by Rips et al. (*235*) by means of correlation increments using only the boiling point of the substance. Whereas the correct boiling point of $FClO_2$ was used, its structure was erroneously assumed to be that of the hypofluorite F—O—Cl=O.

4. Chemical Properties

Chloryl fluoride is stable at ambient temperature in well-passivated and dry containers. Its thermal decomposition in quartz was studied by Schumacher et al. (*24*, *137*). It reaches a measurable rate only above 300°C. The decomposition reaction is monomolecular and its rate is pressure-dependent. The activation energy was calculated to be 45 ± 2 kcal mole^{-1} and the rate constant was determined as $k_\infty = 2.3 \times 10^{13} \times 10^{-45000/4.5T}$ sec^{-1}. The following decomposition mechanism was proposed:

$$FClO_2 \longrightarrow FClO + O$$
$$O + FClO_2 \longrightarrow FClO + O_2$$
$$2FClO \longrightarrow 2ClF + O_2$$

However, based on our present knowledge about FClO (see Section III, A), a more likely decomposition mode for FClO in the above mechanism would be

$$2FClO \longrightarrow ClF + FClO_2$$

The thermal decomposition of $FClO_2$ in Monel was studied by Macheteau and Gillardeau (183). Decomposition to ClF and O_2 was observed at 100°C (2.5% in 144 hr) and 200°C (10% in 235 hr), but a temperature $\geqslant$250°C was required for rate measurements. It was found that the decomposition is of first order and monomolecular at temperatures up to 285°C. At 300°C the reaction becomes second-order. The calculated rate constants and half-life times are summarized in Table XVI. The

TABLE XVI

THERMAL DECOMPOSITION OF $FClO_2$ IN MONEL[a]

Temp. (°C)	Initial press. of $FClO_2$ (mm)	Average rate constant ($\sec^{-1}$)	Half-life
250	52	6.8×10^{-6}	20 hr
250	101	8.5×10^{-6}	22 hr 30 min
270	52	1.8×10^{-5}	10 hr 30 min
285	52	2.8×10^{-5}	6 hr 40 min

[a] Data from Macheteau and Gillardeau (183).

average activation energy between 250° and 285°C was found to be 23.7 kcal mole^{-1}. The results at temperatures >300°C agree with those reported by Schumacher et al. (137) for the quartz reactor. Glass is only slowly attacked by $FClO_2$ at room temperature, but traces of HF or H_2O catalyze the reaction (90, 265). Chloryl fluoride reacts with water (9, 36) and anhydrous nitric acid (51) according to

$$2FClO_2 + H_2O \longrightarrow 2HF + 2ClO_2 + \tfrac{1}{2}O_2$$

and

$$2FClO_2 + 2HONO_2 \longrightarrow 2HF + 2ClO_2 + N_2O_5 + \tfrac{1}{2}O_2$$

Both reactions are relatively slow and do not go to completion in several hours at room temperature (9, 36, 51). In addition, some of the ClO_2

formed can decompose to Cl_2 and O_2 and the nascent oxygen can oxidize $FClO_2$ to $FClO_3$ which is resistant to hydrolysis:

$$FClO_2 + O \longrightarrow FClO_3$$

These results differ from the previous report by Schmeisser and Fink (255) that the reaction between $FClO_2$ and $HONO_2$ proceeds at $-30°C$ according to (46)

$$2FClO_2 + 2HONO_2 \longrightarrow NO_2ClO_4 + ClO_2 + NO_2 + 2HF$$

The statement made in Gmelin (122) and attributed to Bode and Klesper (31) that $FClO_2$ hydrolyzes to $FClO_3$ and H_2, is obviously incorrect. Hydrolysis of $FClO_2$ with base (253, 264, 296) proceeds as follows:

$$FClO_2 + 2OH^- \longrightarrow ClO_3^- + F^- + H_2O$$

Traces of H_2O in $FClO_2$ generate a red-brown color (256) which is probably due to ClO_2. With NH_3 it ignites at $-78°C$ and the end products are NH_4Cl and NH_4F (99). The observation of a weak band at 1052 cm^{-1} in the Raman spectra of $FClO_2$ in dilute HF solutions in addition to strong bands due to $FClO_2$, was interpreted (283) in terms of the equilibrium:

$$HF + FClO_2 \rightleftarrows ClO_2^+ + HF_2^-$$

With HCl, chloryl fluoride reacts (255) at $-110°C$ according to

$$HCl + FClO_2 \longrightarrow HF + ClO_2 + \tfrac{1}{2}Cl_2$$

With the stronger reducing agent HBr, it reacts explosively at $-110°C$ (99). With $HOSO_2F$, at $-78°C$ (99) it forms the stable ClO_2OSO_2F,

$$FClO_2 + HOSO_3F \longrightarrow HF + ClO_2OSO_2F$$

but with $HOSO_2Cl$ at $-90°C$, only the decomposition products of the analogous ClO_2OSO_2Cl, i.e., SO_3, ClO_2, and Cl_2, are obtained. With anhydrous $HOClO_3$, the following reaction occurs (87, 252):

$$FClO_2 + HOClO_3 \longrightarrow HF + ClO_2OClO_3$$

Sulfur trioxide, at $-10°C$ in $CFCl_3$ solution, undergoes an insertion reaction to yield the orange solid (mp = 27°C) ClO_2OSO_2F (254). The same compound was also obtained (296) in the absence of a solvent:

$$FClO_2 + SO_3 \longrightarrow ClO_2OSO_2F$$

With the strong reducing agent SO_2, chloryl fluoride reacts explosively at $-40°C$ (99). When $FClO_2$ and I_2O_5 are combined at $-196°C$, then warmed to $-50°$ to $-20°C$, I_2O_5 is dissolved with formation of O_2, IF_5, ClO_2, Cl_2O_6, and Cl_2O_7 (294).

Chloryl fluoride is a fluorinating agent and a moderately strong oxidizer. Thus it can fluorinate AsF_3 to the pentafluoride (294):

$$3FClO_2 + AsF_3 \longrightarrow ClO_2^+AsF_6^- + 2ClO_2$$

Sulfur tetrafluoride is oxidized by $FClO_2$ at 50°–300°C to yield a mixture of SF_6, SF_4O, and SF_2O_2 (4). Similarly, N_2F_4 is fluorinated at 30°C to give a mixture to NF_3, FNO_2, and FNO (223). Uranium tetrafluoride can be oxidized by $FClO_2$ to UF_5 and UF_6, the latter step requiring a reaction temperature between 50° and 150°C (27). Metal chlorides are converted by $FClO_2$ into metal fluorides, most of which can form ClO_2^+-containing salts when an excess of $FClO_2$ is used. Typical examples are $SbCl_5$, $SnCl_4$, and $TiCl_4$ which are converted to $ClO_2^+SbF_6^-$, $(ClO_2^+)_2SnF_6^{2-}$, and $(ClO_2^+)_2TiF_6^{2-}$, respectively. Aluminum trichloride is converted to AlF_3 (99, 255). Oxides, such as I_2O_5 (see above), SiO_2, Sb_2O_5, and B_2O_3 can be converted by $FClO_2$ at −10°C to SiF_4, $ClO_2^+SbF_6^-$, and $ClO_2^+BF_4^-$, respectively (87). At 50°–100°C, UO_2F_2 reacts only slowly with $FClO_2$, but at 150°C with contact times of 30 min, UF_6, Cl_2, and O_2 are formed (178, 179) according to

$$4FClO_2 + UO_2F_2 \longrightarrow UF_6 + 2Cl_2 + 5O_2$$

Only one reaction was reported in which $FClO_2$ was oxidized from the penta- to the heptavalent state (49, 52, 69). The powerful oxidizer PtF_6 was required to obtain the following reaction:

$$2FClO_2 + 2PtF_6 \longrightarrow ClF_2O_2^+PtF_6^- + ClO_2^+PtF_6^-$$

Chloryl fluoride was converted to ClF_3O by UV-photolysis of systems containing mixtures such as $FClO_2$–F_2, $FClO_2$–ClF, $FClO_2$–ClF_3, and $FClO_2$–ClF_5 (228, 240). These reactions probably do not involve a direct oxygen–fluorine exchange in $FClO_2$, since ClF_3O can be synthesized by the same technique either directly from the three elements or from ClF_3 and oxygen (228).

Chloryl fluoride, like most of the other known chlorine fluorides and oxyfluorides, possesses amphoteric character. Owing to its weak and polar (p—π*)σ Cl—F bond (see Section II, C), it exhibits a much stronger tendency to form adducts with Lewis acids than with Lewis bases. The adducts with Lewis acids result in salts containing ClO_2^+ cations, and those with bases result in $ClO_2F_2^-$ salts. Both ions are discussed in detail in Sections III, G and H, respectively.

G. Chloryl Cation

Although the chloryl cation does not contain a ClF bond and, therefore, in a strict sense does not belong to the family of the chlorine fluoride oxides, it was included in this review since it is a true derivative of $FClO_2$.

The existence of $FClO_2$ adducts with BF_3, AsF_5, PF_5, SbF_5, SiF_4, SO_3, and TaF_5 was first reported in 1954 by Schmeisser and Ebenhöch (*87, 254*) and Woolf (*296*). In 1957, Schmeisser and Fink obtained (*99, 255*) adducts with TiF_4 and SnF_4. In 1958, Clark and Emeleus described (*73*) the existence of a VF_5 adduct, more recently Christe (*52*) obtained a PtF_5 and IrF_5 adduct, and Yeats and Aubke (*298a*) prepared $ClO_2^+ [AsF_5(SO_3F)]^-$ from ClO_2SO_3F and AsF_5.

In a previous review (*253*) the adducts of $FClO_2$ with the stronger Lewis acids, such as AsF_5 or SbF_5, were considered to be ionic and to contain ClO_2^+ cations. However, the corresponding BF_3 and PF_5 adducts were assumed to be molecular adducts. In 1968, Carter *et al.* (*44*) reported evidence for the existence of solvated ClO_2^+ ions in HSO_3F solution. Since then, vibrational spectroscopy has successfully been used to establish the ionic nature of solid $ClO_2^+AsF_6^-$ (*43, 66*), $ClO_2^+BF_4^-$ (*66, 155, 157*), $ClO_2^+SbF_6^- \cdot xSbF_5$ (*42, 43, 155, 157*), $ClO_2^+ClO_4^-$ (*221*), $ClO_2^+PtF_6^-$, and $ClO_2^+IrF_6^-$ (*52*).

1. Syntheses and Properties

Salts containing the ClO_2^+ cation can be prepared either by direct combination of $FClO_2$ with the corresponding perfluorinated Lewis acid with (*254*) or without a solvent (*43, 66, 73, 155, 209, 296*), by the interaction of $FClO_2$ with oxides (*87*), chlorides (*99, 255*), and lower (*294*) or higher (*52*) oxidation state fluorides, or by interaction of the perfluorinated Lewis acid with chlorine oxides (*210, 247*). The latter reactions, however, produce nonvolatile XF_3O as a by-product:

$$5Cl_2O + 3XF_5 \longrightarrow 2ClO_2^+XF_6^- + XF_3O + 4Cl_2 \quad (X = As, Sb)$$

$$5Cl_2O + 7SbF_5 \longrightarrow 2ClO_2^+Sb_3F_{16}^- + SbF_3O + 4Cl_2$$

$$5ClO_2 + 6SbF_5 \longrightarrow 4ClO_2^+SbF_6^- + 2SbF_3O + \tfrac{1}{2}Cl_2$$

$$5ClO_2 + 14SbF_5 \longrightarrow 4ClO_2^+Sb_3F_{16}^- + 2SbF_3O + \tfrac{1}{2}Cl_2$$

$$Cl_2O_6 + 2SbF_5 \longrightarrow ClO_2^+SbF_6^- + SbF_3O + FClO_3$$

Of the above approaches, the direct combination of $FClO_2$ with the corresponding Lewis acid is generally the most convenient. It yields well-defined products, except for cases, such as SbF_5 (*210*) or TaF_5 (*296*), where polyanion formation is possible. From the $FClO_2$–SbF_5 system, depending on the ratio of the starting materials and the reaction conditions, only $ClO_2^+SbF_6^-$, $ClO_2^+Sb_3F_{16}^-$, or a mixture of the two but no $ClO_2^+Sb_2F_{11}^-$, were obtained (*210*). However, single crystals of

$ClO_2^+Sb_2F_{11}^-$ have been obtained by Edwards and Sills (*88a*) by the interaction of $ClF_2^+SbF_6^-$ solutions with glass.

The $FClO_2$ adducts are generally white solids, except for the yellow PtF_6^- and IrF_6^- salts (*52*) and for $FClO_2 \cdot SO_3$ which was reported to be a red-to-pale yellow low-melting solid (*296*). The properties of the latter compound indicate that in the liquid phase it may exist, by analogy with Cl_2O_6 (*221*), in its covalent form, i.e., O_2ClOSO_2F. The ionicity of chloryl fluorosulfate was also discussed (*298*) in a paper dealing with the liquid range of fluorosulfates. The literature reports on the thermal stability of the ClO_2^+ salts are rather sketchy. In addition to the data given in Table IV of Schmeisser's review (*253*), stability data were published only for the BF_3 and the SbF_5 adducts. The $FClO_2 \cdot BF_3$ adduct reaches a dissociation pressure of 1 atm at 44.1°C (*66*), whereas $ClO_2^+SbF_6^-$ (mp = 220–225°C) and $ClO_2^+Sb_3F_{16}^-$ (mp = 50–53°C) are stable up to 300° and 200°C, respectively (*209*). The PtF_6^- and IrF_6^- salts of ClO_2^+ are stable at room temperature (*52*). It should be pointed out that Table IV of Schmeisser (*253*) implies that the thermal stability of the PF_6^- salt is higher than that of the BF_4^-. However, for related cations the reverse is true, and it appears that the data cited might be inaccurate.

X-Ray powder diffraction data have been reported for $ClO_2^+AsF_6^-$ (*66*) and for $ClO_2^+SbF_6^-$ and $ClO_2^+Sb_3F_{16}^-$ (*209*). All the ClO_2^+ salts react violently with organic compounds and water. With stronger Lewis bases, such as NO, NO_2, $ClNO_2$ (*99, 255*), FNO, and FNO_2 (*51, 68*), the following type of displacement reactions can be carried out:

$$ClO_2^+AsF_6^- + NO_x \longrightarrow NO_x^+AsF_6^- + ClO_2$$

$$ClO_2^+AsF_6^- + ClNO_2 \longrightarrow NO_2^+AsF_6^- + ClO_2 + \tfrac{1}{2}Cl_2$$

$$ClO_2^+PtF_6^- + FNO_x \longrightarrow NO_x^+PtF_6^- + FClO_2$$

2. Molecular Structure

The ClO_2^+ cation has been well characterized by vibrational spectroscopy (*42, 43, 66, 155, 157*). Characteristic frequencies and intensities for ClO_2^+ are summarized in Table XVII. The observed ^{35}Cl–^{37}Cl isotopic shifts were used to calculate the bond angle of ClO_2^+. It was shown that the cation is sharply bent and that the bond angle approximates 120° (*66, 155*). Force constants were computed as a function of the ClO_2^+ bond angle (*66, 155*) and the preferred set of constants is included in Table XVII. The value of 8.96 mdyn/Å obtained (*66*) for the ClO-stretching force constant of ClO_2^+ demonstrates that the ClO bond has double-bond character (see Table I).

TABLE XVII

Characteristic Frequencies[a] and Internal Force Constants[b] of ClO_2^+

Obsd. freq. (cm^{-1}) and intensities		Assignment in point group C_{2v}		
IR	Raman			
1296.4 m	1296.4 (1)	$\nu_3(B_1)$	ν_{as}	$^{35}ClO_2$
1282.6 mw	—	$\nu_3(B_1)$	ν_{as}	$^{37}ClO_2$
1043.7 mw	1044.4 (10)	$\nu_1(A_1)$	ν_s	$^{35}ClO_2$
1038.3 w	1039.1 (4)	$\nu_1(A_1)$	ν_s	$^{37}ClO_2$
521.0 m	521.3 (3)	$\nu_2(A_1)$	δ	$^{35}ClO_2$
517 sh	—	$\nu_2(A_1)$	δ	$^{37}ClO_2$

$f_r = 8.96 \pm 0.06$ mdyn/Å
$f_{rr} = -0.45 \pm 0.13$ mdyn/Å
$f_{r\alpha} = 0.24 \pm 0.13$ mdyn/Å
$f_\alpha = 0.82 \pm 0.03$ mdyn/Å

[a] Taken for $ClO_2^+AsF_6^-$ from Christe et al. (66).
[b] Calculated for ⦨ OClO = 120°.

These conclusions concerning the structure of ClO_2^+ were recently confirmed by Edwards and Sills (88a) who carried out a crystal structure determination for $ClO_2^+Sb_2F_{11}^-$. They found the ClO_2^+ ion to be V-shaped, with an O-Cl-O angle of 122° and a mean Cl-O bond length of 1.31 Å.

H. Difluorochlorate(V) Anion

The existence of difluorochlorates of sodium, potassium, and barium was reported in 1965 by Mitra (195). However, this claim was met by skepticism since the reported synthesis involved the use of 40% aqueous hydrofluoric acid. In a subsequent paper (196), Mitra withdrew his claim. In 1969, Huggins and Fox reported (141, 142) the synthesis of $CsClF_2O_2$ from CsF and $FClO_2$, and a subsequent spectroscopic study by Christe and Curtis showed (54) that the vibrational spectrum of the adduct is consistent with a $ClF_2O_2^-$ anion of symmetry C_{2v}.

1. Synthesis and Properties

The synthesis of $CsClF_2O_2$ can be readily achieved by the interaction of dry CsF with excess $FClO_2$ at room temperature (141, 142). In the original work (*141, 142*), activated CsF was used which was obtained by vacuum pyrolysis of the CsF–hexafluoroacetone complex. The conversion of CsF to $CsClF_2O_2$ was 87%. When ordinary CsF (dried by fusion in a platinum crucible and powdered) was used (*54*), the conversion of CsF to $CsClF_2O_2$ was 73%.

The $CsClF_2O_2$ adduct is a white solid, stable at 25°C. Vacuum pyrolysis at 80°–100°C yields CsF and $FClO_2$ (*142*), demonstrating that the formation reaction is reversible. It fumes in moist air and reacts explosively with water (*142*). Controlled hydrolysis (*54*) proceeds according to

$$ClF_2O_2^- + H_2O \longrightarrow ClO_3^- + 2HF$$

2. Structure

The nature of the $CsClF_2O_2$ adduct was established (*54*) by vibrational spectroscopy. The observed spectra were consistent with a $ClF_2O_2^-$ anion possessing the following structure of symmetry C_{2v}:

$$\left[\begin{array}{c} F \\ | \\ -Cl {\displaystyle {\atop \diagdown}}^{\displaystyle \nearrow O}_{\displaystyle \searrow O} \\ | \\ F \end{array} \right]$$

(XII)

The observed bands and their assignments are summarized in Table XVIII. A normal coordinate analysis was carried out (*54*) for $ClF_2O_2^-$ assuming the following geometry: $R(ClO) = 1.43$ Å, $r(ClF) = 1.79$ Å, $\alpha(\angle OClO) = 120°, \beta(\angle OClF) = 90°$, and $(\angle FClF) = 180°$. The actual bond angles are expected to deviate slightly from this ideal geometry owing to increased repulsion from the free valence electron pair on Cl (see Section II, A). The internal force constants of $ClF_2O_2^-$ are summarized in Table XIX. As can be seen from Table I and the general discussion in Section II, C, the ClO bonds in $ClF_2O_2^-$ have double-bond character and the ClF bonds are as expected, semi-ionic 3 center–4 electron bonds. The polarity of the latter is increased further by the formal negative charge and the high degree of oxygen substitution. The combination of these effects results in the lowest ClF-stretching force constant value found to date for any ClF bond. As demonstrated for several other oxyfluoride anions (see Section III, D), the negative charge in $ClF_2O_2^-$ resides mainly on the ligands having the highest electronegativity, i.e., on the fluorine, and not on the oxygen atoms.

TABLE XVIII

Vibrational Spectrum of $Cs^+ClF_2O_2^-$ and Its Assignment[a]

Obsd. freq. (cm^{-1}) and intensities		Assignment for XO_2F_2 in point group C_{2v}	Approx. description of vibration
IR	Raman		
1225 } vs 1191	1221 (0.8)	$\nu_8(B_2)$	$\nu_{as}(XO_2)$
1070 s	⎧ 1076 (10) ⎨ 1064 ⎩ 1055	$\nu_1(A_1)$	$\nu_s(XO_2)$
559 m	559 (1.2)	$\nu_2(A_1)$	$\delta_s(XO_2)$
510 vs, br		$\nu_6(B_1)$	$\nu_{as}(XF_2)$
	480 (1), br	$\nu_5(A_2)$?	τ
330–370 m	{ 363 (10) 337 (8)	$\nu_3(A_1)$ $\nu_7(B_1), \nu_9(B_2)$	$\nu_s(XF_2)$ $\delta_{rock}, \delta_{wag}$
	198 (0.7)	$\nu_4(A_1)$	$\delta_s(XF_2)$

[a] Data from Christe and Curtis (54).

TABLE XIX

Force Constants of $ClF_2O_2^-$ [a,b]

f_R	8.3	f_β	1.2
f_{RR}	0.1	$f_{\beta\beta'}$	0.57
f_r	1.6	$f_{\beta\beta}$	0.1
f_{rr}	−0.1	$f_{r\beta} - f_{r\beta'}$	0.3
f_α	1.95		

[a] Data from Christe and Curtis (54).

[b] Stretching force constants in mdyn/Å, deformation constants in mdyn Å/radian^2, and stretch–bend interactions in mdyn/radian.

I. Chlorine Trifluoride Dioxide

A compound having the empirical composition $(ClF_3O_2)_n$ was reported in 1962 by Streng and Grosse (128, 276, 278). It was obtained by the interaction of either Cl_2, ClF, or HCl with O_2F_2 between −154°C and −143°C or by UV photolysis of ClF_3 and O_2 mixtures at −78°C. Both methods produced the same product, a violet unstable solid, which irreversibly decomposed above −78°C. In a subsequent study of the infrared and visible spectra of these products, Gardiner and Turner (*108, 109*) proposed the structure F_2ClOOF for the violet compound. However, both the synthetic and the spectroscopic studies are not convincing and further work is required to establish the composition and structure of this violet species.

A well-defined and characterized compound, having the composition ClF_3O_2 and showing no resemblance to Streng and Grosse's violet compound, was reported in 1972 by Christe (*50*). This work is an excellent example for the perfection of handling techniques for extremely reactive oxidizers. Thus the physical, chemical, and spectroscopic properties of ClF_3O_2 and of its $ClF_2O_2^+$ adducts were determined from a total of 2.2 mmol of material. The fact that ClF_3O_2 as a powerful oxidizer is readily reduced to $FClO_2$ which cannot be removed from ClF_3O_2 by simple fractionation (see below), rendered the handling of this compound puarticlarly difficult.

1. Synthesis and Properties

The synthesis of ClF_3O_2 is best described by the following reaction sequence:

$$2FClO_2 + 2PtF_6 \longrightarrow ClF_2O_2^+PtF_6^- + ClO_2^+PtF_6^-$$

Several side reactions compete with this reaction and the yield of $ClF_2O_2^+$ varies greatly with slight changes in the reaction conditions (*52, 68*). The ClF_3O_2 is then displaced from its $ClF_2O_2^+$ salt according to

$$ClF_2O_2^+PtF_6^- + ClO_2^+PtF_6^- + 2FNO_2 \longrightarrow 2NO_2^+PtF_6^- + ClF_3O_2 + FClO_2$$

Chloryl fluoride is slightly less volatile than ClF_3O_2, and, therefore, most of it can be removed from ClF_3O_2 by fractional condensation in a −112°C trap. The remaining $FClO_2$, however, has to be removed by complexing with BF_3:

$$ClF_3O_2 + FClO_2 + 2BF_3 \longrightarrow ClF_2O_2^+BF_4^- + ClO_2^+BF_4^-$$

Since $ClF_2O_2^+BF_4^-$ is stable (*69*) at 20°C, whereas $ClO_2^+BF_4^-$ is not (*66*), the latter can be pumped away at 20°C. The resulting pure $ClF_2O_2^+BF_4^-$ is then treated with an excess of FNO_2 and the evolved ClF_3O_2 and unreacted FNO_2 are readily separated by fractional condensation through a series of −126° and −196°C traps:

$$ClF_2O_2^+BF_4^- + FNO_2 \longrightarrow NO_2^+BF_4^- + ClF_3O_2$$

The overall yield of pure ClF_3O_2 based on the PtF_6 used in step 1 was found to be about 10 mole%.

Pure ClF_3O_2 is colorless as a gas or liquid and white as a solid. Some of its measured (68) physical properties are summarized in Table XX. Near its melting point the vapor pressure above liquid ClF_3O_2 was found to be reproducibly lower than expected from the vapor pressure curve given in Table XX. This indicates that close to the melting point some ordering effect occurs in the liquid.

The measured vapor density of ClF_3O_2 indicates that no appreciable association occurs in the gas phase. Its relatively low boiling point and Trouton constant imply little association in the liquid phase. This prediction is confirmed by the vibrational spectra of the liquid and the neat solid which exhibit only minor frequency shifts when compared to

TABLE XX

Some Properties of ClF_3O_2[a]

Property	Value
Melting point	$-81.2°C$
Boiling point	$-21.58°C$
ΔH_{vap}	5.57 kcal mole^{-1}
Trouton constant	22.13 e.u.
Vapor pressure	$\text{Log } P(\text{mm}) = 7.719 - \dfrac{1217.2}{T\ (°K)}$

[a] Data from Christe and Wilson (68).

the spectra of the gas and the matrix-isolated solid. This finding is somewhat surprising since both ClF_3 (102) and ClF_3O (55, 226) show a pronounced tendency to associate in the liquid and solid state through bridges involving the axial fluorine atoms.

The thermodynamic properties were computed with the molecular geometry and vibrational frequencies given below assuming an ideal gas at 1 atm pressure and using the harmonic-oscillator rigid-rotor approximation. These properties are given for the range 0°–2000°K in the Appendix (Table AII).

Chlorine trifluoride dioxide resembles chlorine fluorides and oxyfluorides in its corrosive and oxidizing properties. It must be handled in systems consisting of corrosion-resistant metals, Teflon, or sapphire. It appears to be marginally stable in a well-passivated system at ambient temperature. It is a strong oxidative fluorinator as evidenced by its tendency to fluorinate metal surfaces to metal fluorides with $FClO_2$ formation. It reacts explosively with organic materials and care must be taken to avoid such combinations. The hydrolysis of ClF_3O_2 was not quantitatively studied; however, on one occasion a slight leak in an infrared gas cell containing ClF_3O_2 resulted in the formation of $FClO_3$ and HF indicating the following reaction.

$$ClF_3O_2 + H_2O \longrightarrow FClO_3 + 2HF$$

Chlorine trifluoride dioxide forms stable adducts with strong Lewis acids, such as BF_3, AsF_5, or PtF_5 (*49, 68, 69*). These adducts have ionic structures containing the $ClF_2O_2^+$ cation (see Section III, I, 2). The high stability of these adducts can be explained by the change from the energetically unfavorable trigonal-bipyramidal structure of ClF_3O_2 to the more favorable tetrahedral $ClF_2O_2^+$ configuration (see Section II, D). Contrary to ClF_3 (*295*), but by analogy with ClF_3O (*64*), it does not form stable adducts with FNO or FNO_2 at temperatures as low as $-78°C$. This was demonstrated by the various displacement reactions where ClF_3O_2 and unreacted FNO or FNO_2 could be readily removed from the reactor at $-78°C$. With the stronger base, CsF, it did not form a stable adduct but decomposed to $FClO_2$ and F_2. However, only relatively small amounts of ClF_3O_2 were available for the complex formation study with CsF, and the possibility of preparing salts such as $Cs^+ClF_4O_2^-$ under more favorable reaction conditions cannot entirely be ruled out.

2. *Molecular Structure*

Vibrational (*57*) and ^{19}F NMR (*68*) spectroscopy were used to establish for ClF_3O_2 the following structure of symmetry C_{2v}, which according to semi-empirical linear combination of atomic orbitals–molecular orbitals (LCAO-MO) self-consistent field (SCF) calculations (*239*) is most stable:

$$F-\underset{\underset{F}{|}}{\overset{\overset{F}{|}}{Cl}}\begin{matrix}\nearrow O \\ \searrow O\end{matrix}$$

(XIII)

The ^{19}F NMR spectrum of liquid ClF_3O_2 was measured in the temperature range $-20°$ to $-80°C$. It showed at all temperatures one partially resolved signal centered at -413 ppm below the external standard $CFCl_3$. The observed signal is in excellent agreement with an AB_2 pattern with $J/\nu_0\delta = 1.0$ and $J_{FF} = 443$ Hz. The low chemical shift of -413 ppm for ClF_3O_2 is in excellent agreement with a heptavalent chlorine fluoride, and the fluorine–fluorine coupling constant of 443 Hz observed for ClF_3O_2 is similar to that of 421 Hz observed for the structurally related ClF_3 (61). Additional support for the above structure was derived from the fact that the B_2 part of the AB_2 pattern occurs downfield from the A part as expected for the axial fluorine atoms in a trigonal bipyramidal arrangement (120, 200).

The infrared spectra of gaseous, solid, and matrix-isolated ClF_3O_2 and the Raman spectra of gaseous and liquid ClF_3O_2 were reported (57) and are summarized in Table XXI. The observed data are in excellent

TABLE XXI

VIBRATIONAL SPECTRUM OF ClF_3O_2 AND ITS ASSIGNMENT IN POINT GROUP C_{2v} [a]

IR (cm^{-1})	Ra (cm^{-1})	Assignment for ClF_3O_2 in point group C_{2v}		Approx. description of mode
1093 s	1093 (4) p	A_1	ν_1	Sym ClO_2 str
683 m	683 (10) p		ν_2	ClF_{eq} str
519 w	520 (8) p		ν_3	ClO_2 scissor
487 vw	487 (6) p		ν_4	Sym $F_{ax}ClF_{ax}$ str
287 w	285 (1)		ν_5	$F_{ax}ClF_{ax}$ scissor in ClF_3 plane
(417)[b]	402 (0+)	A_2	ν_6	Torsion
695 vs		B_1	ν_7	Antisym $F_{ax}ClF_{ax}$ str
592 s	586 (0+)		ν_8	ClO_2 wag
372 w			ν_9	Antisym $F_{eq}ClF_{2ax}$ def in ClF_3 plane
1327 vs	1320 (0+)	B_2	ν_{10}	Antisym ClO_2 str
531 m	530 (1)		ν_{11}	ClO_2 rock
c	222 (1)		ν_{12}	$F_{ax}ClF_{ax}$ scissor out of ClF_3 plane

[a] Data from Christe and Curtis (57).
[b] Observed only for solid ClF_3O_2.
[c] Below frequency range of spectrometer used.

agreement with the preceding model (XIII) of symmetry C_{2v}. A normal coordinate analysis was carried out for ClF_3O_2 assuming the following geometry: $D(ClO) = 1.40$ Å, $R(ClF_{eq}) = 1.62$ Å, $r(ClF_{ax}) = 1.72$ Å, $\alpha(OClO) = 130°$, $\beta(F_{eq}ClF_{ax}) = \delta(OClF_{ax}) = 90°$, and $\gamma(OClF_{eq}) = 115°$, based on the observed geometries of ClF_3 and $FClO_3$ and a correlation between ClO bond length and stretching frequency. The deviation of the OClO bond angle from the ideal 120° was estimated by comparison with the known geometries of SF_4O and $FClO_3$. The force constants thus obtained are summarized in Table XXII. The value of the ClO-stretching force constant (9.23 mdyn/Å) is in excellent agreement with that of

TABLE XXII

Internal Force Constants of ClF_3O_2 [a,b]

$f_D = 9.23$	$f_{\beta\beta} = 0.09$
$f_R = 3.35$	$f_{r\beta} = f_{r\beta'} = 0.10$
$f_r = 2.70$	$f_{r\delta} = -f_{r\delta'} = 0.25$
$f_\alpha = 1.41$	$f_{D\alpha} = 0.61$
$f_\beta = 1.40$	$f_{\beta\delta} = -f_{\beta\delta'} = -0.16$
$f_\gamma = 1.33$	$f_{\delta\delta} = -f_{\delta\delta'} = -0.34$
$f_\delta = 1.30$	$f_{\delta\delta''} = -0.17$
$f_{DD} = -0.09$	$f_{\gamma\gamma} = -0.30$
$f_{rr} = -0.04$	$f_{R\alpha} = -0.37$

[a] Data from Christe and Curtis (57).
[b] Stretching constants in mdyn/Å, deformation constants in mdyn Å/radian², and stretch–bend interaction constants in mdyn/radian.

9.37 mdyn/Å found for ClF_3O (55) and the general valence force field values of 9.07 and 8.96 mdyn/Å reported for $FClO_2$ (270) and ClO_2^+, (66) respectively. The values of the ClF-stretching force constants are comparable to those previously reported for the related pseudotrigonal bipyramidal molecules ClF_3 (102) and ClF_3O (55) and are summarized in Table XXIII. In all three molecules, the stretching force constant of the equatorial ClF bond is significantly higher than that of the two axial bonds, although their relative difference decreases with increasing oxidation state of the central atom. The difference in bond strength between equatorial and axial bonds implies significant contributions from semi-ionic 3 center–4 electron bonds to the axial ClF bonds. This bonding scheme has been discussed in detail for the related pseudotrigonal bipyramidal ClF_2^- anion in Section II, C.

TABLE XXIII

ClF Stretching Force Constants of ClF_3O_2 Compared to Those of Pseudotrigonal Bipyramidal ClF_3O, ClF_3, ClF_2^-, and $ClF_2O_2^-$

Compound	f_R (mdyn/Å)	f_r (mdyn/Å)	f_{rr} (mdyn/Å)	$(f_R-f_r)/f_R$	Ref.
ClF_3	4.2	2.7	0.36	0.36	(102)
ClF_3O	3.2	2.3	0.26	0.26	(55)
ClF_3O_2	3.4	2.7	−0.04	0.19	(57)
ClF_2^-	—	2.4	0.17	—	(63)
$ClF_2O_2^-$	—	1.6	−0.1	—	(54)

Inspection of Table XXIII also reveals that the value of f_r does not depend exclusively on the oxidation state of the central atom. Obviously, formal negative charges (as in the anions) and increasing oxygen substitution facilitate the formation of semi-ionic bonds and, hence, counteract the influence of the oxidation state of the central atom. It is interesting to note that the relative contribution from semi-ionic bonding (see Section II, C) to the axial ClF bonds $[= (f_R - f_r)/f_R]$ decreases from ClF_3 to ClF_3O and ClF_3O_2 (see Table XXIII). This can be attributed to the decreasing electron density around the central atom with increasing oxidation state, thus making it more difficult to release electron density to the axial fluorine ligands as required for the formation of semi-ionic bonds.

The bonding in ClF_3O_2 might be described by the following approximation (53). The bonding of the three equatorial ligands, ignoring the second bond of the Cl=O double bond, is mainly due to an sp^2 hybrid, whereas the bonding of the two axial ClF bonds involves one delocalized p-electron pair of the chlorine atom for the formation of a semi-ionic 3 center–4 electron pσ bond.

J. DIFLUOROPERCHLORYL CATION

The existence of the $ClO_2F_2^+$ cation in the form of its PtF_6^- salt was reported in 1972 by Christe (49). In a subsequent paper (69), a full account was given of the synthesis and properties of the PtF_6^-, AsF_6^-, and BF_4^- salts of $ClO_2F_2^+$.

1. Synthesis and Properties

It was found (52) that PtF_6 and $FClO_2$, when combined at $-196°C$ and allowed to warm up slowly to $25°C$, interacted according to

$$2FClO_2 + 2PtF_6 \longrightarrow ClO_2F_2^+PtF_6^- + ClO_2^+PtF_6^-$$

The yield of $ClO_2F_2^+$ was not 50% as expected from the foregoing equation, but generally about 25% owing to the competing reaction

$$2FClO_2 + 2PtF_6 \longrightarrow 2ClO_2^+PtF_6^- + F_2$$

In some of the experiments, small amounts of $ClF_6^+PtF_6^-$ or ClF_5 and $FClO_3$ were observed, depending on the exact reaction conditions. The formation of some $FClO_3$ is not surprising since it is known that $FClO_2$ readily interacts with nascent oxygen to yield $FClO_3$ (9, 36, 51).

Attempts to suppress the competing reaction by changing the reaction conditions (rapid warm-up from $-196°$ to $-78°C$ and completion of the reaction at $-78°C$) resulted on one occasion in an entirely different course for the reaction:

$$6FClO_2 + 6PtF_6 \longrightarrow 5ClO_2^+PtF_6^- + ClF_6^+PtF_6^- + O_2$$

Further modification of the reaction conditions (rapid warm-up of the $FClO_2$–PtF_6 mixture from $-196°$ to either $-78°$ or $25°C$ and completion of the reaction at $25°C$) did not produce detectable amounts of either $ClO_2F_2^+$ or $ClF_6^+PtF_6^-$, but only $ClO_2^+PtF_6^-$ and ClF_5, F_2, and O_2. This indicates that the nature of the reaction products is more influenced by the warm-up rate of the starting materials from $-196°$ to about $-78°C$ than by the final reaction temperature. Slow warm-up favors the formation of $ClO_2F_2^+$, whereas rapid warm-up yields ClF_6^+ or ClF_5 and F_2 (52).

The BF_4^- and AsF_6^- salts were prepared (69) as follows:

$$ClO_2^+PtF_6^- + ClO_2F_2^+PtF_6^- + 2FNO_2 \longrightarrow FClO_2 + ClF_3O_2 + 2NO_2^+PtF_6^-$$

Unreacted FNO_2 and some of the $FClO_2$ could be separated from ClF_3O_2 by fractional condensation. The remaining $FClO_2$ was separated from ClF_3O_2 by complexing with BF_3. Since the resulting $ClO_2^+BF_4^-$ has a dissociation pressure (66) of 182 mm at $22.1°C$ while $ClO_2F_2^+BF_4^-$ is stable, the former salt could be readily removed by pumping at $20°C$. Conversion of $ClO_2F_2^+BF_4^-$ to the corresponding AsF_6^- salt was accomplished through displacement of BF_4^- by the stronger Lewis acid AsF_5:

$$ClO_2F_2^+BF_4^- + AsF_5 \longrightarrow ClO_2F_2^+AsF_6^- + BF_3$$

All three salts, $ClO_2F_2^+PtF_6^-$, $ClO_2F_2^+AsF_6^-$, and $ClO_2F_2^+BF_4^-$, are solids, stable at $25°C$, and react violently with water or organic materials. The PtF_6^- compound is canary yellow, whereas those of AsF_6^- and BF_4^- are white.

The salts dissolve in anhydrous HF without decomposing. They are crystallinic in the solid state, and the X-ray powder diffraction patterns of $ClO_2F_2^+BF_4^-$ and $ClO_2F_2^+AsF_6^-$ have been reported (69). The pattern of the former was tentatively indexed on the basis of an orthorhombic unit cell with $a = 5.45$, $b = 7.23$, and $c = 13.00$ Å. Assuming four molecules per unit cell and neglecting contributions from the highly charged central atoms to the volume, a plausible average volume of 16 Å^3 per F or O atom was obtained.

The thermal stability of $ClO_2F_2^+BF_4^-$ is higher than that of $ClO_2^+BF_4^-$ (66), $ClF_2^+BF_4^-$ (259), or other similar salts. The pronounced tendency of ClF_3O_2 to form stable adducts with Lewis acids is in good agreement with the correlations between the stability of an adduct and the structure of the parent molecule and its ions (see Section II, D). Thus, tetrahedral $ClO_2F_2^+$ (see below) should be energetically much more favorable than trigonal bipyramidal ClF_3O_2.

2. Molecular Structure

The structure of $ClO_2F_2^+$ salts was established by ^{19}F NMR and vibrational spectroscopy (69).

In the ^{19}F NMR spectrum of $ClF_2O_2^+PtF_6^-$ in anhydrous HF, a broad singlet at −310 ppm relative to external $CFCl_3$ was tentatively assigned (61) to $ClF_2O_2^+$. Subsequent studies (69) of $ClF_2O_2^+BF_4^-$ and $ClF_2O_2^+AsF_6^-$ confirmed the original assignment. The spectrum of $ClF_2O_2^+BF_4^-$ in HF showed a strong temperature dependence. At 30°C it consisted of a single peak at 185 ppm relative to external $CFCl_3$. With decreasing temperature the peak at first became broader and then separated at about 0°C into three signals at −301 ($ClO_2F_2^+$), 146 (BF_4^-), and 194 ppm (HF) which became narrower with further decrease in temperature. The observed peak area ratio of approximately 2:1 for the 146- and −301-ppm signals confirmed their assignment to BF_4^- and $ClO_2F_2^+$, respectively, and proved the ionic nature of the $ClF_3O_2 \cdot BF_3$ adduct in HF solution.

The spectrum of $ClF_2O_2^+AsF_6^-$ in HF (which was acidified with AsF_5) consisted of two resonances at −307 ($ClO_2F_2^+$) and 105 ppm (HF, AsF_5, AsF_6^-), respectively. Rapid exchange among HF, AsF_5, and AsF_6^- preempted the measurement of the $ClO_2F_2^+$ to AsF_6^- peak area ratio (69).

The vibrational spectra of the BF_4^-, AsF_6^-, and PtF_6^- salts of $ClO_2F_2^+$ were recorded for both the solids and HF solutions (69). It was shown that all three salts are ionic containing, in addition to the anions, a common cation. The vibrational spectrum of this cation closely resembled that of SO_2F_2 indicating a pseudotetrahedral structure of symmetry C_{2v}. The observed frequencies together with the stretching force constants obtained from Cl isotopic shifts are listed in Table XXIV. Inspection of Table I shows that $ClO_2F_2^+$ possesses the highest value known for a ClO-stretching force constant. This is not surprising, since the central atom in $ClO_2F_2^+$ has a high oxidation state (+VII), highly electronegative ligands, and a formal positive charge (cation). The influence of these factors on f_{ClO} was discussed in Section II, C. By analogy with ClF_2O^+ (58), the only other known species exhibiting a f_{ClO} value of similar magnitude, contributions from the resonance structure,

$$\text{(XIV)} \quad \longleftrightarrow \quad \text{(XV)}$$

TABLE XXIV

Observed Frequencies, Approximate Description of Modes, and Most Important Internal Force Constants Computed to Fit the Observed ^{35}Cl and ^{37}Cl Isotopic Shifts and Assuming Two Different Bond Angles of $ClO_2F_2^+$ [a]

Assignment		Obs. freq. (cm^{-1})	Approx. description of mode
A_1	ν_1	1241	$\nu_{sym}(ClO_2)$
	ν_2	756	$\nu_{sym}(ClF_2)$
	ν_3	514	$\delta_{sym}(ClO_2)$
	ν_4	390	$\delta_{sym}(ClF_2)$
A_2	ν_5	390	τ
B_1	ν_6	1479	$\nu_{asym}(ClO_2)$
	ν_7	530	$\delta_{rock}(ClO_2)$
B_2	ν_8	830	$\nu_{asym}(ClF_2)$
	ν_9	514	$\delta_{rock}(ClF_2)$

	$\angle OClO, \angle FClF$, deg	
	124, 96	114, 105
f_D (mdyn/Å) (ClO)	12.20	12.04
f_{DD} (mdyn/Å)	−0.46	−0.66
f_R (mdyn/Å) (ClF)	4.40	4.53
f_{RR} (mdyn/Å)	−0.32	0.03

[a] Data from Christe et al. (69).

might be invoked to explain the high f_{ClO} value. The value of the ClF-stretching force constant (4.46 mdyn/Å) falls within the range expected for a predominantly covalent ClF bond in a cation having a central atom with a +VII oxidation state.

K. Perchloryl Fluoride

Perchloryl fluoride,

$$\underset{O}{\overset{F}{\underset{\|}{O=Cl=O}}}$$

(XVI)

the acyl fluoride of perchloric acid, was first obtained by Bode and Klesper in 1951 (30) by the action of F_2 on $KClO_3$ at $-40°C$, but believed to be ClO_2OF. In 1952 it was prepared by Engelbrecht and Atzwanger (91) by electrolysis of $NaClO_4$ in anhydrous HF and was correctly identified. In the mid-fifties it became commercially available from Pennsalt Chemical Corporation and can be purchased in research quantities from Ozark Mahoning Company. Owing to its remarkably low reactivity and high specific impulse (see Section III, K, 5), it received considerable interest as a rocket propellant oxidizer, resulting in a rather thorough study of its properties. Unfortunately, its high vapor pressure (53 atm at $T_c = 95°C$) and coefficient of expansion rendered it inferior to other oxidizer candidates. Owing to its relative inertness (it hydrolyzes only slowly in water), it has found use as a fluorinating agent in organic chemistry. In addition to the general reviews, listed in the Introduction, and brief reviews in Japanese (205) and Chinese (48), reviews that are devoted exclusively to $FClO_3$ have been published by Pennsalt (222), Gall (106), and Khutoretskii et al. (158). The inertness of $FClO_3$ is due to its energetically favorable pseudotetrahedral configuration, its highly covalent and strong Cl—F bond (see Section II, C), and its extremely small dipole moment of 0.023 D. Combined, these properties give it a high kinetic stability in spite of its low thermodynamic stability ($\Delta H°_{f298} = -5.7$ kcal mole^{-1}).

1. Synthesis

Perchloryl fluoride can be prepared by electrolysis of a saturated solution of $NaClO_4$ in anhydrous HF with a current efficiency of 10% (91, 92).

Fluorination of solid $KClO_3$ by F_2 (30, 31) produces $FClO_3$, $FClO_2$, ClF, Cl_2O_6, Cl_2, and O_2 (89, 92). The yields of $FClO_3$ were about 45% based on the F_2 used (92). When the fluorination was carried out below $-20°C$, yields of $FClO_3$ as high as 60% were obtained (265). The fluorination of $NaClO_3$ with F_2 can also be carried out in aqueous solution at $25°-75°C$ resulting in a 50% yield of $FClO_3$ (299; see also 125). Replacement of F_2 by other fluorinating agents, such as ClF_3, BrF_3, or SbF_5, gives mainly $FClO_2$ and Cl_2 and only low yields of $FClO_3$ (92). Purification of crude $FClO_3$ by washing of the products condensible at $-196°C$ with an alkaline $Na_2S_2O_3$ solution produces material containing less than 1.5% of impurities (92).

The thermal decomposition of Cl_2O_7 at 100°C in the presence of F_2 produces a mixture of $FClO_3$ and $FClO_2$ in a yield of about 75% (98). Similarly, the fluorination of either Cl_2O_6 or Cl_2O_7 with SbF_5 produces $FClO_3$ in high yield (210) according to

$$Cl_2O_6 + 2SbF_5 \longrightarrow ClO_2^+SbF_6^- + SbF_3O + FClO_3$$

and

$$Cl_2O_7 + nSbF_5 \longrightarrow SbF_3O \cdot (SbF_5)_{n-1} + 2FClO_3$$

The fluorination of NO_2ClO_4 by ClF_3 at room temperature results in the formation of $FClO_3$ and smaller amounts of $FClO_2$, ClO_2, and $ClNO_2$ (25). Perchloryl fluoride is also formed by the interaction of $FClO_2$ with nascent oxygen (9, 36, 51) and in the reaction of gaseous ClF_3 with UO_2, U_3O_8, and UO_3 (149) and with UO_2F_2 (263), or by the reaction of $ClF_2^+BiF_6^-$ with metal oxides (78). Xenon dioxide tetrafluoride, XeO_2F_4, is capable of oxidizing either ClF_3 or ClF_5 to $FClO_3$ (143). Almost quantitative yields of $FClO_3$ and $R_f C\underset{F}{\overset{O}{\diagdown}}$ can be obtained by the alkali metal fluoride-catalyzed decomposition of the corresponding $R_f CF_2 OClO_3$ at slightly elevated temperatures (249).

The most convenient and commercially attractive methods for preparing $FClO_3$ involve the fluorination of perchlorates. Heating of $KClO_4$ to 70°–120°C in an excess of SbF_5 produces $FClO_3$ in 50% yield (90). The yield of $FClO_3$ can be increased to 90% and the reaction temperature can be lowered to 20°–50°C, when a mixture of HF–SbF_5 is used (292, 293). Slightly lower yields were obtained when the HF solvent was replaced by AsF_3, IF_5, or BrF_5.

Most of the commercial processes are based on the use of $HOSO_2F$. This method was proposed in 1956 by Barth-Wehrenalp (20). Evolution of $FClO_3$ starts at 50°C and goes to completion at 85°–110°C. The yields of $FClO_3$ vary from 50 to 80% (20, 22, 162, 163) and, if necessary, the $HOSO_2F$ can be regenerated (22). If desired, the reaction can be carried out in glass apparatus. The influence of certain additives on the yield of $FClO_3$ was studied (81). The addition of 5 to 25% of SbF_3 to the $HOSO_2F$ increases the yield of $FClO_3$ to 90% and higher but hinders the regeneration of $HOSO_2F$. The addition of HF–BF_3 increases the $FClO_3$ yield to 85% but requires elevated pressure. Zinc, aluminum, silver, and lead fluorides were found to decrease the yield of $FClO_3$.

The highest yield of perchloryl fluoride (97%) was achieved with a mixture of fluorosulfonic acid and SbF_5 as fluorinating medium. Potassium, sodium, lithium, magnesium, barium, calcium, and silver perchlorates and perchloric acid itself undergo the reaction. Commercial reagents are used and their additional purification is not necessary; unlike all the previous methods the preparation of perchloryl fluoride by this method can be carried out at room temperature. At high temperature (100°–135°C) the reaction time is 1–10 min in all, which allows the process to be carried out continuously in a packed column. The purity of product obtained after the usual purification reaches 98% and over; air and carbon dioxide are present as trace impurities (23).

The exact mechanism of the reaction between ClO_4^- and superacids has as yet not been established, although numerous comments on it were published (19, 21, 167, 253, 292, 297). Based on our present understanding of superacid chemistry (67, 118, 216) and of the complex formation of $FClO_3$ (see Section III, K, 4), a mechanism involving ClO_3^+ as an intermediate is very unlikely. Furthermore, the high yields of $FClO_3$ (up to 97%) would be surprising in view of the expected instability of ClO_3^+. In our opinion, other mechanisms, such as the one shown, involving protonated perchloric acid (166) are more plausible:

$$4HF + 2SbF_5 \longrightarrow 2H_2F^+ + 2SbF_6^-$$
$$2H_2F^+ + ClO_4^- \longrightarrow H_2OClO_3^+ + 2HF$$
$$H_2OClO_3^+ + HF \longrightarrow FClO_3 + H_3O^+$$
$$\overline{ClO_4^- + 3HF + 2SbF_5 \longrightarrow FClO_3 + H_3O^+ + 2SbF_6^-}$$

2. Molecular Structure

The structural parameters of $FClO_3$ were determined by Clark, Beagley, and Cruickshank (72) by gas-phase electron diffraction. The molecule has symmetry C_{3v} and the following bond angles and distances:

$$\angle (OClF) \quad 103.0°$$
$$r(Cl-F) \quad 1.610 \text{ Å}$$
$$r(Cl-O) \quad 1.402 \text{ Å}$$
$$\angle (OClO) \quad 115.1°$$

(XVII)

Owing to its small dipole moment, $FClO_3$ exhibits only a very weak microwave spectrum (*171, 173*). Since only the $J = 4 \to 5$, $K = 3$ and the $J = 6 \to 7$, $K = 3$ and $K = 6$ transitions were observed, a complete structure determination was not possible. However, the estimated geometry and dipole moment are in good agreement with the exact values measured by other methods. Table XXV lists the frequency values and constants that were obtained. The rotational constants, B_0, are in good agreement with the values obtained from the high-resolution infrared spectrum (*184*) of the 549- and 589-cm^{-1} fundamentals.

The dipole moment of $FClO_3$ was determined by dielectric relaxation measurements (*192*) as 0.023 ± 0.003 D and from the $J_{11 \to 12}$ transition in a resonant cavity at 126196 MHz (*101*) as 0.025 ± 0.003 D. This low dipole moment indicates that the electronegativity of F and the ClO_3

TABLE XXV

FREQUENCY VALUES AND CONSTANTS FOR PERCHLORYL FLUORIDE

Transitions and constants	$F^{35}ClO_3$ (MHz)	$F^{37}ClO_3$ (MHz)
$J = 4 \to 5, K = 3$	52585.97 ± 0.05	52560.4 ± 0.3
$J = 6 \to 7, K = 3$	73619.40 ± 0.05	73583.94 ± 0.05
$J = 6 \to 7, K = 6$	73618.72 ± 0.05	—
B_0	5258.692 ± 0.005	5256.149 ± 0.005
D_J	0.0014 ± 0.0002	
D_{JK}	0.0018 ± 0.0003	
eqQ	-19.2 ± 0.5	-15.4 ± 1.5

group are comparable, thus resulting in a high degree of covalency for the Cl—F bond.

The ^{19}F NMR spectrum of $FClO_3$, according to Brownstein (*41*) consists of a partially resolved quartet ($J_{ClF} \sim 310$ Hz) of equal intensity at $\phi = -241.5$ ppm. The lack of rapid quadrupole relaxation indicates a highly symmetric electric field around the central atom in good agreement with the small dipole moment observed for $FClO_3$ (see above). The temperature dependence of the ^{19}F NMR spectrum of $FClO_3$ was studied by Bacon et al. (*13*). An expression for the line broadening was derived, and a value of 1.0 kcal mole^{-1} was obtained for the activation energy of molecular reorientation. A value of 278 ± 5 Hz was calculated for $J35_{ClF}$. According to Agahigian et al. (*1*), the ^{19}F resonance of $FClO_3$ occurs at $\phi = -287$ ppm, but measurements in our laboratory indicate that this value is inaccurate. A value of $\phi = -252.9 \pm 2$ ppm was found by us for liquid $FClO_3$ at $-120°C$. The ^{35}Cl and ^{19}F NMR spin-lattice relaxation times and rotational diffusion in liquid $FClO_3$ were measured by Maryott et al. (96, 191) using pulse techniques.

The mass spectrum of $FClO_3$ was measured (82, 138, 234). The vertical ionization potential and the F—ClO_3 bond dissociation energy were found to be 13.6 ± 0.2 eV and ~ 60 kcal mole^{-1}, respectively. The average ClO bond dissociation energy and the heat of formation were estimated (82) to be 60 and -5.3 kcal mole^{-1}, respectively.

The UV absorption spectrum of $FClO_3$ was reported by Sicre and Schumacher (264) and Pilipovich et al. (228).

The vibrational spectrum of $FClO_3$ has been well characterized. The infrared spectrum was thoroughly analyzed by Lide and Mann (174) and

TABLE XXVI

VIBRATIONAL SPECTRUM OF GASEOUS $FClO_3$ AND ITS ASSIGNMENT FOR POINT GROUP C_{3v}

		Assignment	Infrared[a] (cm^{-1})	Raman[b] (cm^{-1})
A_1	ν_1	sym ClO_3 stretch	1061 s	1062.8, 1060.9[c] vs, p
	ν_2	ClF stretch	717 s, 707 m	716.8, 706.6 s, p
	ν_3	sym ClO_3 deform.	549 w	548.8 m, p
E	ν_4	asym ClO_3 stretch	1315 vs	1314 w
	ν_5	asym ClO_3 deform.	589 m	573 w
	ν_6	rocking	405 w	414 w

[a] Data from Lide and Mann (174).
[b] Data from Claassen and Appelman (71).
[c] Splittings are due to ^{35}Cl and ^{37}Cl isotopes.

two of the fundamentals (ν_3 and ν_5) were studied at high resolution by Madden and Benedict (184). The Raman spectra of the liquid and of the gas were reported by Powell et al. (233) and Dunlap et al. (85) and by Claassen and Appelman (71), respectively. The observed fundamentals together with their assignment are summarized in Table XXVI and are in excellent agreement with a molecule of symmetry C_{3v}. The infrared spectrum of $FClO_3$ has also been reported by Engelbrecht et al. (92), Pennsalt (222), Smith et al. (271), and Karelin et al. (154). A correlation of ClO-stretching frequencies (236) and force constants with bond lengths and bond orders was given by Robinson (237); however, his plots and assumptions must be thoroughly updated before being used. Absolute infrared intensities were reported for $FClO_3$ by Kharitonov et al. (157). Quantum mechanical studies of the atomic, bond, and molecular polarizabilities were carried out by Nagarajan and Redmon (204). Numerous force fields (107, 140, 154, 157, 201, 204, 245, 273) were computed for $FClO_3$, but owing to the lack of sufficient experimental data, no unique solution was obtained. Values of about 9.4 and 3.9 mdyn/Å for the ClO- and the ClF-stretching force constants, respectively, appear to us most reasonable. Mean square amplitudes of vibration of $FClO_3$ were calculated by Müller et al. (201, 203) and Nagarajan and Redmon (204). Müller et al. (201) have also computed the Coriolis zeta constants for $FClO_3$; however, their values differ significantly from those given by Hoskins (140). Molecular reorientation in liquid $FClO_3$ was studied by Sunder and co-workers (279a) using Raman spectroscopy.

The high-resolution photoelectron spectrum of $FClO_3$ was studied by DeKock et al. (80). The results from this study, including ab initio SCF MO calculations of the electronic structures, are summarized in Tables XXVII and XXVIII. These calculations indicate considerable participation by 3d orbitals of the Cl atom, although they tend to overestimate the importance of 3d orbitals in bonding by correcting for some inade-

TABLE XXVII

Ionization Data for Perchloryl Fluoride[a,b]

Band No.	Adiabatic i.p. (eV)	Vertical i.p. (eV)	Vibrational spacing (cm^{-1})	Vibrational assignment	Orbital assignment
1	13.04 (1)	—	370 (40)	ν_3 or ν_4	$6b_2$
	13.57 (2)	—	475 (60)	ν_3	$2a_2$
2	14.85 (1)	15.181 (6)	340 (16)	ν_4	$6b_1$
	15.181 (6)	15.307 (6)	1025 (30)	ν_1	$11a_1$
3	16.676 (5)	16.676 (5)	1135 (16)	ν_1	$5b_2$
			805 (30)	ν_2	
			510 (20)	ν_3	
4	18.07 (3)	18.31 (2)	—	—	$5b_1$
5	19.175 (7)	19.390 (4)	850 (30)	ν_2	$4b_2$
			485 (40)	ν_3	
	19.699 (7)	19.807 (7)	855 (30)	ν_2	$9a_1$
			500 (20)	ν_2	
6	—	21.7 (1)	—	—	$4b_1$
7	—	24.2 (1)	—	—	$8a_1$
		Ground state	1269	ν_1	
			848	ν_2	
			544	ν_3	
			384	ν_4	

[a] Data from DeKock et al. (80).
[b] Standard deviations are given in parentheses after each quantity.

TABLE XXVIII

CALCULATED EIGENVALUES AND PERCENTAGE CHARACTER OF VALENCE MOLECULAR ORBITALS FOR $FClO_3$[a]

Orbital	Eigenvalue (eV)	Atomic character (%)						
		Chlorine orbital			Oxygen orbital		Fluorine orbital	
		3d	3s	3p	2s	2p	2s	2p
$1a_2$	−12.9	—	—	—	—	100	—	—
$7e$	−14.1	10.5	—	—	—	78.0	—	11.0
$10a_1$	−15.4	12.6	—	1.9	1.1	44.3	—	39.8
$6e$	−16.0	16.5	—	—	5.3	74.0	—	3.9
$5e$	−18.6	9.2	—	1.8	1.3	18.7	—	68.9
$4e$	−21.5	4.3	—	24.8	19.1	38.0	—	13.3
$9a_1$	−23.3	1.1	4.1	34.1	16.8	29.4	2.6	11.4
$8a_1$	−26.2	—	18.2	10.0	34.2	12.3	7.6	17.2
$3e$	−40.6	2.9	—	24.5	66.7	5.0	—	—
$7a_1$	−43.0	2.4	3.9	8.1	14.0	1.9	68.7	—
$6a_1$	−47.6	—	44.7	—	29.2	6.5	17.5	1.6

[a] Data from DeKock et al. (80)

quacy in the s and p bases. Results of MO calculations were also reported by Hillier et al. (129, 139) and Ionov and Ionova (146). The latter authors calculated the electron density distribution in $FClO_3$ as Q_{Cl} = +0.83, Q_O = −0.23, and Q_F = −0.14, using the geometry, the ionization potential of the molecule and of the free atoms, and the orbital exponents of the Slater functions as input data.

X-Ray diffraction data were reported by Tallman et al. (280, 281) for solid $FClO_3$ at liquid air temperature. The data were indexed in terms of a tetragonal unit cell with a = 7.66 and c = 5.31 Å, Z = 4, and d = 2.18 gm/cm^3. Barberi (16, 17) has shown that solid $FClO_3$ exists between its melting point and −196°C in only one solid phase. Based on entropy calculations, Koehler and Giauque (160) suggested that there is a high degree of disorder in the arrangement of the F and O atoms in crystallinic $FClO_3$.

3. Physical Properties

Some of the physical properties of $FClO_3$ are summarized in Table XXIX. In the Appendix (Table AIII), the temperature dependence of some of the thermodynamic properties is given (147). In addition to these data, the viscosity of gaseous $FClO_3$ between 50 and 150°C was reported

(218). Some thermodynamic properties of $FClO_3$ were calculated (235) using only the boiling point of the compound and correlation increments.

Perchloryl fluoride is white as a solid and colorless as a liquid and gas. It possesses a characteristic sweetish odor (92). Its toxicity is moderate and comparable to that of CH_2CHCN or Cl_2. Tests on mice showed an acute vapor toxicity (LD_{50}) of 630 ppm at 4-hr exposure time. Exposure of monkeys to 40 ppm $FClO_3$ in air for 3 months resulted in enlarged spleens and lungs together with some evidence of red cell destruction (222).

The dielectric strength of $FClO_3$ is outstanding and over a broad pressure range is about 30% higher than that of SF_6. During irradiation with ^{60}Co γ-rays, the dielectric strength decreased only by 5% (46). The correlation between negative-ion formation and electric breakdown of $FClO_3$ was studied by Hickam and Berg (138) by mass spectroscopy. Perchloryl fluoride has been used as an insulator in high-voltage systems.

Perchloryl fluoride was reported (185, 195, 205, 206, 222) to be sparingly soluble (1–3 gm/liter at 1 atm and 25°C) in a wide variety of polar and nonpolar solvents, such as aqueous solutions, alcohols, ketones, esters, ethers, and aromatic and halogenated solvents. However, more recent measurements by Golub et al. (124) show that these solubilities are substantially (several-fold) higher. When working with larger amounts of $FClO_3$ in organic solvents, all necessary precautions should be taken since mixtures of this kind are potentially explosive. Hammond et al. (132–134) have extensively studied the extremely weak electron acceptor–donor (ball-plane) interactions between $FClO_3$ and aromatic hydro- and fluorocarbons. Several inorganic acid halides, $HOSO_2F$, PCl_3, $POCl_3$, SO_2Cl_2, $SOCl_2$, $TiCl_4$, and $SiCl_4$ dissolve gaseous $FClO_3$ to the extent of 20–30 gm/liter at 25°C and 1 atm pressure (106).

Liquid perchloryl fluoride is a typical nonpolar solvent. Most inorganic and organic salts are insoluble in it. Conversely, most covalent, essentially nonpolar substances, boiling within about 50°C of perchloryl fluoride, are completely miscible, e.g., chlorine, boron trifluoride, sulfur hexafluoride, silicon tetrafluoride, phosgene, nitrous oxide, chlorine trifluoride, chlorofluorocarbons, silicon tetrachloride, sulfuryl chloride, dinitrogen tetroxide, and thionyl chloride (106).

Blends of perchloryl fluoride with halogen fluorides are homogeneous and stable. When these are used as storable liquid oxidizers for rocket propulsion, the halogen fluoride usually confers hypergolicity, increased density, and lowered vapor pressure; whereas the perchloryl fluoride provides oxygen needed for efficient combustion of carbon in the fuel or of certain metal additives. The mixtures are thermally stable and their

CHLORINE OXYFLUORIDES 379

TABLE XXIX

SOME PHYSICAL PROPERTIES OF $FClO_3$

Property	Value	Ref.
Melting point	$-147.75°C$	(92, **160**[a])
Boiling point	$-46.67°C$	(30, 92, **160**)
T_{crit}	$95.17°C$	(92, 100, **148**)
P_{crit}	53.0 atm	(148)
Crit. density	0.637 gm cm^{-3}	(**92**, 100)
Crit. molar volume	161 cm^3	(92)
Vapor pressure (for $T = -109°$ to $-44°C$)	$\log P(\text{mm}) = -1652.3/T(°K) - 8.62625 \log T + 0.0046098 T + 28.44780$	(92, 148, **160**)
Density of solid ($-190°C$)	2.19 gm cm^{-3}	(281)
Density of liquid (for $T = -142°$ to $-39°C$)	$\rho(\text{gm cm}^{-3}) = 2.266 - 1.603 \times 10^{-3} T - 4.080 \times 10^{-6} T^2 (°K)$	(92, 100, **148**, 222)
(for $T = 29.9°$ and $53.8°C$)	$\rho(\text{gm cm}^{-3}) = 1.390$ and 1.276	(268)
Viscosity of liquid (for $T = -77°$ to $54°C$)	$\log \eta = 299\ T^{-1} - 1.755$ (centipoise)	(222, **268**)
Surface tension (for $T = -75.2°$ to $-55.6°C$)	24.1 to 21.3 dyn cm^{-1}	(268)
ΔH_{fusion}	0.9163 kcal mole^{-1}	(16, 17, **160**)
ΔS_{fusion}	7.12 e.u.	(16, 17)
ΔS_{vap} ($-46.67°C$)	4.619 kcal mole^{-1}	(30, 92, 148, **160**)
Trouton constant	20.395 e.u.	(30, 92, **160**)
ΔH^0_{f298} (g)	-5.7 kcal mole^{-1}	(15–18, 82, 204, **291**)
ΔG^0_{f298} (g)	11.5 kcal mole^{-1}	(291)
S^0_{298}	66.65 e.u.	(147, 148, 172, 208, **291**)
$C_{p\ 298}$ (g)	15.517 e.u.	(147, **160**, 204, 291)
$C_{p\ 298}$ (l)	27.19 e.u.	(**150**, 160)
Specific heat ratio, C_p/C_v, gas at $25°C$	1.12	(174, **189**)

[a] Bold face reference number indicates reference from which the listed value is quoted.

compatibility with container materials is determined mainly by the halogen fluoride. The density and vapor pressure of perchloryl fluoride–chlorine trifluoride blends have been summarized in tables by Gall (*106*). The miscibility and compatibility of $FClO_3$ at low temperatures was studied by Streng (*277*) for O_2, O_3, O_2F_2, ClF, ClF_3, SF_4, SF_6, CF_3Cl, and C_4H_{10}.

4. Chemical Properties

Owing to its pseudotetrahedral configuration, its highly covalent strong Cl—F bond, and low dipole moment, $FClO_3$ possesses high kinetic stability in spite of ΔH_f^0 being only -5.7 and ΔG_f^0 being positive (11.5 kcal mole^{-1}). This is reflected in its high thermal stability and its reluctance to hydrolyze. It is not shock-sensitive and at room temperature is relatively inert. At elevated temperature, however, or under conditions supplying a sufficient amount of activation energy, it is a powerful oxidizer (*211*).

Perchloryl fluoride is thermally stable up to about 400°C. The thermal decomposition of $FClO_3$ in quartz at pressures between 5 and 930 mm and temperatures between 465° and 495°C was studied by Gatti *et al.* (*112*). They found that the decomposition reaction,

$$2FClO_3 \longrightarrow 2ClF + 3O_2$$

is unimolecular and homogeneous with an activation energy of 58.4 ± 2 kcal mole^{-1}. The rate constant at 495.4°C was found to be $k = 9.25 \times 10^{-4}$ sec^{-1} and the following decomposition mechanism was suggested:

$$FClO_3 \longrightarrow FClO_2 + O$$
$$O + FClO_3 \longrightarrow FClO_2 + O_2$$
$$FClO_2 \longrightarrow ClF + O_2$$

The decomposition kinetics were also calculated by Usmanov and Magarra (*287*) using a dimensionless molecular transfer equation. Perchloryl fluoride can be heated almost to the softening point of glass without explosion (*92*).

Hydrolysis of $FClO_3$ is very slow even at 250°–300°C (*92*). For quantitative hydrolysis, heating of $FClO_3$ with concentrated aqueous hydroxide solution to 300°C in a sealed tube is required:

$$FClO_3 + 2NaOH \longrightarrow NaClO_4 + NaF + H_2O$$

For quantitative analysis, $FClO_3$ can conveniently be reduced at 25°C by an alcoholic solution of KOH resulting in dissolved KF and a precipitate of $KClO_4$ (*222*).

The reaction of $FClO_3$ with metallic sodium or potassium starts only at ~300°C, although it proceeds vigorously (*92*). At room temperature $FClO_3$ is unreactive with a considerable number of gases, liquids, and solids. Again, however, if sufficient activation energy, such as heating to 100°–300°C, is supplied, violent reactions usually occur. With reducing agents, oxides, fluorides, and chlorides are formed. Typical examples are H_2, N_2O, H_2S, SO_2, SCl_2, PCl_3, CaC_2, KCN, NaI, KSCN, $CH_2\!=\!CCl_2$, and hydrocarbons (*122, 158, 222*). Using dilute mixtures, the H_2S–$FClO_3$ reaction can be controlled and the following products are obtained (*222*):

$$3FClO_3 + 4H_2S \longrightarrow 4SO_2 + 3HF + 3HCl + H_2O$$

In the spectra of H_2S–$FClO_3$ and H_2–$FClO_3$ flames, bands due to S_2, SO_2, OH and to ClO, OH, respectively, were observed (*177*). With HCl at 200°–300°C, the following gas-phase reaction occurs:

$$FClO_3 + 7HCl \longrightarrow HF + 4Cl_2 + 3H_2O$$

Many inorganic ions are oxidized by $FClO_3$ in aqueous solution (*112*). The oxidation rate often depends on the pH of the solution and the temperature. For example: the oxidation of KI in the presence of $NaHCO_3$ is barely detectable; in caustic soda, a slow oxidation occurs; and in 0.1 M mineral acid one observes (*92*) quantitative reaction within 4 hr according to

$$FClO_3 + 8I^- + 6H^+ \longrightarrow Cl^- + F^- + 4I_2 + 3H_2O$$

Other ions oxidized by $FClO_3$ include NO_2^-, SO_3^{2-}, and CN^- which are converted to NO_3^-, SO_4^{2-}, and NCO^-, respectively (*106, 122, 222*).

Whereas $FClO_3$ is rather inert toward most compounds, including gaseous NH_3, at room temperature it reacts (*92, 186, 187*) easily with liquid NH_3 at $-78°C$ or its aqueous solutions:

$$FClO_3 + 3NH_3 \longrightarrow NH_4F + NH_4NHClO_3$$

The reaction is complete in several hours and, in liquid NH_3, it is greatly accelerated by $NaNH_2$ (*186, 187*). From the ammonium perchloryl amide, which could not be isolated in pure form, the corresponding Ag^+, Cs^+, and K^+ salts and K_2NClO_3 and Cs_2NClO_3 have been obtained. These salts, especially when dry, are impact- and friction-sensitive.

Perchloryl fluoride does not attack glass at moderate temperature, but decomposes at 25°C on contact with activated SiO_2 or Al_2O_3, particularly in the presence of small amounts of H_2O. With other surface-active materials, such as charcoal, ignition may take place. However, there is no reaction at room temperature with synthetic zeolites. It passes freely through a 4 Å molecular sieve, but is completely absorbed by a 5 Å molecular sieve (188). Most combustible substances in contact with liquid $FClO_3$ form shock-sensitive explosive compositions. Generally, metal oxides, fluorides, or chlorides do not react with $FClO_3$ at temperatures up to 400°C (186). Lalande reported (164) that $FClO_3$ oxidizes UF_4 to UF_6. However, a subsequent study by Rude et al. (241) showed that an intermediate uranium oxyfluoride that disproportionates to UF_6 and UO_2F_2 is formed. Photolysis of mixtures of $FClO_3$ with F_2 or ClF_5 produces ClF_3O (228, 240).

Perchloryl fluoride shows no tendency to form adducts with either strong Lewis acids or bases. This behavior has been rationalized in Section II, D. The binary systems of $FClO_3$ with BF_3, PF_5, AsF_5, SbF_5, or SO_3 were studied by Lang (167), at Pennsalt (224), and by Nikitina and Rosolovskii (209). Similarly, at Pennsalt (224) no evidence was found for complexing of $FClO_3$ with either CsF or FNO_2.

Anhydrous $FClO_3$ does not corrode most of the common metals, but, in the presence of moisture, slow hydrolysis may occur causing corrosion (40, 122, 127, 222). The compatibility of various elastomers with 1:1 mixtures of $FClO_3$ and N_2F_4 was studied by Green et al. (126) and Grigger et al. (127).

In reactions with organic compounds, $FClO_3$ behaves as either an oxidant or a 1- or 2-center electrophile which, depending on the reaction conditions, can be used for the introduction of either fluorine, a ClO_3 group, or both fluorine and oxygen. A large number of publications have appeared on this subject and have been extensively reviewed by Khutoretskii et al. (158). Additional general information can be found in Refs. (106, 122, 169, 262, and 284). Since a systematic coverage of this subject is beyond the scope of this review, we give examples only of the most important type of reactions, in addition to references to some of the more recent publications not covered in the previous reviews.

Since $FClO_3$ is highly susceptible to nucleophilic attack at the chlorine atom, it reacts readily with anions. These reactions are relatively well-understood, and Sheppard has proposed (261) a general mechanism for these reactions by which the most nucleophilic center in the anion (oxygen or other heteroatom related to carbon) always attacks the chlorine and never the more electronegative fluorine. For localized nucleophiles (such as alkoxides), simple fluoride ion displacement occurs, but, for the mesomeric ions (ambient electrophiles), an intramolecular (cyclic) transfer of F^- can occur in the intermediate to give a C—F bond. The high energy gained by the formation of the C—F bond provides a strong driving force for this fluoride transfer, and fluorine never has to achieve a highly unfavorable energy state with positive charge. This mechanism explains why phenyllithium reacts with $FClO_3$ to give perchloryl benzene, whereas 2-lithiothiophene gives 2-fluorothiophene in high yield (257):

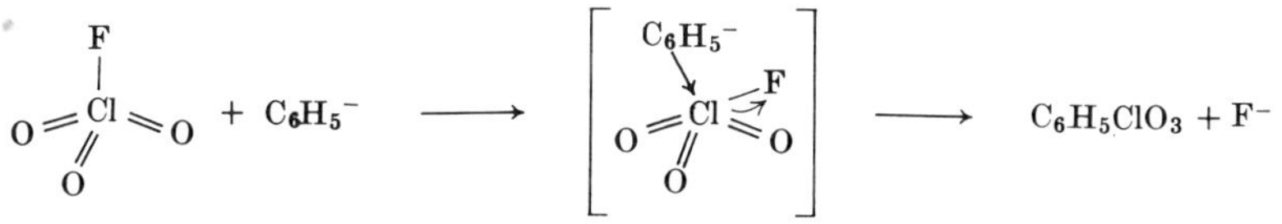

$$\text{O=}\overset{\mathrm{F}}{\underset{\mathrm{O}}{\mathrm{Cl}}}\text{=O} + \mathrm{C_6H_5^-} \longrightarrow \left[\text{O=}\overset{\mathrm{C_6H_5^-}}{\underset{\mathrm{O}}{\mathrm{Cl}}}\overset{\mathrm{F}}{=}\mathrm{O}\right] \longrightarrow \mathrm{C_6H_5ClO_3} + \mathrm{F^-}$$

and

$$\text{O=}\overset{\mathrm{F}}{\underset{\mathrm{O}}{\mathrm{Cl}}}\text{=O} + \mathrm{C_4H_3S^-} \longrightarrow \left[\text{thiophene-Cl(O)(O)(F) intermediate}\right]^- \longrightarrow \text{2-F-thiophene} + \mathrm{ClO_3^-}$$

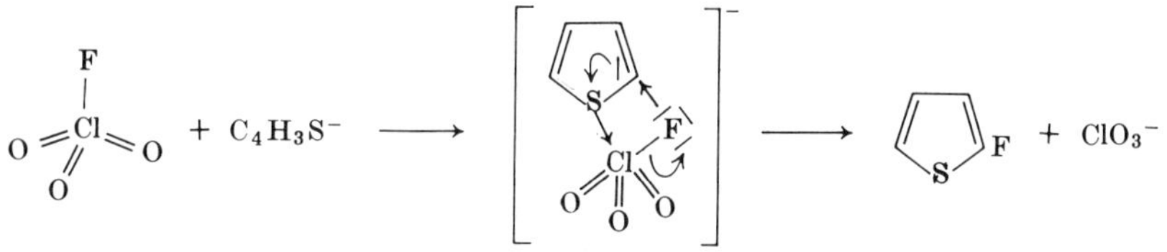

Compounds having a cyclic double bond conjugated with an aromatic ring are capable of reacting with $FClO_3$ to give α-fluoroketones. This type of reaction was named oxofluorination and in it $FClO_3$ acts as a 2-center electrophile as shown for indene (207):

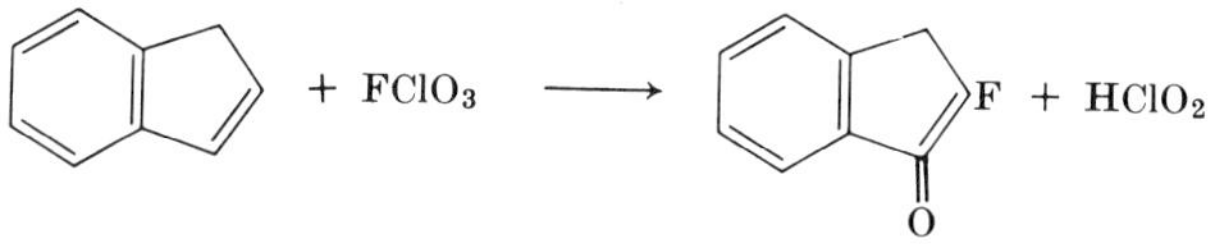

indene + $FClO_3$ ⟶ 2-fluoro-1-indanone + $HClO_2$

In the presence of Friedel–Crafts catalysts, such as $AlCl_3$, the $FClO_3$ can be used for introducing a ClO_3 group (perchlorylation) into an aromatic ring (144):

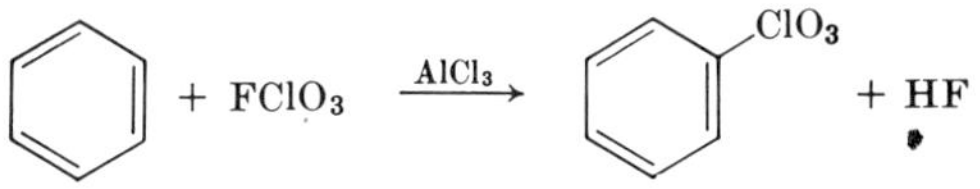

benzene + $FClO_3$ $\xrightarrow{AlCl_3}$ $C_6H_5ClO_3$ + HF

Hydrogenolysis (258) of perchloryl aromatic compounds yields ArH and not ArOH, thus confirming the presence of a C—Cl bond. Another useful reaction of $FClO_3$ involves the replacement of the active hydrogens of methylene compounds by fluorine (145, 262, 284). A typical example is the fluorination of malonic esters:

$$\mathrm{CH_2(COOR)_2} \xrightarrow{+\mathrm{FClO_3}} \mathrm{CF_2(COOR)_2}$$

Since $FClO_3$ is a very mild fluorinating agent, it has found widespread use for the selective fluorination of compounds such as steroids.

The reaction of cyclic amines with $FClO_3$ is similar to that of $FClO_3$ with NH_3 (see above). For example, the following reaction takes place with piperidine (110):

$$2 \langle NH \rangle + FClO_3 \longrightarrow \langle NClO_3 \rangle + \langle NH_2^+F^- \rangle$$

(For additional recent publications dealing with the use of $FClO_3$ as a reagent for the synthesis of organic compounds, see Refs. *94, 103, 111, 114–116, 153, 159, 212, 213, 217, 250, 286, 289* and *290*.)

5. Uses

The most thoroughly studied application of $FClO_3$ is its use as an oxidant. The spectra of fuel–$FClO_3$ flames were studied (*177*), and the flame speed in mixtures of CH_4 with air and $FClO_3$ was measured (*131*). The H_2–$FClO_3$ flame was found (*251*) to be readily controllable with a low background and useful as an excitation source for flame photometry. For rocket propulsion, the performance of either neat $FClO_3$ or combinations with other oxidizers, such as halogen fluorides (*11, 26*), was studied and typical performance data (*106*) are given in Table XXX. Small amounts of ClF_3 can be added to neat $FClO_3$ to provide self-ignition. The performance of $FClO_3$ as an oxidizer is similar to that of N_2O_4 (*106, 136*), and the burning rate of solid propellants is increased by $FClO_3$ (*267*). It has also been proposed to use an acetylene–$FClO_3$ torch

TABLE XXX

Performance of Selected Storable Liquid Oxidizers for Rocket Propulsion[a]

Oxidizer	Fuel	Specific impulse[b] (sec)	Density impulse (gm sec/cm³)
$FClO_3$	UDMH[c]	290	337
ClF_3	UDMH	279	382
65 ClF_3/35 $FClO_3$	UDMH	288	386
$FClO_3$	LiH solid[d]	273	337
ClF_3	LiH solid	288	436
88 ClF_3/12 $FClO_3$	LiH solid	291	433
$FClO_3$	N_2H_4	295	358
ClF_3	N_2H_4	292	436
N_2O_4	N_2H_4	291	354

[a] Data from Gall (*106*).
[b] Pound force × sec/lb mass; shifting equilibrium; pressure ratio 1000:14.7.
[c] Unsymmetrical dimethylhydrazine.
[d] LiH, 85%; organic binder, 15%.

for cutting and welding of metals, in the Sterling cycle engine, in high-pressure gas generation for turbine drive, in fuel cells, and in explosives similar to Sprengel liquid O_2-carbon powder combinations (105), and as a deodorant in aerosol sprays (170). However, the latter application appears very doubtful in view of the substantial toxicity of $FClO_3$ (see above).

The use of $FClO_3$ as a chemical reagent for the introduction of fluorine or a ClO_3 group has been discussed in detail in Section III, K, 4 and is of special value for the synthesis of fluorine-containing steroids. The polymerization of ethylene under a pressure of hundreds of atmospheres and a temperature of about 200°C in the presence of $FClO_3$ has been patented (135).

There are patents on the use of $FClO_3$ as a heat transfer medium in refrigeration (165) and as an insecticide-fungicide (123). Owing to its ability to absorb intensively slow electrons (138), $FClO_3$ can be used as a gaseous insulator. Its dielectric properties are superior to those of SF_6, and it hardly deteriorates on exposure to γ-irradiation (104).

General information on shipping, handling, safety, etc., of $FClO_3$ can be found in Gall's review (106).

L. CHLORINE FLUORIDE OXIDE RADICALS

Very little is known about chlorine fluoride oxide radicals. Although the formation of the $FClO_3^-$ radical anion in the reaction of $FClO_3$ with nucleophilic agents has been postulated (286), it has not been isolated and characterized.

The only well-known species is the $FClO^+$ radical cation. The ESR spectrum of this species was first reported by Olah and Comisarow (214, 215) for both the ClF_3-SbF_5 and the ClF_5-SbF_5 system. However, the spectrum was incorrectly interpreted in terms of a ClF^+ radical cation. Eachus, Slight, and Symons (86) suggested that the observed spectrum is due to $FClO^+$ and not to ClF^+. This conclusion was supported by Christe and Muirhead (62) who showed that, in the pure ClF_3-SbF_5 and ClF_5-SbF_5 systems, this species could not be observed but was generated by impurities in the starting materials. Additional evidence for this species containing oxygen was obtained by Gillespie and Morton (121) who investigated the reaction of ClF and of Cl_2-ClF with the superacid medium HSO_3F-SbF_5-SO_3. It was shown that the addition of H_2O to solutions of $ClF_2^+SbF_6^-$ in SbF_5 strongly enhanced the ESR signal attributed to ClF^+ by Olah and Comisarow. They suggested that the species was due either to $FClO^+$ or $FClO_2^+$, although their attempts to detect ^{17}O hyperfine splitting in a sample treated with enriched water were unsuccessful. The conclusive identification of this species was recently reported by Morton and Preston (*199*). By using ^{17}O substitution techniques, they succeeded in proving that the species contains 1 oxygen atom and is best described as $FClO^+$. This radical cation is characterized by its g value of 2.0059 and the following hyperfine interactions: $a_{17} = 18.0$, $a_{19} = 20.4$, and $a_{35} = 12.9$ G.

During a matrix-isolation infrared study of the F_2–Cl_2O and ClF–O_3 systems, a new species was observed by Andrews *et al.* (*5*) at 733.8 cm^{-1} which was tentatively assigned to the $ClF_2O \cdot$ radical. However, more data are needed for the positive identification of this species.

M. MISCELLANEOUS

The ClO_3F^{2-} anion has been reported by Mitra and Ray (*197*). However, in our opinion their claim is almost certainly incorrect.

A compound of the empirical composition FCl_2O_6 was claimed by DeGuevara (*79*). The following self-explanatory abstract of this patent was found in *Chemical Abstracts* and, we are confident, will be enjoyed by the more knowledgeable readers:

> The title compd. which is claimed to be novel is prepd. by reacting in a hermetically sealed flask 500 ml. 55° Be H_2SO_4, $Cs(OCl)_2$ 5–10, $KClO_2$ 9–20, $KClO_3$ 10–20, and $Mg(ClO_4)_2$ 10–20 g. Cl and a Cl oxide are given off, washed, and collected as a stabilized aq. soln. Simultaneously, F is produced from CaF_2 and H_2SO_4 and washed and dried. The F is passed into the stabilized aq. soln. of FCl_2O_6, which is suitable for use as an antiseptic, preservative, and purifier in the food, wine, perfume, and water industries and as a humectant and bleach for textiles.

IV. Appendix: Tables of Thermodynamic Properties for Some Chlorine Oxyfluorides

TABLE AI

THERMODYNAMIC PROPERTIES FOR ClF_3O GAS[a]

$T(°K)$	C_p^0 [cal/(mole deg)]	$H^0-H_0^0$ (kcal/mole)	$-(F°-H_0^0)/T$ [cal/(mole deg)]	S^0 [cal/(mole deg)]
0	0	0	0	0
100	9.721	0.837	49.255	57.624
200	14.932	2.072	55.613	65.971
298.15	18.593	3.732	60.159	72.675
300	18.646	3.766	60.237	72.790
400	20.875	5.751	64.108	78.486
500	22.260	7.913	67.478	83.305
600	23.160	10.187	70.470	87.448
700	23.771	12.536	73.159	91.067
800	24.200	14.936	75.602	94.271
900	24.512	17.372	77.838	97.141

TABLE AI—continued

$T(°K)$	C_p^0 [cal/(mole deg)]	$H^0-H_0^0$ (kcal/mole)	$-(F°-H_0^0)/T$ [cal/(mole deg)]	S^0 [cal/(mole deg)]
1000	24.744	19.835	79.900	99.736
1100	24.921	22.319	81.813	102.103
1200	25.059	24.818	83.595	104.277
1300	25.168	27.330	85.265	106.288
1400	25.256	29.851	86.834	108.156
1500	25.328	32.380	88.314	109.901
1600	25.387	34.916	89.715	111.538
1700	25.437	37.458	91.044	113.078
1800	25.479	40.003	92.309	114.533
1900	25.514	42.553	93.516	115.912
2000	25.545	45.106	94.668	117.221

[a] Data from Christe and Curtis (55).

TABLE AII

THERMODYNAMIC PROPERTIES FOR ClF_3O_2 GAS[a]

$T(°K)$	C_p^0 [cal/(mole deg)]	$H^0-H_0^0$ (kcal/mole)	$-(F°-H_0^0)/T$ [cal/(mole deg)]	S^0 [cal/(mole deg)]
0	0	0	0	0
100	10.127	0.847	48.967	57.437
200	16.511	2.179	55.516	66.411
298.15	21.256	4.049	60.375	73.956
300	21.327	4.089	60.459	74.088
400	24.384	6.386	64.711	80.675
500	26.362	8.930	68.484	86.344
600	27.685	11.636	71.881	91.275
700	28.599	14.453	74.968	95.615
800	29.251	17.347	77.795	99.479
900	29.727	20.298	80.400	102.953
1000	30.085	23.289	82.816	106.105
1100	30.360	26.312	85.066	108.985
1200	30.574	29.359	87.171	111.637
1300	30.745	32.425	89.148	114.091
1400	30.883	35.507	91.012	116.375
1500	30.995	38.601	92.775	118.509
1600	31.089	41.705	94.447	120.513
1700	31.167	44.818	96.036	122.400
1800	31.233	47.938	97.551	124.183
1900	31.289	51.064	98.997	125.873
2000	31.337	54.196	100.382	127.480

[a] Data from Christe and Curtis (57).

TABLE AIII: Thermodynamic Properties for $FClO_3$ Gas[a]

$T(°K)$	C_p^0	S^0	$-(F° - H_{298}^0)T$	$H° - H_{298}^0$	ΔH_f^0	ΔF_f^0	$\text{Log } K_P$
	cal mole^{-1} deg^{-1}			kcal mole^{-1}			
0	0.000	0.000	Infinite	−3.178	−3.034	−3.034	Infinite
100	8.462	54.278	78.032	−2.375	−3.996	1.097	−2.397
200	12.073	61.160	67.968	−1.362	−4.715	6.493	−7.095
298	15.517	66.653	66.653	0.000	−5.120	12.090	−8.861
300	15.573	66.749	66.653	0.029	−5.125	12.196	−8.884
400	18.152	71.602	67.297	1.722	−5.298	18.002	−9.835
500	20.000	75.863	68.593	3.635	−5.312	23.831	−10.416
600	21.319	79.633	70.125	5.704	−5.226	29.653	−10.801
700	22.271	82.994	71.728	7.886	−5.076	35.455	−11.069
800	22.967	86.016	73.328	10.150	−4.884	41.231	−11.263
900	23.487	88.753	74.893	12.474	−4.665	46.983	−11.408
1000	23.883	91.249	76.405	14.843	−4.426	52.710	−11.519
1100	24.189	93.540	77.860	17.248	−4.175	58.410	−11.604
1200	24.430	95.655	79.256	19.679	−3.914	64.090	−11.672
1300	24.624	97.619	80.594	22.132	−3.649	69.746	−11.725
1400	24.780	99.449	81.876	24.602	−3.382	75.380	−11.767
1500	24.909	101.164	83.106	27.087	−3.112	80.998	−11.801
1600	25.016	102.775	84.285	29.584	−2.843	86.596	−11.828
1700	25.105	104.294	85.418	32.090	−2.574	92.179	−11.850
1800	25.181	105.731	86.507	34.604	−2.311	97.744	−11.867
1900	25.246	107.094	87.555	37.125	−2.051	103.297	−11.881
2000	25.301	108.391	88.564	39.653	−1.795	108.833	−11.892
2100	25.349	109.626	89.538	42.185	−1.544	114.359	−11.901
2200	25.391	110.807	90.478	44.723	−1.300	119.871	−11.908
2300	25.428	111.936	91.387	47.264	−1.060	125.374	−11.913
2400	25.461	113.019	92.266	49.808	−0.828	130.868	−11.917
2500	25.489	114.059	93.117	52.356	−0.605	136.347	−11.919
2600	25.515	115.059	93.942	54.906	−0.386	141.823	−11.921

CHLORINE OXYFLUORIDES

T						
2800	25.558	116.952	95.518	60.013	0.027	152.747	−11.922
2900	25.577	117.849	96.273	62.570	0.223	158.199	−11.922
3000	25.593	118.716	97.007	65.129	0.411	163.641	−11.921
3100	25.608	119.556	97.721	67.689	0.593	169.082	−11.920
3200	25.622	120.369	98.416	70.250	0.767	174.513	−11.918
3300	25.635	121.158	99.093	72.813	0.935	179.937	−11.916
3400	25.646	121.923	99.753	75.377	1.095	185.360	−11.914
3500	25.657	122.667	100.398	77.942	1.249	190.776	−11.912
3600	25.666	123.390	101.026	80.508	1.396	196.192	−11.910
3700	25.675	124.093	101.640	83.075	1.537	201.600	−11.907
3800	25.683	124.778	102.240	85.643	1.673	207.009	−11.905
3900	25.691	125.445	102.827	88.212	1.803	212.404	−11.902
4000	25.698	126.096	103.400	90.782	1.926	217.803	−11.900
4100	25.704	126.730	103.961	93.352	2.043	223.200	−11.897
4200	25.711	127.350	104.511	95.922	2.156	228.592	−11.894
4300	25.716	127.955	105.049	98.494	2.264	233.986	−11.892
4400	25.721	128.546	105.577	101.066	2.366	239.370	−11.889
4500	25.726	129.124	106.093	103.638	2.464	244.755	−11.886
4600	25.731	129.690	106.600	106.211	2.558	250.143	−11.884
4700	25.735	130.243	107.097	108.784	2.646	255.525	−11.881
4800	25.739	130.785	107.585	111.358	2.731	260.904	−11.879
4900	25.743	131.316	108.064	113.932	2.812	266.277	−11.876
5000	25.747	131.836	108.534	116.507	2.889	271.652	−11.873
5100	25.750	132.346	108.996	119.082	2.961	277.029	−11.871
5200	25.753	132.846	109.450	121.657	3.030	282.399	−11.868
5300	25.756	133.336	109.896	124.232	3.097	287.779	−11.866
5400	25.759	133.818	110.335	126.808	3.158	293.140	−11.863
5500	25.762	134.290	110.766	129.384	3.220	298.518	−11.861
5600	25.765	134.755	111.190	131.960	3.274	303.881	−11.859
5700	25.767	135.211	111.608	134.537	3.327	309.248	−11.857
5800	25.769	135.659	112.019	137.114	3.378	314.621	−11.855
5900	25.771	136.099	112.423	139.691	3.425	319.982	−11.852
6000	25.773	136.533	112.821	142.268	3.470	325.349	−11.850

[a] JANAF Thermochemical Tables (147).

Acknowledgments

The authors are indebted to the Office of Naval Research, Power Branch, for their continued interest in and funding of halogen oxidizer research at Rocketdyne and to Dr. L. R. Grant for helpful discussions, and to Mrs. C. Mirras for typing the manuscript.

References

1. Agahigian, H., Gray, A. P., and Vickers, G. D., *Can. J. Chem.* **40**, 157 (1962).
2. Alexakos, L. G., and Cornwell, C. D., *J. Chem. Phys.* **41**, 2098 (1964).
3. Alexandre, M., and Rigny, P., *Can. J. Chem.* **52**, 3676 (1974).
4. Allied Chemical Corporation, unpublished results on Contract No. DA-30-069-ORD-2638 (Sept. 1959–Sept. 1964).
5. Andrews, L., Chi, F. K., and Arkell, A., *J. Amer. Chem. Soc.* **96**, 1997 (1974).
6. Arvia, A. J., and Aymonino, P. J., *Spectrochim. Acta* **19**, 1449 (1963).
7. Arvia, A. J., Basualdo, W. H., and Schumacher, H. J., *Angew. Chem.* **67**, 616 (1955).
8. Arvia, A. J., Basualdo, W. H., and Schumacher, H. J., *Z. Anorg. Allg. Chem.* **286**, 58 (1956).
9. Aubert, J., Bougon, R., and Carles, M., *Commis. Energ. At.* [*Fr.*], *Rapp.* **CEA-R-3282** (1967).
10. Axworthy, A. E., Mueller, K. H., and Wilson, R. D., "Photochemistry of Interest as Rocket Propellants," Final Report on Contract No. AFOSR-TR-73-2183 (1973).
11. Ayers, O. E., and Huskins, C. W., U.S. Patent 3,717,997 (1973).
12. Aymonino, P. J., Sicre, J. E., and Schumacher, H. J., *J. Chem. Phys.* **22**, 756 (1954).
13. Bacon, J., Gillespie, R. J., and Quail, J. W., *Can. J. Chem.* **41**, 3063 (1963).
14. Baran, E. J., *Z. Chem.* **13**, 391 (1973).
15. Barberi, P., *Bull. Inform. Sci. Technol. CEA* No. 180, p. 55 (1973).
16. Barberi, P., Ph.D. Thesis, University of Provence, Aix Marseille, France (1974).
17. Barberi, P., *5th Eur. Symp. Fluorine Chem.* Paper I-7 (1974).
18. Barberi, P., and Carre, J., *5th Eur. Symp. Fluorine Chem.* Paper I-4 (1974).
19. Barr, J., Gillespie, R. J., and Thompson, R. C., *Inorg. Chem.* **3**, 1149 (1964).
20. Barth-Wehrenalp, G., *J. Inorg. Nucl. Chem.* **2**, 266 (1956).
21. Barth-Wehrenalp, G., *J. Inorg. Nucl. Chem.* **4**, 374 (1957).
22. Barth-Wehrenalp, G., U.S. Patent 2,942,948 (1960).
23. Barth-Wehrenalp, G., and Mandell, H., U.S. Patent 2,942,949 (1960); German Patent 1,076,640 (1960).
24. Basualdo, W. H., and Schumacher, H. J., *Angew. Chem.* **67**, 231 (1955).
25. Beardell, A. W., and Grelecki, C. J., U.S. Patent 3,404,958 (1968).
26. Beighley, C. M. *Missiles Rockets* 30 (1960).
27. Benoit, R., Besnard, G., Hartmanshenn, O., Luce, M., Mougin, J., and Pelisse, J., *Commis. Energ. At.* [*Fr.*], *Rapp.* **CEA-R-3963** (1970).
28. Biemann, K., "Mass Spectrometry," p. 161. McGraw-Hill, New York 1962.
29. Bilham, J., and Linnett, J. W., *Nature (London)* **301**, 1323 (1964).
30. Bode, H., and Klesper, E., *Z. Anorg. Allg. Chem.* **266**, 275 (1951).
31. Bode, H., and Klesper, E., *Angew. Chem.* **66**, 605 (1954).
32. Bougon, R., *Bull. Inform. Sci. Technol. CEA* No. 161, p. 9 (1971).

33. Bougon, R., *C. R. Acad. Sci., Ser. C* **274**, 696 (1972).
34. Bougon, R., Bui Huy, T., Cadet, A., Charpin, P., and Rousson, R., *Inorg. Chem.* **13**, 690 (1974).
35. Bougon, R., Bui Huy, T., and Charpin, P., *Inorg. Chem.* **14**, 1822 (1975).
36. Bougon, R., Carles, M., and Aubert, J., *C. R. Acad. Sci., Ser. C* **265**, 179 (1967).
37. Bougon, R., Isabey, J., and Plurien, P., *C. R. Acad. Sci., Ser. C* **271**, 1366 (1970).
38. Bougon, R., Isabey, J., and Plurien, P., *C. R. Acad. Sci., Ser. C* **273**, 415 (1971).
39. Bougon, R., Isabey, J., and Plurien, P., French Patent 2,110, 555 (1972).
40. Boyd, W. K., Berry, W. E., and White, E. L., Rept. No. AD 613553, NASA Accession No. N65-24361 (1965).
41. Brownstein, S., *Can. J. Chem.* **38**, 1597 (1960).
42. Carter, H. A., and Aubke, F., *Can. J. Chem.* **48**, 3456 (1970).
43. Carter, H. A., Johnson, W. M., and Aubke, F., *Can. J. Chem.* **47**, 4619 (1969).
44. Carter, H. A., Qureshi, A. M., and Aubke, F., *Chem. Commun.* p. 1461 (1968).
45. Carter, H. A., Ruddick, J. N., Sams, J. R., and Aubke, F., *Inorg. Nucl. Chem. Lett.* **11**, 29 (1975).
46. Chapman, J. J., and Frisco, L. J., *Pap., 111th Meet. Electrochem. Soc.* (1957).
47. Chi, F. K., and Andrews, L., *J. Phys. Chem.* **77**, 3062 (1973).
48. Ching-Yung Tao, *Hua Hsueh Tung Pao* **5**, 44 (1962).
49. Christe, K. O., *Inorg. Nucl. Chem. Lett.* **8**, 453 (1972).
50. Christe, K. O., *Inorg. Nucl. Chem. Lett.* **8**, 457 (1972).
51. Christe, K. O., *Inorg. Chem.* **11**, 1220 (1972).
52. Christe, K. O., *Inorg. Chem.* **12**, 1580 (1973).
53. Christe, K. O., *XXIVth Int. Congr. Pure Appl. Chem.*, Vol. IV, 115 (1974).
54. Christe, K. O., and Curtis, E. C., *Inorg. Chem.* **11**, 35 (1972).
55. Christe, K. O., and Curtis, E. C., *Inorg. Chem.* **11**, 2196 (1972).
56. Christe, K. O., and Curtis, E. C., *Inorg. Chem.* **11**, 2209 (1972).
57. Christe, K. O., and Curtis, E. C., *Inorg. Chem.* **12**, 2245 (1973).
58. Christe, K. O., Curtis, E. C., and Schack, C. J., *Inorg. Chem.* **11**, 2212 (1972).
59. Christe, K. O., Curtis, E. C., and Schack, C. J., *Spectrochim. Acta, Part A* **31**, 1035 (1975).
60. Christe, K. O., Curtis, E. C., and Wilson, R. D., *7th Int. Symp. Fluorine Chem.*, Paper I-25 (1973).
61. Christe, K. O., Hon, J. F., and Pilipovich, D., *Inorg. Chem.* **12**, 84 (1973).
62. Christe, K. O., and Muirhead, J. S., *J. Amer. Chem. Soc.* **91**, 7777 (1969).
63. Christe, K. O., Sawodny, W., and Guertin, J. P., *Inorg. Chem.* **6**, 1159 (1967).
64. Christe, K. O., Schack, C. J., and Pilipovich, D., *Inorg. Chem.* **11**, 2205 (1972).
65. Christe, K. O., Schack, C. J., Pilipovich, D., Curtis, E. C., and Sawodny, W., *Inorg. Chem.* **12**, 620 (1973).
66. Christe, K. O., Schack, C. J., Pilipovich, D., and Sawodny, W., *Inorg. Chem.* **8**, 2489 (1969).
67. Christe, K. O., Schack, C. J., and Wilson, R. D., *Inorg. Chem.* **14**, 2224 (1975).
68. Christe, K. O., and Wilson, R. D., *Inorg. Chem.* **12**, 1356 (1973).
69. Christe, K. O., Wilson, R. D., and Curtis, E. C., *Inorg. Chem.* **12**, 1358 (1973).
70. Christe, K. O., Wilson, R. D., and Schack, C. J., *Inorg. Nucl. Chem. Lett.* **11**, 161 (1975).

71. Claassen, H. H., and Appelman, E. H., *Inorg. Chem.* **9**, 622 (1970).
72. Clark, A. H., Beagley, B., and Cruickshank, D. W. J., *Chem. Commun.* p. 14 (1968).
73. Clark, H. C., and Emeleus, H. J., *J. Chem. Soc., London* p. 190 (1958).
74. Cooper, T. D., Dost, F. N., and Wang, C. H., *J. Inorg. Nucl. Chem.* **34**, 3564 (1972).
75. Cotton, F. A., and Wilkinson, G., "Advanced Inorganic Chemistry," 3rd ed. Wiley, New York, 1972.
76. Coulson, C. A., *J. Chem. Soc., London* p. 1442 (1964).
77. Dadieu, A., Damm, R., and Schmidt, E. W., "Raketentreibstoffe." Springer-Verlag, Berlin and New York, 1968.
78. Dale, J. W., U.S. Patent 3,663,183 (1972).
79. DeGuevara, M. L., French Patent 1,583,711 (1966).
80. DeKock, R. L., Lloyd, D. R., Hillier, I. H., and Saunders, V. R., *Proc. Roy. Soc., Ser. A* **328**, 401 (1972).
81. Dess, H., U.S. Patent 2,982,618 (1961).
82. Dibeler, V. H., Reese, R. M., and Mann, D. E., *J. Chem. Phys.* **27**, 176 (1957).
83. Dow Chemical Co., Report No. T-0009-4Q-68, Quarterly Technical Report No. 4 under Contract F04611-67-C-0009 (1969).
84. Downs, A. J., and Adams, C. J., *in* "Comprehensive Inorganic Chemistry" (J. C. Bailar *et al.*, eds.), Vol. II, pp. 1386–1396. Pergamon, Oxford, 1973.
85. Dunlap, J. L., and Jones, E. A., *Spectrosc. Mol.* **9**, 32 (1960).
86. Eachus, R. S., Sleight, T. P., and Symons, M. C. R., *Nature (London)* **222**, 769 (1969).
87. Ebenhöch, F. L., Ph.D. Thesis, University of Munich, Germany (1954).
88. Edwards, A. J., and Sills, R. J. C., *J. Chem. Soc., A* p. 2697 (1970).
88a. Edwards, A. J., and Sills, R. J. C., *J. Chem. Soc., Dalton Trans.* p. 1726 (1974).
89. Engelbrecht, A., *Angew. Chem.* **66**, 442 (1954).
90. Engelbrecht, A., U.S. Patent 2,942,947 (1960).
91. Engelbrecht, A., and Atzwanger, H., *Monatsh. Chem.* **83**, 1087 (1952).
92. Engelbrecht, A., and Atzwanger, H., *J. Inorg. Nucl. Chem.* **2**, 348 (1956).
93. Engelbrecht, A., Mayr, O., Ziller, G., and Schandara, E., *Monatsh. Chem.* **105**, 796 (1974).
94. Erashko, V. I., Sankov, B. G., Shevelev, S. A., and Fainzilberg, A. A., *Izv. Akad. Nauk SSSR, Ser. Khim.* p. 344 (1973).
95. Farrar, R. L., Jr., Report No. K-1416. Office of Technical Services, U.S. Dept. of Commerce, Washington, D.C., 1960.
96. Farrar, T. C., Maryott, A. A., and Malmberg, M. S., *Ber. Bunsenges. Phys. Chem.* **75**, 246 (1971).
97. Faust, J. P., Jache, A. W., and Klanica, A. J., U.S. Patent 3,545,924 (1970); French Patent 1,497,123 (1967).
98. Figini, R. V., Goloccia, E., and Schumacher, H. J., *Z. Phys. Chem. (Frankfurt am Main)* [N.S.] **14**, 32 (1958).
99. Fink, W., Ph.D. Thesis, University of Munich, Germany (1956).
100. Francis, A. W., *Chem. Eng. Sci.* **10**, 37 (1959).
101. Frenkel, L., Smith, W., and Gallagher, J. J., *J. Chem. Phys.* **45**, 2251 (1966).
102. Frey, R. A., Redington, R. L., and Aljibury, A. L. K., *J. Chem. Phys.* **54**, 344 (1971).
103. Fridland, S. V., Dmitrieva, N. V., Vigalok, I. V., Zykova, T. V., and Salakhutdinov, R. A., *Zh. Obshch. Khim.* **43**, 572 (1973).

104. Gall, J. F., U.S. Patent 3,038,955 (1956).
105. Gall, J. F., U.S. Patent 3,066,058 (1962).
106. Gall, J. F., in "Kirk Othmer Encyclopedia of Chemical Technology" 2nd ed., Vol. 9, p. 598. Wiley (Interscience), New York, 1966.
107. Gans, P., *J. Mol. Struct.* **12**, 411 (1972).
108. Gardiner, D. J., *J. Fluorine Chem.* **3**, 226 (1973).
109. Gardiner, D. J., and Turner, J. J., *6th Int. Fluorine Symp.* Paper C-13, (1971).
110. Gardner, D. M., Helitzer, R., and Mackley, C., *J. Org. Chem.* **29**, 3738 (1964).
111. Gardiner, D. M., Helitzer, R., and Rosenblatt, D. H., *J. Org. Chem.* **32**, 1115 (1967).
112. Gatti, R., Sicre, J. E., and Schumacher, H. J., *Z. Phys. Chem. (Frankfurt am Main)* [N.S.] **23**, 164 (1960); *Angew. Chem.* **69**, 638 (1957).
113. Gatti, R., Staricco, E., Sicre, J. E., and Schumacher, H. J., *Z. Phys. Chem. (Frankfurt am Main)* [N.S.] **35**, 343 (1962); see also Ghibaudi, E., Sicre, J. E., and Schumacher, H. J., *ibid* **90**, 95 (1974).
114. Gensler, W. J., Ahmed, W. A., and Leeding, M. V., *J. Org. Chem.* **33**, 4279 (1968).
115. Gershon, H., Renwick, J. A. A., Wynn, W. K., and Ascoli, R. D., *J. Org. Chem.* **31**, 916 (1966).
116. Gershon, H., Schulman, S. G., and Spevack, A. D., *J. Med. Chem.* **10**, 536 (1967).
117. Gillardeau, J., and Macheteau, Y., French Patent 1,527,112 (1968).
118. Gillespie, R. J., *Accounts Chem. Res.* **1**, 202 (1968).
119. Gillespie, R. J., "Molecular Geometry." Van Nostrand-Reinhold, Princeton, New Jersey, 1972.
120. Gillespie, R. J., Landa, B., and Schrobilgen, G. J., *Chem. Commun.* p. 1543 (1971).
121. Gillespie, R. J., and Morton, M. J., *Inorg. Chem.* **11**, 591 (1972).
122. Gmelin's "Handbuch der Anorganischen Chemie," Syst. No. 6, Part B, No. 2. Verlag Chemie, Weinheim, 1969.
123. Goebel, M., U.S. Patent 2,913,366 (1959).
124. Golub, V. B., Khutoretskii, V. M., Besprozvannyi, M. A., Temchenko, V. G., and Antipenko, G. L., *Zh. Prikl. Khim. (Leningrad)* **44**, 679 (1971).
125. Grakauskas, V., French Patent 1,360,968 (1964).
126. Green, J., Levine, N. B., and Sheehan, W., *Rubber Chem. Technol.* **39**, 1222 (1966).
127. Grigger, J. C., and Miller, H. C., WADD Tech. Rep. 61-54. U.S. Air Force, Wright-Patterson Air Force Base, Ohio, 1961.
128. Grosse, A. V., and Streng, A. G., U.S. Patent 3,285,842 (1966).
129. Guest, M. F., and Hillier, I. H., *Int. J. Quantum Chem.* **6**, 967 (1972).
130. Hach, R. J., and Rundle, R. E., *J. Amer. Chem. Soc.* **73**, 4321 (1951).
131. Halpern, C., *J. Res. Nat. Bur. Stand., Sect. A* **65**, 513 (1961).
132. Hammond, P. R., *J. Chem. Soc., A* 3826 (1971).
133. Hammond, P. R., and Lake, R. R., *Chem. Commun.* p. 987 (1968).
134. Hammond, P. R., and Lake, R. R., *J. Chem. Soc., A* p. 3819 (1971).
135. Hardwike, N., U.S. Patent 2,947,738 (1960).
136. Hendel, F., and Cavecche, E., *Chem. Eng.* **67**, 93 (1960).
137. Heras, M. J., Aymonino, P. J., and Schumacher, H. J., *Z. Phys. Chem. (Frankfurt am Main)* [N.S.] **22**, 161 (1959).
138. Hickam, W., and Berg, D., *J. Chem. Phys.* **29**, 517 (1958).

139. Hillier, I. H., and Saunders, V. R., *Chem. Commun.* p. 1183 (1970).
140. Hoskins, L. C., *J. Chem. Phys.* **50**, 1130 (1969).
141. Huggins, D. K., and Fox, W. B., U.S. Patent 3,423,168 (1969).
142. Huggins, D. K., and Fox, W. B., *Inorg. Nucl. Chem. Lett.* **6**, 337 (1970).
143. Huston, J. L., *J. Amer. Chem. Soc.* **93**, 5255 (1971).
144. Inman, C., Oesterling, R., and Tyczkowski, E., *J. Amer. Chem. Soc.* **80**, 5286 (1958).
145. Inman, C., Oesterling, R., and Tyczkowski, E., *J. Amer. Chem. Soc.* **80**, 6533 (1958).
146. Ionov, S. P., and Ionova, G. V., *Russ. J. Inorg. Chem.* **14**, 886 (1969).
147. JANAF Thermochemical Tables, PB-168370 (1965).
148. Jarry, R. L., *J. Phys. Chem.* **61**, 498 (1957).
149. Jarry, R. L., and Davis, W., U.S. *At. Energy Comm.* **K-847** (1951).
150. Jarry, R. L., and Fritz, J. J., *Chem. Eng. Data Ser.* **3**, 34 (1958).
151. Johnson, G. K., Smith, P. N., and Hubbard, W. N., *J. Chem. Thermodyn.* **5**, 793 (1973).
152. Kalman, O. F., private communication (1967).
153. Kamlet, M. J., U.S. Patent 3,624,129 (1971).
154. Karelin, A. J., Ionov, S. P., and Ionova, G. V., *Zh. Strukt. Khim.* **11**, 454 (1970).
155. Karelin, A. I., Nikitina, Z. K., Kharitonov, Y. Y., and Rosolovskii, V. Y., *Russ. J. Inorg. Chem.* **15**, 480 (1970).
156. Kemmit, R. D. W., and Sharp, D. W. A., *Advan. Fluorine Chem.* **4**, 242 (1965).
157. Kharitanov, Y. Y., Karelin, A. I., and Rosolovskii, V. Ya., *J. Mol. Struct.* **19**, 545 (1973).
158. Khutoretskii, V. M., Okhlobystina, L. V., and Fainzilberg, A. A., *Usp. Khim.* **36**, 377 (1967).
159. Khutoretskii, V. M., Okhlobystina, L. V., and Fainzilberg, A. A., *Izv. Akad. Nauk SSSR, Ser. Khim.* p. 387 (1970).
160. Koehler, J. K., and Giauque, W. F., *J. Amer. Chem. Soc.* **80**, 2659 (1958).
161. Krishna Pillai, M. G., and Curl, R. F., Jr., *J. Chem. Phys.* **37**, 2921 (1962).
162. Kwasnik, W., *in* "Handbook of Preparative Inorganic Chemistry" (G. Brauer, ed.), 2nd ed., Vol. 1, pp. 165–166. Academic Press, New York, 1963.
163. Lalande, W., U.S. Patent 2,982,617 (1961); German Patent 1,026,285 (1958).
164. Lalande, W., U.S. Patent 3,086,842 (1963).
165. Lalande, W., and Gall, J., U.S. Patent 2,998, 388 (1956).
166. Lang, K., Diploma Thesis, University of Munich, Germany (1955).
167. Lang, K., Ph.D. Thesis, University of Munich, Germany (1956).
168. Lau, C., and Passmore, J., *Inorg. Chem.* **13**, 2278 (1974).
169. Lawless, E. W., and Smith, I. C., "Inorganic High-Energy Oxidizers." Dekker, New York, 1968.
170. Laycock, T. B., and Tucker, N. B., U.S. Patent 3,328,312 (1967).
171. Lide, D. R., *J. Chem. Phys.* **43**, 3767 (1965).
172. Lide, D. R., and Mann, D. E., *Nat. Bur. Stand. (U.S.) Rep.* **4399** (1955).
173. Lide, D. R., and Mann, D. E., *J. Chem. Phys.* **25**, 595 (1956).
174. Lide, D. R., and Mann, D. E., *J. Chem. Phys.* **25**, 1128 (1956).
175. Lindahl, C. B., U.S. Patent 3,709,982 (1973).
176. Lindahl, C. B., Schack, C. J., and Pilipovich, D., U.S. Patent 3,701,630 (1972).
177. Lodwig, R. M., and Margrave, J. L., *Combust. Flame* **3**, 147 and 249 (1959).

CHLORINE OXYFLUORIDES

178. Luce, M., and Hartmanshenn, O., *J. Inorg. Nucl. Chem.* **29**, 2823 (1967).
179. Luce, M., and Hartmanshenn, O., *Commis. Energ. At.* [*Fr.*], *Rapp.* **3210** (1967).
180. Lustig, M., and Shreeve, J. M., *Advan. Fluorine Chem.* **7**, 175 (1973).
181. Lynton, H., and Passmore, J., *Can. J. Chem.* **49**, 2539 (1971).
182. Macheteau, Y., and Gillardeau, J., *Bull. Soc. Chim. Fr.* p. 4075 (1967).
183. Macheteau, Y., and Gillardeau, J., *Bull. Soc. Chim. Fr.* p. 1819 (1969).
184. Madden, R. P., and Benedict, W. S., *J. Chem. Phys.* **25**, 594 (1956).
185. Magerlein, B., Pike, J., Jackson, R., Vandenberg, G., and Kagan, F., *J. Org. Chem.* **29**, 2982 (1964).
186. Mandell, H. C., in "Kirk Othmer Encyclopedia of Chemical Technology" 1st ed., 2nd Suppl., Wiley (Interscience), New York, 1960.
187. Mandell, H. C., and Barth-Wehrenalp, G., *J. Inorg. Nucl. Chem.* **12**, 90 (1959).
188. Mandell, H. C., and Barth-Wehrenalp, G., U.S. Patent 3,140,934 (1964).
189. Margrave, J. L., and Wendt, R. P., *J. Chem. Phys.* **31**, 857 (1959).
190. Martin, D., and Tantot, G., *5th Eur. Symp. Fluorine Chem.*, Paper I-12 (1974).
191. Maryott, A. A., and Farrar, T. C., *J. Chem. Phys.* **54**, 64 (1971).
192. Maryott, A. A., and Kryder, S. J., *J. Chem. Phys.* **27**, 1211 (1957).
193. McDowell, C. A., Raghunathan, P., and Tait, J. C., *J. Chem. Phys.* **59**, 5858 (1973).
194. Meinert, H., and Gross, U., *Z. Chem.* **9**, 455 (1969).
195. Mitra, G., *Z. Anorg. Allg. Chem.* **340**, 110 (1965).
196. Mitra, G., *Z. Anorg. Allg. Chem.* **368**, 336 (1969).
197. Mitra, G., and Ray, A., *Sci. Cult.* **21**, 179 (1956).
198. Monsanto Research Corporation, Quarterly Technical Summary Report No. 2, MRB-2022-Q2, Contract No. AF 04(611)-8520 (1963).
199. Morton, J. R., and Preston, K. F., *Inorg. Chem.* **13**, 1786 (1974).
200. Muetterties, E. L., Mahler, W., Packer, K. J., and Schmutzler, R., *Inorg. Chem.* **3**, 1298 (1964).
201. Müller, A., Krebs, B., Fadini, A., Glemser, O., Cyvin, S. J., Brunvoll, J., Cyvin, B. N., Elvebredd, I., Hagen, G., and Vizi, B., *Z. Naturforsch. A* **23**, 1656 (1968).
202. Müller, A., and Nagarajan, G., *Z. Anorg. Allg. Chem.* **349**, 87 (1966).
203. Müller, A., Peacock, C. J., Schulze, H., and Heidborn, U., *Mol. Struct.* **3**, 252 (1969).
204. Nagarajan, G., and Redmon, M. J., *Monatsh. Chem.* **103**, 1406 (1972).
205. Nakanishi, S., *J. Jap. Chem.* **13**, 864 (1959).
206. Nathan, A., Magerlein, B., and Hogg, J., *J. Org. Chem.* **24**, 1517 (1959).
207. Neeman, M., and Osawa, Y., *J. Amer. Chem. Soc.* **85**, 232 (1963).
208. Neugebauer, C. A., and Margrave, J. L., *J. Amer. Chem. Soc.* **79**, 1338 (1957).
209. Nikitina, Z. K., and Rosolovskii, V. Ya., *Izv. Akad. Nauk SSSR, Ser. Khim.* p. 750 (1972).
210. Nikitina, Z. K., and Rosolovskii, V. Ya., *Izv. Akad. Nauk SSSR, Ser. Khim.* p. 273 (1973).
211. Oesterling, R. E., and Tyczkowski, E. A., *J. Amer. Chem. Soc.* **80**, 5286 (1958).
212. Okhlobystina, L. V., and Khutoretskii, V. M., *Izv. Akad. Nauk SSSR, Ser. Khim.* p. 1188 (1969).
213. Okhlobystina, L. V., Khutoretskii, V. M., and Fainzilberg, A. A., *Izv. Akad. Nauk SSSR, Ser. Khim.* p. 1487 (1971).

214. Olah, G. A., and Comisarow, M. B., *J. Amer. Chem. Soc.* **90**, 5033 (1968).
215. Olah, G. A., and Comisarow, M. B., *J. Amer. Chem. Soc.* **91**, 2172 (1969).
216. Olah, G. A., White, A. M., and O'Brien, D. H., *Chem. Rev.* **70**, 561 (1970).
217. Osawa, Y., and Neeman, M., *J. Org. Chem.* **32**, 3055 (1967).
218. Ostero, J., *Commis. Energ. At.* [*Fr.*], Note **CEA-N-1293** (1970).
219. Parent, C. R., and Gerry, M. C. L., *Chem. Commun.* p. 285 (1972).
220. Parent, C. R., and Gerry, M. C. L., *J. Mol. Spectrosc.* **49**, 343 (1974).
221. Pavia, A. C., Pascal, J. L., and Potier, A., *C. R. Acad. Sci., Ser. C* **272**, 1495 (1971).
222. Pennsalt Chemicals Corporation, "Perchloryl Fluoride," New Products Booklet No. DC-1819 (1957).
223. Pennsalt Chemicals Corporation, unpublished results on Contract No. AF 33 (616)-6532 (April 1960–March 1962).
224. Pennsalt Chemicals Corporation, unpublished results on Contract No. AF 04 (611)-8518 (Nov. 1963–Dec. 1964).
225. Pilipovich, D., U.S. Patent 3,707,413 (1972).
226. Pilipovich, D., Lindahl, C. B., Schack, C. J., Wilson, R. D., and Christe, K. O., *Inorg. Chem.* **11**, 2189 (1972).
227. Pilipovich, D., Maya, W., Lawton, E. A., Bauer, H. F., Sheehan, D. F., Ogimachi, N. N., Wilson, R. D., Gunderloy, F. C., and Bedwell, V. E., *Inorg. Chem.* **6**, 1918 (1967).
228. Pilipovich, D., Rogers, H. H., and Wilson, R. D., *Inorg. Chem.* **11**, 2192 (1972).
229. Pilipovich, D., and Schack, C. J., U.S. Patent 3,692,476 (1972).
230. Pilipovich, D., and Wilson, R. D., U.S. Patent 3,697,394 (1972); British Patent 1,278,684 (1972).
231. Pilipovich, D., Wilson, R. D., and Bauer, H. F., U.S. Patent 3,733,392 (1972); British Patent 1,278,863 (1972).
232. Pimentel, G. C., *J. Chem. Phys.* **10**, 446 (1951).
233. Powell, F., and Lippincott, E. R., *J. Chem. Phys.* **32**, 1883 (1960).
234. Reese, R., Dibeler, V., and Mohler, F., *J. Res. Nat. Bur. Stand.* **57**, 367 (1957).
235. Rips, S. M., Zercheninov, A. N., and Pankratov, A. V., *Russ. J. Phys. Chem.* **43**, 208 (1969).
236. Robinson, E. A., *Can. J. Chem.* **41**, 173 (1963).
237. Robinson, E. A., *Can. J. Chem.* **41**, 3021 (1963).
238. Robinson, E. A., Lavery, D. S., and Weller, S., *Spectrochim. Acta, Part A* **25**, 151 (1968).
239. Rode, B. M., and Engelbrecht, A., *Chem. Phys. Lett.* **16**, 26 (1972).
240. Rogers, H. H., and Pilipovich, D., U.S. Patent 3,718,557 (1973).
241. Rude, H., Benoit, R., and Hartmanshenn, O., *Commis. Energ. At.*[*Fr.*], Rapp. **CEA-R-4205** (1971).
242. Ruff, O., and Krug, H., *Z. Anorg. Allg. Chem.* **190**, 270 (1930).
243. Rundle, R. E., *J. Amer. Chem. Soc.* **85**, 112 (1963).
244. Ryss, I. G., "The Chemistry of Fluorine and its Inorganic Compounds." State Publ. House Sci., Tech. Chem. Lit. Moscow, 1956 [translated by the U.S. Atomic Energy Comm. (AEC-tr-3927)].
245. Sawodny, W., Fadini, A., and Ballein, K., *Spectrochim. Acta* **21**, 995 (1965).
246. Schack, C. J., Lindahl, C. B., Pilipovich, D., and Christe, K. O., *Inorg. Chem.* **11**, 2201 (1972).
247. Schack, C. J., and Pilipovich, D., *Inorg. Chem.* **9**, 387 (1970).
248. Schack, C. J., and Pilipovich, D., U.S. Patent 3,777,901 (1973).

249. Schack, C. J., Pilipovich, D., and Hon, J. F., *Inorg. Chem.* **12**, 897 (1973).
250. Schlosser, M., and Heinz, G., *Chem. Ber.* **102**, 1944 (1969).
251. Schmauch, G., and Servass, E., *Appl. Spectrosc.* **12**, 98 (1958); *Anal. Chem.* **30**, 1160 (1958).
252. Schmeisser, M., *Angew. Chem.* **67**, 493 (1955).
253. Schmeisser, M., and Brändle, K., *Advan. Inorg. Chem. Radiochem.* **5**, 41 (1963).
254. Schmeisser, M., and Ebenhöch, F. L., *Angew. Chem.* **66**, 230 (1954).
255. Schmeisser, M., and Fink, W., *Angew. Chem.* **69**, 780 (1957).
256. Schmitz, H., and Schumacher, H. J., *Z. Anorg. Allg. Chem.* **249**, 238 (1942).
257. Schuetz, R. D., Taft, D. D., O'Brien, J. P., Shea, J. L., and Mork, H. M., *J. Org. Chem.* **28**, 1420 (1963).
258. Scott, F. L., and Oesterling, R. E., *J. Org. Chem.* **25**, 1688 (1960).
259. Selig, H., and Shamir, J., *Inorg. Chem.* **3**, 294 (1964).
260. Seppelt, K., *Z. Anorg. Allg. Chem.* **406**, 287 (1974).
261. Sheppard, W. A., *Tetrahedron Lett.* p. 83 (1969).
262. Sheppard, W. A., and Sharts, C. M., "Organic Fluorine Chemistry." Benjamin, New York, 1969.
263. Shrewsberry, R. C., and Williamson, E. L., *J. Inorg. Nucl. Chem.* **28**, 2535 (1966).
264. Sicre, J. E., and Schumacher, H. J., *Z. Anorg. Allg. Chem.* **286**, 232 (1956).
265. Sicre, J. E., and Schumacher, H. J., *Angew. Chem.* **69**, 266 (1957).
266. Siebert, H., "Anwendungen der Schwingungsspektroskopie in der Anorganischen Chemie." Springer-Verlag, Berlin and New York, 1966.
267. Silla, H., Burwasser, H., and Calcote, H. F., *U.S. Dep. Comm., Off. Tech. Serv.* AD **25890** (1960).
268. Simkin, J., and Jarry, R. L., *J. Phys. Chem.* **61**, 503 (1957).
269. Sinke, G., private communication (1967).
270. Smith, D. F., Begun, G. M., and Fletcher, W. H., *Spectrochim. Acta* **20**, 1763 (1964).
271. Smith, E. A., Steinbach, F. C., and Beu, K. E., *U.S. At. Energy Comm., Rep.* **GAT T-687** (1959).
272. So, S. P., and Chau, F. T., *Z. Phys. Chem. (Frankfurt am Main)* [N.S.] **84**, 241 (1973).
273. So, S. P., and Chau, F. T., *Z. Phys. Chem. (Frankfurt am Main)* [N.S.] **85**, 69 (1973).
274. Spratley, R. D., and Pimentel, G. C., *J. Amer. Chem. Soc.* **88**, 2394 (1966).
275. Streng, A. G., *Chem. Rev.* **63**, 607 (1963).
276. Streng, A. G., *J. Amer. Chem. Soc.* **85**, 1380 (1963).
277. Streng, A. G., *J. Chem. Eng. Data* **16**, 357 (1971).
278. Streng, A. G., and Grosse, A. V., *Advan. Chem. Ser* **36**, 159 (1962).
279. Stull, R. D., *Ind. Eng. Chem.* **39**, 545 (1947).
279a. Sunder, S., Hallin, K. E., and McClung, R. E. D., *J. Chem. Phys.* **61**, 2920 (1974).
280. Tallman, R., Ph.D. Thesis, University of Wisconsin, Madison (1960); *Diss. Abstr.* **20**, 4293 (1960).
281. Tallman, R., Wampler, D., and Margrave, J. L., *J. Inorg. Nucl. Chem.* **21**, 38 (1961).
282. Tantot, G., Ph.D. Thesis, University of Paris (1974).

283. Tantot, G., and Bougon, R., *5th Eur. Symp. Fluorine Chem.* Paper I–11 (1974).
284. Titov, Y. A., Reshotova, I. G., and Akhram, A. A., *Reakts. Metody Issled. Org. Soedin.* **15**, 7 (1966).
285. Toeniskoetter, R. H., and Gortsema, F. P., Final Report under Contract No. DA-31-124-ARO(D)-77 Union Carbide Corp., 1965.
286. Tyurikov, V. A., Okhlobystina, L. V., Shapiro, B. I., Khutoretskii, V. M., Fainzilberg, A. A., and Syrkin, Y. K., *Izv. Akad. Nauk SSSR, Ser. Khim.* p. 2373 (1972).
287. Usmanov, A. G., and Magarra, R. I., *Russ. J. Phys. Chem.* **36**, 1454 (1962).
288. Veyre, R., Quenault, M., and Eyraud, C., *C. R. Acad. Sci., Ser. C* **268**, 1480 (1969).
289. Vigalok, I. V., Il'yasov, A. V., and Levin, Ya. A., *Zh. Obshch. Khim.* **39**, 715 (1969).
290. Vigalok, I. V., and Ostrovskaya, A. V., *Zh. Obshch. Khim.* **41**, 1410 (1971).
291. Wagman, D. C., Evans, W. H., Halow, J., Parker, V. B., Bailey, S. M., and Schumm, R. H., *Nat. Bur. Stand. (U.S.) Tech. Note* **270**–1 (1965).
292. Wamser, C. A., Fox, W. B., Gould, D., and Sukornick, B., *Inorg. Chem.* **7**, 1933 (1968).
293. Wamser, C. A., Sukornick, B., Fox, W. B., and Gould, D., *Inorg. Syn.* **14**, 29 (1973).
294. Weiss, R., Ph.D. Thesis, Technical University, Aachen, Germany (1959).
295. Whitney, E. D., MacLaren, R. O., Hurley, T. J., and Fogle, C. E., *J. Amer. Chem. Soc.* **86**, 4340 (1964).
296. Woolf, A. A., *J. Chem. Soc., London* p. 4113 (1954).
297. Woolf, A. A., *J. Inorg. Nucl. Chem.* **3**, 250 (1956).
298. Woolf, A. A., *J. Chem. Soc., A* p. 401 (1967).
298a. Yeats, P. A., and Aubke, F., *J. Fluor. Chem.* **4**, 243 (1974).
299. Yodis, A. W., and Cunningham, W. J., *U.S.* Patent 3,375,072 (1968).
300. Züchner, K., and Glemser, O., *Angew. Chem.* **84**, 1147 (1972).

SUBJECT INDEX

A

Addition reactions, *see* specific compounds
Arsenic compounds, mass spectra of, 253, 254

B

Barbituric acid, 187
Benzvalene, 41, 57
Bismuth clusters, 50
2,4-Bis(trifluoromethylmercapto)ureti-
    dine-1,3-dione, reactions of, 160, 161
Borabenzenes, 42
Boranes, 2, *see also* specific types
    bonding in, 7–16
    electron deficient, 7
    hydrocarbons having similar pattern, 35–42
    interatomic distances in, 45
    localized bonds, 7–10
    polyhedron-edge bond orders, 44
    structural patterns, 3–7
    structure, 67
arachno-Boranes, 6, 7
    comparison of, 83
    MO diagram, 14
    polyhedra, 5
    skeletal bonding electron pairs, 13–15
    structure of iron complex, 15
closo-Boranes, 5, 6
    comparison of, 82
    MO diagram for, 12
    polyhedra, 3
    skeletal bonding electron pairs, 10–13
nido-Boranes, 6
    comparison of, 82
    MO diagram, 14
    polyhedra, 4
    skeletal bonding electron pairs, 13–15
Borazines, 42
    ion clusters, 266
    rearrangements in, 262

Boron compounds, mass spectra of, 250, 251
Boron trihalides, mass spectra of, 250

C

^{13}C shifts in NMR, 210–214
Carboranes, 2, *see also* specific types
    analogs and derivatives, 97–132
    BH group-substituted, 78
        CNPR theory, 125–127
    bond distances in, 46
    bonding in, 7–16
    bridge hydrogen, 90–93
    carbaaza-substituted, 80
    carbathia-substituted, 80
    CH group-substituted, 78
        CNPR theory, 125–127
    coordination number pattern recognition theory (CNPR) of structure, 67–137
    deltahedra and deltahedral fragments, 69, 70
    electron deficient, 7
    endohydrogens, 90–93
    heteroatom analogs, CNPR theory, 125–129
    heteroatoms donating two electrons, CNPR theory, 129–132
    Lowry–Brønsted acidities of, 132–136
    reactions of, 48
    structural patterns, 3–7
    structural rules, 85–97
    structures of, 68, 69
    transition element group (TEG)-substituted, 79
        CNPR theory, 127–129
arachno-Carboranes, 6
    [B_3H_9] family of, 76
        CNPR theory, 117–119
    B_4H_{10} family of, 76
        CNPR theory, 119, 120
    B_5H_{11} family of, 76
        CNPR theory, 120, 121

SUBJECT INDEX

arachno-Carboranes—*continued*
  B_6H_{12} family of, 76
    CNPR theory, 121
  [B_7H_{13}] family of, 76
    CNPR theory, 121, 122
  B_8H_{14} family of, 77
    CNPR theory, 122
  B_9H_{15} family of, 77
    CNPR theory, 122–124
  [$B_{10}H_{16}$] family of, 78
    CNPR theory, 124
  [$B_{11}H_{17}$] family of, 78
    CNPR theory, 124, 125
  CNPR theory, 116–125
  comparison of, 83
  skeletal bonding electron pairs, 13–15
closo-Carboranes, 5, 6
  CNPR theory, 97
  comparison of, 82
  polyhedra, 3
  rearrangement-prone CNPR theory, 98
  skeletal bonding electron pairs, 10–13
  stable, CNPR theory, 98
  unstable and/or unknown, 71
    CNPR theory, 98–100
nido-Carboranes, 6
  B_5H_9 family of, 72
    CNPR theory, 105
  B_6H_{10} family of, 72
    CNPR theory, 101, 102
  [B_7H_{11}] family of, 72
    CNPR theory, 109, 110
  B_8H_{12} family of, 73
    CNPR theory, 110, 111
  [B_9H_{13}] family of, 73
    CNPR theory, 106–109
  $B_{10}H_{14}$ family of, 74
    CNPR theory, 102–105
  [$B_{11}H_{15}$] family of, 75
    CNPR theory, 111–113
  [$B_{12}H_{16}$] family of, 75
    CNPR theory, 113–116
  CNPR theory, 100–116
  comparison of, 82
  skeletal bonding electron pairs, 13–15
CF_3SNCO, reactions of, 159
Chlorination reactions, 149
Chlorine fluoride(s)
  geometry of, 320–322
  oxygenation of, 349

Chlorine fluoride oxide radicals, 385, 386
Chlorine monofluoride, addition reaction of, 147–150
Chlorine monofluoride oxide, 328–330
  force field of, 329, 330
  infrared spectrum of, 328, 329
  stretching force constants for, 330
  synthesis of, 328
Chlorine nitrate, fluorination of, 332
Chlorine oxides, fluorination of, 348
Chlorine oxyfluorides, 319–389, *see also* specific compounds
  adduct formation, 327, 328
  amphoteric nature of, 327, 328
  bond lengths, 326
  bond strengths, 323–327
  geometry of, 320–323
  ligand distribution, 323
  reactivity of, 327, 328
  stretching force constants, 324–327
Chlorine pentafluoride oxide, 345, 346
Chlorine trifluoride dioxide, 361–367
  bonding in, 366, 367
  internal force constants of, 366
  molecular structure of, 364–367
  properties of, 362–364
  stretching force constants of, 366
  synthesis of, 362–364
  thermodynamic data for, 387
  vibrational spectra of, 364, 365
Chlorine trifluoride oxide, 331–340
  chemical properties of, 337–340
  internal force constants, 335
  molecular structure of, 334–336
  physical properties of, 336, 337
  reactions of, 338, 339
  stretching force constants, 336
  synthesis of, 331–334
  thermodynamic data for, 386, 387
  vibrational spectra of, 334
Chloryl cation, 356–359
  internal force constants of, 359
  molecular structure of, 358, 359
  properties of, 357, 358
  synthesis of, 357, 358
  vibrational spectra of, 358, 359
Chloryl fluoride, 347–356
  chemical properties of, 353–356
  molecular structure of, 349–352

SUBJECT INDEX

physical properties of, 352, 353
reactions of, 356
synthesis of, 347–349
thermal decomposition of, 354, 355
vapor pressures of, 353
vibrational spectra of, 349–352
Cluster compounds, *see also* specific compounds
   cage opening, 47, 48
   interatomic distances in, 42–47
   mixed, 20–34
      skeletal bond pairs, 30–32
      skeletal electron-counting procedures, 20–23
   reactions of, 47–50
   skeletal bond orders, 44
   structural and bonding patterns in, 1–59
Cobalt-carbonyl acetylene complex, 22
Cobalt-carbonyl clusters, 16, 17, 19, 20
Cobalt mixed cluster compounds, 31–34
Coordination number pattern recognition theory (CNPR), 67–137, *see also* Carboranes
   boron considerations, 94, 95
   bridge and endohydrogen considerations, 90–93, 132–136
   carbon and other heteroelements, 93, 94
   coordination number range, 87
   deltahedron–deltahedral fragment hypothesis, 85, 86
   history of, 69–85
   primary, secondary, and tertiary expressions, 95–97
   structural preference of various moieties, 86–90
   violations of various rules, 88, 89
Cornwell effect, 206, 207
Cubane, 57
Cuneane, 57
Cyclocatenasulfur, charge-transfer complexes, 305
Cyclodecasulfur (S_{10}), 301
Cyclododecasulfur (S_{12}), 301, 302
   configuration of, 290
Cycloheptasulfur (S_7), 294, 295
Cyclohexasulfur (S_6), 288, 293, 294
Cycloicosasulfur (S_{20}), 303, 304
Cyclooctadecasulfur (S_{18}), 302, 303
Cyclooctasulfur (S_8), 287, 296–300
   configuration of, 290

monoclinic β-sulfur, 297, 299
   structure, 298
monoclinic γ-sulfur, 299, 300
   structure, 299
orthorhombic α-sulfur, 297
   structure, 298
Cyclopentadiene-metal complexes, 40, 41
Cycloundecasulfur (S_{11}), 301

D

Dewar benzene, 57
Dicarba-*closo*-carboranes
   rearrangement-prone, 71
   stable, 71
Difluorochlorate(V) anion, 359–361
   force constants of, 360, 361
   properties of, 359, 360
   structure of, 360, 361
   synthesis of, 359, 360
   vibrational spectra of, 360, 361
Difluorooxychloronium(V) cation, 340–343
   force constants, 341, 342
   molecular structure, 340–342
   properties of, 342, 343
   synthesis of, 340
   vibrational spectra of, 340, 341
Difluoroperchloryl cation, 367–370
   force constants, 370
   molecular structure of, 369, 370
   properties of, 367–369
   synthesis of, 367–369
   vibrational spectra of, 369, 370
Disilanes, mass spectra of, 251
Disulfanes, preparation of, 166
Dyes, 177

F

^{19}F shielding parameters, 219–225
Fluorides, binary, ^{19}F shielding parameters, 220–225
Fluorination reactions, 146, 147
Fungicides, 177

G

Gas chromatography and mass spectroscopy (GC/MS) of inorganic and organometallic compounds, 273–276

Germanium chlorides, mass spectra of, 248, 249
Germanium compounds, mass spectra of, 251, 252
Gillespie–Nyholm valence shell electron-pair repulsion theory, 325
Gold clusters, 52, 53
Group IV compounds, mass spectra of, 251, 253
Group V compounds, mass spectra of, 252, 253
Group VI compounds, mass spectra of, 254

H

Heterocycles, rearrangements in B–N and B–O, 262, 263
Hydrides, binary, proton-shielding parameters for, 219
Hydrocarbon(s)
　clusters, 57–59
　conforming to borane pattern, 35–42
Hydroquinone, 179

I

Insecticides, 177
Iridium-carbonyl clusters, 16, 17
Iron-carbonyl clusters, 16, 17, 19, 20
Iron 8-hydroxyquinolates, mass spectra of, 245
Iron mixed cluster compounds, 30–34
Iron-nitrosyl clusters, 55
Iron tricarbonyl pentafluorophenyl complexes, mass spectra of, 240
Isotope abundances in mass spectra, 264–267

L

Larmor's theorem, 202
Lowry–Brønsted acidity
　of bridge hydrogens, 135
　of carboranes, 132–136

M

Magnetic nuclei
　natural abundance of, 198, 199
　periodic table of, 198–202
　proton-shielding parameters for binary hydrides, 219
　receptivity of, 198, 199
　variations in ^{19}F shielding parameters, 220–225
Mass analyzers, 231, 232
Mass spectra, see also Mass spectroscopy
　effect of electron beam ionization energy on, 245–247
　　of inlet temperature on, 242
　　of instrument background on, 243, 244
　　of ion-molecule reactions on, 244
　　of metal parts on, 242, 243
　　of sample size on, 243, 244
　　of source on, 242
　　of volatility on, 244, 245
　of fragments, 239–242
　of molecular ions, 239–242
　of negative ions, 267
　of polymers, 239–242
Mass spectrometers, 232–237
　chemical ionization sources, 233
　electrohydrodynamic ionization source, 235
　electron impact sources, 232, 233
　field ionization/field desorption, 234, 235
　inlet systems, 235–237
　　batch, 236
　　direct insertion, 235, 236
　　gas/liquid probe, 236
　　GC/MS interfaces, 236, 237
Mass spectroscopy, 229–276, see also Mass spectra, Mass spectrometers
　coupled gas chromatography (GC/MS) and, 273–276
　fragmentations, 248–257
　high-resolution studies, 268–270
　of inorganic and organometallic compounds, 229–276
　instrumentation, 231–237
　isotope abundances, 264–267
　low- and medium-resolution studies, 247–268
　metastable-ion techniques, 270–273
　negative ions, 267
　rearrangements, 257–264
　sample handling, 237–239
　synthetic models, 268
Mercury compounds, mass spectra of, 254
Metal carbonyl(s), reactions with sulfeny chlorides, 188, 189

Metal-carbonyl clusters, 2, 16–20, *see also* specific metals
  octahedral, 16–18, 20
  reactions of, 49, 50
  not related to borane patterns, 53–56
  skeletal bonding electrons, 20–23
  triangular and tetrahedral, 16, 17
Metal β-diketonates, mass spectra of, 254, 255
Metal fluorides
  bond formation, 257–262
  fluorination with, 146, 147
Metal halide(s), mass spectra of, 247–250
Metal halide clusters, 51, 52
Metal-hydrocarbon complexes
  arachno structures of, 36, 37
  bond distances in, 46
  borane pattern in, 35–42
  closo structures of, 36
  nido structures of, 36, 37
  reactions of, 48, 49
Metal pentafluorides, mass spectra of, 249, 250
Metal phthalocyanines
  FD spectra, 234
  mass spectra, 255
Metalloboranes, 23–30
  with metal atom in bridging position, 29
closo-Metalloboranes, 24
nido-Metalloboranes, 25
Metallocarboranes, 23–30
  from smaller carboranes, 28
closo-Metallocarboranes, 24
nido-Metallocarboranes, 25
Metastable ions, in mass spectroscopy, 270–273

N

Nickel-carbonyl clusters, 20
Nickel-mixed cluster compounds, triple-decker sandwich structure, 39
Nickel tetracarbonyl, mass spectroscopy of, 229
Nuclear magnetic resonance, *see also* Nuclear magnetic shielding
  measurement of, and periodic table, 198–202
Nuclear magnetic shielding, 197–225
  ab initio calculations, 214
  absolute, 215–218
    methods of measurement, 217, 218
    scales, 218
    spin-rotation interaction, 215–217
  atom-plus-ligand local-term approximation, 209, 210
  atomic local-term approximation, 207–209
  Cornwell effect, 206, 207
  electronegativity correlations, 214, 215
  molecular shielding terms, 202–206
  periodicity in, 218–225
  physical models of, 202–215
  (r^{-1}) dependence, 203–205
    variation with atomic number, 205
  σ_d, 202–204
    variation with atomic number, 204
  substituent effects in, 210–214
  theory of, 202–215

O

Organometallic compounds
  GC/MS of, 273–276
  mass spectra of, 250–254
  mass spectroscopy of, 229–276
Organosulfur rings, 308
Orotic acid, 187
Osmium-carbonyl clusters, 16, 17, 20
Osmium mixed cluster compounds, 30–34
Oxygen-sulfur rings, 306, 307

P

Perchloryl fluoride, 371–385
  chemical properties, 380–384
  dipole moment, 374
  eigenvalues for, 377
  frequency values, 374
  ionization data for, 376
  molecular structure of, 373–377
  physical properties of, 377–380
  reactions of, 380, 381, 383
  for rocket propulsion, 384
  synthesis of, 371–373
  thermodynamic data for, 388, 389
  uses of, 384, 385
  valence molecular orbitals for, 377
  vibrational spectra, 374–377
Perfluorohalogenoolefins, addition to, 153

Perfluorohalogenoorganodisulfanes, chlorolysis of, 150–152
Perfluorohalogenoorganomercaptides, reactions of, 152
Perfluorohalogenoorganomercapto groups, characteristics of, 189, 190
Perfluorohalogenoorganosulfenyl bromides, 155–157
Perfluorohalogenoorganosulfenyl chlorides
 addition of chlorine to perhalogenothiocarbonyl compounds, 147–150
  of sulfur chlorides to perfluorohalogenoolefins, 153
 chlorolysis of perfluorohalogenoorganodisulfanes, 150–152
 fluorination with metal fluorides or HF, 146, 147
 hydrolysis of, 154
 photolysis reactions, 153, 154
 preparation of, 146–154
 properties of, 154, 155
 reactions of perfluorohalogenoorganothiols or -mercaptides with chlorine, 152
Perfluorohalogenoorganosulfenyl fluorides
 chemical shifts for, 144
 coupling constants for, 144
 preparation of, 144, 145
 properties of, 145
Perfluorohalogenoorganosulfenyl halides, 143–190, *see also* specific compounds
 reactions of, 157–189
  with alkanes, 175
  with alkenes, 175
  with alkynes, 176
  with amides, 172
  with amines, 168–172
  with ammonia, 167, 168
  with aromatics, 177–188
  with arsines, 172
  with carbonyl compounds, 173–175
  with heteroaromatics, 177–188
  with nitriles, 176
  with perfluorohalogenothioketones, 165–167
  with pseudohalides, 157–162
  with silver perfluorohalogenocarboxylates, 163, 164
  with sulfinates, 173
  with thioalcohols, 173

Perfluorohalogenoorganothiols, reactions of, 152
Perfluorohalogenothioketones, 165–167
Perhalogenothiocarbonyl compounds, 147–150
Periodic table of magnetic nuclei, 198–202
Pesticides, mass spectra of, 253
Pharmaceuticals, 177
Phosphorus compounds, mass spectra of, 252, 253
Photosulfur, 305
Phthalocyanines, FD spectra, 234
Platinum-carbonyl clusters, 54
Polycatenasulfur, 304, 305
Polysilanes, mass spectra of, 251
Prismane, 57

R

Ramsey shielding terms, *see* Nuclear magnetic shielding
Rhodium-carbonyl carbide clusters, 55
Rhodium-carbonyl clusters, 16–20
Rocket propulsion oxidizers, 384, 385
Ruthenium carbide complexes, mass spectra of, 240, 241
Ruthenium-carbonyl clusters, 16–19

S

Sampling techniques in mass spectroscopy, 237–239
 air- and moisture-sensitive compounds, 237–239
Selenium-sulfur rings, 307
Silver perfluorohalogenocarboxylates, 163, 164
Silver pseudohalides, 157, 162
Spin-rotation interactions, 215–217
Sulfenyl chlorides
 conversions of, 188, 189
 cyclizations of, 188, 189
 reactions of, 188, 189
Sulfenylthiocyanates, 161
Sulfur
 allotropes of, 291–314, *see also* specific forms
  incompletely characterized, 308–314
  with less than six atoms, 291–293

SUBJECT INDEX

    from liquid sulfur, 312, 313
    from solid sulfur, 312
    from solutions of cyclooctasulfur, 311
      of other sulfur compounds, 311, 312
    structural data for solid, 309
    from sulfur vapor, 313, 314
  elemental
    S–S bond configurations, 289
    S–S bond parameters, 288, 289
    structural parameters, 288–291, 296
    structures of, 287–314, see also specific forms
  fibrous, see Polycatenasulfur
  ions, 305, 306
  rings containing other elements, 306–308
Sulfur chlorides, addition to perfluorohalogenoolefins, 153
Sulfur cluster, 56, 57
Sulfur polymers, 308

T

Tellurium-sulfur rings, 307, 308
Tetrafluorooxychlorate(V) anion, 343–345
  force constants of, 345
  molecular structure of, 343–345
  properties of, 343
  synthesis of, 343
  vibrational spectra of, 344
Tetrasulfur tetranitride, 57
Thiaformaldehyde, 308
Thiophenes, 183, 194
Thiozone (S_3), 292
Tin pentafluorophenyl derivatives, mass spectra of, 240
Transition metal complexes, mass spectra of, 254–257

U

Uracil, reactions of, 185, 186

CONTENTS OF PREVIOUS VOLUMES

Volume I

Mechanisms of Redox Reactions of Simple Chemistry
H. Taube

Compounds of Aromatic Ring Systems and Metals
E. O. Fischer and H. P. Fritz

Recent Studies of the Boron Hydrides
William N. Lipscomb

Lattice Energies and Their Significance in Inorganic Chemistry
T. C. Waddington

Graphite Intercalation Compounds
W. Rüdorff

The Szilard-Chalmers Reaction in Solids
Garman Harbottle and Norman Sutin

Activation Analysis
D. N. F. Atkins and A. A. Smales

The Phosphonitrilic Halides and Their Derivatives
N. L. Paddock and H. T. Searle

The Sulfuric Acid Solvent System
R. J. Gillespie and E. A. Robinson

AUTHOR INDEX—SUBJECT INDEX

Volume 2

Stereochemistry of Ionic Solids
J. D. Dunitz and L. E. Orgel

Organometallic Compounds
John Eisch and Henry Gilman

Fluorine-Containing Compounds of Sulfur
George H. Cady

Amides and Imides of the Oxyacids of Sulfur
Margot Becke-Goehring

Halides of the Actinide Elements
Joseph J. Katz and Irving Sheft

Structures of Compounds Containing Chains of Sulfur Atoms
  Olav Foss

Chemical Reactivity of the Boron Hydrides and Related Compounds
  F. G. A. Stone

Mass Spectrometry in Nuclear Chemistry
  H. G. Thode, C. C. McMullen, and K. Fritze

AUTHOR INDEX—SUBJECT INDEX

Volume 3

Mechanisms of Substitution Reactions of Metal Complexes
  Fred Basolo and Ralph G. Pearson

Molecular Complexes of Halogens
  L. J. Andrews and R. M. Keefer

Structures of Interhalogen Compounds and Polyhalides
  E. H. Wiebenga, E. E. Havinga, and K. H. Boswijk

Kinetic Behavior of the Radiolysis Products of Water
  Christiane Ferradini

The General, Selective, and Specific Formation of Complexes by Metallic Cations
  G. Schwarzenbach

Atmospheric Activities and Dating Procedures
  A. G. Maddock and E. H. Willis

Polyfluoroalkyl Derivatives of Metalloids and Nonmetals
  R. E. Banks and R. N. Haszeldine

AUTHOR INDEX—SUBJECT INDEX

Volume 4

Condensed Phosphates and Arsenates
  Erich Thilo

Olefin, Acetylene, and π-Allylic Complexes of Transition Metals
  R. G. Guy and B. L. Shaw

Recent Advances in the Stereochemistry of Nickel, Palladium, and Platinum
  J. R. Miller

The Chemistry of Polonium
  K. W. Bagnall

The Use of Nuclear Magnetic Resonance in Inorganic Chemistry
  E. L. Muetterties and W. D. Phillips

Oxide Melts
  J. D. Mackenzie

AUTHOR INDEX—SUBJECT INDEX

Volume 5

The Stabilization of Oxidation States of the Transition Metals
  R. S. Nyholm and M. L. Tobe

Oxides and Oxyfluorides of the Halogens
  M. Schmeisser and K. Brandle

The Chemistry of Gallium
  N. N. Greenwood

Chemical Effects of Nuclear Activation in Gases and Liquids
  I. G. Campbell

Gaseous Hydroxides
  O. Glemser and H. G. Wendlandt

The Borazines
  E. K. Mellon, Jr., and J. J. Lagowski

Decaborane-14 and Its Derivatives
  M. Frederick Hawthorne

The Structure and Reactivity of Organophosphorus Compounds
  R. F. Hudson

AUTHOR INDEX—SUBJECT INDEX

Volume 6

Complexes of the Transition Metals with Phosphines, Arsines, and Stibines
  G. Booth

Anhydrous Metal Nitrates
  C. C. Addison and N. Logan

Chemical Reactions in Electric Discharges
Adli S. Kana'an and John L. Margrave

The Chemistry of Astatine
A. H. W. Aten, Jr.

The Chemistry of Silicon–Nitrogen Compounds
U. Wannagat

Peroxy Compounds of Transition Metals
J. A. Connor and E. A. V. Ebsworth

The Direct Synthesis of Organosilicon Compounds
J. J. Zuckerman

The Mössbauer Effect and Its Application in Chemistry
E. Fluck

AUTHOR INDEX—SUBJECT INDEX

Volume 7

Halides of Phosphorus, Arsenic, Antimony, and Bismuth
L. Kolditz

The Phthalocyanines
A. B. P. Lever

Hydride Complexes of the Transition Metals
M. L. H. Green and D. L. Jones

Reactions of Chelated Organic Ligands
Quintus Fernando

Organoaluminum Compounds
Roland Köster and Paul Binger

Carbosilanes
G. Fritz, J. Grobe, and D. Kummer

AUTHOR INDEX—SUBJECT INDEX

Volume 8

Substitution Products of the Group VIB Metal Carbonyls
Gerard R. Dobson, Ingo W. Stolz, and Raymond K. Sheline

Transition Metal Cyanides and Their Complexes
  B. M. Chadwick and A. G. Sharpe

Perchloric Acid
  G. S. Pearson

Neutron Diffraction and Its Application in Inorganic Chemistry
  G. E. Bacon

Nuclear Quadrupole Resonance and Its Application in Inorganic Chemistry
  Masaji Kubo and Daiyu Nakamura

The Chemistry of Complex Aluminohydrides
  E. C. Ashby

AUTHOR INDEX—SUBJECT INDEX

Volume 9

Liquid–Liquid Extraction of Metal Ions
  D. F. Peppard

Nitrides of Metals of the First Transition Series
  R. Juza

Pseudohalides of Group IIIB and IVB Elements
  M. F. Lappert and H. Pyszora

Stereoselectivity in Coordination Compounds
  J. H. Dunlop and R. D. Gillard

Heterocations
  A. A. Woolf

The Inorganic Chemistry of Tungsten
  R. V. Parish

AUTHOR INDEX—SUBJECT INDEX

Volume 10

The Halides of Boron
  A. G. Massey

Further Advances in the Study of Mechanisms of Redox Reactions
  A. G. Sykes

Mixed Valence Chemistry—A Survey and Classification
  Melvin B. Robin and Peter Day

AUTHOR INDEX—SUBJECT INDEX—CUMULATIVE TOPICAL INDEX FOR VOLUMES 1–10

Volume 11

Technetium
  K. V. Kotegov, O. N. Pavlov, and V. P. Shvedov

Transition Metal Complexes with Group IVB Elements
  J. F. Young

Metal Carbides
  William A. Frad

Silicon Hydrides and Their Derivatives
  B. J. Aylett

Some General Aspects of Mercury Chemistry
  H. L. Roberts

Alkyl Derivatives of the Group II Metals
  B. J. Wakefield

AUTHOR INDEX—SUBJECT INDEX

Volume 12

Some Recent Preparative Chemistry of Protactinium
  D. Brown

Vibrational Spectra of Transition Metal Carbonyl Complexes
  Linda M. Haines and M. H. Stiddard

The Chemistry Complexes Containing 2,2'-Bipyridyl, 1,10-Phenanthroline, or 2,2', 6',2''-Terpyridyl as Ligands
  W. R. McWhinnie and J. D. Miller

Olefin Complexes of the Transition Metals
  H. W. Quinn and J. H. Tsai

Cis and Trans Effects in Cobalt(III) Complexes
  J. M. Pratt and R. G. Thorp

AUTHOR INDEX—SUBJECT INDEX

Volume 13

Zirconium and Hafnium Chemistry
  E. M. Larsen

Electron Spin Resonance of Transition Metal Complexes
  B. A. Goodman and J. B. Raynor

Recent Progress in the Chemistry of Fluorophosphines
  John F. Nixon

Transition Metal Clusters with Π-Acid Ligands
  R. D. Johnston

AUTHOR INDEX—SUBJECT INDEX

Volume 14

The Phosphazotrihalides
  M. Bermann

Low Temperature Condensation of High Temperature Species as a Synthetic Method
  P. L. Timms

Transition Metal Complexes Containing Bidentate Phosphine Ligands
  W. Levason and C. A. McAuliffe

Beryllium Halides and Pseudohalides
  N. A. Bell

Sulfur–Nitrogen–Fluorine Compounds
  O. Glemser and R. Mews

AUTHOR INDEX—SUBJECT INDEX

Volume 15

Secondary Bonding to Nonmetallic Elements
  N. W. Alcock

Mössbauer Spectra of Inorganic Compounds: Bonding and Structure
  G. M. Bancroft and R. H. Platt

Metal Alkoxides and Dialkylamides
  D. C. Bradley

Fluoroalicyclic Derivatives of Metals and Metalloids
  W. R. Cullen

The Sulfur Nitrides
  H. G. Heal

AUTHOR INDEX—SUBJECT INDEX

Volume 16

The Chemistry of Bis(trifluoromethyl)amino Compounds
  H. G. Ang and Y. C. Syn

Vacuum Ultraviolet Photoelectron Spectroscopy of Inorganic Molecules
  R. L. DeKock and D. R. Lloyd

Fluorinated Peroxides
  Ronald A. De Marco and Jean'ne M. Shreeve

Fluorosulfuric Acid, Its Salts, and Derivatives
  Albert W. Jache

The Reaction Chemistry of Diborane
  L. H. Long

Lower Sulfur Fluorides
  F. Seel

AUTHOR INDEX—SUBJECT INDEX

Volume 17

Inorganic Compounds Containing the Trifluoroacetate Group
  C. D. Garner and B. Hughes

Homopolyatomic Cations of the Elements
  R. J. Gillespie and J. Passmore

Use of Radio-Frequency Plasma in Chemical Synthesis
  S. M. L. Hamblyn and B. G. Reuben

Copper(I) Complexes
  F. H. Jardine

Complexes of Open-Chain Tetradentate Ligands Containing Heavy Donor Atoms
  C. A. McAuliffe

The Functional Approach to Ionization Phenomena in Solutions
  U. Mayer and V. Gutmann

Coordination Chemistry of the Cyanate, Thiocyanate, and Selenocyanate Ions
  A. H. Norbury

SUBJECT INDEX